78 Springer Series i

Edited by Manuel Cardona

Springer Series in Solid-State Sciences

Editors: M. Cardona P. Fulde K. von Klitzing H.-J. Queisser

Managing Editor: H. K. V. Lotsch

Volumes 1–89 are listed at the end of the book

90 **Earlier and Recent Aspects of Superconductivity**
Editors: J. G. Bednorz and K. A. Müller

91 **Electronic Properties of Conjugated Polymers III**
Basic Models and Applications
Editors: H. Kuzmany, M. Mehring, and S. Roth

92 **Physics and Engineering Applications of Magnetism**
Editors: Y. Ishikawa and N. Miura

93 **Quasicrystals**
Editors: T. Fujiwara and T. Ogawa

94 **Electronic Conduction in Oxides**
By N. Tsuda, K. Nasu, A. Yanase, and K. Siratori

T. Inui Y. Tanabe Y. Onodera

Group Theory and Its Applications in Physics

With 72 Figures

Springer-Verlag
Berlin Heidelberg New York London
Paris Tokyo Hong Kong Barcelona

Professor Dr. Teturo Inui †
Professor Dr. Yukito Tanabe

Japan Women's University, 2-8-1, Mejirodai, Bunkyo-ku, Tokyo 112, Japan

Professor Dr. Yositaka Onodera

Department of Physics, School of Science and Technology,
Meiji University, Tama-ku, Kawasaki 214, Japan

Series Editors:

Professor Dr., Dres. h. c. Manuel Cardona
Professor Dr., Dr. h. c. Peter Fulde
Professor Dr. Klaus von Klitzing
Professor Dr. Hans-Joachim Queisser

Max-Planck-Institut für Festkörperforschung, Heisenbergstrasse 1,
D-7000 Stuttgart 80, Fed. Rep. of Germany

Managing Editor: Dr. Helmut K. V. Lotsch

Springer-Verlag, Tiergartenstrasse 17, D-6900 Heidelberg, Fed. Rep. of Germany

---

Title of the original Japanese edition: Ouyou gun ron – Gun hyougen to butsuri gaku

---

ISBN-13: 978-3-540-60445-7 e-ISBN-13: 978-3-642-80021-4
DOI: 10.1007/978-3-642-80021-4

Softcover reprint of the hardcover 1st edition 1990

Typesetting: Macmillan India Ltd., India
2154/3150-543210 – Printed on acid-free paper

Preface to the English Edition

This book has been written to introduce readers to group theory and its applications in atomic physics, molecular physics, and solid-state physics.

The first Japanese edition was published in 1976. The present English edition has been translated by the authors from the revised and enlarged edition of 1980. In translation, slight modifications have been made in Chaps. 8 and 14 to update and condense the contents, together with some minor additions and improvements throughout the volume.

The authors cordially thank Professor J. L. Birman and Professor M. Cardona, who encouraged them to prepare the English translation.

Tokyo, January 1990

T. Inui · Y. Tanabe
Y. Onodera

Preface to the Japanese Edition

As the title shows, this book has been prepared as a textbook to introduce readers to the applications of group theory in several fields of physics.

Group theory is, in a nutshell, the mathematics of symmetry. It has three main areas of application in modern physics. The first originates from early studies of crystal morphology and constitutes a framework for classical crystal physics. The analysis of the symmetry of tensors representing macroscopic physical properties (such as elastic constants) belongs to this category. The second area was enunciated by E. Wigner (1926) as a powerful means of handling quantum-mechanical problems and was first applied in this sense to the analysis of atomic spectra. Soon, H. Bethe (1929) found applications of group theory in the understanding of the electronic structures of molecules and crystals. Nobody will deny the great influence of group theory since then on the development and success of modern atomic, molecular and solid-state physics. The third area concerns applications in the physics of elementary particles. Here group theory serves as the guiding principle in investigating the mathematical structure of the equations governing the fields of particles. Of these three aspects, the present book is concerned with the second.

In writing this book, the authors had in mind as readers those students and research workers who want to learn group theory out of theoretical interest. However, the authors also intended that the book be of value to those research workers who want to apply group-theoretical methods to solve their own problems in chemical or solid-state physics. Accordingly, care has been taken to provide sufficient details of the calculations required to derive the final results as well as practical applications, not to mention detailed accounts of the fundamental concepts involved. In particular, a number of practical examples and problems have been included so that they may arouse the readers' interest and help deepen their understanding.

For the completion of the present book, the encouragement and patience of Mr. K. Endo, editor at Syokabo, have been invaluable. For the publication, the assistance rendered by Mr. S. Makiya (also of Syokabo) was essential. The authors wish to take this opportunity to express their sincere thanks.

Tokyo, October 1976

T. Inui · Y. Tanabe
Y. Onodera

Preface to the Japanese Edition

Contents

Sections marked with an asterisk may be omitted on a first reading.

List of Mathematical Symbols

$\mathbb{R}$	The set of all real numbers
$\mathbb{C}$	The set of all complex numbers
$\sim$	Homomorphic
$\cong$	Isomorphic
$\mathscr{G}/\mathscr{H}$	Factor group
$\{\boldsymbol{k}\}$	Star
$\mathscr{G}(\boldsymbol{k})$	Group of the wavevector $\boldsymbol{k}$
$\mathscr{G}_0(\boldsymbol{k})$	Point group of the wavevector $\boldsymbol{k}$
$\doteq$	Equal modulo reciprocal lattice vectors
$A^\dagger$	Hermitian conjugate matrix, $(A^\dagger)_{ij} = A_{ji}^*$
${}^tA$	Transposed matrix, $({}^tA)_{ij} = A_{ji}$
$D\downarrow\mathscr{H}$	Subduced representation
$\Delta\uparrow\mathscr{G}$	Induced representation
θ	Time-reversal operator
$\mathfrak{S}_n$	Symmetric group of degree n
$A \in \mathscr{G}$	A is a member of the set $\mathscr{G}$
$\mathscr{G}_1 \cap \mathscr{G}_2$	Intersection of sets $\mathscr{G}_1$ and $\mathscr{G}_2$
$[D\times D]$	Symmetric product representation
$\{D\times D\}$	Antisymmetric product representation
$\{D(S) \mid S \in \mathscr{H}\}$	The set of matrices $D(S)$ satisfying the condition $S \in \mathscr{H}$

1. Symmetry and the Role of Group Theory

Any student of science knows nowadays that the basic units of materials are atoms and molecules and that these microscopic building blocks aggregate together to form macroscopic bodies. In early days, chemists tried to understand the binding of molecules in chemical reactions – for example, carbon and oxygen molecules reacting to form carbon oxide – by imagining that each molecule had its own key or hook to catch other molecules with. This primitive model was later replaced by Lewis's octet or valence model (1916), which led to a successful explanation of the saturation of valence. In 1919, Kossel reached an (even quantitative) understanding of the growth of beautiful crystals, such as rock salt, by his theory of valence as heteropolar bonding before the advent of quantum mechanics (1925). With the development of quantum physics, quantitative treatments have been developed for the energy-level structures of atoms, molecules and solids and radiative processes involving them. Homopolar binding, which was beyond the realm of classical physics, was also given an explanation by Heitler–London theory as originating from quantum-mechanical resonance. It should also be remarked that characteristic features of metallic binding are now well understood as a new mechanism of cohesion. Most readers will already be familiar with these facts, to some extent.

Now, what are the *fundamental reasons* for the success of the above theories for level structures of atoms, molecules and solids and for varieties of bonding phenomena? In our opinion, they are not to be sought in the concrete models such as those primitive keys, hooks and valence lines that were later replaced by the quantitative spatial dependence of bond wavefunctions, but are to be found in the fact that *these physical systems are provided with a certain symmetry and the theories were able to reflect it correctly*. Here also lies the reason for the fact that group theory has become a central mathematical tool for dealing with symmetry and that its applications in physics, which the present book is mostly concerned with, have led to rich and fruitful consequences.

Keeping in mind the fact that treatments based on symmetry do not depend upon details of the model, let us digress for a while from the invisible world of atoms and molecules and turn our eyes to the symmetry of more familiar figures in our world. We study the symmetry of patterns seen in Japanese family crests, a heritage of the Japanese culture.

The designs of the three crests shown in Figs. 1.1–3 are based on leaves of water plantain, which grows at the waterside. In the pattern of Fig. 1.1, two leaves are placed symmetrically with respect to the central line MM'. If we put a vertical mirror along MM', the mirror image of the pattern will precisely cover the original pattern. The pattern is said to be invariant under the *mirror*

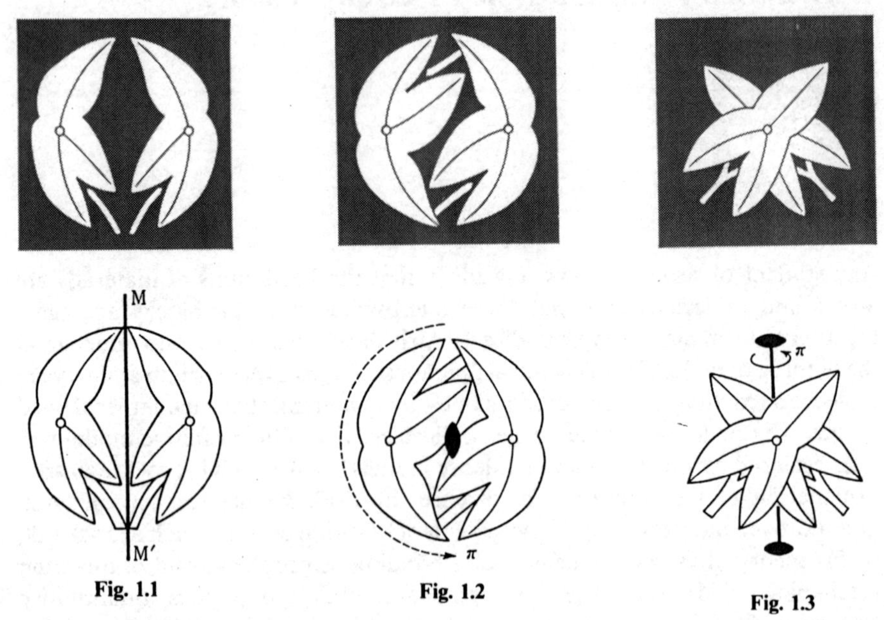

Fig. 1.1. Embracing Leaves of Water Plantain

Fig. 1.2. Chasing Leaves of Water Plantain. The symbol at the center of the lower figure denotes the twofold axis of rotation

Fig. 1.3. Crossing Leaves of Water Plantain

reflection. The operation of reflection is usually denoted by σ, which originates from the first letter of the German word *Spiegelung.*

In the pattern of Fig. 1.2, a counterclockwise 180° rotation about the vertical axis through the center of the figure will bring the right leaf onto the left and vice versa so that the rotated pattern covers the original one. In other words, the pattern of Fig. 1.2 has 180° rotation as its covering operation. We denote it by $R(\pi)$, R standing for the rotation and π representing the angle of rotation in radians. The axis is called the twofold axis, because a further rotation through the same angle in the same sense after the operation of $C_2 = R(\pi)$ brings the pattern back into the original position. We express this fact by writing $C_2 C_2 \equiv C_2^2 = E$, where E is the notation for the identity operation, coming from the German word *Einheit*. (Note that two successive operations are expressed as a product, the second being put to the left of the first.)

In this case, a clockwise rotation through the same angle, 180°, also brings the pattern into the same position as attained by $R(\pi)$, which means that C_2 and its inverse operation $C_2^{-1} = R(-\pi)$ are identical: $C_2^{-1} = C_2$. This then leads to the identities $C_2^2 = C_2^{-1} C_2 = C_2 C_2^{-1} = E$ in accordance with the geometrical considerations given above. The two operations E and C_2 satisfy the product relations

$$C_2^2 = E \ , \quad C_2 E = E C_2 = C_2 \ , \quad E^2 = E \ , \tag{1.1}$$

so that they are closed within the set $\{E, C_2\}$. This means that the set satisfies the group axiom to be stated later in Sect. 2.1. That is, the set $\{E, C_2\}$ constitutes a group called C_2, the cyclic group of order 2. The pattern of Fig. 1.2 is thus a geometrical realization of the abstract group C_2.

If we write E for the operation that leaves the original position intact also in the case of mirror symmetry, we have $\sigma^2 = E$, because reflection of the mirror image reproduces the original pattern. Thus we find

$$\sigma^2 = E , \quad \sigma E = E\sigma = \sigma . \tag{1.2}$$

These relations show that the set $\{E, \sigma\}$ has the same structure as C_2. The crest pattern in Fig. 1.1 is therefore another realization of the group C_2.

A glance at the crest given in Fig. 1.3 will convince us that none of the operations mentioned above (except E) qualify as a covering operation for this pattern. If we imagine, however, that the same pattern is printed on the back of the paper, we come across another kind of operation. Suppose we perform a 180° rotation about the longitudinal axis through the center of the figure as depicted in the lower part of Fig. 1.3. The pattern will come out of the paper during the rotation process but will eventually return to the same plane and the rotated image, although now turned over, will exactly cover the original pattern. This rotation is called *Umklappung* (turning over) and is denoted by C'_2. Here again, we have

$$(C'_2)^2 = E , \quad C'_2 E = EC'_2 = C'_2 , \tag{1.3}$$

as in (1.1, 2), so that the pattern of Fig. 1.3 provides yet another realization of the group C_2.

If we write G for C_2, σ, or C'_2, we note that all three patterns share the symmetry characterized by the relations

$$G^2 = E , \quad GE = EG = G . \tag{1.4}$$

In general, two groups are said to be isomorphic when one-to-one correspondence can be established between the elements as well as their products. Isomorphism is expressed by the notation $\cong$, so that

$$\{E, \sigma\} \cong \{E, C_2\} \cong \{E, C'_2\} \tag{1.5}$$

in the present example.

So far we have relied on our intuition to study the symmetry operations for the three crests. Another effective means of treating more complicated figures or objects is to examine the coordinate transformation associated with the covering operations. Let us briefly review how this is done in our present examples.

For the pattern of Fig. 1.1 we choose the line MM' as the y-axis with the x-axis perpendicular to it in the plane of the paper. When a point $P(x, y)$ on the pattern is carried over to the point $P'(x', y')$ by the mirror reflection σ, we have the relation

$$x' = -x , \quad y' = y .$$

If this is interpreted as a transformation of the column vector $\begin{bmatrix} x \\ y \end{bmatrix}$ into the vector $\begin{bmatrix} x' \\ y' \end{bmatrix}$, we obtain

$$\begin{bmatrix} x' \\ y' \end{bmatrix} = \begin{bmatrix} -1 & 0 \\ 0 & 1 \end{bmatrix} \begin{bmatrix} x \\ y \end{bmatrix},$$

which suggests that the mirror reflection σ can be represented by the matrix $\hat{\sigma}$:

$$\hat{\sigma} = \begin{bmatrix} -1 & 0 \\ 0 & 1 \end{bmatrix}.$$

In the case of the pattern of Fig. 1.2, the coordinates (x', y') of the image point P' generated by the 180° rotation C_2 are given by

$$x' = -x, \quad y' = -y ,$$

or

$$\begin{bmatrix} x' \\ y' \end{bmatrix} = \begin{bmatrix} -1 & 0 \\ 0 & -1 \end{bmatrix} \begin{bmatrix} x \\ y \end{bmatrix},$$

so that the matrix corresponding to C_2 is given by

$$\hat{C}_2 = \begin{bmatrix} -1 & 0 \\ 0 & -1 \end{bmatrix}.$$

If we calculate the squares of the matrices $\hat{\sigma}$ and $\hat{C}_2$, the result is the 2×2 unit matrix $\hat{E}$ as seen below:

$$(\hat{\sigma})^2 = \begin{bmatrix} -1 & 0 \\ 0 & 1 \end{bmatrix} \begin{bmatrix} -1 & 0 \\ 0 & 1 \end{bmatrix} = \begin{bmatrix} 1 & 0 \\ 0 & 1 \end{bmatrix} = \hat{E} ,$$

$$(\hat{C}_2)^2 = \begin{bmatrix} -1 & 0 \\ 0 & -1 \end{bmatrix} \begin{bmatrix} -1 & 0 \\ 0 & -1 \end{bmatrix} = \begin{bmatrix} 1 & 0 \\ 0 & 1 \end{bmatrix} = \hat{E} .$$

We thus find that these 2×2 matrices satisfy equations entirely isomorphic to (1.1, 2), which hold for the geometric operations σ, C_2 and E. When every element G of a group $\mathscr{G}$ has a corresponding matrix $\hat{G}$ and isomorphism holds between $\mathscr{G}$ and the group $\hat{\mathscr{G}}$ of the set of matrices $\hat{G}$, $\hat{\mathscr{G}}$ is called a representation of the group $\mathscr{G}$. Details of the representation theory of a group will be given in Chap. 4.

In a similar way, for the pattern of Fig. 1.3, we have the coordinate transformation due to C_2' as

$$x' = -x , \quad y' = y , \quad z' = -z .$$

Since y remains unchanged, we put

$$\begin{bmatrix} x' \\ z' \end{bmatrix} = \begin{bmatrix} -1 & 0 \\ 0 & -1 \end{bmatrix} \begin{bmatrix} x \\ z \end{bmatrix}$$

so that the matrix

$$\hat{C}_2' = \begin{bmatrix} -1 & 0 \\ 0 & -1 \end{bmatrix}$$

corresponds to the *Umklappung*. Needless to say, we have $(\hat{C}_2')^2 = \hat{E}$ in correspondence with $(C_2')^2 = E$.

We have sketched the gist of the representation theory of groups referring to very simple symmetrical objects. These considerations will serve as a miniature model of the subject to be developed and explored fully in Chap. 4.

1.1 Arrangement of the Book

The organization of the present book can be gathered from the table of contents together with Fig. 1.4. Broadly speaking, chapters up to Chap. 6 are devoted to general theories concerning groups, their representations and applications in quantum mechanics. Chapter 7 and subsequent chapters deal with important groups and their applications in physics. Since these latter chapters have been prepared so that they can be read fairly independently of the others, readers already familiar with general theories may proceed directly to any one of them according to their own interest. Newcomers to the subject who want to learn group theory and its applications using the present book may set the point groups and their applications (Chaps. 8–10) as their first goal. Sections marked with * in Chaps. 2–6 are not prerequisite to attaining this goal. A reader who

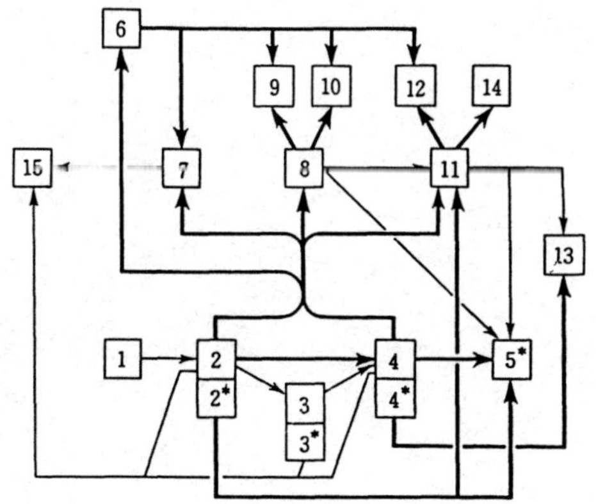

Fig. 1.4. Map showing the interrelation between chapters. The relation M → N indicates that subjects treated in Chap. M are assumed to be known in Chap. N. The bold lines signify a close relationship. The symbol M* stands for the sections of Chap. M marked *

starts with Chap. 1, reads through Chaps. 2–6 skipping sections marked * and reaches Chap. 10 following the bold lines of the map will be rewarded by a first view of the theory and physical applications of the symmetry groups in outline. For applications in solid-state physics, Chaps. 11 and 12 are indispensable.

The text is interspersed with exercises to help readers confirm their understanding. Some of them are, however, intended to supplement the text. Readers are therefore advised at least to try to understand the meaning of the exercises, even when they feel they are much too difficult.

2. Groups

This chapter is devoted to the mathematical definition of a group and related concepts. The examples of groups found throughout the chapter are intended to enable beginners to become acquainted with the concepts of groups. The asterisked sections (Sects. 2.7, 9–11) concern more advanced applications of group theory and may be skipped on a first reading, with the reader returning to them later as occasion arises.

2.1 Definition of a Group

By a *group* $\mathscr{G}$, we mean a set of distinct *elements* $G_1, G_2, \ldots, G_g$ such that for any two elements G_i and G_j, an operation called the group multiplication ($\circ$) is defined[1,2] which satisfies the following four requirements (the group axioms):

G1: The set $\mathscr{G}$ is closed under multiplication: For any two elements G_i and G_j of $\mathscr{G}$, their unique product $G_j \circ G_i$ also belongs to $\mathscr{G}$.

G2: The associative law holds:

$$G_k \circ (G_j \circ G_i) = (G_k \circ G_j) \circ G_i \ .$$

G3: There exists in $\mathscr{G}$ an element G_1 which satisfies

$$G_1 \circ G = G \circ G_1 = G$$

for any element $G \in \mathscr{G}$. Such an element G_1 is called the *unit element* or the *identity element*; it will hereafter be denoted by E:

$$E \circ G = G \circ E = G \qquad \text{(existence of the unit element).}$$

G4: For any element $G \in \mathscr{G}$, there exists an element G^- which satisfies

$$G^- \circ G = G \circ G^- = E \qquad \text{(existence of inverse elements).}$$

We call G^- the *inverse element* of G. In the following we write it as G^{-1}.

The elements G_i are sometimes called *group elements*, particularly when we wish to emphasize that they are members of the group $\mathscr{G}$. Groups having an

---

[1] From Sect. 2.2 onward, we omit the product symbol $\circ$ and write simply G_jG_i for $G_j \circ G_i$.

[2] In physical applications, the group elements G_i represent various operators. The product $G_j \circ G_i$ means "first operate with G_i, and then operate with G_j." Note that the operations are performed from right to left.

infinite number of elements are called *infinite groups*, while groups having a finite number of elements are *finite groups*. The total number of elements in a finite group is the *order of the group*.

It is assumed that the commutative law does not necessarily hold, but note the following:

G5: If any two elements G_i and G_j of a given group $\mathscr{G}$ commute, i.e., if

$$G_j \circ G_i = G_i \circ G_j \qquad \text{(commutative law)}$$

holds, then such a group $\mathscr{G}$ is said to be a *commutative group* or an *Abelian group*.

Exercise 2.1. Show that the set of elements $\{E, G\}$, where $G \circ G = G^2 = E$, satisfies the group axioms G1–G4, i.e., the set $\{E, G\}$ constitutes a group of order two.

Let G be an element of the group $\mathscr{G}$ and E the unit element. The smallest integer p which satisfies the equation $G^p = E$ is called the *order of the element G*.

Exercise 2.2. Show that the set $C_n = \{C, C^2, \ldots, C^{n-1}, C^n = E\}$, in which $C^k \circ C^l = C^{k+l}$, constitutes a group. This group C_n is called the *cyclic group* of order n.

Exercise 2.3. Show that cyclic groups are commutative.

2.1.1 Multiplication Tables

The structure of a group becomes manifest when we construct the multiplication table (or group table). To set up the table, we place the group elements $G_1, G_2, \ldots, G_i, \ldots, G_g$ in the top row and in the leftmost column, as shown in Table 2.1, and then put the product $G_j \circ G_i$ at the intersection of the G_j row and G_i column.

2.1.2 Generating Elements

In the case of a cyclic group, every element in it may be expressed as the power of a single element. In general, if every element of a given group $\mathscr{G}$ is expressible as

Table 2.1. Construction of a multiplication table

G_j \ G_i	G_1	G_2	…	G_i	…	G_g
G_1	$G_1 \circ G_1$	$G_1 \circ G_2$	…	$G_1 \circ G_i$	…	$G_1 \circ G_g$
G_2	$G_2 \circ G_1$	$G_2 \circ G_2$	…	$G_2 \circ G_i$	…	$G_2 \circ G_g$
$\vdots$	$\vdots$	$\vdots$		$\vdots$		$\vdots$
G_j	$G_j \circ G_1$	$G_j \circ G_2$	…	$G_j \circ G_i$	…	$G_j \circ G_g$
$\vdots$	$\vdots$	$\vdots$		$\vdots$		$\vdots$
G_g	$G_g \circ G_1$	$G_g \circ G_2$	…	$G_g \circ G_i$	…	$G_g \circ G_g$

the product of a smaller number of distinct elements, we call those elements the generating elements (or generators) of $\mathcal{G}$. Choice of the generating elements is not unique in general.

Exercise 2.4. Construct the multiplication table for the cyclic groups C_3 and C_4.

Exercise 2.5. Let X and Y be elements of order two. Show that if X and Y commute, i.e., $X \circ Y = Y \circ X$, the set $V = \{E, X, Y, X \circ Y\}$ constitutes a group. This is called the four group and has the two generating elements X and Y.

Exercise 2.6. Show that the four group has the multiplication table given in Table 2.2 if we write Z for $X \circ Y$ in Exercise 2.5.

Exercise 2.7. Demonstrate that if we designate the rotations through 180° about the x-, y- and z-axes as C_{2x}, C_{2y} and C_{2z}, then the set $D_2 = \{E, C_{2x}, C_{2y}, C_{2z}\}$ constitutes a group, and its multiplication table has the same structure as that of the four group.

*2.1.3 Commutative Groups

In commutative groups, it is convenient to use the addition symbol + instead of the product symbol ∘. The above-mentioned five axioms (including the commutative law) may then be written as follows:

A1: The set $\mathcal{A}$ is closed under addition +: For any elements A_i and A_j in the set $\mathcal{A}$, the unique sum $A_i + A_j$ always exists in $\mathcal{A}$.

A2: The associative law holds:

$$A_k + (A_j + A_i) = (A_k + A_j) + A_i \ .$$

A3: There exists an element 0 in $\mathcal{A}$ which satisfies the relation

$$A + 0 = 0 + A = A \qquad \text{(existence of the zero element)}$$

for any element $A \in \mathcal{A}$. Here, the unit element 0 is called the *zero element*.

A4: For any element $A \in \mathcal{A}$, there exists an element $(-A) \in \mathcal{A}$ which satisfies

$$(-A) + A = A + (-A) = 0 \qquad \text{(existence of inverse elements).}$$

The element $(-A)$ is the inverse element of A.

Table 2.2. Multiplication table of the four group V

G_j \ G_i	E	X	Y	Z
E	E	X	Y	Z
X	X	E	Z	Y
Y	Y	Z	E	X
Z	Z	Y	X	E

* The asterisked sections may be skipped on a first reading.

A5: For any two elements A_i, $A_j \in \mathscr{A}$, the commutative law

$$A_i + A_j = A_j + A_i$$

holds.

A set $\mathscr{A}$ that fulfills the above five axioms is called an *additive group*. Additive groups are nothing other than commutative groups in which the product operation is understood to be addition. The set of all real numbers $\mathbb{R}$ forms an additive group under the ordinary meaning of addition. Similarly, the set of all complex numbers $\mathbb{C}$ forms an additive group.

The set $\mathbb{R}$ is closed with respect to addition. What about with respect to multiplication? A real nonzero number f_i has a reciprocal, $f_i^{-1} = 1/f_i$, but the reciprocal does not exist for $f_i = 0$. Remove then the zero element from $\mathbb{R}$ and define the set $\mathbb{R}^* \equiv \mathbb{R} - \{0\}$. The set $\mathbb{R}^*$ now satisfies the four group axioms G1–G4 and the commutative law G5 for ordinary multiplication. Therefore, $\mathbb{R}^*$ constitutes a commutative group. Its unit element is the real number 1. Furthermore, for the combined operations of addition and multiplication, two types of distributive law hold:

$$f_k(f_j + f_i) = f_k f_j + f_k f_i \ ,$$

$$(f_k + f_j) f_i = f_k f_i + f_j f_i \ .$$

To sum up, for any two elements f_i and f_j in $\mathbb{R}$, the sum $f_i + f_j$ and product $f_i f_j$ are defined; the set $\mathbb{R}$ is an additive group with the zero element 0; the set $\mathbb{R}^*$ is a commutative group with the unit element 1; and the distributive laws hold. Such a set $\mathbb{R}$, in general, is called a field. The set of all complex numbers $\mathbb{C}$ also forms a field.

2.2 Covering Operations of Regular Polygons

An example of a group may be obtained by considering the *covering operations* (*symmetry operations*) of an equilateral triangle. Figure 2.1 shows a fixed equilateral triangle 123 on which a congruent triangle $\alpha\beta\gamma$ is allowed to rotate. We now rotate the triangle $\alpha\beta\gamma$ and seek the positions where the two congruent triangles cover each other exactly. As the rotation angle ϕ increases from zero, the first covering takes place at $\phi = 2\pi/3$ (Fig. 2.2a); the corresponding rotation will be denoted by $R(2\pi/3)$. Next we proceed to the second covering position $\phi = 2 \times 2\pi/3$, drawn in Fig. 2.2b. If we write C_3 for $R(2\pi/3)$, then we have $C_3^2 = R(4\pi/3)$ since $R(4\pi/3)$ is obtained by repeating C_3. Because of the indeterminacy of the rotation angle ϕ by multiples of 2π, the second covering position (Fig. 2.2b) may also be obtained by a reverse rotation $\phi = -2\pi/3$, which means $C_3^2 = R(-2\pi/3) = C_3^{-1}$. Increasing ϕ by $2\pi/3$ once again from the second position (b), we obtain $C_3^3 = R(2\pi) = R(0)$, consistent with the fact that

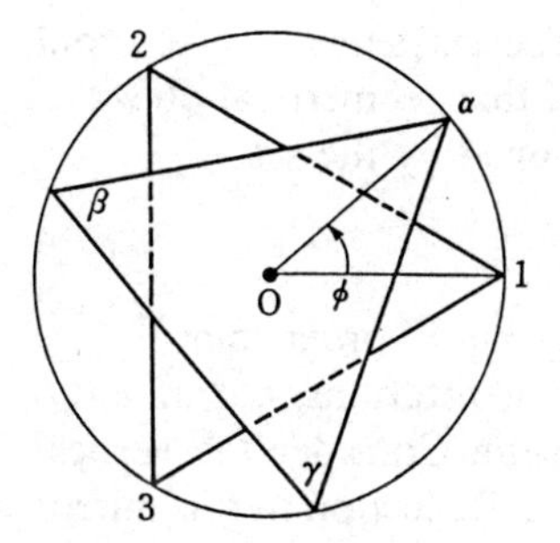

Fig. 2.1. The triangle $\alpha\beta\gamma$ rotates anticlockwise on the base triangle 123

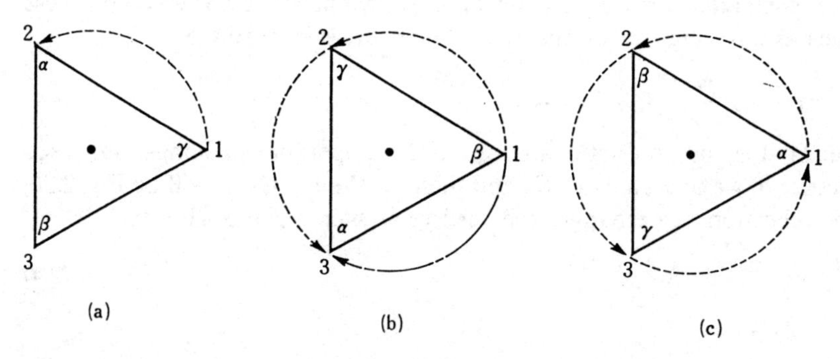

Fig. 2.2. Effects of the rotations (a) C_3, (b) C_3^2, and (c) C_3^3 on the triangle $\alpha\beta\gamma$

the third covering position coincides with the original position $\phi = 0$:

$$C_3 C_3^2 = C_3^3 = E ,$$

where E is the identity operation, which leaves the triangle $\alpha\beta\gamma$ as it stands (Fig. 2.2c).

Including the identity rotation $E = R(0)$ as a member of the covering operations, we have three covering operations E, C_3, and $C_3^2 = C_3^{-1}$. The set of these operations

$$\mathrm{C}_3 = \{E, C_3, C_3^{-1}\} \tag{2.1}$$

is closed, if we consider multiplication to mean successive operations. The set C_3 has the unit element E, the generating element C_3 and its inverse element C_3^{-1}, and satisfies the group axioms. Therefore, it constitutes a group identical to the cyclic group of order three.

In a similar manner, we can discuss the rotational symmetry of a square about its center. The first covering takes place at $\phi = 2\pi/4$ and $R(\pi/2)$ is the corresponding operation, $C_4 = R(\pi/2)$. The second covering position is given by $C_4^2 = C_2 = R(\pi)$ and the third by $C_4^3 = R(3\pi/2) = R(-\pi/2) = C_4^{-1}$. At the fourth step, the turning square comes back to the starting position, $\phi = 2\pi$, giving the relation $C_4^4 = C_2^2 = R(2\pi) = R(0) = E$. The existence of a fourfold rotation axis determines the symmetry properties of this geometrical object.

The rotational symmetry of the square is determined by the set

$$C_4 = \{E, C_4, C_2, C_4^{-1}\} , \tag{2.2}$$

which constitutes the group identical to the cyclic group of order four.

We have so far limited the covering operations to rotations, but an equilateral triangle has another kind of symmetry element. Consider the vertical mirror plane σ_1 through the straight line O1 (Fig. 2.3). Reflection in this mirror plane brings the triangle $\alpha\beta\gamma$ into coincidence with the base triangle 123. We have three such reflections, σ_1, σ_2 and σ_3, as shown in Fig. 2.3. If we count these reflections as covering operations, then the set of six operations

$$C_{3v} = \{E, C_3, C_3^{-1}, \sigma_1, \sigma_2, \sigma_3\} \tag{2.3}$$

is closed. That is, the product of any two of these operations belongs to this set. For instance, if we operate with C_3 and then σ_1, the net result will be Fig. 2.3b, since the reflection σ_1 exchanges the vertices at sites 2 and 3. Hence,

$$\sigma_1 C_3 = \sigma_2 . \tag{2.4}$$

Similarly, we have

$$\sigma_1 \sigma_2 = C_3 . \tag{2.5}$$

Carrying out all the product calculations in this way, we obtain the multiplication table shown in Table 2.3.

Exercise 2.8. Verify that if two mirror planes σ_1 and σ_2 form an angle θ, the product operation $\sigma_1 \sigma_2$ is the rotation $R(2\theta)$ whose rotation axis is the intersection of the two mirror planes. In particular, for $\theta = \pi/3$ we have (2.5).

We have discussed properties of the covering operations by relying on geometric intuition. The structure of the resulting groups can be inspected by

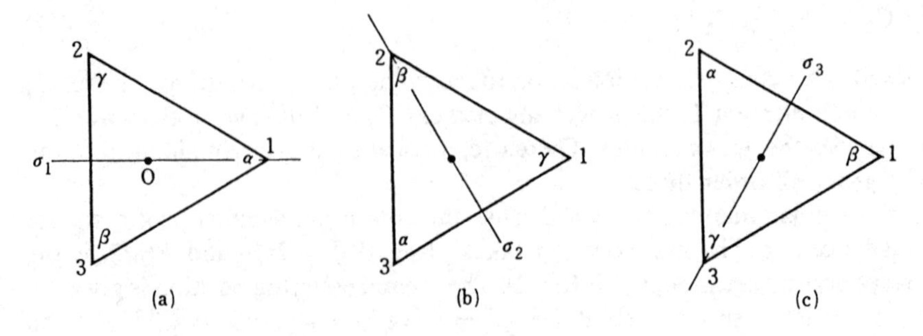

Fig. 2.3. Effects of the reflections (**a**) σ_1, (**b**) σ_2, and (**c**) σ_3 on the triangle $\alpha\beta\gamma$

Table 2.3. Multiplication table of the group C_{3v}

G_j \ G_i	E	C_3	C_3^{-1}	σ_1	σ_2	σ_3
E	E	C_3	C_3^{-1}	σ_1	σ_2	σ_3
C_3	C_3	C_3^{-1}	E	σ_3	σ_1	σ_2
C_3^{-1}	C_3^{-1}	E	C_3	σ_2	σ_3	σ_1
σ_1	σ_1	σ_2	σ_3	E	C_3	C_3^{-1}
σ_2	σ_2	σ_3	σ_1	C_3^{-1}	E	C_3
σ_3	σ_3	σ_1	σ_2	C_3	C_3^{-1}	E

analytic means as well. Consider the transformation of coordinates x and y by the mirror reflection σ_1, which sends the point $P(x, y)$ to $P'(x', y')$, where

$$x' = x ,$$
$$y' = -y ,$$

or, in vector form,

$$\begin{bmatrix} x' \\ y' \end{bmatrix} = \begin{bmatrix} 1 & 0 \\ 0 & -1 \end{bmatrix} \begin{bmatrix} x \\ y \end{bmatrix} .$$

This means that the effect of the mirror reflection σ_1 may be represented by the matrix

$$\hat{\sigma}_1 = \begin{bmatrix} 1 & 0 \\ 0 & -1 \end{bmatrix} ,$$

For rotations $R(\phi)$, it is convenient to use polar coordinates and write $x = r\cos\alpha$, $y = r\sin\alpha$ and $x' = r\cos(\alpha + \phi)$, $y' = r\sin(\alpha + \phi)$, from which we obtain the relation

$$\begin{bmatrix} x' \\ y' \end{bmatrix} = \hat{R}(\phi) \begin{bmatrix} x \\ y \end{bmatrix} , \tag{2.6}$$

where

$$\hat{R}(\phi) = \begin{bmatrix} \cos\phi & -\sin\phi \\ \sin\phi & \cos\phi \end{bmatrix} . \tag{2.7}$$

Thus the effect of the rotation $R(\phi)$ may be represented by the above transformation matrix. We obtain the following six matrices corresponding to the six group elements (2.3):

$$\hat{E} = \begin{bmatrix} 1 & 0 \\ 0 & 1 \end{bmatrix}, \quad \hat{C}_3 = \begin{bmatrix} -\frac{1}{2} & -\frac{\sqrt{3}}{2} \\ \frac{\sqrt{3}}{2} & -\frac{1}{2} \end{bmatrix}, \quad \hat{C}_3^{-1} = \begin{bmatrix} -\frac{1}{2} & \frac{\sqrt{3}}{2} \\ -\frac{\sqrt{3}}{2} & -\frac{1}{2} \end{bmatrix},$$

$$\hat{\sigma}_1 = \begin{bmatrix} 1 & 0 \\ 0 & -1 \end{bmatrix}, \quad \hat{\sigma}_2 = \begin{bmatrix} -\frac{1}{2} & -\frac{\sqrt{3}}{2} \\ -\frac{\sqrt{3}}{2} & \frac{1}{2} \end{bmatrix}, \quad \hat{\sigma}_3 = \begin{bmatrix} -\frac{1}{2} & \frac{\sqrt{3}}{2} \\ \frac{\sqrt{3}}{2} & \frac{1}{2} \end{bmatrix}. \tag{2.8}$$

For instance, the matrix $\hat{C}_3$ is obtained by putting $\phi = 2\pi/3$ in (2.7). The set of these six matrices obeys the multiplication table for the group C_{3v} (Table 2.3) under the usual matrix multiplication. For example, we have

$$\hat{\sigma}_1 \hat{C}_3 = \begin{bmatrix} 1 & 0 \\ 0 & -1 \end{bmatrix} \begin{bmatrix} -\frac{1}{2} & -\frac{\sqrt{3}}{2} \\ \frac{\sqrt{3}}{2} & -\frac{1}{2} \end{bmatrix} = \begin{bmatrix} -\frac{1}{2} & -\frac{\sqrt{3}}{2} \\ -\frac{\sqrt{3}}{2} & \frac{1}{2} \end{bmatrix} = \hat{\sigma}_2 . \tag{2.9}$$

Exercise 2.9. Besides the fourfold rotations (2.2), the square has four mirror planes $\sigma_x, \sigma_y, \sigma_d$ and $\sigma_{d'}$ shown in Fig. 2.4. Show that the set

$$C_{4v} = \{E, C_4, C_2, C_4^{-1}, \sigma_x, \sigma_y, \sigma_d, \sigma_{d'}\} \tag{2.10}$$

constitutes a group with the multiplication table given in Table 2.4.

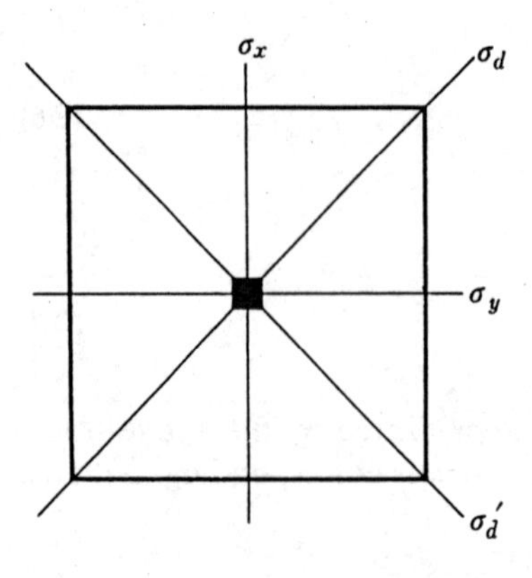

Fig. 2.4. Covering operations of a square. The filled square at the center signifies the fourfold rotation axis. σ_x and σ_y stand for the reflections in the planes perpendicular to the x- and y-axes

Table 2.4. Multiplication table of the group C_{4v}

	E	C_4	C_2	C_4^{-1}	σ_x	σ_y	σ_d	σ_d'
E	E	C_4	C_2	C_4^{-1}	σ_x	σ_y	σ_d	σ_d'
C_4	C_4	C_2	C_4^{-1}	E	σ_d'	σ_d	σ_x	σ_y
C_2	C_2	C_4^{-1}	E	C_4	σ_y	σ_x	σ_d'	σ_d
C_4^{-1}	C_4^{-1}	E	C_4	C_2	σ_d	σ_d'	σ_y	σ_x
σ_x	σ_x	σ_d	σ_y	σ_d'	E	C_2	C_4	C_4^{-1}
σ_y	σ_y	σ_d'	σ_x	σ_d	C_2	E	C_4^{-1}	C_4
σ_d	σ_d	σ_y	σ_d'	σ_x	C_4^{-1}	C_4	E	C_2
σ_d'	σ_d'	σ_x	σ_d	σ_y	C_4	C_4^{-1}	C_2	E

2.3 Permutations and the Symmetric Group

In the case of equilateral triangles discussed in the preceding section, the covering operations relocated the vertices α, β and γ of the rotatable triangle $\alpha\beta\gamma$. Therefore, the covering operations may also be interpreted as *permutations* of the three objects α, β and γ.

When, more generally, we have n objects, we have $n!$ permutations to relocate them on n sites. The set of such $n!$ permutations will form a group called the *permutation group* or the *symmetric group* of degree n, which is denoted by $\mathfrak{S}_n$, using $\mathfrak{S}$, the gothic capital letter S. The notation for permutations is defined as follows: if the permutation P relocates the object on the site p_i to the site i, then P is designated as

$$P = \begin{pmatrix} 1 & 2 & \dots & k & \dots & n \\ p_1 & p_2 & \dots & p_k & \dots & p_n \end{pmatrix} . \tag{2.11}$$

Note that the order of the columns is unimportant, for example

$$\begin{pmatrix} 1 & 2 & \dots & k & \dots & n \\ p_1 & p_2 & \dots & p_k & \dots & p_n \end{pmatrix} = \begin{pmatrix} 1 & k & \dots & 2 & \dots & n \\ p_1 & p_k & \dots & p_2 & \dots & p_n \end{pmatrix} .$$

With the above definition of the permutation symbol, the mirror reflection σ_2 shown in Fig. 2.3b may be interpreted as the permutation $P = \begin{pmatrix} 1 & 2 & 3 \\ 3 & 2 & 1 \end{pmatrix}$, since it exchanges the objects on the sites 1 and 3. Similarly, the reflection σ_1 of Fig. 2.3a corresponds to the permutation $Q = \begin{pmatrix} 1 & 2 & 3 \\ 1 & 3 & 2 \end{pmatrix}$, which exchanges the objects on the sites 2 and 3. If we operate with P and then Q, the net result will be $QP = \begin{pmatrix} 1 & 2 & 3 \\ 3 & 1 & 2 \end{pmatrix}$, since $\alpha\beta\gamma$ is replaced by $\gamma\alpha\beta$ (Fig. 2.5). If we note that P may be

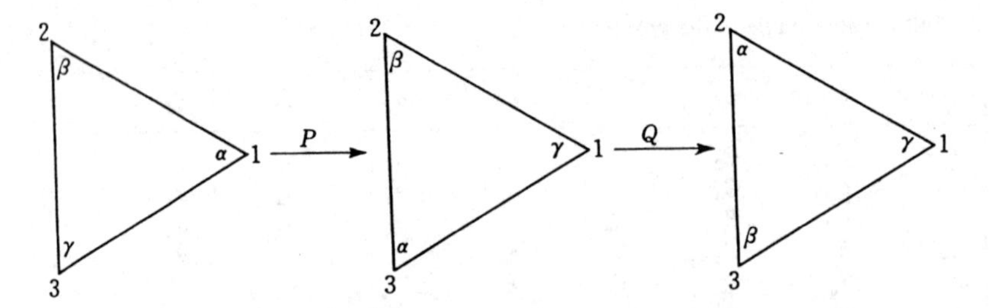

Fig. 2.5. Effect of the permutation P followed by another permutation Q

rewritten as $P = \begin{pmatrix} 1 & 2 & 3 \\ 3 & 2 & 1 \end{pmatrix} = \begin{pmatrix} 1 & 3 & 2 \\ 3 & 1 & 2 \end{pmatrix}$, the product QP of the two permutations can be evaluated as

$$QP = \begin{pmatrix} 1 & 2 & 3 \\ 1 & 3 & 2 \end{pmatrix}\begin{pmatrix} 1 & 2 & 3 \\ 3 & 2 & 1 \end{pmatrix} = \begin{pmatrix} 1 & 2 & 3 \\ 1 & 3 & 2 \end{pmatrix}\begin{pmatrix} 1 & 3 & 2 \\ 3 & 1 & 2 \end{pmatrix} = \begin{pmatrix} 1 & 2 & 3 \\ 3 & 1 & 2 \end{pmatrix} . \tag{2.12}$$

In general, the product of two permutations is

$$\begin{aligned} &\begin{pmatrix} 1 & 2 & \dots & k & \dots & n \\ q_1 & q_2 & \dots & q_k & \dots & q_n \end{pmatrix}\begin{pmatrix} 1 & 2 & \dots & k & \dots & n \\ p_1 & p_2 & \dots & p_k & \dots & p_n \end{pmatrix} \\ &\quad = \begin{pmatrix} 1 & 2 & \dots & k & \dots & n \\ q_1 & q_2 & \dots & q_k & \dots & q_n \end{pmatrix}\begin{pmatrix} q_1 & q_2 & \dots & q_k & \dots & q_n \\ p_{q_1} & p_{q_2} & \dots & p_{q_k} & \dots & p_{q_n} \end{pmatrix} \\ &\quad = \begin{pmatrix} 1 & 2 & \dots & k & \dots & n \\ p_{q_1} & p_{q_2} & \dots & p_{q_k} & \dots & p_{q_n} \end{pmatrix} . \end{aligned} \tag{2.13}$$

Exercise 2.10. Place the four aces from a pack of cards (spades, hearts, diamonds, clubs) in front of you and carry out the permutations $P = \begin{pmatrix} 1 & 2 & 3 & 4 \\ 3 & 1 & 4 & 2 \end{pmatrix}$ and $Q = \begin{pmatrix} 1 & 2 & 3 & 4 \\ 1 & 3 & 4 & 2 \end{pmatrix}$ in succession. Verify that the result may also be obtained by using the multiplication rule (2.13).

Example: $\mathfrak{S}_3$. $\begin{pmatrix} 1 & 2 & 3 \\ 1 & 2 & 3 \end{pmatrix} = E$ is the unit element of the symmetric group $\mathfrak{S}_3$ of degree three. We introduce the following abbreviated notation for the permutations:

$$(1\;3\;2) \equiv \begin{pmatrix} 1 & 2 & 3 \\ 3 & 1 & 2 \end{pmatrix} , \quad (1\;2\;3) \equiv \begin{pmatrix} 1 & 2 & 3 \\ 2 & 3 & 1 \end{pmatrix} ,$$

$$(2\;3) \equiv \begin{pmatrix} 1 & 2 & 3 \\ 1 & 3 & 2 \end{pmatrix} , \quad (1\;3) \equiv \begin{pmatrix} 1 & 2 & 3 \\ 3 & 2 & 1 \end{pmatrix} , \quad (1\;2) \equiv \begin{pmatrix} 1 & 2 & 3 \\ 2 & 1 & 3 \end{pmatrix} .$$

The set of these six permutations is closed and constitutes the group $\mathfrak{S}_3$. Products of these permutations may be evaluated using (2.13). For example, we find

$$(1\ 3\ 2)^2 = \begin{pmatrix} 1 & 2 & 3 \\ 3 & 1 & 2 \end{pmatrix}\begin{pmatrix} 1 & 2 & 3 \\ 3 & 1 & 2 \end{pmatrix} = \begin{pmatrix} 1 & 2 & 3 \\ 3 & 1 & 2 \end{pmatrix}\begin{pmatrix} 3 & 1 & 2 \\ 2 & 3 & 1 \end{pmatrix}$$

$$= \begin{pmatrix} 1 & 2 & 3 \\ 2 & 3 & 1 \end{pmatrix} = (1\ 2\ 3)\ .$$

Similarly,

$$(1\ 2\ 3)^2 = (1\ 3\ 2)\ .$$

$$(1\ 3\ 2)(1\ 2\ 3) = \begin{pmatrix} 1 & 2 & 3 \\ 3 & 1 & 2 \end{pmatrix}\begin{pmatrix} 1 & 2 & 3 \\ 2 & 3 & 1 \end{pmatrix}$$

$$= \begin{pmatrix} 1 & 2 & 3 \\ 3 & 1 & 2 \end{pmatrix}\begin{pmatrix} 3 & 1 & 2 \\ 1 & 2 & 3 \end{pmatrix} = \begin{pmatrix} 1 & 2 & 3 \\ 1 & 2 & 3 \end{pmatrix} = E\ ,$$

$$(1\ 2\ 3)(1\ 3\ 2) = E\ .$$

Thus the subset $\{E, (1\ 3\ 2), (1\ 2\ 3)\}$ forms a subgroup. In addition, from (2.12), we have

$$(2\ 3)(1\ 3) = (1\ 3\ 2)\ .$$

As a result of similar product calculations, we obtain the multiplication table shown in Table 2.5. More details of the symmetric group will be discussed in Chap. 15.

2.4 The Rearrangement Theorem

Theorem: Let $\mathscr{G} = \{G_1, G_2, \ldots, G_g\}$ be a group of order g. Multiplying every element of $\mathscr{G}$ on the right by an arbitrary element G in $\mathscr{G}$, we obtain the set

Table 2.5. Multiplication table of the group $\mathfrak{S}_3$

	E	(1 3 2)	(1 2 3)	(2 3)	(1 3)	(1 2)
E	E	(1 3 2)	(1 2 3)	(2 3)	(1 3)	(1 2)
(1 3 2)	(1 3 2)	(1 2 3)	E	(1 2)	(2 3)	(1 3)
(1 2 3)	(1 2 3)	E	(1 3 2)	(1 3)	(1 2)	(2 3)
(2 3)	(2 3)	(1 3)	(1 2)	E	(1 3 2)	(1 2 3)
(1 3)	(1 3)	(1 2)	(2 3)	(1 2 3)	E	(1 3 2)
(1 2)	(1 2)	(2 3)	(1 3)	(1 3 2)	(1 2 3)	E

$$\mathscr{G}G = \{G_1 G, G_2 G, \ldots, G_g G\} , \tag{2.14}$$

where each element of $\mathscr{G}$ appears once and only once.

Proof: Pick an element $G_i \in \mathscr{G}$, and multiply it from the right by G^{-1}, whose existence is guaranteed by the axiom G4. The product $G_i G^{-1}$ must be equal to some element G_k of $\mathscr{G}$ according to the axiom G1, and we have $G_i = G_k G$; the right-hand side is a member of the set $\mathscr{G}G$. Therefore, every element G_i appears in the set $\mathscr{G}G$. Moreover, it is certain that G_i does not appear twice in $\mathscr{G}G$, for if it did, the same element would appear in the forms $G_k G$ and $G_l G$. From this we would have $G_k = G_l$ by postmultiplying by G^{-1}, contrary to the assumption that the group elements are distinct. □

The rearrangement theorem holds for the set $G\mathscr{G} = \{GG_1, GG_2, \ldots, GG_g\}$ as well. According to this theorem, in every row and every column of the multiplication table, each group element appears once and only once. The theorem may also be stated as follows: Let f be an arbitrary function that takes group elements G_i as its argument. Then for any element $G \in \mathscr{G}$ there holds the relation

$$\sum_{i=1}^{g} f(GG_i) = \sum_{i=1}^{g} f(G_i G) = \sum_{i=1}^{g} f(G_i) . \tag{2.15}$$

2.5 Isomorphism and Homomorphism

2.5.1 Isomorphism

As was suggested in Sect. 2.3, there holds the following one-to-one correspondence between the group C_{3v} of the regular triangle and the symmetric group $\mathfrak{S}_3$ of degree three:

$$E \longleftrightarrow E, \quad C_3 \longleftrightarrow (1\ 3\ 2), \quad C_3^{-1} \longleftrightarrow (1\ 2\ 3) ,$$
$$\sigma_1 \longleftrightarrow (2\ 3), \quad \sigma_2 \longleftrightarrow (1\ 3), \quad \sigma_3 \longleftrightarrow (1\ 2) .$$

If we compare the two multiplication tables (Tables 2.3 and 2.5), we find that they have the same structure under the above correspondence. The two groups C_{3v} and $\mathfrak{S}_3$ are then said to be isomorphic.

The general definition of *isomorphism* is as follows: If there exists a *one-to-one correspondence* between elements G of a group $\mathscr{G}$ and elements G' of another group $\mathscr{G}'$ such that to a multiplication $G_i G_j = G_k$ in $\mathscr{G}$ there corresponds $G_i' G_j' = G_k'$ in $\mathscr{G}'$, then $\mathscr{G}$ and $\mathscr{G}'$ are *isomorphic* and we write

$$\mathscr{G} \cong \mathscr{G}' .$$

In terms of this symbol, the above example of isomorphism may be expressed as

$$C_{3v} \cong \mathfrak{S}_3 .$$

Mathematically, isomorphic groups are considered to be identical since they have the same structure. If we generalize the one-to-one correspondence of isomorphism to n-to-one correspondence, we reach the concept of homomorphism.

2.5.2 Homomorphism

For two given groups $\mathscr{G}$ and $\mathscr{G}'$, let f be a mapping that maps group elements G of $\mathscr{G}$ onto G' of $\mathscr{G}'$; that is, $G' = f(G)$. If the relation

$$f(G_i G_j) = f(G_i) f(G_j) \tag{2.16}$$

holds for any two elements G_i and G_j of $\mathscr{G}$, then f is called a *homomorphic mapping*. The two groups $\mathscr{G}$ and $\mathscr{G}'$ related by a homomorphic mapping are said to be *homomorphic* and this relation is written as

$$\mathscr{G} \sim \mathscr{G}' .$$

Example: $C_{3v} \sim C_2$, where $C_2 = \{E, C\}$, $C^2 = E$. The elements of the group C_{3v} can be made to correspond to the two elements of C_2 as follows:

$$E, C_3, C_3^{-1} \xrightarrow{f} E , \quad \sigma_1, \sigma_2, \sigma_3 \xrightarrow{f} C . \tag{2.17}$$

As may be readily seen from the multiplication table (Table 2.3), this mapping fulfills the relation (2.16) for all elements, so C_{3v} and C_2 are homomorphic. In this example, three elements in the group C_{3v} are mapped onto a single element in C_2.

Homomorphism between two groups signifies n-to-one correspondence between the elements of the two groups. In particular, when the mapping f is one-to-one and satisfies the homomorphism condition (2.16), it is an isomorphic mapping.

Exercise 2.11. Using the relation (2.16), show that $f(E)$ is the identity element of $\mathscr{G}'$ and that $f(G^{-1})$ is equal to the inverse element of $f(G)$.

2.5.3 Note on Mapping

A mapping $f: \mathscr{A} \to \mathscr{B}$, which maps the set $\mathscr{A}$ onto the set $\mathscr{B}$, is defined by a rule (or a function, a transformation) that associates an element A of $\mathscr{A}$ with an element B of $\mathscr{B}$. The element B is the *image* of A, while A is the *inverse image* of B. When every element of $\mathscr{B}$ has a corresponding inverse image in $\mathscr{A}$, such a mapping f is an *onto-mapping*. The homomorphic mapping defined above is an onto-mapping.

2.6 Subgroups

A subgroup $\mathscr{H}$ of a group $\mathscr{G}$ is a subset of $\mathscr{G}$ that is itself a group under the multiplication defined in the mother group $\mathscr{G}$. Both the single element $\{E\}$ and the group $\mathscr{G}$ itself are trivial subgroups of $\mathscr{G}$. The other subgroups, if any, are *proper subgroups*.

A nonempty subset $\mathscr{H}$ of a group $\mathscr{G}$ is a subgroup if and only if the following two conditions are fulfilled:

1) $H_i, H_j \in \mathscr{H} \longrightarrow H_i H_j \in \mathscr{H}$.

2) $H \in \mathscr{H} \longrightarrow H^{-1} \in \mathscr{H}$.

Proof: If $\mathscr{H}$ is a group, 1 and 2 hold. Conversely, suppose both 1 and 2 hold, then we can show that $\mathscr{H}$ satisfies all the requirements for a group. According to 1, the axiom G1 of Sect. 2.1 is satisfied. The associative law holds since $\mathscr{H}$ is a subset of $\mathscr{G}$, while 2 guarantees the axiom G4. From 1 and 2, the element $HH^{-1} = E$ is included in $\mathscr{H}$. Thus the four group axioms are satisfied by $\mathscr{H}$. □

Example: C_{3v} has the following four proper subgroups:

$$\{E, \sigma_1\}, \{E, \sigma_2\}, \{E, \sigma_3\}, \{E, C_3, C_3^{-1}\} . \tag{2.18}$$

Exercise 2.12. Find the eight proper subgroups of the group C_{4v} of Exercise 2.9.

*2.7 Cosets and Coset Decomposition

As has been mentioned in the previous section, $\mathscr{H} = \{E, \sigma_1\}$ is a subgroup of the group C_{3v}. If we multiply the elements of $\mathscr{H}$ with σ_2 on the right, we obtain the set

$$\mathscr{H}\sigma_2 = \{\sigma_2, \sigma_1\sigma_2\} = \{\sigma_2, C_3\} .$$

Similarly, we have

$$\mathscr{H}\sigma_3 = \{\sigma_3, \sigma_1\sigma_3\} = \{\sigma_3, C_3^{-1}\} .$$

We see that the six elements of C_{3v} are just exhausted by the three subsets, or by the three *right cosets*, so that

$$C_{3v} = \mathscr{H} + \mathscr{H}\sigma_2 + \mathscr{H}\sigma_3 , \quad \mathscr{H} = \{E, \sigma_1\} . \tag{2.19}$$

The relation (2.19) is called the *right coset decomposition* of C_{3v} with respect to the subgroup $\mathscr{H}$. One can also carry out the left coset decomposition

$$C_{3v} = \mathscr{H} + \sigma_2\mathscr{H} + \sigma_3\mathscr{H} , \quad \mathscr{H} = \{E, \sigma_1\} . \tag{2.20}$$

in terms of the *left cosets*,

$$\sigma_2\mathscr{H} = \{\sigma_2, C_3^{-1}\} , \quad \sigma_3\mathscr{H} = \{\sigma_3, C_3\} .$$

The general process for obtaining the coset decomposition is as follows: Let $\mathscr{G}$ be a group of order g having a proper subgroup $\mathscr{H}$ of order h. Take some element G_2 of $\mathscr{G}$ which does not belong to the subgroup $\mathscr{H}$, and make a right coset $\mathscr{H}G_2$. If $\mathscr{H}$ and $\mathscr{H}G_2$ do not exhaust the group $\mathscr{G}$, take some element G_3 of $\mathscr{G}$ which appears in neither $\mathscr{H}$ nor $\mathscr{H}G_2$, and make a right coset $\mathscr{H}G_3$. Continue making right cosets $\mathscr{H}G_i$ in this way. If $\mathscr{G}$ is a finite group, all the elements of $\mathscr{G}$ will be exhausted in a finite number of steps, so we obtain the right coset decomposition

$$\mathscr{G} = \mathscr{H}G_1 + \mathscr{H}G_2 + \cdots + \mathscr{H}G_l , \quad G_1 \equiv E . \tag{2.21}$$

The elements G_i are called *coset representatives.* Different cosets $\mathscr{H}G_i$ and $\mathscr{H}G_j$ $(i \neq j)$ have no elements in common. (Otherwise, we would have $H_1G_i = H_2G_j$ for some elements H_1 and H_2 belonging to the group $\mathscr{H}$, then $G_j = H_2^{-1}H_1G_i$, which means G_j is a member of the coset $\mathscr{H}G_i$, contrary to the definition of right cosets. □)

Since every coset $\mathscr{H}G_i$ consists of h distinct elements, the equality $g = hl$ must hold. Hence, the order of the mother group $\mathscr{G}$ is divisible by the order of the subgroup $\mathscr{H}$. The integer $l = g/h$ is called the *index* of $\mathscr{H}$ in $\mathscr{G}$. When g is a prime number, h must be equal to either g or unity, so, the group whose order is a prime number has no proper subgroups.

We can also decompose $\mathscr{G}$ into the left cosets $G_i'\mathscr{H}$:

$$\mathscr{G} = G_1'\mathscr{H} + G_2'\mathscr{H} + \cdots + G_l'\mathscr{H} , \quad G_1' \equiv E . \tag{2.22}$$

The numbers of cosets appearing in (2.21 and 22) are equal, although their contents may be different, as the example given in (2.19, 20) shows.

Exercise 2.13. Derive the right and left coset decompositions of C_{3v} with respect to the proper subgroups (2.18).

Exercise 2.14. Prove that groups whose order is equal to a prime number are cyclic.

2.8 Conjugate Elements; Classes

An element B of the group $\mathscr{G}$ is said to be *conjugate* to A if there exists a group element G such that $B = GAG^{-1}$. We also say in this case that B is obtained from A through transformation by G.

If B is conjugate to A, then A is conjugate to B. If B is conjugate to A and C is conjugate to B, then C is conjugate to A, because from $B = GAG^{-1}$ and $C = G'BG'^{-1}$ it follows that $C = (G'G)A(G'G)^{-1}$.

The set of all elements that are conjugate to each other is called a *conjugate class* or simply a *class*. By this definition, different classes have no elements in common. A class is determined once some representative element A of it is given. Thus, the elements generated by the sequence

$$A,\ G_2AG_2^{-1},\ G_3AG_3^{-1},\ \ldots,\ G_gAG_g^{-1} \tag{2.23}$$

belong to the same class. Note that in (2.23) the same element can appear several times. By a suitable choice of A's, the elements of $\mathscr{G}$ are *classified* into classes. In particular, when we choose $A = E$ in (2.23), we have no elements other than E in the sequence (2.23). Therefore, in any group, the unit element E forms a class $\mathscr{C}_1$ by itself.

We shall explain below the process of classification for the group C_{4v} considered in Exercise 2.9. By use of the multiplication table for C_{4v} (Table 2.4), we calculate every member of the sequence (2.23) for the five elements E, C_2, C_4, σ_x, and σ_d. The result is given in Table 2.6.

In the first and second rows, we observe that the unit element E and the twofold rotation C_2 respectively form the classes $\mathscr{C}_1$ and $\mathscr{C}_2$ by themselves. In the third row, two elements C_4 and C_4^{-1} constitute a third class $\mathscr{C}_3$. In the fourth and fifth rows, $\{\sigma_x, \sigma_y\}$ and $\{\sigma_d, \sigma_d'\}$ form the classes $\mathscr{C}_4$ and $\mathscr{C}_5$. Altogether, we find that the elements of C_{4v} are classified into five classes, see Table 2.7.

We can calculate the sequence (2.23) almost by rote, referring to the multiplication table. Such a calculation is indeed straightforward, but tedious.

Table 2.6. Calculation of the sequence (2.23)

A	$G\ A\ G^{-1}$ \ G	E	C_4	C_2	C_4^{-1}	σ_x	σ_y	σ_d	σ_d'
E	$G\ E\ G^{-1}$	E	E	E	E	E	E	E	E
C_2	$G\ C_2\ G^{-1}$	C_2	C_2	C_2	C_2	C_2	C_2	C_2	C_2
C_4	$G\ C_4\ G^{-1}$	C_4	C_4	C_4	C_4	C_4^{-1}	C_4^{-1}	C_4^{-1}	C_4^{-1}
σ_x	$G\ \sigma_x\ G^{-1}$	σ_x	σ_y	σ_x	σ_y	σ_x	σ_x	σ_y	σ_y
σ_d	$G\ \sigma_d\ G^{-1}$	σ_d	σ_d'	σ_d	σ_d'	σ_d'	σ_d'	σ_d	σ_d

Table 2.7. The classes of C_{4v}

Class	Elements in the class	Number of elements h_k
$\mathscr{C}_1$	E	1
$\mathscr{C}_2$	C_2	1
$\mathscr{C}_3$	C_4, C_4^{-1}	2
$\mathscr{C}_4$	σ_x, σ_y	2
$\mathscr{C}_5$	σ_d, σ_d'	2

Intuitive considerations described below will help in such a classification process. Figure 2.6 shows how the fourfold rotation C_4 transforms the mirror reflection σ_x. If we follow the effect of the product operation $C_4\sigma_x C_4^{-1}$ from right to left, points on the plane are moved as shown in Fig. 2.6, with the result

$$C_4\sigma_x C_4^{-1} = \sigma_y \ . \tag{2.24}$$

It should be emphasized that this conjugate relation follows because the rotation C_4 brings the σ_x mirror plane to the σ_y mirror plane. The relation between σ_d and σ_d' is the same. On the other hand, the group C_{4v} has no elements that bring the σ_x plane to the σ_d plane, so σ_x and σ_d cannot be conjugate in the group C_{4v}. On the basis of these geometric considerations, we find that conjugate elements are geometrically equivalent operations. Thus, the above classification for the group C_{4v} can be carried out more easily.

Exercise 2.15. Show that the elements of the group C_{3v} can be classified into the three classes

$$\mathscr{C}_1 = \{E\} \ , \quad \mathscr{C}_2 = \{C_3, C_3^{-1}\} \ , \quad \mathscr{C}_3 = \{\sigma_1, \sigma_2, \sigma_3\} \ . \tag{2.25}$$

Exercise 2.16. Demonstrate that in a commutative group every element constitutes a class by itself. Consequently, the number of the classes is equal to the order of the group.

Exercise 2.17. Prove that elements belonging to the same class have the same order.

*2.9 Multiplication of Classes

Let $\mathscr{C}_k$ be a class of the group $\mathscr{G}$ consisting of h_k distinct elements. If we transform the elements of the class $\mathscr{C}_k$ with an arbitrary element G, the resulting set $G\mathscr{C}_k G^{-1}$ coincides with $\mathscr{C}_k$ itself,

$$G\mathscr{C}_k G^{-1} = \mathscr{C}_k \ . \tag{2.26}$$

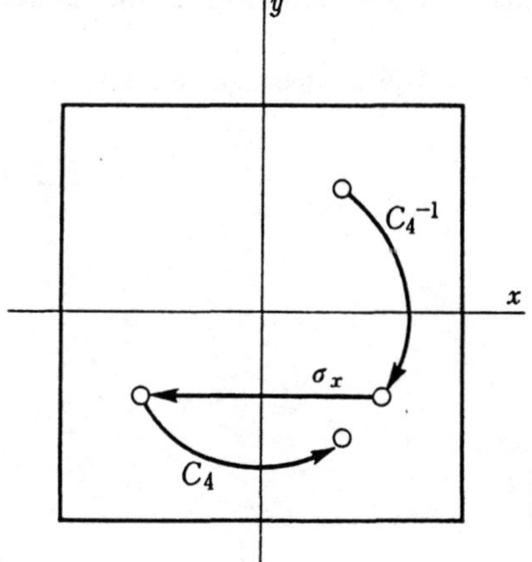

Fig. 2.6. The reflection σ_x as transformed by the fourfold rotation C_4

(By the definition of a class, the elements of the transformed set $G\mathscr{C}_k G^{-1}$ should belong to the class $\mathscr{C}_k$. Since $G\mathscr{C}_k G^{-1}$ consists of h_k distinct elements, it must coincide with $\mathscr{C}_k$ as a set.□)

When we take several classes together to form a set

$$\mathscr{C} = \sum_k a_k \mathscr{C}_k \tag{2.27}$$

with nonnegative integers a_k, the set $\mathscr{C}$ satisfies

$$G\mathscr{C}G^{-1} = \mathscr{C} , \tag{2.28}$$

for any group element G.

The converse is also true. A set $\mathscr{C}$ of group elements satisfying (2.28) must have a structure like (2.27); that is, it must include group elements in complete classes. (Subtract all complete classes from both sides of (2.28). Then the residual set $\mathscr{C}'$ satisfies the same equation $G\mathscr{C}'G^{-1} = \mathscr{C}'$ for any G, if it is a nonempty set at all. Since this equality is assumed to hold for any G_i $(i = 1, 2, \ldots, g)$, $\mathscr{C}'$ includes all the transformed sets $G_i\mathscr{C}'G_i^{-1}$. Such a $\mathscr{C}'$ must contain all mutually conjugate elements. In other words, it must consist of classes.□)

Having established the basic properties of classes, we next consider class multiplication. The product $\mathscr{C}_i\mathscr{C}_j$ of two classes $\mathscr{C}_i$ and $\mathscr{C}_j$ is defined as the set that consists of the products of the elements of $\mathscr{C}_i$ and $\mathscr{C}_j$. Note that the same element can appear several times in the product $\mathscr{C}_i\mathscr{C}_j$. In that event, it should be counted independently every time it appears.

Consider, for example, the classes (2.25) of the group C_{3v}. Their products can be calculated using the multiplication table:

$$\mathscr{C}_2\mathscr{C}_2 = (C_3 + C_3^{-1})(C_3 + C_3^{-1}) = 2E + C_3 + C_3^{-1} = 2\mathscr{C}_1 + \mathscr{C}_2 ,$$

$$\mathscr{C}_2\mathscr{C}_3 = (C_3 + C_3^{-1})(\sigma_1 + \sigma_2 + \sigma_3) = 2(\sigma_1 + \sigma_2 + \sigma_3) = 2\mathscr{C}_3 .$$

From these and other products, we obtain the multiplication table for the classes shown in Table 2.8.

As the above example indicates, the product $\mathscr{C}_i\mathscr{C}_j$ consists of classes,

$$\mathscr{C}_i\mathscr{C}_j = \sum_k c_{ij}^k \mathscr{C}_k , \tag{2.29}$$

Table 2.8. Class multiplications in the group C_{3v}

	$\mathscr{C}_1$	$\mathscr{C}_2$	$\mathscr{C}_3$
$\mathscr{C}_1$	$\mathscr{C}_1$	$\mathscr{C}_2$	$\mathscr{C}_3$
$\mathscr{C}_2$	$\mathscr{C}_2$	$2\mathscr{C}_1 + \mathscr{C}_2$	$2\mathscr{C}_3$
$\mathscr{C}_3$	$\mathscr{C}_3$	$2\mathscr{C}_3$	$3\mathscr{C}_1 + 3\mathscr{C}_2$

where the nonnegative integers c_{ij}^k are called the *class constants.* This relation signifies that the class $\mathscr{C}_k$ appears c_{ij}^k times in the product of the classes $\mathscr{C}_i$ and $\mathscr{C}_j$. The proof of (2.29) is straightforward if we note

$$G\mathscr{C}_i\mathscr{C}_jG^{-1} = G\mathscr{C}_iG^{-1}G\mathscr{C}_jG^{-1} = \mathscr{C}_i\mathscr{C}_j$$

for an arbitrary group element G.

Exercise 2.18. Construct a table similar to Table 2.8 for the group C_{4v}.

Exercise 2.19. The inverse elements of the h_j elements constituting a class $\mathscr{C}_j$ form a class by themselves, which will be denoted by $\mathscr{C}_{j'}$. Show that

$$c_{ij}^1 = \begin{cases} h_i, & \text{when } \mathscr{C}_i = \mathscr{C}_{j'} \\ 0, & \text{otherwise} \,, \end{cases} \tag{2.30}$$

where $\mathscr{C}_1$ is the class consisting only of the unit element.

Exercise 2.20. Show that $\mathscr{C}_i\mathscr{C}_j = \mathscr{C}_j\mathscr{C}_i$.

*2.10 Invariant Subgroups

Let $\mathscr{H}$ be a subgroup of the group $\mathscr{G}$. If we transform the elements of $\mathscr{H}$ with an element G of $\mathscr{G}$, the set of those elements

$$G\mathscr{H}G^{-1}$$

also turns out to be a subgroup of $\mathscr{G}$. It is called a *conjugate subgroup* of $\mathscr{H}$, and is isomorphic to $\mathscr{H}$. (The above set $G\mathscr{H}G^{-1}$ satisfies conditions 1 and 2 for subgroups mentioned in Sect. 2.6. The product of two elements GH_iG^{-1} and GH_jG^{-1} in the set $G\mathscr{H}G^{-1}$ is

$$(GH_iG^{-1})(GH_jG^{-1}) = GH_iH_jG^{-1} = GH_kG^{-1} \,,$$

which belongs to $G\mathscr{H}G^{-1}$, where $H_k = H_iH_j$. The inverse element of GH_iG^{-1} is $GH_i^{-1}G^{-1}$.)

Since a subgroup does not necessarily contain all mutually conjugate elements, $\mathscr{H}$ is in general different from $G\mathscr{H}G^{-1}$. However, if the subgroup $\mathscr{H}$ satisfies the relation

$$G\mathscr{H}G^{-1} = \mathscr{H} \tag{2.31}$$

for all elements $G \in \mathscr{G}$, then $\mathscr{H}$ is called an *invariant subgroup* of $\mathscr{G}$. An invariant subgroup is also called a *normal subgroup* or *normal divisor*.

As has been proved in Sect. 2.9, the set of group elements satisfying (2.31) for all G contains the elements in classes. Therefore invariant subgroups must be composed of classes.

We illustrate this for the group C_{3v}, which has the four proper subgroups (2.18). The six elements of the group are classified into the three classes (2.25). The subgroup $\{E, \sigma_1\}$ is not invariant, since it lacks σ_2 and σ_3, which are conjugate to σ_1. Similarly, $\{E, \sigma_2\}$ and $\{E, \sigma_3\}$ are not invariant subgroups. The last subgroup $C_3 = \{E, C_3, C_3^{-1}\}$ consists of two classes $\mathscr{C}_1 + \mathscr{C}_2$; it is an invariant subgroup of C_{3v}.

Exercise 2.21. Find invariant subgroups of the group C_{4v} among the eight proper subgroups obtained in Exercise 2.12.

Equation (2.31) can be modified to

$$G\mathscr{H} = \mathscr{H}G \ . \tag{2.32}$$

This means that the left coset is identical with the right coset as a set, i.e., an invariant subgroup is a subgroup whose right and left cosets are identical. Therefore, in the coset decomposition with respect to an invariant subgroup we need not worry about the difference between right and left cosets.

*2.11 The Factor Group

Let us consider the coset decomposition of the group $\mathscr{G}$ of order g with respect to its invariant subgroup $\mathscr{N}$ of order n:

$$\mathscr{G} = \mathscr{N}G_1 + \mathscr{N}G_2 + \cdots + \mathscr{N}G_l \ , \qquad G_1 = E \ . \tag{2.33}$$

Now, products of the elements belonging to the cosets $\mathscr{N}G_i$ and $\mathscr{N}G_j$ may be written in the form

$$(N_pG_i)(N_qG_j) = N_pG_iN_qG_i^{-1}G_iG_j \ . \tag{2.34}$$

Since $\mathscr{N}$ is an invariant subgroup, $G_iN_qG_i^{-1}$ belongs to $\mathscr{N}$, therefore (2.34) can be expressed as

$$(N_pG_i)(N_qG_j) = N_rG_iG_j \ , \tag{2.35}$$

where $N_r = N_pG_iN_qG_i^{-1}$ is an element of $\mathscr{N}$. The right-hand side of (2.35) is an element belonging to the coset $\mathscr{N}G_iG_j$. Thus the products of two elements taken from the cosets $\mathscr{N}G_i$ and $\mathscr{N}G_j$ belong to the coset $\mathscr{N}G_iG_j$. We write this relation as

$$(\mathscr{N}G_i)(\mathscr{N}G_j) = \mathscr{N}G_iG_j \ . \tag{2.36}$$

If we take the left-hand side of (2.36) as a multiplication of two cosets, (2.36) means that a product of two cosets turns out to be a coset. The cosets themselves therefore form a group under the "multiplication" defined in this way. The elements of this "group", called the *factor group*, are cosets themselves. The factor group of $\mathscr{G}$ with respect to the invariant subgroup $\mathscr{N}$ is denoted by

$$\mathscr{G}/\mathscr{N} \ .$$

The order of the factor group is equal to the index $l = g/n$ of $\mathscr{N}$ in $\mathscr{G}$.

Exercise 2.22. Show that the cosets satisfy the group axioms under the multiplication rule (2.36). Note that the unit element of the factor group is the subgroup $\mathscr{N}$ itself and that the inverse of the coset $\mathscr{N}G_i$ is the coset $\mathscr{N}G_i^{-1}$.

Example: Factor group C_{3v}/C_3. Coset decomposition of the group C_{3v} with respect to its invariant subgroup C_3 reads

$$C_{3v} = C_3 + C_3\sigma_1 \ , \tag{2.37}$$

where

$$C_3 = \{E, C_3, C_3^{-1}\} \quad C_3\sigma_1 = \{\sigma_1, \sigma_3, \sigma_2\} \ .$$

The products of these cosets may be calculated using (2.36) and Table 2.3, and we obtain the multiplication table of the factor group C_{3v}/C_3 given in Table 2.9. In this case, $g = 6$ and $n = 3$, from which we have the order of the factor group $l = g/n = 2$. This factor group is isomorphic to the cyclic group of order two.

Exercise 2.23. Find the coset decomposition of the group C_{4v} with respect to its invariant subgroup $C_2 = \{E, C_2\}$, and construct the multiplication table of the factor group C_{4v}/C_2.

The above result may be rephrased in terms of the concept of homomorphic mapping. By means of a mapping f: $\mathscr{G} \to \mathscr{G}/\mathscr{N}$, the group elements G_i are mapped onto the cosets $f(G_i) = \mathscr{N}G_i$. Then from (2.36), we have

$$f(G_i)f(G_j) = f(G_iG_j) \ ,$$

which means that f is a homomorphic mapping, and hence $\mathscr{G}$ is homomorphic to the factor group $\mathscr{G}/\mathscr{N}$,

$$\mathscr{G} \sim \mathscr{G}/\mathscr{N} \ . \tag{2.38}$$

In other words, correspondence between the group elements and the cosets is homomorphic when $\mathscr{N}$ is an invariant subgroup.

Table 2.9. Multiplication table of the factor group C_{3v}/C_3

	C_3	$C_3\sigma_1$
C_3	C_3	$C_3\sigma_1$
$C_3\sigma_1$	$C_3\sigma_1$	C_3

*2.11.1 The Kernel

Let f be a homomorphic mapping that maps a group $\mathscr{G}$ onto a group $\mathscr{G}'$ such that

$$f(G_1 G_2) = f(G_1) f(G_2)$$

for $G_1, G_2 \in \mathscr{G}$. The set of all elements which are mapped onto the unit element E' of $\mathscr{G}'$,

$$\mathscr{K} = \{G \mid G \in \mathscr{G}, \quad f(G) = E'\}$$

is called the kernel of the mapping f.

Exercise 2.24. Show that the kernel $\mathscr{K}$ of a homomorphic mapping f is an invariant subgroup of $\mathscr{G}$.
Hint: Show that (2.31) and the two conditions of Sect. 2.6 are fulfilled.

*2.11.2 Homomorphism Theorem

Let $f: \mathscr{G} \to \mathscr{G}'$ be a homomorphic mapping from a group $\mathscr{G}$ onto a group $\mathscr{G}'$ and let the kernel of f be denoted by $\mathscr{K}$, which is an invariant subgroup of $\mathscr{G}$. If we define the mapping $\bar{f}: \mathscr{G}/\mathscr{K} \to \mathscr{G}'$ by $\bar{f}(\mathscr{K} G_i) = f(G_i)$, then $\bar{f}$ is an isomorphic mapping, and hence

$$\mathscr{G}/\mathscr{K} \cong \mathscr{G}' \ .$$

Proof: From (2.16 and 36), the mapping $\bar{f}$ is homomorphic,

$$\begin{aligned} \bar{f}(\mathscr{K} G_i)\bar{f}(\mathscr{K} G_j) &= f(G_i) f(G_j) = f(G_i G_j) \\ &= \bar{f}(\mathscr{K} G_i G_j) = \bar{f}(\mathscr{K} G_i \mathscr{K} G_j) \ . \end{aligned}$$

To prove the isomorphism, we have to show a one-to-one correspondence. The mapping $\bar{f}$ transforms two different cosets $\mathscr{K} G_i$, $\mathscr{K} G_j$ $(G_i \neq G_j)$ to different elements $f(G_i)$, $f(G_j)$ of $\mathscr{G}'$. If $f(G_i) = f(G_j)$, then we would have

$$f(G_i G_j^{-1}) = f(G_i) f(G_j^{-1}) = f(G_i) f(G_j)^{-1} = E' \ ,$$

and $G_i G_j^{-1}$ would be an element of the kernel $\mathscr{K}$, which means G_i would be a member of the coset $\mathscr{K} G_j$, contrary to the definition of cosets. □

2.12 The Direct-Product Group

If we have two groups $\mathscr{A} = \{A_1 \equiv E_{\mathscr{A}}, A_2, \ldots, A_g\}$ and $\mathscr{B} = \{B_1 \equiv E_{\mathscr{B}}, B_2, \ldots, B_l\}$ of order g and l, and all elements of $\mathscr{A}$ commute with those of $\mathscr{B}$,

$$A_i B_j = B_j A_i \ , \qquad (2.39)$$

then we can prove that the gl pairs of such elements A_iB_j will form a group. We name this the *direct-product group* of $\mathscr{A}$ and $\mathscr{B}$ and write it as $\mathscr{A} \times \mathscr{B}$.

The unit element E of the direct-product group $\mathscr{A} \times \mathscr{B}$ is $E = E_{\mathscr{A}} E_{\mathscr{B}}$. The product of two elements A_iB_j and $A_{i'}B_{j'}$ of $\mathscr{A} \times \mathscr{B}$ becomes

$$(A_iB_j)(A_{i'}B_{j'}) = A_iA_{i'}B_jB_{j'} = A_{i''}B_{j''} ,$$

using (2.39). Here $A_{i''}$ and $B_{j''}$ are elements of $\mathscr{A}$ and $\mathscr{B}$, so that this product is an element of $\mathscr{A} \times \mathscr{B}$. Thus the axiom G1 is satisfied. The associative law holds by (2.39). $A_i^{-1}B_j^{-1}$ is inverse to A_iB_j.

As an example of a direct-product group, let us consider again the symmetry of an equilateral triangle. If we do not distinguish between the two faces of the triangle, it has an additional symmetry element – the mirror reflection σ_h in the horizontal plane. Since repeated reflection leads to the identity operation, we have $\sigma_h^2 = E$. Hence, the group

$$C_{1h} = \{E, \sigma_h\} \tag{2.40}$$

is a cyclic group of order two.

Now the horizontal reflection σ_h commutes with the six elements (2.3) of the group C_{3v}, and hence the complete symmetry of the triangle is described by the direct-product group

$$C_{3v} \times C_{1h} = \begin{Bmatrix} E, C_3, C_3^{-1}, \sigma_1, \sigma_2, \sigma_3 \\ \sigma_h, C_3\sigma_h, C_3^{-1}\sigma_h, \sigma_1\sigma_h, \sigma_2\sigma_h, \sigma_3\sigma_h \end{Bmatrix} . \tag{2.41}$$

This group is named D_{3h}, using the notation explained in Sect. 8.2. Using the result from Exercise 2.8, e find that

$$\sigma_i\sigma_h = U_i , \quad i = 1, 2, 3 ,$$

are rotations through π about the i axis in Fig. 2.3. In terms of these rotations, the elements of the group D_{3h} may be written as

$$C_{3v} \times C_{1h} = \begin{Bmatrix} E, C_3, C_3^{-1}, U_1\sigma_h, U_2\sigma_h, U_3\sigma_h \\ \sigma_h, C_3\sigma_h, C_3^{-1}\sigma_h, U_1, U_2, U_3 \end{Bmatrix} .$$

In this form, the group D_{3h} may also be expressed as the direct product of

$$D_3 = \{E, C_3, C_3^{-1}, U_1, U_2, U_3\}$$

and the group C_{1h} of horizontal reflection,

$$D_{3h} = D_3 \times C_{1h} .$$

Exercise 2.25. Construct the multiplication table of the group D_{3h}. Classify the elements of D_{3h} into classes.

Exercise 2.26. Find all the covering operations of a square without distinguishing between its two faces.

3. Vector Spaces

Having learned the elements of group theory, we are now ready to proceed to the theory of group representations. This chapter is intended to provide the necessary mathematical background about vectors and to review related subjects, which will help in understanding Chap. 4. Readers familiar with vectors are advised to run quickly over this chapter.

3.1 Vectors and Vector Spaces

As is well known in ordinary three-dimensional space (three-dimensional euclidean space), a *vector* $\boldsymbol{u}$ is defined by specifying its direction and length (magnitude). Vectors have the following fundamental properties:

(1) The sum of two vectors is also a vector. In the ordinary three dimensions, the sum is constructed according to the parallelepiped rule.
(2) Multiplication of a real number c with a vector $\boldsymbol{u}$ gives a vector $c\boldsymbol{u}$, which is parallel (if $c > 0$) or antiparallel (if $c < 0$) to the original vector $\boldsymbol{u}$.

The set of vectors constitutes the *vector space* $\mathscr{V}$. The most remarkable property of vectors is their linearity:

$$\boldsymbol{u}_1, \boldsymbol{u}_2 \in \mathscr{V} \rightarrow c_1\boldsymbol{u}_1 + c_2\boldsymbol{u}_2 \in \mathscr{V} \ .$$

In fact, a vector space is called a *linear space* as well.

*3.1.1 Mathematical Definition of a Vector Space

Strictly speaking, the mathematical definition of the vector space reads as follows.

The set $\mathscr{V}$ consisting of vectors $\{\boldsymbol{u}_i\}$ is called a vector space if it satisfies the following ten axioms.

V1: The set $\mathscr{V}$ is closed under addition. That is, for any two vectors $\boldsymbol{u}_i$ and $\boldsymbol{u}_j$ in $\mathscr{V}$, the unique sum $\boldsymbol{u}_i + \boldsymbol{u}_j$ also belongs to $\mathscr{V}$.

V2: The associative law

$$\boldsymbol{u}_i + (\boldsymbol{u}_j + \boldsymbol{u}_k) = (\boldsymbol{u}_i + \boldsymbol{u}_j) + \boldsymbol{u}_k$$

holds.

V3: The set $\mathscr{V}$ contains the zero element 0, which satisfies

$$\boldsymbol{u} + 0 = 0 + \boldsymbol{u} = \boldsymbol{u}$$

for any $\boldsymbol{u}$ in $\mathscr{V}$.

V4: For any vector $\boldsymbol{u}$ in $\mathscr{V}$, there exists a vector $-\boldsymbol{u}$ in $\mathscr{V}$ such that

$$\boldsymbol{u} + (-\boldsymbol{u}) = (-\boldsymbol{u}) + \boldsymbol{u} = 0 \ .$$

V5: The commutative law

$$\boldsymbol{u}_i + \boldsymbol{u}_j = \boldsymbol{u}_j + \boldsymbol{u}_i$$

holds.

The above five axioms are nothing other than the axioms for additive groups mentioned in Sect. 2.1. In addition, we have five more.

V6: The product $c\boldsymbol{u}$ of any vector $\boldsymbol{u}$ in $\mathscr{V}$ and any scalar quantity (real or complex number) c also belongs to $\mathscr{V}$.

V7: $(c_1 c_2)\boldsymbol{u} = c_1(c_2\boldsymbol{u})$ (associative law).

V8: $c(\boldsymbol{u}_i + \boldsymbol{u}_j) = c\boldsymbol{u}_i + c\boldsymbol{u}_j$

V9: $(c_1 + c_2)\boldsymbol{u} = c_1\boldsymbol{u} + c_2\boldsymbol{u}$ (distributive laws)

V10: $1\boldsymbol{u} = \boldsymbol{u}$.

According as the scalars c are real or complex numbers, the space $\mathscr{V}$ is called the *real vector space* or *complex vector space.*

3.1.2 Basis of a Vector Space

Take n vectors $\boldsymbol{a}_1, \boldsymbol{a}_2, \ldots, \boldsymbol{a}_n$ from the vector space $\mathscr{V}$ and construct an appropriate linear combination

$$c_1\boldsymbol{a}_1 + c_2\boldsymbol{a}_2 + \cdots + c_n\boldsymbol{a}_n$$

with the coefficients $c_1, c_2, \ldots, c_n$. If the homogeneous equation

$$c_1\boldsymbol{a}_1 + c_2\boldsymbol{a}_2 + \cdots + c_n\boldsymbol{a}_n = 0 \tag{3.1}$$

has no solution except for the trivial one

$$c_1 = 0 \ , \quad c_2 = 0, \ldots , \quad c_n = 0 \ , \tag{3.2}$$

we say that the n vectors $\boldsymbol{a}_1, \boldsymbol{a}_2, \ldots, \boldsymbol{a}_n$ are *linearly independent*. Conversely, if the above equation has a nontrivial solution, i.e., other than (3.2), the vectors are *linearly dependent.*

If we can find at most n independent vectors in the vector space $\mathscr{V}$, we call n the *dimension* of this vector space. To emphasize the dimension n of the vector space $\mathscr{V}$, the notation $\mathscr{V}^n$ is sometimes used.

Take n independent vectors $\boldsymbol{a}_1, \boldsymbol{a}_2, \ldots, \boldsymbol{a}_n$ from $\mathscr{V}^n$ and let $\boldsymbol{u}$ be an arbitrary vector in $\mathscr{V}^n$. Then the $n+1$ vectors $\boldsymbol{u}, \boldsymbol{a}_1, \boldsymbol{a}_2, \ldots, \boldsymbol{a}_n$ must be linearly dependent, and the equation

$$c\boldsymbol{u} + c_1\boldsymbol{a}_1 + c_2\boldsymbol{a}_2 + \cdots + c_n\boldsymbol{a}_n = 0 \tag{3.3}$$

for the scalars $c, c_1, c_2, \ldots, c_n$ must have a non-trivial solution with $c \neq 0$. (In the case $c = 0$, we would have the trivial solution (3.2) because of the linear independence of the vectors $\boldsymbol{a}_i$.) Dividing (3.3) through by c, we obtain

$$\boldsymbol{u} = u_1\boldsymbol{a}_1 + u_2\boldsymbol{a}_2 + \cdots + u_n\boldsymbol{a}_n \ . \tag{3.4}$$

In this way, any vector $\boldsymbol{u}$ in $\mathscr{V}^n$ can be expressed as a linear combination of n linearly independent vectors $\boldsymbol{a}_1, \boldsymbol{a}_2, \ldots, \boldsymbol{a}_n$. Such a set of vectors $\{\boldsymbol{a}_1, \boldsymbol{a}_2, \ldots, \boldsymbol{a}_n\}$ is called a *basis* of the vector space $\mathscr{V}^n$, and the coefficients $u_1, u_2, \ldots, u_n$ of (3.4), the *components* of vector $\boldsymbol{u}$ with respect to this basis.

Note that the choice of the basis is not unique. It has only to satisfy the requirement of linear independence.

3.2 Transformation of Vectors

In this section, we study the transformation of vectors in a vector space $\mathscr{V}^n$, starting from the $n = 2$ case. In two dimensions, it is natural to choose the unit vectors $\boldsymbol{i}$ and $\boldsymbol{j}$ in the x and y directions as the basis vectors, as shown in Fig. 3.1. Any vector $\boldsymbol{r}$ in this space may be expressed as

$$\boldsymbol{r} = x\boldsymbol{i} + y\boldsymbol{j} \tag{3.5}$$

in terms of its components x and y.

Now we rotate the basis vectors $\boldsymbol{i}$ and $\boldsymbol{j}$ through an angle α about the vertical axis. Then $\boldsymbol{i}$ and $\boldsymbol{j}$ are transformed to $\boldsymbol{i}'$ and $\boldsymbol{j}'$, respectively, which are related to $\boldsymbol{i}$ and $\boldsymbol{j}$ through

$$\begin{aligned} \boldsymbol{i}' &= \boldsymbol{i}\cos\alpha + \boldsymbol{j}\sin\alpha \ , \\ \boldsymbol{j}' &= -\boldsymbol{i}\sin\alpha + \boldsymbol{j}\cos\alpha \ . \end{aligned} \tag{3.6}$$

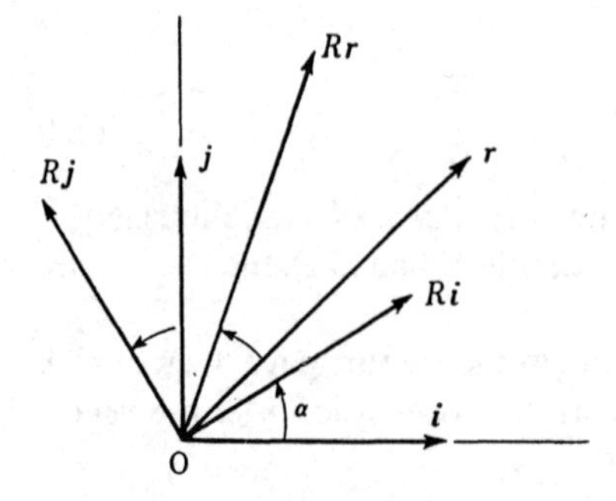

Fig. 3.1. Basis vectors $\boldsymbol{i}, \boldsymbol{j}$ of two-dimensional space and the effect of rotation R on them

Using the identity

$$[\boldsymbol{i},\boldsymbol{j}]\begin{bmatrix}\cos\alpha & -\sin\alpha\\ \sin\alpha & \cos\alpha\end{bmatrix}$$
$$= [\boldsymbol{i}\cos\alpha + \boldsymbol{j}\sin\alpha,\ -\boldsymbol{i}\sin\alpha + \boldsymbol{j}\cos\alpha]\ , \tag{3.7}$$

they can be written as the single equation

$$[\boldsymbol{i}',\boldsymbol{j}'] = [\boldsymbol{i},\boldsymbol{j}]\hat{R}\ , \tag{3.8}$$

where $\hat{R}$ is the 2×2 orthogonal matrix

$$\hat{R} = \hat{R}(\alpha) = \begin{bmatrix}\cos\alpha & -\sin\alpha\\ \sin\alpha & \cos\alpha\end{bmatrix}. \tag{3.9}$$

Note that the matrix $\hat{R}$ postmultiplies a row vector in (3.7, 8). Because $\boldsymbol{i}'(\boldsymbol{j}')$ is obtained from $\boldsymbol{i}(\boldsymbol{j})$ by the rotation, we write $\boldsymbol{i}' = R\boldsymbol{i}$ ($\boldsymbol{j}' = R\boldsymbol{j}$), using the rotation operator R. Then we can write (3.8) as

$$[R\boldsymbol{i}, R\boldsymbol{j}] = [\boldsymbol{i},\boldsymbol{j}]\hat{R}\ . \tag{3.10}$$

Now consider how the vector $\boldsymbol{r}$ is transformed by this rotation R. As R is a linear operator, we have

$$\begin{aligned}\boldsymbol{r}' &= R\boldsymbol{r} = R(x\boldsymbol{i} + y\boldsymbol{j}) = xR\boldsymbol{i} + yR\boldsymbol{j} = x\boldsymbol{i}' + y\boldsymbol{j}'\\ &= x(\boldsymbol{i}\cos\alpha + \boldsymbol{j}\sin\alpha) + y(-\boldsymbol{i}\sin\alpha + \boldsymbol{j}\cos\alpha)\\ &= (x\cos\alpha - y\sin\alpha)\boldsymbol{i} + (x\sin\alpha + y\cos\alpha)\boldsymbol{j}\ .\end{aligned}$$

However, $\boldsymbol{r}'$ is also given by

$$\boldsymbol{r}' = x'\boldsymbol{i} + y'\boldsymbol{j}$$

in terms of its components x' and y'. Equating the coefficients of $\boldsymbol{i}$ and $\boldsymbol{j}$, we obtain

$$x' = x\cos\alpha - y\sin\alpha\ , \qquad y' = x\sin\alpha + y\cos\alpha\ , \tag{3.11}$$

or

$$\begin{bmatrix}x'\\ y'\end{bmatrix} = \hat{R}\begin{bmatrix}x\\ y\end{bmatrix}, \tag{3.12}$$

using the rotation matrix $\hat{R}$ defined by (3.9). Note the difference between the transformation properties of the basis vectors and the vector components: the basis vectors $\boldsymbol{i},\boldsymbol{j}$ transform according to (3.8), while the vector components x, y transform according to (3.12).

The above considerations on the transformation of vectors are readily generalized to the case of higher dimensions. In three dimensions, the rotation operator R is expressible in terms of three angular parameters (Euler angles) α, β, and γ, and the corresponding rotation matrix $\hat{R}$ is given by a 3×3 matrix (see (7.7)).

Let us now proceed to the case of n-dimensional vector space $\mathscr{V}^n$. Consider a linear operator T which operates on vectors in $\mathscr{V}^n$. We assume that, for any vector $\boldsymbol{u}$ in $\mathscr{V}^n$, the transformed vector $T\boldsymbol{u}$ also belongs to $\mathscr{V}^n$. The linearity of T implies

$$T(\boldsymbol{u} + \boldsymbol{v}) = T\boldsymbol{u} + T\boldsymbol{v} \ , \qquad T(c\boldsymbol{u}) = cT\boldsymbol{u} \ , \tag{3.13}$$

where $\boldsymbol{u}, \boldsymbol{v} \in \mathscr{V}^n$ and c is an arbitrary complex number.

Suppose a vector $\boldsymbol{u}$ is expressed as

$$\boldsymbol{u} = u_1\boldsymbol{a}_1 + u_2\boldsymbol{a}_2 + \cdots + u_k\boldsymbol{a}_k + \cdots + u_n\boldsymbol{a}_n$$

in terms of basis vectors $\boldsymbol{a}_k, k = 1, 2, \ldots, n$. When T operates on $\boldsymbol{a}_k$, the transformed vector $T\boldsymbol{a}_k$, being a member of $\mathscr{V}^n$, must be expressible as a linear combination of the basis vectors,

$$T\boldsymbol{a}_k = \sum_{i=1}^{n} \boldsymbol{a}_i T_{ik} \ , \qquad k = 1, 2, \ldots, n \ . \tag{3.14}$$

Introducing the row vector $[\boldsymbol{a}_1, \boldsymbol{a}_2, \ldots, \boldsymbol{a}_n]$ and the transformation matrix

$$\hat{T} = [T_{ik}] = \begin{bmatrix} T_{11} & T_{12} & \ldots & T_{1n} \\ T_{21} & T_{22} & \ldots & T_{2n} \\ \cdot & \cdot & \ldots & \cdot \\ \cdot & \cdot & \ldots & \cdot \\ \cdot & \cdot & \ldots & \cdot \\ T_{n1} & T_{n2} & \ldots & T_{nn} \end{bmatrix} , \tag{3.15}$$

we may write (3.14) in a single equation,

$$[T\boldsymbol{a}_1, T\boldsymbol{a}_2, \ldots, T\boldsymbol{a}_n] = [\boldsymbol{a}_1, \boldsymbol{a}_2, \ldots, \boldsymbol{a}_n]\hat{T} \ . \tag{3.16}$$

If we express the transformed vector $\boldsymbol{u}'$ in the same basis, we have

$$\boldsymbol{u}' = u'_1\boldsymbol{a}_1 + u'_2\boldsymbol{a}_2 + \cdots + u'_i\boldsymbol{a}_i + \cdots + u'_n\boldsymbol{a}_n \ . \tag{3.17}$$

On the other hand, as $\boldsymbol{u}'$ is obtained by operating T on $\boldsymbol{u}$, we have

$$\begin{aligned} \boldsymbol{u}' &= T\boldsymbol{u} = T\sum_k u_k\boldsymbol{a}_k = \sum_k u_k T\boldsymbol{a}_k \\ &= \sum_k u_k \sum_i \boldsymbol{a}_i T_{ik} = \sum_i \left(\sum_k T_{ik} u_k \right) \boldsymbol{a}_i \ . \end{aligned} \tag{3.18}$$

Equating the coefficients of $\boldsymbol{a}_i$ in this relation with those in (3.17), we find

$$u_i' = \sum_k T_{ik} u_k \ , \qquad i = 1, 2, \ldots, n \ . \tag{3.19}$$

If we use the transformation matrix $\hat{T}$ of (3.15), this system of equations can be written as

$$\begin{bmatrix} u_1' \\ u_2' \\ \vdots \\ u_n' \end{bmatrix} = \hat{T} \begin{bmatrix} u_1 \\ u_2 \\ \vdots \\ u_n \end{bmatrix} \ . \tag{3.20}$$

In conclusion, if the basis $\{\boldsymbol{a}_1, \boldsymbol{a}_2, \ldots, \boldsymbol{a}_n\}$ transforms according to (3.16), the transformation of the corresponding vector components is given by (3.20). The action of the operator T is determined by the transformation matrix $\hat{T}$.

Let us next consider the effect of successive transformations. Operating with S on $T\boldsymbol{a}_k$, we obtain

$$\begin{aligned} ST\boldsymbol{a}_k &= S \sum_j \boldsymbol{a}_j T_{jk} = \sum_i \boldsymbol{a}_i \sum_j S_{ij} T_{jk} \\ &= \sum_i \boldsymbol{a}_i (\hat{S}\hat{T})_{ik} \ . \end{aligned} \tag{3.21}$$

As for the transformation of vector components, from

$$u_j' = \sum_k T_{jk} u_k \quad \text{and} \quad u_i'' = \sum_j S_{ij} u_j' \ ,$$

we obtain

$$\begin{aligned} u_i'' &= \sum_j S_{ij} \sum_k T_{jk} u_k = \sum_k \left(\sum_j S_{ij} T_{jk} \right) u_k \\ &= \sum_k (\hat{S}\hat{T})_{ik} u_k \ . \end{aligned}$$

This may be written in the form

$$\begin{bmatrix} u_1'' \\ u_2'' \\ \vdots \\ u_n'' \end{bmatrix} = \hat{S}\hat{T} \begin{bmatrix} u_1 \\ u_2 \\ \vdots \\ u_n \end{bmatrix} \ . \tag{3.22}$$

If the determinant of the transformation matrix $\hat{T}$ is not zero, the inverse matrix $\hat{T}^{-1}$ exists. Such a transformation is called invertible, regular or non-singular. In this case, we can solve (3.20) and obtain

$$\begin{bmatrix} u_1 \\ u_2 \\ \vdots \\ u_n \end{bmatrix} = \hat{T}^{-1} \begin{bmatrix} u'_1 \\ u'_2 \\ \vdots \\ u'_n \end{bmatrix} . \tag{3.23}$$

If we change the basis from $\{\boldsymbol{a}_1, \boldsymbol{a}_2, \ldots, \boldsymbol{a}_n\}$ to $\{\boldsymbol{a}'_1, \boldsymbol{a}'_2, \ldots, \boldsymbol{a}'_n\}$ by the transformation

$$\boldsymbol{a}'_k = T\boldsymbol{a}_k = \sum_j \boldsymbol{a}_j T_{jk} \tag{3.24}$$

with a nonsingular matrix $\hat{T} \equiv [T_{ik}]$, we can solve (3.24) in the form

$$\boldsymbol{a}_i = \sum_h \boldsymbol{a}'_h (\hat{T}^{-1})_{hi} . \tag{3.25}$$

When the transformed basis $\boldsymbol{a}'_k$ undergoes a linear transformation A, we have

$$A\boldsymbol{a}'_k = \sum_j A\boldsymbol{a}_j T_{jk} = \sum_{i,j} \boldsymbol{a}_i A_{ij} T_{jk} . \tag{3.26}$$

Substituting (3.25) for $\boldsymbol{a}_i$, we have

$$\begin{aligned} A\boldsymbol{a}'_k &= \sum_h \boldsymbol{a}'_h \sum_{i,j} (\hat{T}^{-1})_{hi} A_{ij} T_{jk} \\ &= \sum_h \boldsymbol{a}'_h (\hat{T}^{-1}\hat{A}\hat{T})_{hk} = \sum_h \boldsymbol{a}'_h A'_{hk} , \end{aligned}$$

where A'_{hk} is the hk element of the matrix

$$\hat{A}' = \hat{T}^{-1}\hat{A}\hat{T} . \tag{3.27}$$

Equation (3.27) shows how the linear transformation T transforms the matrix elements of the linear operator A.

3.3 Subspaces and Invariant Subspaces

A nonempty subset $\mathscr{V}'$ of the vector space $\mathscr{V}$ is called a *subspace* if it is closed within itself.

More precisely, $\mathscr{V}'$ is a subspace of $\mathscr{V}$ if

$$\boldsymbol{u}, \boldsymbol{v} \in \mathscr{V}' \longrightarrow \boldsymbol{u} + \boldsymbol{v} \in \mathscr{V}'$$

and

$$\boldsymbol{u} \in \mathscr{V}' \longrightarrow c\boldsymbol{u} \in \mathscr{V}'$$

for any complex number c. The above two conditions are sufficient for the subset $\mathscr{V}'$ to satisfy the axioms V1–V10 of the vector space.

If any vector $\boldsymbol{u}$ in a given vector space $\mathscr{V}$ is uniquely expressed in the form

$$\boldsymbol{u} = \boldsymbol{u}' + \boldsymbol{u}'' ,$$

where $\boldsymbol{u}'$ belongs to a subspace $\mathscr{V}'$ and $\boldsymbol{u}''$ belongs to another subspace $\mathscr{V}''$, we say that the vector space $\mathscr{V}$ is decomposed into a *direct sum* of its subspaces $\mathscr{V}'$ and $\mathscr{V}''$, and write

$$\mathscr{V} = \mathscr{V}' + \mathscr{V}'' . \qquad (3.28)$$

$\mathscr{V}''$ is called the *complementary subspace* of $\mathscr{V}'$. If we combine the basis of the subspace $\mathscr{V}'$ with the basis of $\mathscr{V}''$, we obtain the basis for the entire space $\mathscr{V}$. Such a basis of $\mathscr{V}$ is *adapted to the decomposition* (3.28).

If the dimension of $\mathscr{V}'$ is n', and that of $\mathscr{V}''$ is n'', their sum $n = n' + n''$ is equal to the dimension of $\mathscr{V}$.

Let A be a linear operator which operates on vectors in the vector space $\mathscr{V}^n$, and let $\mathscr{V}^{n'}$ be a subspace of $\mathscr{V}^n$. If, for any vector $\boldsymbol{u}$ in $\mathscr{V}^{n'}$, $A\boldsymbol{u}$ remains in $\mathscr{V}^{n'}$, the subspace $\mathscr{V}^{n'}$ is called an *invariant subspace* of $\mathscr{V}^n$ with respect to the operator A. We say that the subspace $\mathscr{V}^{n'}$ remains invariant under the operator A.

Suppose that the vectors $\boldsymbol{u}_1, \boldsymbol{u}_2, \ldots, \boldsymbol{u}_{n'}$, form a basis for the invariant subspace $\mathscr{V}^{n'}$. Since they are closed in $\mathscr{V}^{n'}$ under the operation A, we have

$$A\boldsymbol{u}_k = \sum_{j=1}^{n'} \boldsymbol{u}_j A_{jk} , \qquad k = 1, 2, \ldots, n' . \qquad (3.29)$$

To this set of vectors we add $n - n'$ linearly independent ones $\boldsymbol{u}_{n'+1}$, $\boldsymbol{u}_{n'+2}, \ldots, \boldsymbol{u}_n$ and complete the basis for $\mathscr{V}^n$. Then we see that

$$\begin{aligned} A\boldsymbol{u}_l &= \sum_{j=1}^{n} \boldsymbol{u}_j A_{jl} \\ &= \sum_{j'=1}^{n'} \boldsymbol{u}_{j'} A_{j'l} + \sum_{j''=n'+1}^{n} \boldsymbol{u}_{j''} A_{j''l} \end{aligned} \qquad (3.30)$$

for $\boldsymbol{u}_l$ $(l = n' + 1, \ldots, n)$, so that the transformation matrix has the form

$$\begin{array}{c} 1 \\ 2 \\ \vdots \\ n' \\ n'+1 \\ \vdots \\ n \end{array} \quad \left[\begin{array}{c|c} A_{jk} & A_{j'l} \\ \hline 0 & A_{j''l} \end{array}\right] \quad \begin{array}{l} \left.\begin{array}{c} \\ \\ \end{array}\right\} \mathscr{V}^{n'} \\ \\ \end{array} \left.\begin{array}{c} \\ \\ \\ \\ \end{array}\right\} \mathscr{V}^{n} . \qquad (3.31)$$

All elements in the lower left $(n - n') \times n'$ rectangular part of the matrix are zero.

3.4 Metric Vector Spaces

3.4.1 Inner Product of Vectors

The vector space considered so far contains as yet no important concepts like length (magnitude) and orthogonality of vectors. Introduction of these concepts calls for definition of the inner product (or scalar product) of vectors. The vector space $\mathscr{V}$ equipped with the inner product is called the *metric vector space.*

The *inner product* $(\boldsymbol{u}, \boldsymbol{v})$ of complex vectors $\boldsymbol{u}$ and $\boldsymbol{v}$ should have the following properties:

S1: $(\boldsymbol{u}, \boldsymbol{v}) = (\boldsymbol{v}, \boldsymbol{u})^*$.
S2: $(\boldsymbol{u}, \boldsymbol{v} + \boldsymbol{w}) = (\boldsymbol{u}, \boldsymbol{v}) + (\boldsymbol{u}, \boldsymbol{w})$.
S3: For any complex number c, $(\boldsymbol{u}, c\boldsymbol{v}) = c(\boldsymbol{u}, \boldsymbol{v})$.
S4: $(\boldsymbol{u}, \boldsymbol{u}) \geqq 0$.
S5: $(\boldsymbol{u}, \boldsymbol{u}) = 0$, if and only if $\boldsymbol{u} = 0$.

The asterisk in S1 means the complex conjugate, which is necessitated by the compatibility with S3 and S4. Note that we have $(c\boldsymbol{u}, \boldsymbol{v}) = c^*(\boldsymbol{u}, \boldsymbol{v})$ from S1 and S3. If $(\boldsymbol{u}, \boldsymbol{v}) = 0$, we say that $\boldsymbol{u}$ and $\boldsymbol{v}$ are orthogonal. Since $(\boldsymbol{u}, \boldsymbol{u})$ is nonnegative by definition, we can define a nonnegative quantity

$$|\boldsymbol{u}| = (\boldsymbol{u}, \boldsymbol{u})^{1/2} , \tag{3.32}$$

called the *norm* or *length* of the vector $\boldsymbol{u}$.

3.4.2 Orthonormal Basis

As mentioned in Sect. 3.1.2, a vector space $\mathscr{V}^n$ of dimension n can have a basis set of n linearly independent vectors $\{\boldsymbol{a}_1, \boldsymbol{a}_2, \ldots, \boldsymbol{a}_n\}$. If the inner product is defined in this space, by application of an appropriate linear transformation to this basis (Gram–Schmidt orthogonalization), we can construct an orthonormal basis $\{\boldsymbol{e}_1, \boldsymbol{e}_2, \ldots, \boldsymbol{e}_n\}$, which satisfies

$$(\boldsymbol{e}_i, \boldsymbol{e}_j) = \delta_{ij} , \quad i, j = 1, 2, \ldots, n . \tag{3.33}$$

By means of this basis, any vector $\boldsymbol{u}$ in $\mathscr{V}$ may be expressed as

$$\boldsymbol{u} = u_1\boldsymbol{e}_1 + u_2\boldsymbol{e}_2 + \cdots + u_n\boldsymbol{e}_n . \tag{3.34}$$

Thus, the inner product of two vectors $\boldsymbol{u}$ and $\boldsymbol{v}$ may be written as

$$(\boldsymbol{u}, \boldsymbol{v}) = u_1^* v_1 + u_2^* v_2 + \cdots + u_n^* v_n , \tag{3.35}$$

using their components with respect to this basis. The norm becomes

$$|\boldsymbol{u}| = \left(\sum_i |u_i|^2\right)^{1/2} . \tag{3.36}$$

To sum up, the metric vector space is provided with the inner product. In this space, it is convenient to construct the orthonormal basis (3.33) and express any vector $\boldsymbol{u}$ in the form (3.34).

3.4.3 Unitary Operators and Unitary Matrices

Suppose $\boldsymbol{u}$ and $\boldsymbol{v}$ belong to a metric vector space $\mathscr{V}$. If an operator U transforms $\boldsymbol{u}$ and $\boldsymbol{v}$ as

$$\boldsymbol{u} \to \boldsymbol{u}' = U\boldsymbol{u} , \qquad \boldsymbol{v} \to \boldsymbol{v}' = U\boldsymbol{v} , \tag{3.37}$$

keeping their inner product invariant

$$(\boldsymbol{u}, \boldsymbol{v}) = (U\boldsymbol{u}, U\boldsymbol{v}) , \tag{3.38}$$

then U is said to be *unitary*. The components of (3.37) are written as

$$u_i' = \sum_k U_{ik} u_k , \qquad v_i' = \sum_l U_{il} v_l .$$

Therefore, we have

$$(\boldsymbol{u}', \boldsymbol{v}') = \sum_i u_i'^* v_i' = \sum_{k,l} \left(\sum_i U_{ik}^* U_{il}\right) u_k^* v_l .$$

Since this must be equal to

$$(\boldsymbol{u}, \boldsymbol{v}) = \sum_k u_k^* v_k = \sum_k \sum_l \delta_{kl} u_k^* v_l ,$$

we have

$$\sum_i U_{ik}^* U_{il} = \delta_{kl} , \qquad k, l = 1, 2, \ldots, n \tag{3.39}$$

If we use the transformation matrix $\hat{U}$ whose ik element is U_{ik}, then (3.39) is written as

$$\hat{U}^\dagger \hat{U} = \hat{U}\hat{U}^\dagger = \hat{1} , \tag{3.40}$$

or

$$\hat{U}^\dagger = \hat{U}^{-1} . \tag{3.41}$$

The matrix $\hat{U}^\dagger$ is the *Hermitian conjugate* matrix of $\hat{U}$, satisfying

$$(\hat{U}^\dagger)_{ij} = U^*_{ji} \ . \tag{3.42}$$

Matrices satisfying (3.41) are called *unitary matrices.*

3.4.4 Hermitian Operators and Hermitian Matrices

A matrix $\hat{H}$ whose Hermitian conjugate $\hat{H}^\dagger$ is equal to $\hat{H}$ such that

$$\hat{H}^\dagger = \hat{H} \tag{3.43}$$

is a *Hermitian matrix* or *self-adjoint matrix.* Its elements satisfy

$$H^*_{ji} = H_{ij} \ . \tag{3.44}$$

3.5 Eigenvalue Problems of Hermitian and Unitary Operators

Hermitian and unitary matrices are most useful in physical applications. They have common important properties in the structure of their invariant subspaces. Let a general Hermitian or unitary matrix be represented by $\hat{A}$, and the corresponding linear transformation operator by A. Then we have the following lemma.

Lemma: If A transforms a subspace $\mathscr{S}$ of the vector space $\mathscr{V}$ into itself, then it also transforms the complementary subspace (totally orthogonal subspace) $\mathscr{S}^\perp$ into itself. Thus, both $\mathscr{S}$ and $\mathscr{S}^\perp$ form invariant subspaces by themselves with respect to the operator A.

Proof: If A is Hermitian,

$$\begin{aligned}(A\boldsymbol{u}, \boldsymbol{v}) &= \sum_{i,j} A^*_{ij} u^*_j v_i = \sum_j u^*_j \sum_i A_{ji} v_i \\ &= (\boldsymbol{u}, A\boldsymbol{v}) \ .\end{aligned} \tag{3.45}$$

For an arbitrary vector $\boldsymbol{u}$ in the subspace $\mathscr{S}$, and an arbitrary vector $\boldsymbol{v}$ in $\mathscr{S}^\perp$, we have $(\boldsymbol{u}, \boldsymbol{v}) = 0$. Furthermore, since $A\boldsymbol{u} \in \mathscr{S}$, $A\boldsymbol{u}$ is orthogonal to $\boldsymbol{v}$: $(A\boldsymbol{u}, \boldsymbol{v}) = 0$. Therefore, from (3.45), $(\boldsymbol{u}, A\boldsymbol{v}) = 0$. This means that $A\boldsymbol{v}$ is orthogonal to any vector in $\mathscr{S}$, and hence $A\boldsymbol{v} \in \mathscr{S}^\perp$. Consequently, A transforms any vector $\boldsymbol{v}$ of $\mathscr{S}^\perp$ into $\mathscr{S}^\perp$.

If A is unitary, we have $(A\boldsymbol{u}, A\boldsymbol{v}) = (\boldsymbol{u}, \boldsymbol{v})$. For any vector $\boldsymbol{u}$ of $\mathscr{S}$ and any vector $\boldsymbol{v}$ of $\mathscr{S}^\perp$, $(\boldsymbol{u}, \boldsymbol{v}) = 0$, so that $(A\boldsymbol{u}, A\boldsymbol{v}) = 0$. Because $A\boldsymbol{u}$ is in $\mathscr{S}$, this means $A\boldsymbol{v}$ is orthogonal to $\mathscr{S}$, and $A\boldsymbol{v} \in \mathscr{S}^\perp$ follows. □

Theorem: Every Hermitian or unitary operator A possesses a complete set of eigenvectors.

Proof: For a root $\lambda = \lambda_1$ of the secular equation

$$\det(\hat{A} - \lambda\hat{1}) = \begin{bmatrix} A_{11} - \lambda & A_{12} & \dots & A_{1n} \\ A_{21} & A_{22} - \lambda & \dots & A_{2n} \\ \vdots & \vdots & & \vdots \\ A_{n1} & A_{n2} & \dots & A_{nn} - \lambda \end{bmatrix} = 0 \tag{3.46}$$

the system of equations

$$\begin{aligned} (A_{11} - \lambda_1)c_1 + \quad & A_{12}c_2 + \dots + \quad & A_{1n}c_n = 0 \\ A_{21}c_1 + (A_{22} - \lambda_1)c_2 + \dots + \quad & & A_{2n}c_n = 0 \\ \vdots \quad & & \\ A_{n1}c_1 + \quad & A_{n2}c_2 + \dots + (A_{nn} - \lambda_1)c_n = 0 & \end{aligned} \tag{3.47}$$

has a nontrivial solution $(c_1, c_2, \dots, c_n)$. The column vector

$$\boldsymbol{v}_1 = \begin{bmatrix} c_1 \\ c_2 \\ \vdots \\ c_n \end{bmatrix}$$

is the eigenvector for the eigenvalue equation

$$A\boldsymbol{v}_1 = \lambda_1 \boldsymbol{v}_1 \ . \tag{3.48}$$

By the above lemma, A transforms the subspace $\mathscr{S}_{n-1}^{\perp}$ of all vectors orthogonal to $\boldsymbol{v}_1$ into $\mathscr{S}_{n-1}^{\perp}$ itself. Thus the original eigenvalue problem is reduced to the one with a decreased dimension $n - 1$. By the same procedure, an eigenvector $\boldsymbol{v}_2$ within $\mathscr{S}_{n-1}^{\perp}$ can be found, the secular equation being of degree $n - 1$. Within $\mathscr{S}_{n-1}^{\perp}$, the vectors orthogonal to $\boldsymbol{v}_2$ form a subspace $\mathscr{S}_{n-2}^{\perp}$, and so on. Thus, we find the orthogonal system

$$A\boldsymbol{v}_k = \lambda_k \boldsymbol{v}_k \ , \quad k = 1, 2, \dots, n \ . \tag{3.49}$$

with orthonormality

$$(\boldsymbol{v}_k, \boldsymbol{v}_l) = \delta_{kl} \ . \qquad \square \tag{3.50}$$

If we choose $\boldsymbol{v}_1, \boldsymbol{v}_2, \dots, \boldsymbol{v}_n$ as the basis vectors of $\mathscr{V}^n$, the matrix $\hat{A}$ becomes diagonal,

$$A_{ij} = \lambda_i \delta_{ij} \ . \tag{3.51}$$

Thus the operator A is diagonalized.

Exercise 3.1. Prove that if $\hat{A}$ is a Hermitian matrix, its eigenvalues are real.

Exercise 3.2. Show that if $\hat{A}$ is a unitary matrix, the absolute values of its eigenvalues are equal to 1. In addition, show that if $\hat{A}$ is also symmetric, we can choose the eigenvectors as real.

*3.6 Linear Transformation Groups

As we learned in Sect. 3.2, linear operators T in the vector space $\mathscr{V}^n$ are represented by $n \times n$ matrices

$$\hat{T} = [T_{ik}] \ .$$

When the matrix $\hat{T}$ is not singular ($\det \hat{T} \neq 0$), the inverse matrix $\hat{T}^{-1}$ exists. In this case, consider the result of successive linear transformations T and S whose effects are defined by nonsingular matrices $\hat{S}$ and $\hat{T}$. The result becomes (3.21, 22). Thus, the effect of the successive transformations is described by the product matrix $\hat{S}\hat{T}$. If we regard the operation of forming the matrix product as the multiplication ($\circ$) for group elements explained in Sect. 2.1, the set of nonsingular matrices is closed, the unit element exists as the unit matrix $\hat{1}$, and inverse elements exist as inverse matrices; in this way, the group axioms are fulfilled. Therefore, this set of nonsingular matrices will form a group, which is called the *linear transformation group*. In particular, if their elements are complex numbers, we call the matrix group

$$\mathrm{GL}(n, \mathbb{C}) = \{\hat{A} = [A_{ik}] \,|\, i, k = 1, 2, \ldots, n;\ A_{ik} \in \mathbb{C}, \det \hat{A} \neq 0\}$$

the *complex general linear group* of order n. If the elements A_{ik} are real numbers,

$$\mathrm{GL}(n, \mathbb{R}) = \{\hat{A} = [A_{ik}] \,|\, i, k = 1, 2, \ldots, n;\ A_{ik} \in \mathbb{R}, \det \hat{A} \neq 0\}$$

is the *real general linear group* of order n. In $\mathrm{GL}(n, \mathbb{C})$ and $\mathrm{GL}(n, \mathbb{R})$, the subsets satisfying $\det \hat{A} = 1$ constitute subgroups. We denote them by

$$\mathrm{SL}(n, \mathbb{C}) = \{\hat{A} \in \mathrm{GL}(n, \mathbb{C}) |\, \det \hat{A} = 1\} \ ,$$

$$\mathrm{SL}(n, \mathbb{R}) = \{\hat{A} \in \mathrm{GL}(n, \mathbb{R}) |\, \det \hat{A} = 1\} \ ,$$

and call them the *complex special linear group* and the *real special linear group*.

Furthermore, the set of unitary matrices

$$\mathrm{U}(n) = \{\hat{U} \,|\, U_{ik} \in \mathbb{C} \ , \ \hat{U}\hat{U}^\dagger = \hat{U}^\dagger \hat{U} = \hat{1}\}$$

is called the *unitary group*. The subgroup of $\mathrm{U}(n)$ that satisfies the condition $\det \hat{U} = 1$,

$$\mathrm{SU}(n) = \{\hat{U} \,|\, \hat{U} \in \mathrm{U}(n) \ , \ \det \hat{U} = 1\} \ ,$$

is the *special unitary group*. In $\mathrm{U}(n)$ and $\mathrm{SU}(n)$, the elements U_{ik} are understood to be complex numbers. When they are restricted to real numbers, we obtain the *orthogonal group*

$$\mathrm{O}(n) = \{\hat{O} \mid O_{ik} \in \mathbb{R} \ , \ {}^t\hat{O}\hat{O} = \hat{O}{}^t\hat{O} = \hat{1}\} \ ,$$

and the *special orthogonal group*

$$\mathrm{SO}(n) = \{\hat{O} \mid \hat{O} \in \mathrm{O}(n) \ , \ \det \hat{O} = 1\} \ .$$

$\mathrm{O}(3)$ is the well-known orthogonal transformation group of three-dimensional euclidean space (Sect. 7.6). $\mathrm{SO}(n)$ is the extension to n dimensions of the proper rotation group.

All the groups introduced in this section are also called *classical groups*.

4. Representations of a Group I

In this and the following chapter, we shall develop the theory of representations of finite groups. We begin with the definition of group representations and related fundamental concepts (Sect. 4.1), and follow this with examples of representations (Sects. 4.2, 4.4). Between the examples (Sect. 4.3), effects of symmetry transformation operators on functions are considered. After having become familiar with group representations from these examples, we present the general representation theory (Sects 4.5–13). Several theorems, such as Schur's lemma and orthogonality theorems, appear at this stage. Most of their proofs are given at the end of the chapter (Sect. 4.13).

Readers may jump, on a first reading, from Sect. 4.9 to Chap. 6, which is concerned with the application of representation theory to quantum mechanics. Section 4.12 deals with complex conjugate representations, and should be read before studying time-reversal degeneracies in Chap. 13.

4.1 Representations

Let $\mathscr{G}$ be a finite group of order g consisting of the elements $G_1(=E)$, $G_2, \ldots, G_g$, and let a square matrix $\hat{D}(G_i)$ be associated with each group element G_i. If the matrices satisfy

$$\hat{D}(G_j)\hat{D}(G_i) = \hat{D}(G_k) \tag{4.1}$$

for the corresponding relation of the group elements

$$G_j G_i = G_k \ , \tag{4.2}$$

then the set of the matrices $\hat{D}(G_1), \hat{D}(G_2), \ldots, \hat{D}(G_g)$ is called a *representation*[1] of the group $\mathscr{G}$. Individual matrices $\hat{D}(G_i)$ are called representation matrices, and the size of the matrices is the *dimension* of the representation.

When G_i is the unit element E in (3.1), we have

$$\hat{D}(G_j)\hat{D}(E) = \hat{D}(G_j) \ ,$$

[1] Equation (4.1) means that the mapping D: $G_i \to \hat{D}(G_i)$ is homomorphic, see (2.16). Such a homomorphic mapping in general is called a representation. The symbol D derives from the German word *Darstellung* for representation.

which means

$$\hat{D}(E) = \hat{1} ,$$

i.e., the unit element E is represented by the unit matrix $\hat{1}$. Furthermore, from $G_i G_i^{-1} = E$, we have

$$\hat{D}(G_i)\hat{D}(G_i^{-1}) = \hat{D}(E) = \hat{1} ,$$

and hence

$$\hat{D}(G_i^{-1}) = \hat{D}(G_i)^{-1} ,$$

i.e., the inverse element G_i^{-1} is represented by the inverse matrix $\hat{D}(G_i)^{-1}$.

The correspondence between the group elements and the representation matrices is not necessarily one-to-one. The representation is said to be *faithful* in the case of one-to-one correspondence.

In Sect. 2.2, we saw that the coordinates x and y in two dimensions are transformed according to (2.8) under the symmetry group C_{3v} of the equilateral triangle. In the present context, the matrices of (2.8) may be reproduced as follows:

$$\hat{D}^{(E)}(E) = \begin{bmatrix} 1 & 0 \\ 0 & 1 \end{bmatrix} , \quad \hat{D}^{(E)}(C_3) = \begin{bmatrix} c & -s \\ s & c \end{bmatrix} , \quad \hat{D}^{(E)}(C_3^{-1}) = \begin{bmatrix} c & s \\ -s & c \end{bmatrix} ,$$

$$\hat{D}^{(E)}(\sigma_1) = \begin{bmatrix} 1 & 0 \\ 0 & -1 \end{bmatrix} , \quad \hat{D}^{(E)}(\sigma_2) = \begin{bmatrix} c & -s \\ -s & -c \end{bmatrix} , \quad \hat{D}^{(E)}(\sigma_3) = \begin{bmatrix} c & s \\ s & -c \end{bmatrix} , \tag{4.3}$$

where

$$c = \cos(2\pi/3) = -1/2 , \quad s = \sin(2\pi/3) = \sqrt{3}/2 . \tag{4.4}$$

These six matrices, satisfying (4.1), form a faithful representation of the group C_{3v}. The above representation is said to be a *unitary representation*, because all the representation matrices are unitary. The superscript E in (4.3) is the name given to this representation.

Any group $\mathscr{G}$ always has the *identity representation*, in which every group element G_i is represented simply by 1, regarded as a 1×1 matrix. In the group C_{3v}, the identity representation A_1 is given by

$$D^{(A_1)}(E) = 1 , \quad D^{(A_1)}(C_3) = 1 , \quad D^{(A_1)}(C_3^{-1}) = 1 ,$$

$$D^{(A_1)}(\sigma_1) = 1 , \quad D^{(A_1)}(\sigma_2) = 1 , \quad D^{(A_1)}(\sigma_3) = 1 . \tag{4.5}$$

Its dimension is 1. The group C_{3v} has another one-dimensional representation:

$$\begin{aligned} &D^{(A_2)}(E) = 1 , \quad &D^{(A_2)}(C_3) = 1 , \quad &D^{(A_2)}(C_3^{-1}) = 1 , \\ &D^{(A_2)}(\sigma_1) = -1 , \quad &D^{(A_2)}(\sigma_2) = -1 , \quad &D^{(A_2)}(\sigma_3) = -1 . \end{aligned} \tag{4.6}$$

As will be seen in due course, C_{3v} has only the three independent representations given above. The nomenclature for the representations A_1, A_2 and E will be explained in Sect. 8.4.

4.1.1 Basis for a Representation

Let $\psi_1, \psi_2, \ldots, \psi_d$ be independent elements in a vector space, and suppose that the linear operators[2] G_i operating on them form a group $\mathscr{G}$. The set $\{\psi_1, \psi_2, \ldots, \psi_d\}$ is called a *basis* for a representation, if

$$G_i \psi_\nu = \sum_{\mu=1}^{d} \psi_\mu D_{\mu\nu}(G_i) \ , \quad G_i \in \mathscr{G} \ , \tag{4.7}$$

holds, i.e., if the basis is closed within itself under the operations of the group $\mathscr{G}$. Individual members of the basis are called *partners*, basis vectors, or basis functions.

It is easy to see that the coefficient matrices $D_{\mu\nu}(G_i)$ form a representation. Operating a linear operator G_j on both sides of (4.7), and using again (4.7), we obtain

$$\begin{aligned} G_j G_i \psi_\nu &= \sum_{\mu=1}^{d} G_j \psi_\mu D_{\mu\nu}(G_i) \\ &= \sum_{\lambda=1}^{d} \sum_{\mu=1}^{d} \psi_\lambda D_{\lambda\mu}(G_j) D_{\mu\nu}(G_i) \\ &= \sum_{\lambda=1}^{d} \psi_\lambda [\hat{D}(G_j)\hat{D}(G_i)]_{\lambda\nu} \ . \end{aligned} \tag{4.8}$$

On the other hand, from (4.7) itself, we have

$$G_k \psi_\nu = \sum_{\lambda=1}^{d} \psi_\lambda D_{\lambda\nu}(G_k) \ . \tag{4.9}$$

Comparing (4.9) with (4.8), we obtain (4.1), which demonstrates that the matrices $\hat{D}(G_i)$ really form a representation. Since the νth partner ψ_ν in the basis transforms according to the νth column of the representation matrices, it may be said to belong to the νth column of the representation.

Because of the closure property of the basis mentioned above, the space spanned by the basis is an *invariant subspace*.

---

[2] Linear operators G, by definition, satisfy

$$G(c_1\phi_1 + c_2\phi_2) = c_1 G\phi_1 + c_2 G\phi_2 \ .$$

4.1.2 Equivalence of Representations

When the matrices of two representations D and D' of the group $\mathcal{G}$ are related by a *similarity transformation* (or an *equivalence transformation*)

$$\hat{D}'(G_i) = \hat{T}^{-1}\hat{D}(G_i)\hat{T} , \quad G_i \in \mathcal{G} , \tag{4.10}$$

through a regular matrix $\hat{T}$, the representations D and D' are said to be *equivalent*. Representations that are not equivalent are *inequivalent*. Representations having different dimensions are necessarily inequivalent.

As has been mentioned, the basis $\psi_1, \psi_2, \ldots, \psi_d$ for the representation D satisfies (4.7). A new basis ψ'_ν is now constructed by means of a linear transformation

$$\psi'_\nu = \sum_{\mu=1}^{d} \psi_\mu T_{\mu\nu} , \tag{4.11}$$

which may be solved for ψ_λ:

$$\psi_\lambda = \sum_{\kappa=1}^{d} \psi'_\kappa (\hat{T}^{-1})_{\kappa\lambda} . \tag{4.12}$$

The effect of G_i on the new basis ψ'_ν becomes

$$\begin{aligned} G_i\psi'_\nu &= \sum_\mu G_i\psi_\mu T_{\mu\nu} \\ &= \sum_\lambda \psi_\lambda [\hat{D}(G_i)\hat{T}]_{\lambda\nu} \\ &= \sum_\kappa \psi'_\kappa [\hat{T}^{-1}\hat{D}(G_i)\hat{T}]_{\kappa\nu} \\ &= \sum_\kappa \psi'_\kappa [\hat{D}'(G_i)]_{\kappa\nu} . \end{aligned} \tag{4.13}$$

Consequently, the transformed basis (4.11) belongs to the equivalent representation (4.10). In other words, a linear transformation of the basis results in an equivalent representation.

4.1.3 Reducible and Irreducible Representations

From two representations $D^{(1)}$ and $D^{(2)}$ of the group $\mathcal{G}$, one can construct a representation of larger dimension

$$\hat{D}(G_i) = \begin{bmatrix} \hat{D}^{(1)}(G_i) & 0 \\ 0 & \hat{D}^{(2)}(G_i) \end{bmatrix} . \tag{4.14}$$

Such a representation D, which certainly satisfies (4.1), is the *direct sum* of the two representations,

$$D = D^{(1)} + D^{(2)} . \quad (4.15)$$

The dimension of D is equal to the sum of the dimensions of $D^{(1)}$ and $D^{(2)}$.

A representation D having a structure like (4.14) is said to be reducible. The reducibility, however, may be concealed by an equivalence transformation. To include such cases, any representations which are equivalent to the apparently reducible ones like (4.14) are called *reducible representations.* If no equivalence transformation can achieve the above block-diagonalization, such representations are *irreducible representations.* A reducible representation can be decomposed into a direct sum of irreducible representations. The decomposition is called *reduction* or *irreducible decomposition.* The representations A_1, A_2, and E shown above for the C_{3v} group are all irreducible.

When a reducible representation D is block-diagonalized as (4.14) with an equivalence transformation (4.10), the corresponding basis undergoes the linear transformation (4.11). The new basis ψ'_v then divides itself into two sets (two subspaces), each of which is closed within itself under the operations of $\mathscr{G}$. In general, the space spanned by a basis of a reducible representation can be decomposed into a direct sum of invariant subspaces corresponding to the reduction of that representation. In contrast to this, in the case of an irreducible representation, the space spanned by its basis can never be divided into invariant subspaces. Such a space spanned by a basis of an irreducible representation is called an *irreducible invariant subspace.*

To be more accurate, representations like (4.14) are called *completely reducible*; reducible representations, in general, may have nonvanishing upper right rectangular matrices as shown, for example, in (3.31). In the completely reducible representation (4.14), each of the basis sets for $D^{(1)}$ and $D^{(2)}$ forms an invariant subspace by itself.

In fact, unitary representations are completely reducible. This follows immediately from the lemma of Sect. 3.5. Since any representation of a finite group may be turned into a unitary one by means of an equivalence transformation (Sect. 4.5.1), we will confine ourselves below to unitary representations and understand reducible representations to mean completely reducible representations.

4.2 Irreducible Representations of the Group $C_{\infty v}$

To study an example of representations, let us consider the group $C_{\infty v}$, which is the symmetry group of heteronuclear diatomic molecules. This group consists of rotations $R(\alpha)$ through an arbitrary angle α about the molecular z-axis and mirror reflections in any plane containing the z-axis. As the generating elements of the group we choose the rotations $R(\alpha)$ and the reflection σ_y in the zx-plane, which are related to each other by

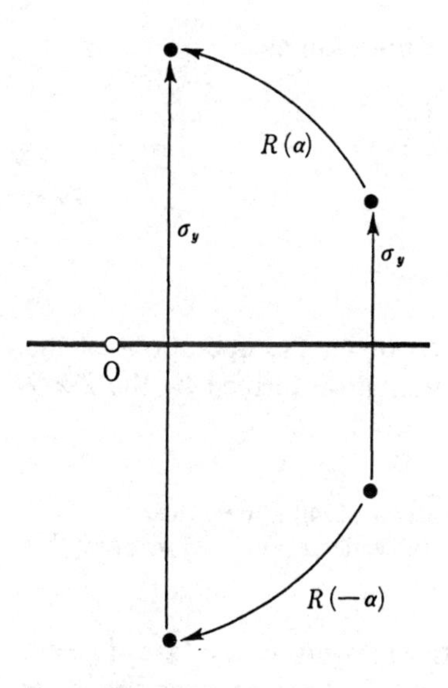

Fig. 4.1. The relation (4.17)

$$R(\alpha)R(\alpha') = R(\alpha + \alpha') , \tag{4.16}$$

$$R(\alpha)\sigma_y = \sigma_y R(-\alpha) , \tag{4.17}$$

as shown in Fig. 4.1.

The rotations $R(\alpha)$ form a subgroup C_∞, whose irreducible representations admit bases v_m that satisfy

$$R(\alpha)v_m = e^{-im\alpha}v_m . \tag{4.18}$$

Since 2π rotation should be the identity operation, we have $R(2\pi) = R(0)$, and the rotational quantum number m is limited to integers.

To consider irreducible representations of the group $C_{\infty v}$, we start with a v_λ with a positive integer λ. For such a v_λ, we have

$$R(\alpha)v_\lambda = e^{-i\lambda\alpha}v_\lambda , \quad \lambda > 0 . \tag{4.19}$$

Operating now with both sides of (4.17) on v_λ and using (4.19), we obtain

$$R(\alpha)\sigma_y v_\lambda = \sigma_y R(-\alpha)v_\lambda = e^{i\lambda\alpha}\sigma_y v_\lambda , \tag{4.20}$$

which means that $\sigma_y v_\lambda$ behaves like v_m with $m = -\lambda$. So we define $v_{-\lambda}$ by

$$v_{-\lambda} \equiv \sigma_y v_\lambda . \tag{4.21}$$

Operating the rotation $R(\alpha)$ and the reflection σ_y on $v_{-\lambda}$ gives

$$R(\alpha)v_{-\lambda} = e^{i\lambda\alpha}v_{-\lambda} , \tag{4.22}$$

$$\sigma_y v_{-\lambda} = \sigma_y^2 v_\lambda = v_\lambda . \tag{4.23}$$

The above relations may be combined in matrix notation as

$$[R(\alpha)v_\lambda, R(\alpha)v_{-\lambda}] = [v_\lambda, v_{-\lambda}]\begin{bmatrix} e^{-i\lambda\alpha} & 0 \\ 0 & e^{i\lambda\alpha} \end{bmatrix},$$
$$[\sigma_y v_\lambda, \sigma_y v_{-\lambda}] = [v_\lambda, v_{-\lambda}]\begin{bmatrix} 0 & 1 \\ 1 & 0 \end{bmatrix}. \tag{4.24}$$

This result shows that the set $\{v_\lambda, v_{-\lambda}\}$ is closed under the operations of the group $C_{\infty v}$. It is therefore a basis for the representation formed by the 2×2 matrices appearing in (4.24).

Exercise 4.1. Show that this representation (the 2×2 matrices in (4.24)) is irreducible. *Hint*: If not, it would be possible to take a suitable linear combination $v = c_1 v_\lambda + c_2 v_{-\lambda}$ such that both $R(\alpha)v$ and $\sigma_y v$ were multiples of v.

Up to this point, we have confined λ to positive integers. In the case of $\lambda = 0$, the representation is not irreducible, because v_λ and $v_{-\lambda}$ become invariant under the rotations $R(\alpha)$. If we take

$$v_0^+ \equiv v_\lambda + v_{-\lambda} \tag{4.25}$$

for $\lambda = 0$, we have

$$R(\alpha)v_0^+ = v_0^+ , \quad \sigma_y v_0^+ = v_0^+ ,$$

showing that v_0^+ is the basis for the identity representation. Another combination

$$v_0^- \equiv v_\lambda - v_{-\lambda} \tag{4.26}$$

for $\lambda = 0$, which behaves like

$$R(\alpha)v_0^- = v_0^- , \quad \sigma_y v_0^- = -v_0^- ,$$

gives another one-dimensional representation. In this way, we have found that the $C_{\infty v}$ group has two one-dimensional representations D_0^+ and D_0^- as well as two-dimensional representations D_λ ($\lambda = 1, 2, 3, \ldots$). In designating molecular electronic levels, these irreducible representations are usually called Σ^+, Σ^-, Π, Δ, Φ, and so on.

The group C_∞, consisting of the rotations $R(\alpha)$, is an invariant subgroup of $C_{\infty v}$, so $C_{\infty v}$ can be decomposed into cosets,

$$C_{\infty v} = C_\infty + \sigma_y C_\infty .$$

In the present section, we have constructed the irreducible representations of $C_{\infty v}$ from those (with basis v_m) of its invariant subgroup C_∞. The general theory for such a procedure is discussed in Chap. 5.

4.3 Effect of Symmetry Transformation Operators on Functions

In Sect. 3.2, we saw that a rotation R in two dimensions transforms the coordinates x and y as (3.11). We wish to see in this section how the same operation transforms functions $f(x, y)$ of the coordinates, and then generalize the result.

Suppose that $f(\boldsymbol{r})$ represents the height at the position $\boldsymbol{r}$ on the map shown in Fig. 4.2a. If we rotate the map about the origin O through an angle α, the position vector $\boldsymbol{r}'$ of the rotated point P is given by

$$\boldsymbol{r}' = R\boldsymbol{r} .$$

The components of the two vectors are related by (3.11).

Let us now consider a new function f' that describes the contours of the rotated map (Fig. 4.2b). For such a function, we write

$$f' = P_R f , \tag{4.27}$$

since it is obtained by applying the rotation R to the function f. The value of the function f' at the point $\boldsymbol{r}' = (x', y')$ is naturally given by $f'(x', y')$. It should be equal to the height $f(x, y)$ before the rotation:

$$f'(x', y') = f(x, y) , \tag{4.28}$$

which may be written as

$$P_R f(x', y') = f(x, y) , \tag{4.29}$$

using (4.27). Here, P_R is an operator that operates on the function f (rather than on the coordinates $\boldsymbol{r}$), and its effect is defined by (4.29).

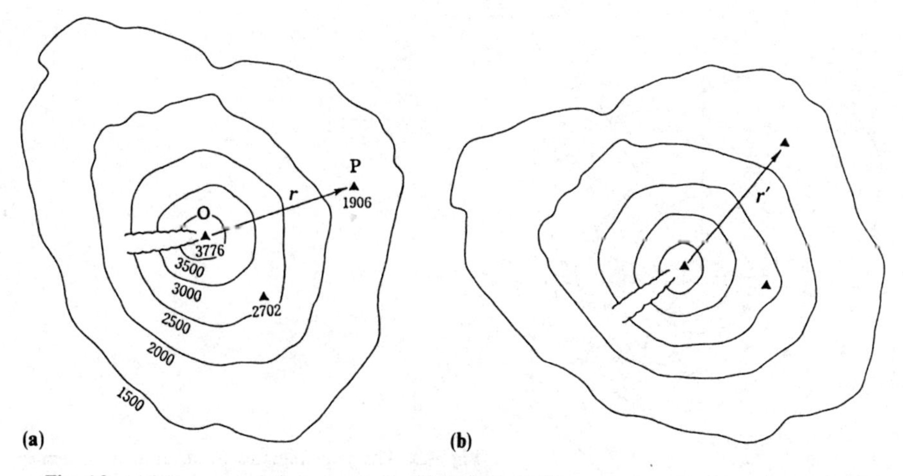

Fig. 4.2. (a) Map around the summit O of Mt. Fuji. (b) The map rotated about the origin O

If we use (3.11) to express x and y in terms of x' and y', use the result in (4.29), and then drop the primes on both sides, we obtain

$$P_R f(x, y) = f(x \cos\alpha + y \sin\alpha,\ y \cos\alpha - x \sin\alpha) \ , \tag{4.30}$$

which explicitly defines the effect of rotational operations P_R on the function $f(x, y)$.

In the above derivation, we rotated the map in the fixed reference frame. Equivalently, one can choose another convention, in which the map is fixed and the reference frame is rotated in the reverse sense (Fig. 4.3). The coordinates of the point P are (x, y) in the fixed reference frame and (x', y') in the rotated frame. They are related through (3.11). Since the height of the mountain does not depend on the choice of the reference frame, we have (4.28).

Exercise 4.2. For the function $f(x, y) = x + \mathrm{i}y$, show that $P_R f(x, y) = \exp(-\mathrm{i}\alpha) f(x, y)$.

Exercise 4.3. Expand the right-hand side of (4.30) in powers of α and show that

$$P_R = \exp(-\mathrm{i}\alpha l_z) \equiv 1 + \sum_{n=1}^{\infty} \frac{1}{n!} (-\mathrm{i}\alpha)^n l_z^n$$

using the angular-momentum operator

$$l_z \equiv -\mathrm{i}\left(x \frac{\partial}{\partial y} - y \frac{\partial}{\partial x}\right) .$$

The definition (4.29) may be generalized to three-dimensional operations. If a symmetry operation R takes the point at $\boldsymbol{r}$ to $\boldsymbol{r}' = R\boldsymbol{r}$, then the value of the new function $P_R f$ at $\boldsymbol{r}'$ must be equal to the value of f at $\boldsymbol{r}$:

$$P_R f(\boldsymbol{r}') = f(\boldsymbol{r}) \ . \tag{4.31}$$

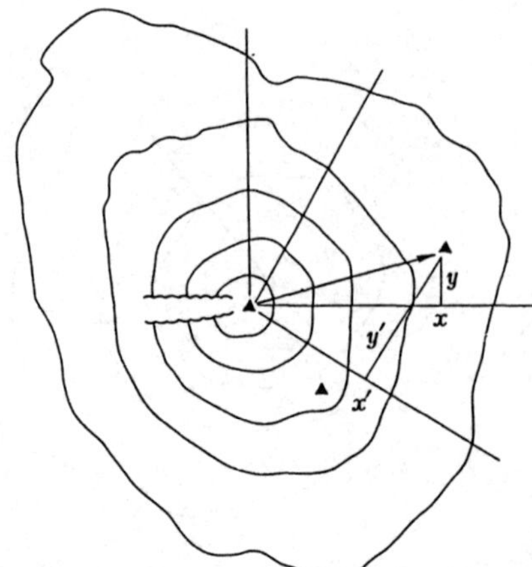

Fig. 4.3. The map remains fixed, and the reference frame is rotated in the reverse sense

If we replace $\boldsymbol{r}$ on the right-hand side by $R^{-1}\boldsymbol{r}'$ and drop the primes on both sides, we obtain

$$P_R f(\boldsymbol{r}) = f(R^{-1}\boldsymbol{r}) \,, \tag{4.32}$$

which means that the value of the transformed function $P_R f$ at a point $\boldsymbol{r}$ is equal to the value of f at $\boldsymbol{r}_0 = R^{-1}\boldsymbol{r}$ which R sends to $\boldsymbol{r}$.

For complicated symmetry operations R as in space groups, the notation P_R becomes unwieldy. So, from the following section on, we shall write simply R for P_R. In this notation, (4.32) is written as

$$Rf(\boldsymbol{r}) = f(R^{-1}\boldsymbol{r}) \,. \tag{4.33}$$

Let us consider next the product of two operations P_R and P_S. The product $P_S P_R$ is defined as the successive operations

$$(P_S P_R)f \equiv P_S(P_R f) \,. \tag{4.34}$$

For the symmetry operation S as well, from (4.32), we have

$$P_S g(\boldsymbol{r}) = g(S^{-1}\boldsymbol{r})$$

for an arbitrary function g. If we put

$$g = P_R f$$

for simplicity, the value of the function (4.34) at $\boldsymbol{r}$ becomes

$$(P_S P_R)f(\boldsymbol{r}) = P_S\, g(\boldsymbol{r}) = g(S^{-1}\boldsymbol{r}) \,,$$

which may be written as

$$(P_S P_R)f(\boldsymbol{r}) = P_R f(S^{-1}\boldsymbol{r}) = f(R^{-1}S^{-1}\boldsymbol{r}) \,,$$

where we have used (4.32). On the other hand, for the product $S\,R$ of symmetry operations, we have

$$P_{SR} f(\boldsymbol{r}) = f((S\,R)^{-1}\boldsymbol{r}) = f(R^{-1}S^{-1}\boldsymbol{r}) \,.$$

Comparison of these two expressions gives

$$P_{SR} = P_S P_R \,. \tag{4.35}$$

Because of this relation, replacement of P_R by R as in (4.33) causes no ambiguities.

4.4 Representations of the Group C_{3v} Based on Homogeneous Polynomials

Let us place an equilateral triangle as shown in Fig. 4.4, such that the threefold axis lies in the [111] direction, and consider how the symmetry operations of the group C_{3v} transform functions $f(x, y, z)$ of the coordinates in such a reference frame.

The C_3 rotation, having the sense shown in Fig. 4.4, takes a point (x, y, z) to (z, x, y). Therefore, the function $C_3 f$ is defined by [see (4.31)]

$$C_3 f(z, x, y) = f(x, y, z) \ ,$$

or, when the arguments are rearranged,

$$C_3 f(x, y, z) = f(y, z, x) \ .$$

The effects of the other group elements may be found in a similar way, and the result is shown in Table 4.1.

Table 4.1. Transformation of functions $f(x, y, z)$ by the six operations R of the group C_{3v}

R	$[Rf](x, y, z)$
E	$f(x, y, z)$
C_3	$f(y, z, x)$
C_3^{-1}	$f(z, x, y)$
σ_1	$f(x, z, y)$
σ_2	$f(z, y, x)$
σ_3	$f(y, x, z)$

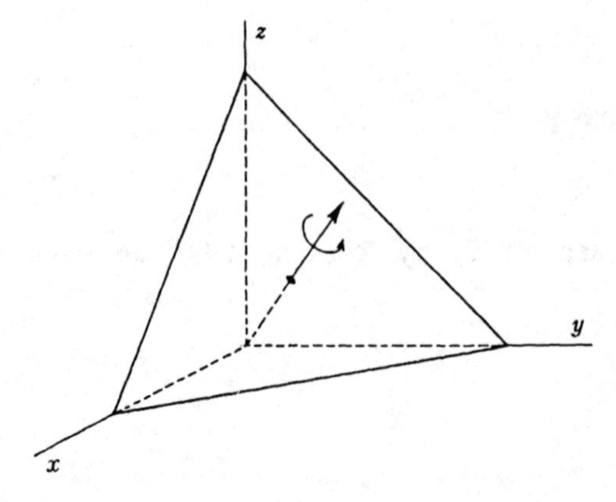

Fig. 4.4. Definition of the coordinate axes for the group C_{3v}

We wish to consider now what representations are obtained from the following three functions:

$$f_1(x, y, z) = x \ ,$$
$$f_2(x, y, z) = y \ ,$$
$$f_3(x, y, z) = z \ . \tag{4.36}$$

The C_3 rotation transforms f_1 as

$$C_3 f_1(x, y, z) = f_1(y, z, x) = y \ ,$$

which means that $C_3 f_1 = f_2$. Similarly, we find

$$C_3 f_2 = f_3 \ , \quad \text{and} \quad C_3 f_3 = f_1 \ .$$

In matrix notation, these relations may be written as

$$[C_3 f_1, C_3 f_2, C_3 f_3] = [f_1, f_2, f_3] \begin{bmatrix} 0 & 0 & 1 \\ 1 & 0 & 0 \\ 0 & 1 & 0 \end{bmatrix} .$$

Similar equations may be found for the other five elements, and we obtain the following six matrices:

$$\hat{D}(E) = \begin{bmatrix} 1 & 0 & 0 \\ 0 & 1 & 0 \\ 0 & 0 & 1 \end{bmatrix} , \quad \hat{D}(C_3) = \begin{bmatrix} 0 & 0 & 1 \\ 1 & 0 & 0 \\ 0 & 1 & 0 \end{bmatrix} ,$$

$$\hat{D}(C_3^{-1}) = \begin{bmatrix} 0 & 1 & 0 \\ 0 & 0 & 1 \\ 1 & 0 & 0 \end{bmatrix} , \quad \hat{D}(\sigma_1) = \begin{bmatrix} 1 & 0 & 0 \\ 0 & 0 & 1 \\ 0 & 1 & 0 \end{bmatrix} ,$$

$$\hat{D}(\sigma_2) = \begin{bmatrix} 0 & 0 & 1 \\ 0 & 1 & 0 \\ 1 & 0 & 0 \end{bmatrix} , \quad \hat{D}(\sigma_3) = \begin{bmatrix} 0 & 1 & 0 \\ 1 & 0 & 0 \\ 0 & 0 & 1 \end{bmatrix} , \tag{4.37}$$

which form a three-dimensional representation of the group C_{3v} with the basis $\{f_1, f_2, f_3\}$.

This representation is reducible. As is apparent from the geometrical construction of Fig. 4.4, the linear combination

$$f^{(A_1)} = \frac{1}{\sqrt{3}}(f_1 + f_2 + f_3) \tag{4.38}$$

is invariant under the six operations, and provides a basis for the identity representation A_1. The other two independent functions

$$f_1^{(E)} = \frac{1}{\sqrt{6}}(2f_1 - f_2 - f_3) ,$$
$$f_2^{(E)} = \frac{1}{\sqrt{2}}(f_2 - f_3) , \tag{4.39}$$

transform according to (4.3), and hence form a basis for the irreducible representation E. Consequently, the representation (4.37) may be reduced into the direct sum $A_1 + E$. Actually, if we apply the linear transformation

$$[f^{(A_1)}, f_1^{(E)}, f_2^{(E)}] = [f_1, f_2, f_3]\hat{T}$$

with the matrix

$$\hat{T} = \begin{bmatrix} \frac{1}{\sqrt{3}} & \frac{2}{\sqrt{6}} & 0 \\ \frac{1}{\sqrt{3}} & -\frac{1}{\sqrt{6}} & \frac{1}{\sqrt{2}} \\ \frac{1}{\sqrt{3}} & -\frac{1}{\sqrt{6}} & -\frac{1}{\sqrt{2}} \end{bmatrix} ,$$

then the equivalence transformation (4.10) generates block-diagonalized representation matrices,

$$\hat{D}'(G_i) = \begin{bmatrix} \hat{D}^{(A_1)}(G_i) & 0 \\ 0 & \hat{D}^{(E)}(G_i) \end{bmatrix} , \quad G_i \in C_{3v} . \tag{4.40}$$

For example, for $G_i = C_3$, we have

$$\hat{D}'(C_3) = \begin{bmatrix} 1 & 0 & 0 \\ 0 & c & -s \\ 0 & s & c \end{bmatrix}$$

with the constants c and s defined by (4.4).

Exercise 4.4. Verify that the functions (4.39) transform according to (4.3).

When we proceed to the homogeneous polynomials of degree two,

$$\begin{aligned}&x^2,\ y^2,\ z^2\ ,\\&yz,\ zx,\ xy\ ,\end{aligned}\tag{4.41}$$

the functions on the upper and lower rows are closed within themselves under the C_{3v} group. For the functions on the upper row, we may put

$$f_1 = x^2\ , \quad f_2 = y^2\ , \quad f_3 = z^2\ .$$

We then obtain the same result as the first-order polynomials. For the functions on the lower row as well, with

$$f_1 = yz\ , \quad f_2 = zx\ , \quad f_3 = xy$$

we obtain the same irreducible decomposition. As a result, we have $2A_1 + 2E$ from the polynomials (4.41). The irreducible representation A_2 appears first in the third degree.

Exercise 4.5. Show that

$$f^{(A_2)}(x, y, z) = x(y^2 - z^2) + y(z^2 - x^2) + z(x^2 - y^2)$$

is a basis for the irreducible representation A_2.

4.5 General Representation Theory

This section concerns itself with general theorems on the representation matrices. The proofs of the theorems are mostly given at the end of this chapter (Sect. 4.13), to avoid interference with the mainstream of the text.

We shall confine ourselves below to a finite group $\mathcal{G}$ of order g. The results remain valid for infinite compact groups as well.[3]

4.5.1 Unitarization of a Representation

Theorem: A representation of the finite group $\mathcal{G}$ can be brought into a unitary representation by means of an equivalence transformation.

Proof: See Sect. 4.13.1. □

This theorem guarantees that any representation is equivalent to a unitary representation, so that we can restrict ourselves to unitary representations without loss of generality.

---

[3] Among the linear transformation groups mentioned in Sect. 3.6, U(n), SU(n), O(n), and SO(n) are compact.

4.5.2 Schur's First Lemma

For two given irreducible representations $D^{(1)}$ and $D^{(2)}$ of dimensions m and n respectively, a rectangular matrix $\hat{M}$ that satisfies

$$\hat{D}^{(1)}(G)\hat{M} = \hat{M}\hat{D}^{(2)}(G) , \quad G \in \mathscr{G} , \tag{4.42}$$

for any group element G of $\mathscr{G}$ must be either

(1) the zero matrix (all the elements vanish), or
(2) a square matrix ($m = n$) and det $M \neq 0$.

Proof: See Sect. 4.13.2. □

Case (1) is the trivial solution of (4.42). In case (2), nonsingularity of the matrix $\hat{M}$ guarantees the existence of the inverse matrix $\hat{M}^{-1}$, and hence we have

$$\hat{M}^{-1}\hat{D}^{(1)}(G)\hat{M} = \hat{D}^{(2)}(G) , \tag{4.43}$$

which shows the equivalence of $D^{(1)}$ and $D^{(2)}$. In the case that $D^{(1)}$ and $D^{(2)}$ are inequivalent, the matrix $\hat{M}$ satisfying (4.42) is restricted to the zero matrix.

4.5.3 Schur's Second Lemma

A matrix $\hat{M}$ that commutes with all the representation matrices $\hat{D}(G)$

$$\hat{D}(G)\hat{M} = \hat{M}\hat{D}(G) , \quad G \in \mathscr{G} , \tag{4.44}$$

of a representation D of the group $\mathscr{G}$ must be a multiple of the unit matrix $\hat{1}$,

$$\hat{M} = c\hat{1} ,$$

if D is irreducible. Here, c is an arbitrary complex number.

Proof: See Sect. 4.13.3. □

Exercise 4.6. Verify that Schur's second lemma holds for the irreducible representation E of the C_{3v} group given in (4.3). Show further that the three-dimensional representation (4.37) has a commutative matrix $\hat{M}$ other than the unit matrix. What does this imply?

4.5.4 The Great Orthogonality Theorem

Matrices of an irreducible unitary representation $D^{(\alpha)}$ satisfy the following orthogonality relations:

$$\sum_G D_{ij}^{(\alpha)}(G)^* D_{kl}^{(\alpha)}(G) = \frac{g}{d_\alpha}\delta_{ik}\delta_{jl} . \tag{4.45a}$$

The summation on the left-hand side runs over all g elements of the group $\mathcal{G}$, and d_α stands for the dimension of the representation $D^{(\alpha)}$.

In addition, if $D^{(\alpha)}$ and $D^{(\beta)}$ are inequivalent irreducible representations,

$$\sum_G D_{ij}^{(\alpha)}(G)^* D_{kl}^{(\beta)}(G) = 0 \ . \tag{4.45b}$$

The two orthogonality relations may be combined together as

$$\sum_G D_{ij}^{(\alpha)}(G)^* D_{kl}^{(\beta)}(G) = \frac{g}{d_\alpha} \delta_{\alpha\beta}\delta_{ik}\delta_{jl} \ . \tag{4.46}$$

Proof: Let $\hat{B}$ be an arbitrary rectangular matrix with d_α rows and d_β columns, and construct a matrix $\hat{M}$ by means of

$$\hat{M} = \sum_G \hat{D}^{(\alpha)}(G^{-1}) \, \hat{B} \hat{D}^{(\beta)}(G) \ . \tag{4.47}$$

This matrix is now found to satisfy Schur's first lemma. For an arbitrary group element G' of $\mathcal{G}$, we have

$$\hat{D}^{(\alpha)}(G')\hat{M} = \sum_G \hat{D}^{(\alpha)}(G' G^{-1}) \hat{B} \hat{D}^{(\beta)}(G) \ .$$

The rearrangement theorem permits us to replace the summation over G by a summation over $G'' \equiv G G'^{-1}$:

$$\begin{aligned} \hat{D}^{(\alpha)}(G')\hat{M} &= \sum_{G''} \hat{D}^{(\alpha)}(G''^{-1}) \, \hat{B} \hat{D}^{(\beta)}(G'' G') \\ &= \hat{M} \hat{D}^{(\beta)}(G') \ . \end{aligned} \tag{4.48}$$

Since (4.48) holds for any G' in $\mathcal{G}$, we can use Schur's first lemma.

If $D^{(\alpha)}$ and $D^{(\beta)}$ are inequivalent, all the elements of $\hat{M}$ must vanish,

$$\sum_i \sum_k \sum_G D_{ji}^{(\alpha)}(G^{-1}) B_{ik} D_{kl}^{(\beta)}(G) = 0 \ .$$

Since $\hat{B}$ is a completely arbitrary matrix, we may put all $B_{ik} = 0$ except one element $B_{ik} = 1$. We then have

$$\sum_G D_{ji}^{(\alpha)}(G^{-1}) D_{kl}^{(\beta)}(G) = 0 \ .$$

Using unitarity,

$$D_{ji}^{(\alpha)}(G^{-1}) = [\hat{D}^{(\alpha)}(G)^\dagger]_{ji} = D_{ij}^{(\alpha)}(G)^* \ , \tag{4.49}$$

we obtain (4.45b).

To prove (4.45a), we set $\hat{D}^{(\alpha)} = \hat{D}^{(\beta)}$ in (4.47). Then we can use Schur's second lemma because of (4.48), and we have

$$\hat{M} = \sum_G \hat{D}^{(\alpha)}(G^{-1})\, \hat{B}\, \hat{D}^{(\alpha)}(G) = c\,\hat{1} .$$

Putting again all $B_{ik} = 0$ except one $B_{ik} = 1$, we obtain

$$\sum_G D_{ji}^{(\alpha)}(G^{-1})\, D_{kl}^{(\alpha)}(G) = c\,\delta_{jl} . \tag{4.50}$$

To determine the constant c, put $j = l$ and sum over j. The left-hand side yields

$$\begin{aligned}\sum_{j=1}^{d_\alpha} \sum_G D_{ji}^{(\alpha)}(G^{-1})\, D_{kj}^{(\alpha)}(G) &= \sum_G [\hat{D}^{(\alpha)}(G)\,\hat{D}^{(\alpha)}(G^{-1})]_{ki} \\ &= \sum_G D_{ki}^{(\alpha)}(GG^{-1}) = g\,\delta_{ki} ,\end{aligned}$$

while the right-hand side yields

$$c \sum_{j=1}^{d_\alpha} 1 = c\,d_\alpha ,$$

so that we obtain

$$c = (g/d_\alpha)\,\delta_{ik} .$$

Substituting this result into (4.50) and using the unitarity (4.49), we arrive at (4.45a). □

Exercise 4.7. Verify that the representation matrices (4.3) satisfy the great orthogonality relation (4.45a).

The orthogonality relations (4.46) allows the following geometric interpretation. Suppose that

$$(D_{ij}^{(\alpha)}(G_1), D_{ij}^{(\alpha)}(G_2), \ldots, D_{ij}^{(\alpha)}(G_g))$$

and

$$(D_{kl}^{(\beta)}(G_1), D_{kl}^{(\beta)}(G_2), \ldots, D_{kl}^{(\beta)}(G_g))$$

are vectors in a g-dimensional vector space. Equation (4.46) may then be understood to imply that these vectors are mutually orthogonal. The vectors carry three indices – the representation label α and the suffixes i and j. Therefore, the number of such vectors is obtained by summing d_α^2 over the n_r independent irreducible representations. Since the g-dimensional vector space can have at most g mutually orthogonal vectors, we have

$$\sum_{\alpha=1}^{n_r} d_\alpha^2 \leq g . \tag{4.51}$$

As a matter of fact, as will be shown in Sect. 4.10, the equality holds in (4.51). The relation

$$\sum_{\alpha=1}^{n_r} d_\alpha^2 = g \tag{4.52}$$

plays an important part in determining dimensions of irreducible representations.

4.6 Characters

Let $\chi(G)$ denote the trace of the representation matrix $\hat{D}(G)$,

$$\chi(G) = \mathrm{Tr}\{\hat{D}(G)\} = \sum_{i=1}^{d} D_{ii}(G) \ . \tag{4.53}$$

Then the set of the traces $\chi(G_1), \chi(G_2), \ldots, \chi(G_g)$ for all the g group elements constitutes the *character* of the representation D. Characters of irreducible representations are called *irreducible characters* or *simple characters*. For example, the characters of the three irreducible representations A_1, A_2, and E of C_{3v} considered in Sect. 4.1 are shown in Table 4.2. Since the unit element E is represented by the unit matrix, $\chi(E)$ is equal to the dimension d of the representation.

Traces of matrices have the following important property:

$$\mathrm{Tr}\{\hat{T}\hat{S}\} = \mathrm{Tr}\{\hat{S}\hat{T}\} \ , \tag{4.54}$$

which may be confirmed by direct calculation,

$$\begin{aligned} \sum_i (\hat{T}\hat{S})_{ii} &= \sum_i \sum_j T_{ij} S_{ji} \\ &= \sum_j \sum_i S_{ji} T_{ij} = \sum_j (\hat{S}\hat{T})_{jj} \ . \end{aligned}$$

If we set $\hat{S} = \hat{Q}\hat{T}^{-1}$ in (4.54), we obtain

$$\mathrm{Tr}\{\hat{T}\hat{Q}\hat{T}^{-1}\} = \mathrm{Tr}\{\hat{Q}\} \ , \tag{4.55}$$

Table 4.2. Characters of the irreducible representations of the group C_{3v}

Class:	$\mathscr{C}_1$	$\mathscr{C}_2$	$\mathscr{C}_3$
Element:	E	C_3, C_3^{-1}	$\sigma_1, \sigma_2, \sigma_3$
A_1	1	1	1
A_2	1	1	-1
E	2	-1	0

i.e., the trace is invariant with respect to similarity transformations. Characters are now seen to have the following fundamental properties:

(1) Values of $\chi(G)$ are common to all the conjugate elements. Any two conjugate elements G_i and G_j of the group $\mathscr{G}$ are related by

$$G G_i G^{-1} = G_j$$

with some group element G of $\mathscr{G}$. The corresponding relation among the representation matrices

$$\hat{D}(G)\hat{D}(G_i)\hat{D}(G)^{-1} = \hat{D}(G_j) ,$$

together with (4.55), yields

$$\chi(G_i) = \chi(G_j) .$$

Thus, $\chi(G_i)$ takes the same value for all the elements belonging to the same class.

(2) Equivalent representations have the same character. Since equivalent representations D and D' are related to each other through (4.10), their characters are equal owing to (4.55).

4.6.1 First and Second Orthogonalities of Characters

Theorem: *First Orthogonality of Characters.* Characters of irreducible representations satisfy the following orthogonality relation:

$$\sum_G \chi^{(\alpha)}(G)^* \chi^{(\beta)}(G) = g\delta_{\alpha\beta} , \tag{4.56}$$

where $\chi^{(\alpha)}$ and $\chi^{(\beta)}$ stand for the characters of irreducible representations $D^{(\alpha)}$ and $D^{(\beta)}$, and the summation runs over all g group elements.

Proof: Put $i = j$ and $k = l$ in the great orthogonality theorem (4.46), and sum over i and k. □

Since the value of $\chi(G)$ is common to the h_k elements in the class $\mathscr{C}_k$, we may write $\chi(\mathscr{C}_k)$ for it, and replace the sum in (4.56) with a sum over the classes:

$$\sum_{k=1}^{n_c} h_k \chi^{(\alpha)}(\mathscr{C}_k)^* \chi^{(\beta)}(\mathscr{C}_k) = g\delta_{\alpha\beta} , \tag{4.57}$$

where the group $\mathscr{G}$ has n_c classes.

Theorem: *Second Orthogonality of Characters.* Characters of irreducible representations satisfy the following orthogonality relation as well:

$$\sum_{\alpha=1}^{n_r} \chi^{(\alpha)}(\mathscr{C}_i)^* \chi^{(\alpha)}(\mathscr{C}_j) = \delta_{ij}\frac{g}{h_i} . \tag{4.58}$$

Proof: See Sect. 4.13.4. □

Exercise 4.8. Verify that the irreducible characters given in Table 4.2 satisfy the first and second orthogonality relations.

The above two kinds of orthogonalities allow the following geometric interpretation. If we regard

$$(\sqrt{h_1}\,\chi^{(\alpha)}(\mathscr{C}_1), \sqrt{h_2}\,\chi^{(\alpha)}(\mathscr{C}_2), \ldots, \sqrt{h_{n_c}}\,\chi^{(\alpha)}(\mathscr{C}_{n_c}))$$

as a vector in n_c dimensions, the left-hand side of (4.57) is simply the inner product of such vectors. Since the n_c-dimensional vector space can have at most n_c mutually orthogonal vectors, the number of the vectors n_r must not exceed the dimension n_c:

$$n_r \leqq n_c \ .$$

Similarly, if we regard

$$(\chi^{(1)}(\mathscr{C}_i), \chi^{(2)}(\mathscr{C}_i), \ldots, \chi^{(n_r)}(\mathscr{C}_i))$$

as a vector in n_r dimensions, (4.58) is the orthogonality of such vectors. The number of vectors n_c should not exceed the dimension n_r:

$$n_c \leqq n_r \ .$$

From these two requirements, we obtain the important result

$$n_r = n_c \ , \tag{4.59}$$

that is, *the number n_r of inequivalent irreducible representations is equal to the number n_c of classes.*

4.7 Reduction of Reducible Representations

The trace of a reducible representation D of the form (4.14) becomes

$$\chi(G_i) = \chi^{(1)}(G_i) + \chi^{(2)}(G_i) \ .$$

In general, a reducible representation D is a direct sum of irreducible representations $D^{(\alpha)}$,

$$D = \sum_{\alpha} q_\alpha D^{(\alpha)} \ , \tag{4.60}$$

where q_α are nonnegative integers. Since the matrices for such a representation can be block-diagonalized with an equivalence transformation, its character $\chi(G)$ is expressible as

$$\chi(G) = \sum_{\alpha} q_\alpha \chi^{(\alpha)}(G) . \tag{4.61}$$

Now the coefficients q_α of the direct sum can be determined by using the first orthogonality relation (4.56). Multiplying (4.61) by $\chi^{(\alpha)}(G)^*$ and summing over G, we obtain

$$\sum_G \chi^{(\alpha)}(G)^* \chi(G) = \sum_\beta q_\beta \sum_G \chi^{(\alpha)}(G)^* \chi^{(\beta)}(G) = q_\alpha g ,$$

and hence

$$q_\alpha = \frac{1}{g} \sum_G \chi^{(\alpha)}(G)^* \chi(G) , \tag{4.62}$$

or, if the summation is taken over the classes,

$$q_\alpha = \frac{1}{g} \sum_k h_k \chi^{(\alpha)}(\mathscr{C}_k)^* \chi(\mathscr{C}_k) . \tag{4.63}$$

Thus, the reduction (4.60) can be worked out just with the characters, without involving full representation matrices.

As an example of the reduction, let us consider the three-dimensional representation (4.37) of the C_{3v} group, which has the following character:

$$\chi(E) = 3 , \quad \chi(C_3) = \chi(C_3^{-1}) = 0 , \quad \chi(\sigma_1) = \chi(\sigma_2) = \chi(\sigma_3) = 1 .$$

Using (4.62) or (4.63), we obtain

$$q_{A_1} = \tfrac{1}{6}[1\,(1)(3) + 2\,(1)(0) + 3\,(1)(1)] = 1 ,$$

$$q_{A_2} = \tfrac{1}{6}[1\,(1)(3) + 2\,(1)(0) + 3\,(-1)(1)] = 0 ,$$

$$q_{E} = \tfrac{1}{6}[1\,(2)(3) + 2\,(-1)(0) + 3\,(0)(1)] = 1 .$$

Hence, the above three-dimensional representation is reduced to $A_1 + E$.

As this example shows, use of the characters greatly simplifies the reduction procedure. However, we do not yet know what transformation matrix $\hat{T}$ would bring about the block-diagonalization. The method of projection operators (Sect. 6.6) answers this problem. Interested readers are invited to refer to Sect. 6.6, which may be understood with the knowledge of the representation theory developed up to this point.

Exercise 4.9. For the character χ of an arbitrary representation D, show that

$$\sum_G |\chi(G)|^2 = mg ,$$

and that (i) D is irreducible if $m = 1$; (ii) D is reducible if $m \geqq 2$. This result may be used as a criterion for examining whether a given representation is irreducible or not.

4.7.1 Restriction to a Subgroup

Let $\mathscr{H}$ be a subgroup of $\mathscr{G}$. Matrices of an irreducible representation D of the group $\mathscr{G}$ are defined for all the elements G of $\mathscr{G}$. If we collect only the matrices $\hat{D}(H)$ for the elements H of $\mathscr{H}$, then such a set will form a representation of the subgroup $\mathscr{H}$. This representation, obtained by restricting the representation D to the subgroup $\mathscr{H}$, is denoted by $D \downarrow \mathscr{H}$ and is called the *subduced representation.* Even when D is irreducible in $\mathscr{G}$, the subduced representation $D \downarrow \mathscr{H}$ is in general reducible.

Let us take, for instance,

$$\mathscr{G} = \mathrm{C}_{3\mathrm{v}} \ , \quad \text{and} \quad \mathscr{H} = \mathrm{C}_\mathrm{s} = \{E, \sigma_1\} \ .$$

The group C_s has two irreducible representations A′ and A″, which are respectively even and odd with respect to the mirror reflection σ, as shown in Table 4.3. If we restrict the irreducible representations of $\mathrm{C}_{3\mathrm{v}}$ given in Table 4.2 to the subgroup C_s, the resulting subduced representations can be reduced as follows:

$$\mathrm{A}_1 \downarrow \mathscr{H} = \mathrm{A}' \ , \quad \mathrm{A}_2 \downarrow \mathscr{H} = \mathrm{A}'' \ , \quad \mathrm{E} \downarrow \mathscr{H} = \mathrm{A}' + \mathrm{A}'' \ .$$

Such relations are known as *compatibility relations.*

4.8 Product Representations

Let $D^{(\alpha)}$ and $D^{(\beta)}$ be representations with dimensions d_α and d_β. The bases for these representations

$$\varphi_1, \varphi_2, \ldots, \varphi_{d_\alpha} \quad \text{and} \quad \psi_1, \psi_2, \ldots, \psi_{d_\beta}$$

are transformed by the elements G of the group $\mathscr{G}$ as

$$G\varphi_j = \sum_i \varphi_i D_{ij}^{(\alpha)}(G) \ , \quad G\psi_l = \sum_i \psi_k D_{kl}^{(\beta)}(G) \ .$$

Table 4.3. Irreducible representations of the group C_s

	E	σ
A′	1	1
A″	1	−1

Then the products $\varphi_j\psi_l$, which amount to $d_\alpha d_\beta$ in number, are transformed as

$$G(\varphi_j\psi_l) \equiv G\varphi_j G\psi_l = \sum_{ik} \varphi_i\psi_k [\hat{D}^{(\alpha\times\beta)}(G)]_{ik,jl} , \tag{4.64}$$

$$[\hat{D}^{(\alpha\times\beta)}(G)]_{ik,jl} \equiv D_{ij}^{(\alpha)}(G) D_{kl}^{(\beta)}(G) . \tag{4.65}$$

The rows and columns of the matrix $\hat{D}^{(\alpha\times\beta)}(G)$ are specified by double indices. Since (4.64) has the same structure as (4.7), the set $\{\varphi_j\psi_l\}$ forms a basis for the representation $D^{(\alpha\times\beta)}$. In fact, it is easy to see that the matrices defined by (4.65) satisfy the requirement (4.1), or

$$\hat{D}^{(\alpha\times\beta)}(G)\hat{D}^{(\alpha\times\beta)}(G') = \hat{D}^{(\alpha\times\beta)}(GG') \tag{4.66}$$

through the following calculation:

$$\begin{aligned} &[\hat{D}^{(\alpha\times\beta)}(G)\hat{D}^{(\alpha\times\beta)}(G')]_{ik,jl} \\ &\quad = \sum_{\mu\nu} [\hat{D}^{(\alpha\times\beta)}(G)]_{ik,\mu\nu} [\hat{D}^{(\alpha\times\beta)}(G')]_{\mu\nu,jl} \\ &\quad = \sum_{\mu\nu} D_{i\mu}^{(\alpha)}(G) D_{k\nu}^{(\beta)}(G) D_{\mu j}^{(\alpha)}(G') D_{\nu l}^{(\beta)}(G') \\ &\quad = D_{ij}^{(\alpha)}(GG') D_{kl}^{(\beta)}(GG') \\ &\quad = [\hat{D}^{(\alpha\times\beta)}(GG')]_{ik,jl} . \end{aligned}$$

This representation, denoted by $D^{(\alpha)} \times D^{(\beta)}$, is called the *direct-product representation* or the *product representation.* Its dimension is equal to $d_\alpha d_\beta$.

The character $\chi^{(\alpha\times\beta)}$ of the product representation $D^{(\alpha)} \times D^{(\beta)}$ is obtained by putting $i = j$, $k = l$ in (4.65) and summing over i and k,

$$\begin{aligned} \chi^{(\alpha\times\beta)}(G) &= \sum_i \sum_k [\hat{D}^{(\alpha\times\beta)}(G)]_{ik,ik} \\ &= \chi^{(\alpha)}(G)\chi^{(\beta)}(G) . \end{aligned} \tag{4.67}$$

Product representations are reducible in general. If $D^{(\alpha)} \times D^{(\beta)}$ has the irreducible decomposition

$$D^{(\alpha)} \times D^{(\beta)} = \sum_\gamma q_\gamma D^{(\gamma)} , \tag{4.68}$$

then the values of q_γ may be calculated by means of

$$\begin{aligned} q_\gamma &= \frac{1}{g} \sum_G \chi^{(\gamma)}(G)^* \chi^{(\alpha\times\beta)}(G) \\ &= \frac{1}{g} \sum_G \chi^{(\gamma)}(G)^* \chi^{(\alpha)}(G)\chi^{(\beta)}(G) . \end{aligned} \tag{4.69}$$

Here q_γ gives the number of times the representation $D^{(\gamma)}$ appears in the reduction of $D^{(\alpha)} \times D^{(\beta)}$.

For representations $D^{(\gamma)}$ with $q_\gamma \neq 0$, appropriate linear combinations of $\varphi_j \psi_l$ will give a basis for them. When such a basis is written in the form

$$\Psi_m^{(\gamma)p} = \sum_{jl} \varphi_j \psi_l \langle \alpha j \beta l | p \gamma m \rangle \ , \tag{4.70}$$

the coefficients $\langle \alpha j \beta l | p \gamma m \rangle$ are called the *Clebsch–Gordan coefficients.* The suffix m specifies partners in the representation $D^{(\gamma)}$, and p takes on the values $1, 2, \ldots, q_\gamma$.

As an example of the product representation, let us consider the product E × E in the C_{3v} group. The bases $\{\varphi_1, \varphi_2\}$ and $\{\psi_1, \psi_2\}$ are transformed according to the matrices (4.3). The character of the product representation E × E is obtained from Table 4.2 using (4.67),

$$\chi^{(\mathrm{E}\times\mathrm{E})}(\mathscr{C}_1) = 4 \ , \quad \chi^{(\mathrm{E}\times\mathrm{E})}(\mathscr{C}_2) = 1 \ , \quad \chi^{(\mathrm{E}\times\mathrm{E})}(\mathscr{C}_3) = 0 \ .$$

The values of q_γ obtained from (4.69) are

$$q_{\mathrm{A}_1} = 1 \ , \quad q_{\mathrm{A}_2} = 1 \ , \quad q_{\mathrm{E}} = 1 \ ,$$

which means that E × E is reduced as

$$\mathrm{E} \times \mathrm{E} = \mathrm{A}_1 + \mathrm{A}_2 + \mathrm{E} \ . \tag{4.71}$$

In this reduction, (4.70) takes the form

$$\begin{aligned}
\Psi^{(\mathrm{A}_1)} &= \frac{1}{\sqrt{2}}(\varphi_1\psi_1 + \varphi_2\psi_2) \ , \\
\Psi^{(\mathrm{A}_2)} &= \frac{1}{\sqrt{2}}(\varphi_1\psi_2 - \varphi_2\psi_1) \ , \\
\Psi_1^{(\mathrm{E})} &= \frac{1}{\sqrt{2}}(-\varphi_1\psi_1 + \varphi_2\psi_2) \ , \\
\Psi_2^{(\mathrm{E})} &= \frac{1}{\sqrt{2}}(\varphi_1\psi_2 + \varphi_2\psi_1) \ .
\end{aligned} \tag{4.72}$$

4.8.1 Symmetric and Antisymmetric Product Representations

Let us return to the beginning of Sect. 4.8 and consider the case $D^{(\alpha)} = D^{(\beta)}$. Equation (4.64) becomes now

$$G(\varphi_j \psi_l) = \sum_{ik} \varphi_i \psi_k D_{ij}^{(\alpha)}(G) D_{kl}^{(\alpha)}(G) \ .$$

If we construct similar equations for $\varphi_l\psi_j$ and take the sum and difference, we obtain

$$G(\varphi_j\psi_l \pm \varphi_l\psi_j) = \sum_{ik}(\varphi_i\psi_k \pm \varphi_k\psi_i)D_{ij}^{(\alpha)}(G)D_{kl}^{(\alpha)}(G) . \quad (4.73)$$

This means that the functions

$$\Psi_{jl}^{+} \equiv \varphi_j\psi_l + \varphi_l\psi_j ,$$

which number $d_\alpha(d_\alpha + 1)/2$, and

$$\Psi_{jl}^{-} \equiv \varphi_j\psi_l - \varphi_l\psi_j ,$$

which number $d_\alpha(d_\alpha - 1)/2$, are closed within themselves, and form bases for different representations of the group $\mathscr{G}$. The representation $[D^{(\alpha)} \times D^{(\alpha)}]$ based on Ψ_{jl}^{+} is called the *symmetric product representation*, while the other one, $\{D^{(\alpha)} \times D^{(\alpha)}\}$, based on Ψ_{jl}^{-}, is called the *antisymmetric product representation.*

If we note

$$\Psi_{jl}^{\pm} = \pm \Psi_{lj}^{\pm} ,$$

then (4.73) may be written as

$$G\Psi_{jl}^{\pm} = \tfrac{1}{2}\sum_{ik}(D_{ij}^{(\alpha)}(G)D_{kl}^{(\alpha)}(G) \pm D_{kj}^{(\alpha)}(G)D_{il}^{(\alpha)}(G))\Psi_{ik}^{\pm} . \quad (4.74)$$

The coefficients on the right-hand side are interpreted as forming the matrices of the symmetric and antisymmetric product representations. Their characters can therefore be obtained by setting $j = i$, $l = k$ and summing those coefficients over i and k. Thus the character of the symmetric product representation becomes

$$\begin{aligned}\chi^{[\alpha\times\alpha]}(G) &= \tfrac{1}{2}\sum_{ik}(D_{ii}^{(\alpha)}(G)D_{kk}^{(\alpha)}(G) + D_{ki}^{(\alpha)}(G)D_{ik}^{(\alpha)}(G)) \\ &= \tfrac{1}{2}(\chi^{(\alpha)}(G)^2 + \chi^{(\alpha)}(G^2)) , \quad (4.75)\end{aligned}$$

while for the antisymmetric product representation,

$$\chi^{\{\alpha\times\alpha\}}(G) = \tfrac{1}{2}(\chi^{(\alpha)}(G)^2 - \chi^{(\alpha)}(G^2)) . \quad (4.76)$$

In the case of the E × E representation in the C_{3v} group, (4.75 and 76) yield

$$\chi^{[E\times E]}(\mathscr{C}_1) = 3 , \quad \chi^{[E\times E]}(\mathscr{C}_2) = 0 , \quad \chi^{[E\times E]}(\mathscr{C}_3) = 1 ,$$

$$\chi^{\{E\times E\}}(\mathscr{C}_1) = 1 , \quad \chi^{\{E\times E\}}(\mathscr{C}_2) = 1 , \quad \chi^{\{E\times E\}}(\mathscr{C}_3) = -1 .$$

From these characters, we find

$$[E \times E] = A_1 + E , \quad \{E \times E\} = A_2 .$$

4.9 Representations of a Direct-Product Group

Consider a direct product $\mathscr{A} \times \mathscr{B}$ of two groups $\mathscr{A}$ and $\mathscr{B}$ (Sect. 2.12), and suppose that $D^{(a)}$ is a d_a-dimensional representation of the group $\mathscr{A}$, and $D^{(b)}$ a d_b-dimensional representation of the group $\mathscr{B}$. For arbitrary elements A and B of the groups $\mathscr{A}$ and $\mathscr{B}$ respectively, we can construct the representation matrix

$$[\hat{D}^{(a \times b)}(AB)]_{ik,jl} \equiv D_{ij}^{(a)}(A)\, D_{kl}^{(b)}(B) \tag{4.77}$$

for the element AB of the direct-product group $\mathscr{A} \times \mathscr{B}$. The above-defined matrices form a representation of the group $\mathscr{A} \times \mathscr{B}$, as may be verified by a calculation similar to the one shown below (4.66). This representation $D^{(a \times b)}$ has the dimension $d_a d_b$, and its character is given by

$$\chi^{(a \times b)}(AB) = \sum_{ik} [\hat{D}^{(a \times b)}(AB)]_{ik,ik} = \chi^{(a)}(A)\chi^{(b)}(B)\ . \tag{4.78}$$

It is called the *direct-product representation* or the *outer direct-product representation.*[4]

Exercise 4.10. Show that if $D^{(a)}$ and $D^{(b)}$ are irreducible representations of the groups $\mathscr{A}$ and $\mathscr{B}$, then the representation $D^{(a \times b)}$ defined by (4.77) is an irreducible representation of the group $\mathscr{A} \times \mathscr{B}$. (*Hint*: Use the result of Exercise 4.9.) In addition, show that those representations exhaust the irreducible representations of the direct-product group $\mathscr{A} \times \mathscr{B}$. (Use (4.59).)

As an example of the direct-product group, we have already considered $C_{3v} \times C_{1h}$ (Sect. 2.12). The group C_{3v} has three irreducible representations, A_1, A_2, and E, shown in Table 4.2, and the group C_{1h} has two irreducible representations, A′ and A″, shown in Table 4.3. The irreducible representations of the direct-product group $D_{3h} = C_{3v} \times C_{1h}$ are obtained by combining them, resulting in six in all, as shown in Table 4.4.

Table 4.4. Characters of the irreducible representations of the group $D_{3h} = C_{3v} \times C_s$

	E	C_3, C_3^{-1}	$\sigma_1, \sigma_2, \sigma_3$	σ_h	$C_3\sigma_h, C_3^{-1}\sigma_h$	U_1, U_2, U_3
$A_1 \times A' = A'_1$	1	1	1	1	1	1
$A_2 \times A' = A'_2$	1	1	−1	1	1	−1
$E \times A' = E'$	2	−1	0	2	1	0
$A_1 \times A'' = A''_1$	1	1	1	−1	−1	−1
$A_2 \times A'' = A''_2$	1	1	−1	−1	−1	1
$E \times A'' = E''$	2	−1	0	−2	1	0

[4] To avoid confusion, we wish to remind readers that we are here concerned with the direct product of two representations $D^{(a)}$, $D^{(b)}$ of two *different* groups $\mathscr{A}$, $\mathscr{B}$. On the other hand, in the preceding section, we considered representations $D^{(\alpha)}$, $D^{(\beta)}$ of one and the same group $\mathscr{G}$.

*4.10 The Regular Representation

The regular representation is defined by the matrices

$$[\hat{D}^{(\mathrm{reg})}(G)]_{ij} = \delta(G_i^{-1}GG_j) \ , \quad \text{where} \tag{4.79}$$

$$\delta(G) = \begin{cases} 1 & \text{if } G = E \text{ (unit element)} \ , \\ 0 & \text{otherwise} \ . \end{cases}$$

The rows and columns of the matrices are specified by the numbers i and j attached to the group elements, and the ij element of the matrix $\hat{D}^{(\mathrm{reg})}(G)$ is nonvanishing only when $G_i = GG_j$ holds. The dimension of the representation is equal to the order g of the group.

Exercise 4.11. Show that the above-defined matrices $\hat{D}^{(\mathrm{reg})}(G)$ form a representation, i.e.,

$$\hat{D}^{(\mathrm{reg})}(G)\,\hat{D}^{(\mathrm{reg})}(G') = \hat{D}^{(\mathrm{reg})}(GG') \ .$$

The explicit forms of the matrices $\hat{D}^{(\mathrm{reg})}(G)$ can be constructed from the multiplication table. Leave the first column of the multiplication table as it stands and replace the first row by the inverse elements. The resulting table is shown in Table 4.5 for the group $\mathrm{C_{3v}}$. One can then obtain the matrix $\hat{D}^{(\mathrm{reg})}(G)$ by replacing G in the table with 1 and filling the rest with zeros. For instance, if we write 1 for C_3 in Table 4.5, we obtain

$$\hat{D}^{(\mathrm{reg})}(C_3) = \begin{bmatrix} 0 & 0 & 1 & 0 & 0 & 0 \\ 1 & 0 & 0 & 0 & 0 & 0 \\ 0 & 1 & 0 & 0 & 0 & 0 \\ 0 & 0 & 0 & 0 & 1 & 0 \\ 0 & 0 & 0 & 0 & 0 & 1 \\ 0 & 0 & 0 & 1 & 0 & 0 \end{bmatrix}$$

Table 4.5. Multiplication table for the group $\mathrm{C_{3v}}$. The elements in the top row have been replaced by the inverse elements

$\mathrm{C_{3v}}$	E	C_3^{-1}	$(C_3^{-1})^{-1}$	σ_1^{-1}	σ_2^{-1}	σ_3^{-1}
E	E	C_3^{-1}	C_3	σ_1	σ_2	σ_3
C_3	C_3	E	C_3^{-1}	σ_3	σ_1	σ_2
C_3^{-1}	C_3^{-1}	C_3	E	σ_2	σ_3	σ_1
σ_1	σ_1	σ_3	σ_2	E	C_3	C_3^{-1}
σ_2	σ_2	σ_1	σ_3	C_3^{-1}	E	C_3
σ_3	σ_3	σ_2	σ_1	C_3	C_3^{-1}	E

From (4.79), the character of the regular representation is

$$\chi^{(\mathrm{reg})}(G) = \sum_{i=1}^{g} \delta(G_i^{-1} G G_i) = \begin{cases} g & \text{if } G = E \ , \\ 0 & \text{otherwise} \ . \end{cases} \tag{4.80}$$

Since the regular representation is reducible, it may be decomposed as

$$D^{(\mathrm{reg})} = \sum_{\alpha} q_\alpha D^{(\alpha)} \ ,$$

and correspondingly

$$\chi^{(\mathrm{reg})}(G) = \sum_{\alpha} q_\alpha \chi^{(\alpha)}(G) \ . \tag{4.81}$$

The value of q_α can be obtained with use of (4.62),

$$q_\alpha = \frac{1}{g} \sum_{G} \chi^{(\alpha)}(G)^* \chi^{(\mathrm{reg})}(G) = \chi^{(\alpha)}(E)^* = d_\alpha \ .$$

Thus, we find that the multiplicity q_α is equal to the dimension d_α of the representation $D^{(\alpha)}$. Using this result in (4.81), we obtain

$$\sum_{\alpha} d_\alpha \chi^{(\alpha)}(G) = 0 \ , \quad \text{for} \quad G \neq E \tag{4.82}$$

and

$$\sum_{\alpha} d_\alpha^2 = g \quad \text{when} \quad G = E \ . \tag{4.83}$$

*4.11 Construction of Character Tables

To determine the characters of irreducible representations, various relations we have derived are used.

(1) The number of inequivalent irreducible representations n_r is equal to the number of classes n_c:

$$n_r = n_c \ . \tag{4.59}$$

(2) The sum of squares of the dimensions of inequivalent irreducible representations is equal to the order g of the group:

$$d_1^2 + d_2^2 + \cdots + d_{n_r}^2 = g \ . \tag{4.83}$$

(3) First orthogonality:

$$\sum_{k=1}^{n_c} h_k \chi^{(\alpha)}(\mathscr{C}_k)^* \chi^{(\beta)}(\mathscr{C}_k) = g\delta_{\alpha\beta} . \tag{4.57}$$

(4) Second orthogonality:

$$\sum_{\alpha=1}^{n_r} \chi^{(\alpha)}(\mathscr{C}_i)^* \chi^{(\alpha)}(\mathscr{C}_j) = \delta_{ij} \frac{g}{h_i} . \tag{4.58}$$

(5) Relation for characters resulting from class multiplications (proved in Sect. 4.13.4):

$$h_i h_j \chi^{(\alpha)}(\mathscr{C}_i) \chi^{(\alpha)}(\mathscr{C}_j) = d_\alpha \sum_k c_{ij}^k h_k \chi^{(\alpha)}(\mathscr{C}_k) . \tag{4.108}$$

Here, c_{ij}^k denotes the class constants defined in (2.29).

In many cases, (1)–(3) above suffice to determine the characters; (4) may be used as a supplement. However, in some groups, (1)–(4) cannot determine the characters uniquely. In that event, we use (5).

Example. The above rules may be used to construct the character table for the C_{3v} group. From the classification of the $g = 6$ elements (Exercise 2.15), we see that the group has $n_c = 3$ classes. So we have $n_r = 3$ inequivalent irreducible representations from (1). Next, we use (2). The three integers that satisfy $d_1^2 + d_2^2 + d_3^2 = g = 6$ are $d_1 = 1$, $d_2 = 1$, and $d_3 = 2$. So the group has two 1-dimensional representations and one 2-dimensional representation. Since one of the 1-dimensional representations is the identity representation, we obtain the intermediate result shown in Table 4.6. The character for the 1-dimensional representation is easy to determine. Since $C_3^3 = E$, the character of the 1-dimensional representation must satisfy $\chi(C_3)^3 = 1$. In addition, since C_3 and C_3^{-1} belong to the same class, $\chi(C_3) = \chi(C_3^{-1}) = 1/\chi(C_3)$. From these two conditions, we find $\chi(C_3) = 1$. Similarly, we obtain $\chi(\sigma_1) = \pm 1$. Characters of one-dimensional representations are usually determined from these considerations.

The character of the remaining two-dimensional representation is determined from the orthogonality (3) with the two 1-dimensional representations.

Table 4.6. Construction of the character table for the group C_{3v}. The blanks are to be filled

$\mathscr{C}_1$ E	$\mathscr{C}_2$ C_3, C_3^{-1}	$\mathscr{C}_3$ $\sigma_1, \sigma_2, \sigma_3$
1	1	1
1		
2		

Exercise 4.12. Show that irreducible representations of commutative groups are always one-dimensional.

Exercise 4.13. Construct the character table for the group C_{4v}. For C_{4v} see Exercise 2.9 and Table 2.7.

*4.12 Adjoint Representations

For a given representation

$$D: G_i \longrightarrow \hat{D}(G_i) \ ,$$

we can construct the *adjoint representation*

$$\tilde{D}: G_i \longrightarrow {}^t\hat{D}(G_i^{-1}) \ ,$$

where ${}^t\hat{D}(G_i^{-1})$ denotes the transposed matrix of $\hat{D}(G_i^{-1})$. In contrast to the basis ϕ_j for the representation D that is transformed as

$$G_i \phi_j = \sum_{k=1}^{d} \phi_k D_{kj}(G_i)$$

by the elements G_i of the group $\mathcal{G}$, the basis $\tilde{\psi}_j$ for the adjoint representation $\tilde{D}$ is transformed as

$$G_i \tilde{\psi}_j = \sum_{k=1}^{d} \tilde{\psi}_k D_{jk}(G_i^{-1}) \ .$$

The character $\tilde{\chi}$ of the representation $\tilde{D}$ is related to the character χ of D through[5]

$$\tilde{\chi}(G_i) = \chi(G_i^{-1}) = \chi(G_i)^* \ . \tag{4.84}$$

From (4.84), we observe that $\tilde{D}$ is irreducible if D is irreducible, and vice versa.

Exercise 4.14. Show that $\tilde{D}$ forms a representation.

Exercise 4.15. Show that for two given irreducible representations D_1 and D_2, the product representation $\tilde{D}_1 \times D_2$ contains the identity representation only if D_1 is equivalent to D_2. In addition, show that if D is an irreducible representation, then $\tilde{D} \times D$ contains the identity representation only once. Note that the basis of that identity representation is given by

$$\Psi_0 = \sum_{j=1}^{d} \tilde{\psi}_j \phi_j \ . \tag{4.85}$$

---

[5] The first equality follows from the definition. The second equality $\chi(G^{-1}) = \chi(G)^*$ is evident in cyclic groups. In noncyclic groups, the equality may be verified by restricting the representation D to the cyclic subgroup with the generating element G.

When D is a unitary representation, we have

$$\hat{\tilde{D}}(G_i) = {}^t\hat{D}(G_i^{-1}) = \hat{D}(G_i)^* , \tag{4.86}$$

and hence the adjoint representation $\tilde{D}$ means the *complex conjugate representation* D^*. The basis $\tilde{\psi}_j$ for the representation D^* is transformed as

$$G_i\tilde{\psi}_j = \sum_{k=1}^{d} \tilde{\psi}_k D_{kj}(G_i)^* . \tag{4.87}$$

As will be discussed in Chap. 13, the basis $\tilde{\psi}_j$ for D^* is related to the basis ψ_j for D through $\tilde{\psi}_j = \theta\psi_j$, where θ is the time-reversal operator. In the case of orbital functions, θ becomes simply complex conjugation and we have $\tilde{\psi}_j = \psi_j^*$.

An irreducible representation D and its complex conjugate representation D^* can be equivalent (when the character χ is real) or inequivalent (χ is complex). The equivalent case further divides itself into two cases: either *all* the representation matrices can be made real, or that is impossible. The representation D is said to be *real* in the former case and *pseudoreal* in the latter case.

Real representations necessarily have real characters. But the converse is not true: real characters do not always guarantee real representation matrices. To examine whether a given irreducible representation D is real or pseudoreal, the *Frobenius–Schur criterion*

$$\frac{1}{g}\sum_{i=1}^{g}\chi(G_i^2) = \begin{cases} 1 & \text{(a) if } D \text{ is real} , \\ -1 & \text{(b) if } D \text{ is pseudoreal} , \\ 0 & \text{(c) } D, D^* \text{ inequivalent} \end{cases} \tag{4.88}$$

may be used. This criterion is derived in the following way.

The left-hand side of (4.88) may be written as

$$\frac{1}{g}\sum_{i=1}^{g}\chi(G_i^2) = \frac{1}{g}\sum_{i}\sum_{kl} D_{kl}(G_i)\, D_{lk}(G_i) . \tag{4.89}$$

In case (c), where D and D^* are inequivalent, the right-hand side vanishes owing to the great orthogonality relation (4.45b).

When D^* is equivalent to D, they are related by a unitary matrix $\hat{U}$ through

$$\hat{D}(G_i) = \hat{U}\hat{D}(G_i)^*\,\hat{U}^{-1} . \tag{4.90}$$

Repeated use of (4.90) gives

$$\hat{D}(G_i) = \hat{U}\hat{U}^*\,\hat{D}(G_i)(\hat{U}\hat{U}^*)^{-1} ,$$

which means that the matrix $\hat{U}\hat{U}^*$ commutes with all the representation matrices $\hat{D}(G_i)$. By virtue of Schur's second lemma, irreducibility of D requires that

$$\hat{U}\hat{U}^* = \lambda\hat{1} \ . \qquad (4.91)$$

From unitarity of $\hat{U}$, we now have

$$\hat{U} = \lambda^2\hat{U} \ ,$$

which yields $\lambda = \pm 1$. The right-hand side of (4.89) may then be written as

$$\begin{aligned} \frac{1}{g}\sum_{i=1}^{g} \chi(G_i^2) &= \frac{1}{g}\sum_{klmn}\sum_{i} D_{kl}(G_i)\, U_{lm} D_{mn}(G_i)^* \, (\hat{U}^{-1})_{nk} \\ &= \frac{1}{d}\sum_{kl} U_{lk}\, U_{kl}^* = \lambda = \pm 1 \ , \end{aligned} \qquad (4.92)$$

where we have used (4.90) and (4.45a).

In case (a) $\lambda = 1$, $\hat{U}$ is a symmetric unitary matrix and

$$\hat{U}^* = \hat{U}^\dagger = \hat{U}^{-1}$$

holds. Such a matrix can be expressed[6] as a square of a symmetric unitary matrix $\hat{T}$,

$$\hat{U} = \hat{T}^2 \ , \qquad \hat{T}^* = \hat{T}^\dagger = \hat{T}^{-1} \ . \qquad (4.93)$$

Now the equivalence transformation with this matrix $\hat{T}$ gives an equivalent representation

$$\hat{D}'(G_i) = \hat{T}^{-1}\hat{D}(G_i)\hat{T} \ , \qquad (4.94)$$

which is certainly real:

$$\begin{aligned} \hat{D}'(G_i)^* &= (\hat{T}^*)^{-1}\hat{D}(G_i)^*\hat{T}^* = (\hat{T}^*)^{-1}\hat{U}^{-1}\hat{D}(G_i)\hat{U}\hat{T}^* \\ &= \hat{T}^{-1}\hat{D}(G_i)\hat{T} = \hat{D}'(G_i) \ . \end{aligned} \qquad (4.95)$$

Consequently, in case (a), the representation D can be made real.

In case (b) $\lambda = -1$, $\hat{U}$ is an antisymmetric unitary matrix, and no choice of unitary matrix $\hat{T}$ can bring (4.94) into real matrices. If it were possible, we should have

$$\hat{U}\hat{T}^*\hat{T}^{-1} = \omega\hat{1}, \quad |\omega| = 1 \ ,$$

---

[6] This can be achieved in the following way: Symmetric unitary matrices have real eigenvectors (see Exercise 3.2). Therefore, $\hat{U}$ may be diagonalized by an orthogonal transformation, $\hat{U} = \hat{S}\hat{\Omega}\hat{S}^{-1}$, where $\hat{\Omega}$ has diagonal elements with modulus unity. Then $\hat{T} = \hat{S}\hat{\Omega}^{1/2}\hat{S}^{-1}$ satisfies (4.93).

from the third equality of (4.95) and Schur's second lemma. The above result would give

$$\hat{U} = \omega \hat{T}(\hat{T}^*)^{-1} \quad \text{or} \quad \hat{U}\hat{U}^* = |\omega|^2 \hat{1} = \hat{1} ,$$

in contradiction with

$$\hat{U}\hat{U}^* = -\hat{1} . \tag{4.96}$$

We note in passing that the dimension of the representation must be an even number in case (b). When we take the determinant of both sides of (4.96), we have

$$\det \hat{U} \cdot \det \hat{U}^* = (-1)^d .$$

The left-hand side is equal to unity because of the unitarity of $\hat{U}$, and hence d must be an even number. Thus, irreducible representations with odd dimensions fall under the cases (a) or (c).

This completes the derivation of the Frobenius–Schur criterion (4.88). In space groups, this criterion is turned into the Herring criterion (13.48).

Examples: *Case* (a): Most of the irreducible representations of point groups fall into this case. In addition, irreducible representations $D^{(l)}$ of the rotation group with nonnegative integers l fall into this case. Since the basis $Y_{lm}(\theta, \phi)$ satisfies (7.68), the matrix $\hat{U}$ has nonvanishing elements

$$U_{m,-m} = (-1)^m , \quad m = l, l-1, \ldots, -l .$$

This matrix $\hat{U}$ is a symmetric unitary matrix of $2l+1$ dimensions. The representation matrices can be made real by choosing the following $2l+1$ real functions as the basis:

$$Y_{l0}, \quad \frac{1}{\sqrt{2}}(Y_{lm} + Y^*_{lm}), \quad \frac{-i}{\sqrt{2}}(Y_{lm} - Y^*_{lm}) , \quad m = 1, 2, \ldots, l .$$

Case (b): Irreducible representations $D^{(j)}$ of the rotation group with j half of an odd integer fall into this case. For example, when $j = 1/2$, the spin functions $\{\alpha, \beta\}$ provide the basis for the $D^{(1/2)}$ representation, and the time-reversed basis $\{\beta, -\alpha\}$, that for the complex conjugate representation $D^{(1/2)*}$. The matrix

$$\hat{U} = \begin{bmatrix} 0 & -1 \\ 1 & 0 \end{bmatrix}$$

is antisymmetric, and the representation matrices cannot be made real. In single-valued representations, point groups have no examples of this case. An example in space groups will be mentioned in Sect. 13.3.

Case (c): Some point group representations, such as Γ_2 and Γ_3 of the group T, belong to this case.

*4.13 Proofs of the Theorems on Group Representations

*4.13.1 Unitarization of a Representation

Theorem: A representation of the finite group $\mathscr{G}$ can be brought into a unitary representation by means of an equivalence transformation.

Proof: For brevity, we shall write $\hat{A}_i$ for the representation matrix $\hat{D}(G_i)$. We construct a Hermitian matrix $\hat{H}$ from $\hat{A}_i$:

$$\hat{H} = \sum_{i=1}^{g} \hat{A}_i^\dagger \hat{A}_i \ .$$

For an arbitrary group element G_j, we have

$$\begin{aligned}\hat{A}_j^\dagger \hat{H} \hat{A}_j &= \sum_{i=1}^{g} \hat{A}_j^\dagger \hat{A}_i^\dagger \hat{A}_i \hat{A}_j \\ &= \sum_{i=1}^{g} (\hat{A}_i \hat{A}_j)^\dagger (\hat{A}_i \hat{A}_j) \ .\end{aligned}$$

The rearrangement theorem permits us to rewrite this as

$$\hat{A}_j^\dagger \hat{H} \hat{A}_j = \sum_{k=1}^{g} \hat{A}_k^\dagger \hat{A}_k = \hat{H} \ . \tag{4.97}$$

The matrix $\hat{H}$ is positive definite (i.e., its eigenvalues λ_i are positive), because for an arbitrary nonzero vector $\boldsymbol{v}$, we have

$$\begin{aligned}(\boldsymbol{v}, \hat{H}\boldsymbol{v}) &= \sum_i (\boldsymbol{v}, \hat{A}_i^\dagger \hat{A}_i \boldsymbol{v}) \\ &= \sum_i (\hat{A}_i \boldsymbol{v}, \hat{A}_i \boldsymbol{v}) > 0 \ .\end{aligned}$$

Owing to the regularity of the representation matrices $\hat{A}_i$, the inner products never vanish.

The Hermitian matrix $\hat{H}$ can be diagonalized by a unitary transformation

$$\hat{H} = \hat{U} \hat{\Lambda} \hat{U}^{-1} \ . \tag{4.98}$$

The elements λ_i of the diagonal matrix

$$\hat{\Lambda} = [\lambda_1, \lambda_2, \ldots, \lambda_d]$$

are positive. Therefore, we can define the regular matrix

$$\hat{V} = \hat{U} \hat{\Lambda}^{-1/2} \ ,$$

where

$$\hat{\Lambda}^{-1/2} = [\lambda_1^{-1/2}, \lambda_2^{-1/2}, \ldots, \lambda_d^{-1/2}] .$$

The equivalence transformation

$$\hat{A}'_j = \hat{V}^{-1} \hat{A}_j \hat{V} , \quad j = \hat{1}, 2, \ldots, d , \tag{4.99}$$

then gives a unitary representation. In fact, it is straightforward to verify the unitarity relation

$$\hat{A}'^{\dagger}_j \hat{A}'_j = \hat{1} \quad \text{(unit matrix)}$$

using (4.97–99). □

*4.13.2 Schur's First Lemma

For two given irreducible representations $D^{(1)}$ and $D^{(2)}$ of dimensions m and n respectively, a rectangular matrix $\hat{M}$ that satisfies

$$\hat{D}^{(1)}(G)\hat{M} = \hat{M}\hat{D}^{(2)}(G) \tag{4.100}$$

for every group element G of $\mathcal{G}$ must be either

(1) the zero matrix (all the elements vanish), or
(2) a square matrix ($m = n$) with det $\hat{M} \neq 0$.

Proof: We examine the possibilities other than the trivial case (1), and assume $\hat{M} \neq 0$. We further assume $m \geqq n$ for the moment. Now, a basis $\psi_1, \psi_2, \ldots, \psi_m$ for the irreducible representation $D^{(1)}$ is transformed by the group elements G as

$$G\psi_j = \sum_{i=1}^{m} \psi_i D_{ij}^{(1)}(G) , \quad j = 1, 2, \ldots, m . \tag{4.101}$$

Let us then construct a new basis with

$$\phi_l = \sum_{j=1}^{m} \psi_j M_{jl} , \quad l = 1, 2, \ldots, n . \tag{4.102}$$

The new basis ϕ_l is seen to be transformed as

$$G\phi_l = \sum_{k=1}^{n} \phi_k D_{kl}^{(2)}(G) , \tag{4.103}$$

because of (4.100–102), which shows that $\phi_1, \phi_2, \ldots, \phi_n$ form a basis for the irreducible representation $D^{(2)}$. However, by virtue of the property of the irreducible invariant subspace mentioned in Sect. 4.1.3, the linear transformation (4.102) can never diminish the dimension of that invariant subspace. It is therefore impossible that $m > n$. We are then left with the possibility of $m = n$. Now if $m = n$, the matrix $\hat{M}$ must be a regular matrix (det $\hat{M} \neq 0$).

Otherwise, the n basis functions generated by (4.102) would not be linearly independent, and we could then construct a smaller invariant subspace than the one spanned by the original basis ψ_i. This would again violate the irreducibility of the original space.

In the case $m < n$, we replace G by G^{-1} in (4.100) and take the transposed equation,

$$ {}^t\hat{M}\,{}^t\hat{D}^{(1)}(G^{-1}) = {}^t\hat{D}^{(2)}(G^{-1})\,{}^t\hat{M} \ . $$

We can then repeat the above arguments for the basis $\tilde{\psi}_j$ of the adjoint representation (Sect. 4.12) that satisfies

$$ G\tilde{\psi}_j = \sum_{i=1}^{n} \tilde{\psi}_i D_{ji}^{(2)}(G^{-1}) \ . $$

As mentioned in Sect. 4.12, when $D^{(1)}$ and $D^{(2)}$ are irreducible, their adjoint representations are also irreducible. □

*4.13.3 Schur's Second Lemma

A matrix $\hat{M}$ that commutes with all the representation matrices $\hat{D}(G)$ of a representation D of the group $\mathscr{G}$,

$$ \hat{D}(G)\hat{M} = \hat{M}\hat{D}(G) \ , \quad G \in \mathscr{G} \ , \tag{4.104} $$

must be a multiple of the unit matrix $\hat{1}$,

$$ \hat{M} = c\hat{1} \ , $$

if D is irreducible.

Proof: For an arbitrary complex number λ, we have from (4.104),

$$ \hat{D}(G)(\hat{M} - \lambda\hat{1}) = (\hat{M} - \lambda\hat{1})\hat{D}(G) \ . $$

Because of the irreducible D, Schur's first lemma tells us that either (1) $\hat{M} - \lambda\hat{1} = 0$, or (2) $\det(\hat{M} - \lambda\hat{1}) \neq 0$. If we choose λ to be an eigenvalue of the matrix $\hat{M}$, then we have $\det(\hat{M} - \lambda\hat{1}) = 0$, which eliminates the second possibility. Therefore we have $\hat{M} = \lambda\hat{1}$. □

*4.13.4 Second Orthogonality of Characters

Theorem: Characters of irreducible representations satisfy the following orthogonality relation:

$$ \sum_{\alpha=1}^{n_r} \chi^{(\alpha)}(\mathscr{C}_i)^* \chi^{(\alpha)}(\mathscr{C}_j) = \delta_{ij} \frac{g}{h_i} \ , \tag{4.105} $$

where h_i stands for the number of elements in the class $\mathscr{C}_i$.

Proof: The proof of this theorem is divided into two stages. In the first stage, we derive the relation (4.108) that holds for products of the irreducible characters.

As we saw in Sect. 2.9, for any class $\mathscr{C}_k$ of the group $\mathscr{G}$, we have (2.26), which may be rewritten as

$$G\mathscr{C}_k = \mathscr{C}_k G \ , \quad G \in \mathscr{G} \ , \tag{4.106}$$

where G is an arbitrary element of $\mathscr{G}$. Because of the homomorphism between the group and its representation, a relation similar to (4.106) holds for the representation matrices as well. If we define the sum of the representation matrices over the elements in the class $\mathscr{C}_k$ by

$$\hat{\mathscr{C}}_k \equiv \sum_{G \in \mathscr{C}_k} \hat{D}^{(\alpha)}(G) \ ,$$

then we have

$$\hat{D}^{(\alpha)}(G)\hat{\mathscr{C}}_k = \hat{\mathscr{C}}_k \hat{D}^{(\alpha)}(G) \ , \quad G \in \mathscr{G} \ .$$

Since $D^{(\alpha)}$ is irreducible, Schur's second lemma requires that

$$\hat{\mathscr{C}}_k = \lambda \hat{1} \ .$$

To determine λ, we take the trace of both sides,

$$h_k \chi^{(\alpha)}(\mathscr{C}_k) = \lambda d_\alpha \ .$$

Thus, we obtain

$$\hat{\mathscr{C}}_k = \frac{h_k}{d_\alpha} \chi^{(\alpha)}(\mathscr{C}_k) \hat{1} \ . \tag{4.107}$$

Next, we use the relation (2.29) for class products. Again from the homomorphism, we have

$$\hat{\mathscr{C}}_i \hat{\mathscr{C}}_j = \sum_k c_{ij}^k \hat{\mathscr{C}}_k \ .$$

Substitution of (4.107) gives

$$h_i h_j \chi^{(\alpha)}(\mathscr{C}_i) \chi^{(\alpha)}(\mathscr{C}_j) = d_\alpha \sum_k c_{ij}^k h_k \chi^{(\alpha)}(\mathscr{C}_k) \ , \tag{4.108}$$

which completes the first stage of the proof.

In the second stage, we make use of the relations (4.82, 83) derived from the regular representation, which may be combined into

$$\sum_\alpha d_\alpha \chi^{(\alpha)}(\mathscr{C}_k) = g \delta_{k1} \ . \tag{4.109}$$

Here $\mathscr{C}_1$ denotes the class that consists of the single element E. Summing both sides of (4.108) over the irreducible representations α and using (4.109), we obtain

$$h_i h_j \sum_\alpha \chi^{(\alpha)}(\mathscr{C}_i)\chi^{(\alpha)}(\mathscr{C}_j) = g c_{ij}^1 \ . \tag{4.110}$$

If we write $\mathscr{C}_{j'}$ for the class that consists of the elements G^{-1} inverse to the ones G constituting the class $\mathscr{C}_j$, we have

$$\chi^{(\alpha)}(\mathscr{C}_{j'}) = \chi^{(\alpha)}(\mathscr{C}_j)^* \quad \text{and} \quad h_{j'} = h_j \ ,$$

using the second equality of (4.84). Moreover, from (2.30),

$$c_{ij'}^1 = h_i \delta_{ij} \ .$$

Replacing j by j' in (4.110), we arrive at the orthogonality (4.105). □

5. Representations of a Group II

In this chapter, we discuss how to construct irreducible representations of a group from the known representations of its invariant subgroup. This problem reduces to the construction of irreducible ray representations with an appropriate factor system. The treatment will be most easily understood by working through it for some point group. The representation theory for space groups is a typical application of the scheme developed in this chapter.

*5.1 Induced Representations

A group $\mathscr{G}$ having a proper subgroup $\mathscr{H}$ may be decomposed into left cosets of $\mathscr{H}$ as

$$\mathscr{G} = R_1\mathscr{H} + R_2\mathscr{H} + \cdots + R_k\mathscr{H} , \quad R_1 = E . \tag{5.1}$$

Let us write $D^{(\lambda)}$ for the λth irreducible representation of $\mathscr{H}$ and denote its basis set as $\phi_1, \phi_2, \ldots, \phi_d$, so that

$$S\phi_\nu = \sum_\mu \phi_\mu D^{(\lambda)}_{\mu\nu}(S) , \quad S \in \mathscr{H} , \tag{5.2}$$

for the element S belonging to $\mathscr{H}$.

If we define the functions $\phi_{j\nu}$ by

$$\phi_{j\nu} = R_j\phi_\nu , \quad j = 1, 2, \ldots, k ; \quad \nu = 1, 2, \ldots, d , \tag{5.3}$$

they will provide us with a basis set for a kd-dimensional representation of $\mathscr{G}$. In fact, if we operate on $\phi_{j\nu}$ with an element P of $\mathscr{G}$, we find

$$P\phi_{j\nu} = \sum_{i\mu} \phi_{i\mu} D_{i\mu,j\nu}(P) , \tag{5.4}$$

where

$$D_{i\mu,j\nu}(P) = \delta_{ij}(P) D^{(\lambda)}_{\mu\nu}(R_i^{-1}PR_j) , \tag{5.5}$$

$$\delta_{ij}(P) = \begin{cases} 1 & \text{when } PR_j \in R_i\mathscr{H} , \\ 0 & \text{otherwise} . \end{cases} \tag{5.6}$$

The representation D defined by (5.5) is called the *induced representation* of $D^{(\lambda)}$

onto $\mathscr{G}$ and is usually denoted as $D^{(\lambda)} \uparrow \mathscr{G}$. In this section, we will employ the notation $\bar{D}^{(\lambda)}$, for simplicity. If we write $\chi^{(\lambda)}$ for the character of $D^{(\lambda)}$, the character $\bar{\chi}^{(\lambda)}$ of the induced representation $\bar{D}^{(\lambda)}$ follows immediately from (5.5):

$$\bar{\chi}^{(\lambda)}(P) = \sum_{j=1}^{k} \delta_{jj}(P)\chi^{(\lambda)}(R_j^{-1}PR_j) \ . \tag{5.7}$$

The representation $\bar{D}^{(\lambda)}$ is reducible in general. Suppose the irreducible representation $D^{(i)}$ of $\mathscr{G}$ appears $w_{\lambda i}$ times in the reduction of $\bar{D}^{(\lambda)}$. We then have

$$\bar{\chi}^{(\lambda)}(P) = \sum_{i} w_{\lambda i}\chi^{(i)}(P) \ , \qquad P \in \mathscr{G} \ , \tag{5.8}$$

if we write $\chi^{(i)}$ for the character of $D^{(i)}$.

The representation matrices of $D^{(i)}$ are defined naturally for all the elements of $\mathscr{G}$. If we restrict them to the elements S of the subgroup $\mathscr{H}$, we obtain a set of matrices $\{\hat{D}^{(i)}(S) | S \in \mathscr{H}\}$ representing $\mathscr{H}$. The representation of $\mathscr{H}$ obtained in this way is sometimes denoted by $D^{(i)} \downarrow \mathscr{H}$ and is called the *subduced representation* of $D^{(i)}$ onto $\mathscr{H}$.

Now, $D^{(i)}$, though irreducible under $\mathscr{G}$, will be in general reducible under $\mathscr{H}$. If the irreducible representation $D^{(\lambda)}$ of $\mathscr{H}$ is included $w_{i\lambda}$ times in the reduction of $D^{(i)} \downarrow \mathscr{H}$, we have

$$\chi^{(i)}(S) = \sum_{\lambda} w_{i\lambda}\chi^{(\lambda)}(S) \ , \qquad S \in \mathscr{H} \ . \tag{5.9}$$

When $w_{i\lambda} \geqq 1$, we say that $D^{(\lambda)}$ is compatible with $D^{(i)}$ and relations (5.9) are called compatibility relations.

For the integers $w_{\lambda i}$ in (5.8) and $w_{i\lambda}$ in (5.9), we have the *Frobenius reciprocity theorem*:

$$w_{i\lambda} = w_{\lambda i} \ . \tag{5.10}$$

It means that the number of times an irreducible representation $D^{(\lambda)}$ of $\mathscr{H}$ appears in the reduction of the subduced representation $D^{(i)} \downarrow \mathscr{H}$ is equal to the number of times the irreducible representation $D^{(i)}$ of $\mathscr{G}$ is included in the induced representation $D^{(\lambda)} \uparrow \mathscr{G}$.

Exercise 5.1. Prove the Frobenius reciprocity theorem.

Exercise 5.2. Show that $D^{(\lambda)} \uparrow \mathscr{G}$ is irreducible if and only if the irreducible representation $D^{(\lambda)}$ of $\mathscr{H}$ appears in $(D^{(\lambda)} \uparrow \mathscr{G}) \downarrow \mathscr{H}$ only once.

Exercise 5.3. Derive the following expression for $\bar{\chi}^{(\lambda)}(P)$ $(P \in \mathscr{C}_i)$ from (5.7):

$$h_i\bar{\chi}^{(\lambda)}(\mathscr{C}_i) = k \sum_{S \in \mathscr{H} \cap \mathscr{C}_i} \chi^{(\lambda)}(S) \ ,$$

where the sum on the right-hand side is to be taken over the elements S of $\mathscr{H}$ that belong to the class $\mathscr{C}_i$ of $\mathscr{G}$.

Table 5.1. Compatibility relations between the irreducible representations of D_{6h}, D_6 and C_6

D_{6h}	D_6	C_6	(m)
$A_{1g,u}$	A_1	A	(0)
$A_{2g,u}$	A_2		
$B_{1g,u}$	B_1	B	(3)
$B_{2g,u}$	B_2		
$E_{1g,u}$	E_1	E_1	(± 1)
$E_{2g,u}$	E_2	E_2	(± 2)

Exercise 5.4. Examine how $D^{(\lambda)} \uparrow \mathscr{G}$ and $D^{(i)} \downarrow \mathscr{H}$ are reduced when $\mathscr{H} = C_6$ and $\mathscr{G} = D_{6h}$ or D_6. Table 5.1 gives the relevant compatibility relations.

*5.2 Irreducible Representations of a Group with an Invariant Subgroup

Suppose that the group $\mathscr{G}$ of order g has a proper invariant subgroup $\mathscr{H}$ of order h. When $\mathscr{G}$ is a space group, $\mathscr{H}$ may be taken as the translation group. The group $\mathscr{G}$ may then be decomposed into left cosets,

$$\mathscr{G} = T_1\mathscr{H} + T_2\mathscr{H} + \cdots + T_k\mathscr{H} , \quad T_1 = E , \quad g = kh . \tag{5.11}$$

Next, take an irreducible unitary representation Δ of $\mathscr{H}$ whose basis will be written as ψ_ν ($\nu = 1, 2, \ldots, d$). In the case of the translation group, d is unity. From ψ_ν, we construct k sets of functions given by

$$\psi_\nu^{(i)} = T_i\psi_\nu , \quad i = 1, 2, \ldots, k .$$

The ith set $\psi_\nu^{(i)}$ ($\nu = 1, 2, \ldots, d$) will form a basis for the irreducible representation $\Delta^{(i)} = \{\hat{\Delta}^{(i)}(S) | S \in \mathscr{H}\}$ of $\mathscr{H}$, where

$$\hat{\Delta}^{(i)}(S) = \hat{\Delta}(T_i^{-1} S T_i) .$$

Remember that $T_i^{-1} S T_i$ belongs to $\mathscr{H}$, because $\mathscr{H}$ is an invariant subgroup of $\mathscr{G}$. In this way, we have k irreducible representations of $\mathscr{H}$. In general, not all of them may be inequivalent. Let us therefore group them so that equivalent ones are put into the same group. As a result, we can renumber and arrange them in the following manner:

(i) Representations $\Delta^{(ij)}$ ($j = 1, 2, \ldots, m$) with the same i are all equivalent to $\Delta^{(i1)}$, and
(ii) representations $\Delta^{(i1)}$ ($i = 1, 2, \ldots, s$) with different i are all inequivalent, with $k = sm$.

In accordance with this, we renumber T_i as follows:

$$T_{i1} = P_i\ , \quad i = 1, 2, \ldots, s\ ,$$

$$T_{1j} = R_j\ , \quad j = 1, 2, \ldots, m\ ,$$

$$T_{ij} = P_i R_j S_0\ , \quad S_0 \in \mathscr{H}\ .$$

Then we have

$$\hat{\Delta}^{(ij)}(S) = \hat{U}_{ij}^{-1}\hat{\Delta}^{(i1)}(S)\hat{U}_{ij}\ , \quad \hat{\Delta}^{(i1)}(S) = \hat{\Delta}(P_i^{-1}SP_i)\ ,$$

$$\hat{\Delta}(R_j^{-1}SR_j) = \hat{U}_j^{-1}\hat{\Delta}(S)\hat{U}_j\ , \quad \hat{U}_{ij} = \hat{U}_j\hat{\Delta}(S_0)\ ,$$

with unitary matrices $\hat{U}_j$.

Now, the set given by

$$\mathscr{L} = R_1\mathscr{H} + R_2\mathscr{H} + \cdots + R_m\mathscr{H}\ , \quad R_1 = E\ , \quad l = mh\ , \tag{5.12}$$

is a set of elements T which make the representation $\Delta(T^{-1}ST)$ of $\mathscr{H}$ equivalent to $\Delta(S)$ and form a group. The group $\mathscr{L}$ is of course determined by specifying the representation Δ and is called the *little group* of Δ in $\mathscr{G}$. In the case of a space group, the representation Δ is characterized by a wavevector $\mathbf{k}$ and $\mathscr{L}$ corresponds to the group $\mathscr{G}(\mathbf{k})$. It is to be remarked that $\mathscr{H}$ is an invariant subgroup of $\mathscr{L}$ in (5.12).

If we decompose $\mathscr{G}$ into left cosets with respect to the subgroup $\mathscr{L}$, we obtain

$$\mathscr{G} = P_1\mathscr{L} + P_2\mathscr{L} + \cdots + P_s\mathscr{L}\ , \quad P_1 = E\ , \quad g = sl\ . \tag{5.13}$$

We have defined $\mathscr{L}$ as the little group of $\Delta = \Delta^{(11)}$. The little group of $\Delta^{(i1)}$ is provided by $P_i\mathscr{L}P_i^{-1} \equiv \mathscr{L}_i$. The set of representations $\Delta^{(11)}, \Delta^{(21)}, \ldots, \Delta^{(s1)}$ are said to form an *orbit* or a *star*. In the treatment of space groups, each representation $\Delta^{(i1)}$ is specified by the wavevector $\mathbf{k}_i$, and star $\{\mathbf{k}\}$ represents the set $\{\mathbf{k}_1, \mathbf{k}_2, \ldots, \mathbf{k}_s\}$. The little group $\mathscr{L}_i$ corresponds to $\mathscr{G}(\mathbf{k}_i)$.

By making use of the decomposition (5.12), we can derive the induced (*md*-dimensional) representation $\Delta \uparrow \mathscr{L}$ as discussed in the preceding section. However, it is not irreducible in general as a representation of $\mathscr{L}$. What we need here is an irreducible representation D of $\mathscr{L}$, called the *small representation*. Postponing the details to the next section, we assume here we have somehow obtained D. Its dimension will be nd, if Δ appears n times in $D \downarrow \mathscr{H}$. Using the decomposition (5.13), we can derive an induced representation $D \uparrow \mathscr{G} \equiv \bar{D}$. The dimension of this representation is *snd*. In the *ij* block, we have the matrix given by

$$\bar{D}_{ij}(Q) = \delta_{ij}(Q)\hat{D}(P_i^{-1}QP_j)\ , \quad i,j = 1, 2, \ldots, s\ , \tag{5.14}$$

with

$$\delta_{ij}(Q) = \begin{cases} 1 & \text{when } QP_j \in P_i\mathscr{L}\ , \\ 0 & \text{otherwise}\ . \end{cases}$$

When $Q \in \mathscr{L}$, the matrix $\hat{D}(Q)$ appears in the 11 block, while for $Q \in \mathscr{L}_i$ we find $\hat{D}(P_i^{-1}QP_i)$ in its ii block. In particular, when $Q = S \in \mathscr{H}$, the d-dimensional matrix $\hat{\Delta}(S) = \Delta^{(11)}$ appears n times on the diagonal of the 11 block, $\Delta^{(21)} = \hat{\Delta}(P_2^{-1}SP_2)$ n times on the diagonal of the 22 block, and so on. The irreducibility of $\bar{D}$ as a representation of $\mathscr{G}$ follows from Exercise 5.2 of the preceding section.

Exercise 5.5. Prove that $\bar{D}$ is irreducible under $\mathscr{G}$.

When the order k of the factor group $\mathscr{G}/\mathscr{H}$ is prime, the problem becomes simple. We observe that $\mathscr{H} \subseteq \mathscr{L} \subseteq \mathscr{G}$ in general. If $k = sm$ is prime, we must have either

(a) $s = k, m = 1,$ so that $\mathscr{L} = \mathscr{H}$,

or

(b) $s = 1, m = k,$ so that $\mathscr{G} = \mathscr{L}$.

In case (a), $\Delta \uparrow \mathscr{G}$ gives an irreducible representation of $\mathscr{G}$.

In case (b), $\mathscr{G}/\mathscr{H} = \mathscr{L}/\mathscr{H}$ is a cyclic group of order k and we may choose $E, R, R^2, \ldots, R^{k-1}$ as coset representatives in (5.12). Since $\hat{\Delta}(R^{-1}SR) = \hat{U}^{-1}\hat{\Delta}(S)\hat{U}$ for $S \in \mathscr{H}$, we obtain

$$\hat{\Delta}(R^{-k}SR^k) = \hat{U}^{-k}\hat{\Delta}(S)\hat{U}^k = (\hat{\Delta}(R^k))^{-1}\hat{\Delta}(S)\hat{\Delta}(R^k) \tag{5.15}$$

for $R^k \in \mathscr{H}$. This means that $\hat{\Delta}(R^k)\hat{U}^{-k}$ commutes with $\hat{\Delta}(S)$ for all elements S of $\mathscr{H}$. By Schur's lemma, it then follows that

$$\hat{U}^k = \alpha\hat{\Delta}(R^k)$$

with a complex constant α of modulus unity, because both sides are unitary matrices. Since (5.15) holds even if we replace $\hat{U}$ by $\alpha^{-1/k}\hat{U}$, we may as well put

$$\hat{U}^k = \hat{\Delta}(R^k) \tag{5.16}$$

from the beginning. If we determine $\hat{U}$ from this together with $\hat{\Delta}(R^{-1}SR) = \hat{U}^{-1}\hat{\Delta}(S)\hat{U}$, the irreducible representation of $\mathscr{G} = \mathscr{L}$ is given by

$$\hat{D}(S) = \hat{\Delta}(S) , \quad S \in \mathscr{H} ,$$
$$\hat{D}(R) = \hat{U} . \tag{5.17}$$

Note that we are still left with the choice of kth roots of 1 in determining $\hat{U}$ from (5.16). Corresponding to this situation, we obtain k irreducible representations from (5.17).

Let us write $\mathscr{G}_1$ for the largest (proper) invariant subgroup of $\mathscr{G} = \mathscr{G}_0$ ($\mathscr{G}_1 \subset \mathscr{G}_0$), $\mathscr{G}_2$ for the largest invariant subgroup of $\mathscr{G}_1$ ($\mathscr{G}_2 \subset \mathscr{G}_1$) and so on, and suppose this sequence ends at $\mathscr{G}_l \equiv \{E\}$.

When the orders of the factor groups $\mathscr{G}_0/\mathscr{G}_1, \mathscr{G}_1/\mathscr{G}_2, \ldots, \mathscr{G}_{l-1}/\mathscr{G}_l = \mathscr{G}_{l-1}$ are all prime, $\mathscr{G}$ is called a *solvable group.* All the irreducible representations of a solvable group can be obtained by the method given in this section starting from those for the smallest cyclic group $\mathscr{G}_{l-1}$. This means that the problem of constructing small representations of $\mathscr{L}$ has also been solved in the case of solvable groups.

The 230 space groups, and accordingly the 32 point groups, are all known to be solvable.

Exercise 5.6. Apply the method of this section to derive the irreducible representations of C_{3v} from those of the invariant subgroup C_3.

*5.3 Irreducible Representations of Little Groups or Small Representations

When the little group $\mathscr{L}$ is decomposed into left cosets with respect to its invariant subgroup $\mathscr{H}$ as in (5.12), we have

$$\hat{\Delta}(R_j^{-1}SR_j) = \hat{U}_j^{-1}\hat{\Delta}(S)\hat{U}_j \ , \quad j = 2, \ldots, m \ , \quad S \in \mathscr{H} \ , \tag{5.18}$$

because $\Delta(R_j^{-1}SR_j)$ and $\Delta(S)$ are equivalent representations by definition. The fact that $\mathscr{H}$ is an invariant subgroup of $\mathscr{L}$ means that the product of coset representatives R_i and R_j may be written as

$$R_iR_j = R_kS_0 \ , \quad S_0 \in \mathscr{H} \ . \tag{5.19}$$

This, together with (5.18), gives

$$\alpha_{ij}^k\hat{U}_i\hat{U}_j = \hat{U}_k\hat{\Delta}(S_0) \tag{5.20}$$

according to Schur's lemma. For $\hat{U}$ and $\hat{\Delta}$ unitary, $|\alpha_{ij}^k| = 1$. The factor α_{ij}^k is a complex number of modulus unity determined from (5.20) when the $\hat{U}_i$'s are given. Note that we have

$$\alpha_{1j}^k = \delta_{jk} \ , \quad \alpha_{i1}^k = \delta_{ik} \ . \tag{5.21}$$

We shall write $R_{i'}$ for the coset representative that satisfies $R_iR_{i'} = S' \in \mathscr{H}$ for a given R_i. We then find

$$\alpha_{ii'}^1\hat{U}_i\hat{U}_{i'} = \hat{\Delta}(S') \tag{5.22}$$

with

$$\alpha_{ii'}^1 = \alpha_{i'i}^1 \ . \tag{5.23}$$

Suppose we have chosen the irreducible representation D of $\mathscr{L}$ in such a way that it is already fully reduced under $\mathscr{H}$. This means that the matrix $\hat{D}(S)$ for the elements S of $\mathscr{H}$ has d-dimensional representation matrices $\hat{\Delta}(S)$ appearing n times on its diagonal. If we denote the matrix elements of $Q \in \mathscr{L}$ as $D_{\gamma\nu,\gamma'\nu'}(Q)$

with $\gamma, \gamma' = 1, 2, \ldots, n$ and $v, v' = 1, 2, \ldots, d$, we find

$$D_{\gamma v, \gamma' v'}(S) = \delta_{\gamma\gamma'} \Delta_{vv'}(S) , \quad S \in \mathscr{H} . \tag{5.24}$$

Put $S = R_j^{-1} S' R_j$ in this equation and use (5.18). It then follows from Schur's lemma that

$$D_{\gamma v, \gamma' v'}(R_j) = D_{\gamma\gamma'}(j)(\hat{U}_j)_{vv'} \tag{5.25}$$

The factors $D_{\gamma\gamma'}(j)$ define a $n \times n$ matrix $\hat{D}(j)$. The matrices $\hat{D}(i)$ satisfy

$$\hat{D}(i)\hat{D}(j) = \alpha_{ij}^k \hat{D}(k) \tag{5.26}$$

because of (5.19, 20). For R_i and $R_{i'}$ whose product belongs to $\mathscr{H}$, we find

$$\hat{D}(i)\hat{D}(i') = \alpha_{ii'}^1 \hat{1}_n , \tag{5.27}$$

where $\hat{1}_n$ is the $n \times n$ unit matrix.

When we let R_j correspond to the element $R_j\mathscr{H}$ (simply denoted as j hereafter) of the factor group $\mathscr{L}/\mathscr{H} \equiv \mathscr{L}^{\mathrm{I}}$, equations (5.26, 27) are the relations to be satisfied by the "representation matrices" $\hat{D}(i)$, $\hat{D}(j)$, ... in accordance with those between the elements of the group $\mathscr{L}^{\mathrm{I}}$: $ij = k$, $ii' = 1$. This "representation" differs from the ordinary one in that we have factors α_{ij}^k, $\alpha_{ii'}^1$ on the right-hand sides. The set D_0 of matrices $\hat{D}(j)$ satisfying such relations is called a *ray representation* of $\mathscr{L}^{\mathrm{I}}$ with the *factor system* $\{\alpha_{ij}^k\}$. In contrast to the ray representations, ordinary ones with all the factors $\alpha_{ij}^k = 1$ are sometimes called vector representations.

The group $\mathscr{L}$ is occasionally denoted as $\mathscr{L}^{\mathrm{II}}$ with respect to $\mathscr{L}^{\mathrm{I}}$, so that $\mathscr{L}^{\mathrm{I}} \cong \mathscr{L}^{\mathrm{II}}/\mathscr{H}$. Especially, when $\mathscr{H}$ is the center of $\mathscr{L}^{\mathrm{II}}$, the group $\mathscr{L}^{\mathrm{II}}$ is called the covering group of $\mathscr{L}^{\mathrm{I}}$. The center is defined as the set of elements of a group that commute with all the elements of the group. The center is naturally an invariant subgroup of the group.

According to the considerations given above, the problem of obtaining an irreducible representation D of the little group $\mathscr{L}$ reduces to the construction of an irreducible ray representation D_0 of the factor group $\mathscr{L}/\mathscr{H}$. That is, we can obtain $D = \{D_{\gamma v, \gamma' v'}(Q) | Q \in \mathscr{L}\}$ through (5.24, 25), once we know $D_0 = \{D_{\gamma\gamma'}(j)\}$. (Conversely, we can obtain an irreducible *ray* representation D_0 of $\mathscr{L}/\mathscr{H}$, through the irreducible representations D of $\mathscr{L}$ and Δ of $\mathscr{H}$.)

In dealing with the representations of space groups, we choose the translation group $\mathscr{T}$ as $\mathscr{H}$. In the case of symmorphic space groups with no glides and screws, we may take rotational operations of the point group $\mathscr{G}/\mathscr{T}$ as R_j and put all the factors α_{ij}^k to 1. It suffices to consider only vector representations. Even with the nonsymmorphic groups, we can put the factors to 1 as long as the wavevector $\boldsymbol{k}$ lies inside the Brillouin zone (BZ). Only for nonsymmorphic space groups and where $\boldsymbol{k}$ ends on the surface of the BZ, do we have to consider the ray representations. Let us see this by calculating the factor system for this problem.

Let $R_j = \{\beta_j|\tau_j\}$ be the representative element of the left coset decomposition of $\mathscr{L} = \mathscr{G}(\boldsymbol{k})$ with respect to the translation group $\mathscr{T}$. Corresponding to (5.19), we have

$$\{\beta_i|\tau_i\}\{\beta_j|\tau_j\} = \{\beta_i\beta_j|\tau_i + \beta_i\tau_j\} = \{\beta_k|\tau_k\}\{\varepsilon|\boldsymbol{t}_0\} , \tag{5.28}$$

$$\beta_i\beta_j = \beta_k ,$$

$$\tau_i + \beta_i\tau_j = \tau_k + \beta_k\boldsymbol{t}_0 .$$

Since the irreducible representation $\Delta(\{\varepsilon|\boldsymbol{t}_n\}) = \exp(\mathrm{i}\boldsymbol{k}\cdot\boldsymbol{t}_n)$ of $\mathscr{T}$ is one-dimensional, we may put $\hat{U}_j = \hat{U}(\{\beta_j|\tau_j\}) = 1$. Then it follows from (5.20) that

$$\alpha_{ij}^k = \exp(\mathrm{i}\boldsymbol{k}\cdot\boldsymbol{t}_0) = \exp[\mathrm{i}\boldsymbol{k}\cdot(\tau_i + \beta_i\tau_j - \tau_k)] , \tag{5.29}$$

where we have taken into account that $\exp(\mathrm{i}\boldsymbol{k}\cdot\beta\boldsymbol{t}_0) = \exp(\mathrm{i}\beta^{-1}\boldsymbol{k}\cdot\boldsymbol{t}_0) = \exp(\mathrm{i}\boldsymbol{k}\cdot\boldsymbol{t}_0)$ for $\{\beta|\tau\}\in\mathscr{G}(\boldsymbol{k})$. The representation matrices $\hat{\underline{D}}^{\boldsymbol{k}(\Gamma)}$ of $\mathscr{G}(\boldsymbol{k})$ are given by

$$\underline{D}_{\gamma\gamma'}^{\boldsymbol{k}(\Gamma)}(\{\beta_j|\tau_j\}\{\varepsilon|\boldsymbol{t}_n\}) = D_{\gamma\gamma'}^{\boldsymbol{k}(\Gamma)}(j)\exp(\mathrm{i}\boldsymbol{k}\cdot\boldsymbol{t}_n) \tag{5.30}$$

in terms of the irreducible ray representation $D^{\boldsymbol{k}(\Gamma)}$ of $\mathscr{G}(\boldsymbol{k})/\mathscr{T}$ with the factor system given by (5.29). In (5.30), the matrices $\{D^{\boldsymbol{k}(\Gamma)}(j)\}$ satisfy (5.26, 27). *When* $\boldsymbol{k}$ *lies within the BZ*, we can write them as follows:

$$D_{\gamma\gamma'}^{\boldsymbol{k}(\Gamma)}(j) = \underline{D}_{\gamma\gamma'}^{\boldsymbol{k}(\Gamma)}(\{\beta_j|\tau_j\}) = D_{\gamma\gamma'}^{\Gamma}(\beta_j)\exp(\mathrm{i}\boldsymbol{k}\cdot\tau_j) \tag{5.31}$$

using the matrices of an irreducible representation D^{Γ} of the point group isomorphic with $\mathscr{G}(\boldsymbol{k})/\mathscr{T}$. Note that the character $\underline{\chi}^{\boldsymbol{k}(\Gamma)}$ of the irreducible representation of $\mathscr{G}(\boldsymbol{k})$ is related to $\chi^{\boldsymbol{k}(\Gamma)}$ for the irreducible ray representation of $\mathscr{G}(\boldsymbol{k})/\mathscr{T}$ by

$$\underline{\chi}^{\boldsymbol{k}(\Gamma)}(\{\beta_j|\tau_j\}\{\varepsilon|\boldsymbol{t}_n\}) = \chi^{\boldsymbol{k}(\Gamma)}(j)\exp(\mathrm{i}\boldsymbol{k}\cdot\boldsymbol{t}_n) . \tag{5.32}$$

We could as well have chosen U_j so that $U_j = \exp(\mathrm{i}\boldsymbol{k}\cdot\tau_j)$ in the above. The factor system in this case is given by

$$\alpha_{ij}^{\prime k} = \exp[\mathrm{i}\boldsymbol{k}\cdot(\beta_i\tau_j - \tau_j)] = \exp[\mathrm{i}\boldsymbol{K}(\beta_i)\cdot\tau_j] , \tag{5.33}$$

with

$$\boldsymbol{K}(\beta) = \beta^{-1}\boldsymbol{k} - \boldsymbol{k} .$$

The two factor systems (5.29 and 33) are equivalent (Sect. 5.4) and either of them may be adopted. We observe that the vector $\boldsymbol{K}(\beta_i)$ vanishes when $\boldsymbol{k}$ lies inside the BZ. When the vector $\boldsymbol{k}$ ends on the BZ boundary, $\boldsymbol{K}(\beta_i)$ will in general be equal to a reciprocal lattice vector. In such cases, nonvanishing τ_j of screws and glides will bring ray representations into play.

*5.4 Ray Representations

Let $i, j, k, \ldots$ denote the elements of a finite group $\mathscr{L}^1$ of order m. When the relations

$$\hat{D}(i)\hat{D}(j) = \alpha_{ij}^k \hat{D}(k) \tag{5.26}$$

hold corresponding to the multiplication $ij = k$ of the group elements, we say that the set D of matrices $\hat{D}(i)$ form a ray representation of $\mathscr{L}^1$ with the factor system $\alpha = \{\alpha_{ij}^k\}$. In correspondence with the associative law for the group elements,

$$iq = kr = p \ , \quad \text{if} \quad ij = k \quad \text{and} \quad jr = q \ ,$$

the factors have to satisfy the *associativity condition*:

$$\alpha_{iq}^p \alpha_{jr}^q = \alpha_{ij}^k \alpha_{kr}^p \ . \tag{5.34}$$

Conversely, if we have a set of m^2 constants α_{ij}^k satisfying this condition, we can construct a ray representation with this set as a factor system.

The ray representation D' obtained by replacing the matrices $\hat{D}(i)$ with $\hat{D}'(i) = \omega_i \hat{D}(i)$ has the factor system

$$\alpha_{ij}'^k = \frac{\omega_i \omega_j}{\omega_k} \alpha_{ij}^k \ . \tag{5.35}$$

When the factors of the systems α and α' are connected by such relations, the two systems are said to be equivalent and regarded as essentially the same. By making use of (5.34, 35), we can show, without loss of generality, that we may assume (5.21, 23) for any factor system. Besides, as shown below, the values of factors α_{ij}^k may be restricted to the mth roots of unity. As a result, it follows that for a given finite group there exist only a finite number of inequivalent factor systems.

We first note that, when $\alpha = \{\alpha_{ij}^k\}$ and $\alpha' = \{\alpha_{ij}'^k\}$ are factor systems, the product $\alpha\alpha' = \alpha'\alpha = \{\alpha_{ij}^k \alpha_{ij}'^k\}$ is also a factor system. Then the mth power α^m of the system α becomes equivalent to the system $1 = \{1\}$. In fact, we may put (5.34) in matrix form as

$$\hat{\alpha}(i)\hat{\alpha}(j) = \alpha_{ij}^k \hat{\alpha}(k) \tag{5.36}$$

by defining the elements of the $m \times m$ matrix $\hat{\alpha}(i)$ through

$$\alpha_{iq}^p = \alpha_{pq}(i) \ .$$

This then leads to

$$\frac{\det\hat{\alpha}(i)\det\hat{\alpha}(j)}{\det\hat{\alpha}(k)} = (\alpha_{ij}^k)^m \ ,$$

which, when compared with (5.35), shows that α^m is equivalent to 1. If we put $\omega_i = [\det\hat{\alpha}(i)]^{-1/m}$ in (5.35) and go over to $\hat{D}'(i)$, we have $(\alpha'^k_{ij})^m = 1$, so that we may regard any factor α^k_{ij} of the system to be one of the mth roots of unity.

Equation (5.36) shows that the set $\{\hat{\alpha}(i)\}$ constitutes a ray representation of $\mathscr{L}^\mathrm{I}$ with the factor system α. If we put this relation into the form

$$\hat{\alpha}(i)\hat{\alpha}(j) = \hat{\alpha}(k)\alpha_{kj}(i) = \beta_{ik}(j)\hat{\alpha}(k) \tag{5.37}$$

with

$$\beta_{pq}(i) = \alpha^q_{pi} \ , \tag{5.38}$$

we note that $\{\hat{\alpha}(i)\}$ may be called the "regular" (unitary) representation of $\mathscr{L}^\mathrm{I}$. It should be remarked that (5.36, 37) also hold if we replace the matrices $\hat{\alpha}(i)$ by unitary matrices $\hat{\beta}(i)$:

$$\hat{\beta}(i)\hat{\beta}(j) = \alpha^k_{ij}\hat{\beta}(k) \tag{5.39}$$

$$= \hat{\beta}(k)\alpha_{kj}(i) = \beta_{ik}(j)\hat{\beta}(k) \ . \tag{5.40}$$

In fact, the representations $\alpha = \{\hat{\alpha}(i)\}$ and $\beta = \{\hat{\beta}(i)\}$ are equivalent.

In the following, we give several theorems concerning unitary ray representations of a finite group without proof [5.1].

The concepts of equivalence and irreducibility for ray representations are the same as those for the vector representations. Note that the equivalence of ray representations is meaningful only for a common factor system.

Schur's lemma is also valid as in the case of vector representations. For two irreducible ray representations $D^{(\sigma)} = \{\hat{D}^{(\sigma)}(i)\}$ and $D^{(\sigma')} = \{\hat{D}^{(\sigma')}(i)\}$ with the same factor system, the matrix $\hat{M}$ satisfying

$$\hat{M}\hat{D}^{(\sigma)}(i) = \hat{D}^{(\sigma')}(i)\hat{M} \ , \quad i \in \mathscr{L}^\mathrm{I} \tag{5.41}$$

for any i will be either the zero matrix or a square matrix with nonvanishing determinant. In the latter case, the two representations are of the same dimension and are in fact equivalent. If a matrix $\hat{M}$ commutes with *all* the matrices of an irreducible ray representation $D^{(\sigma)}$ as

$$\hat{M}\hat{D}^{(\sigma)}(i) = \hat{D}^{(\sigma)}(i)\hat{M} \ , \quad i \in \mathscr{L}^\mathrm{I} \ , \tag{5.42}$$

$\hat{M}$ is nothing other than a multiple of the unit matrix.

The orthogonality theorem for the irreducible ray representations holds in the same form as before:

$$\sum_i D^{(\sigma)}_{\lambda\mu}(i) D^{(\sigma')}_{\lambda'\mu'}(i)^* = \frac{m}{d_\sigma}\,\delta_{\sigma\sigma'}\delta_{\lambda\lambda'}\delta_{\mu\mu'} \ , \tag{5.43}$$

where d_σ is the dimension of the representation $D^{(\sigma)}$.

The character of a ray representation is defined in the same way by

$$\chi^{(\sigma)}(i) = \sum_{\lambda} D^{(\sigma)}_{\lambda\lambda}(i) .$$

In ray representations, however, the values of characters are not necessarily equal within a class. The situation here may be described as follows.

Let us define a set of complex numbers p_{ij} by

$$p_{ij} = \frac{h_i}{m} \operatorname{Tr}\{\hat{\beta}(i)\hat{\alpha}(j)^*\} = \frac{h_i}{m} \sum_{k,l} \alpha^l_{ki}\alpha^{l*}_{jk} , \tag{5.44}$$

where h_i is the number of elements in the class which the element i belongs to. Then it can be shown that they satisfy

$$p_{ij} = p^*_{ji} , \quad p_{ij}p_{jl} = p_{il} , \tag{5.45}$$

and $p_{ij} = 0$ when i and j belong to different classes. It then follows that, when i and j belong to the same class, either

$$\text{(a)}\ p_{ii} = 1 , \quad |p_{ij}| = 1 , \quad i \neq j , \quad \text{or} \tag{5.46}$$

$$\text{(b)}\ p_{ii} = 0 , \quad p_{ij} = 0 , \quad i \neq j , \tag{5.47}$$

holds, depending on the class in question.

When the elements i and j belong to the same class, characters of i and j are related to each other by

$$\chi^{(\sigma)}(i) = p_{ij}\chi^{(\sigma)}(j) \tag{5.48}$$

for any irreducible ray representation $D^{(\sigma)}$ ($\sigma = 1, 2, \ldots, r$). Thus in the class (b) with $p_{ii} = 0$, all the irreducible characters $\chi^{(\sigma)}(i)$ will vanish for any element in that class. Such classes are called *zero classes*. This will not be the case for the class (a) with $p_{ii} = 1$. In this type of class, it may happen that the values of the character for the elements i and j in that class do not coincide. However, as (5.48) shows, the difference is brought about only through the complex factor p_{ij} with modulus unity. For practical purposes, this does not cause any difficulty. We have only to prepare values for some typical representative element of the class in the character table. Classes of this type are called *ray classes*.

If we write ξ and η for the representative elements of the ray classes, the orthogonality theorems for the characters read

$$\sum_{\xi} h_\xi \chi^{(\sigma)}(\xi)^* \chi^{(\sigma')}(\xi) = m\delta_{\sigma\sigma'} , \tag{5.49}$$

$$\sum_{\sigma=1}^{r} \chi^{(\sigma)}(\xi)^* \chi^{(\sigma)}(\eta) = \frac{m}{h_\xi} \delta_{\xi\eta} . \tag{5.50}$$

The sum on the left-hand side of (5.49) runs over the representative elements ξ of

Table 5.2. Character table for irreducible ray representations

	E	$\mathscr{C}_2 \ldots \mathscr{C}_r$	$\mathscr{C}_{r+1} \ldots \mathscr{C}_t$
$\chi^{(1)}$	d_1	$\times \ldots \times$	$0 \ldots 0$
$\chi^{(2)}$	d_2	$\times \ldots \times$	$0 \ldots 0$
$\vdots$	$\vdots$	$\vdots \quad \vdots$	$\vdots \quad \vdots$
$\chi^{(r)}$	d_r	$\times \ldots \times$	$0 \ldots 0$

ray classes. Since only the characters of ray classes are involved in these orthogonality relations, we find that *the number r of inequivalent irreducible ray representations of a finite group is equal to the number of ray classes.* (The latter of course varies depending upon the factor system employed.)

If we reduce the regular ray representation α given by (5.37), every irreducible ray representation $D^{(\sigma)}$ with dimension d_σ appears d_σ times:

$$\alpha = d_1 D^{(1)} + d_2 D^{(2)} + \cdots + d_r D^{(r)} , \tag{5.51}$$

so that we have

$$m = d_1^2 + d_2^2 + \cdots + d_r^2 . \tag{5.52}$$

In passing, it may be remarked that no one-dimensional irreducible ray representation is possible if at least one zero class exists. This is because we have $m = \sum_\xi h_\xi$ from (5.49) in the case of a one-dimensional representation, whereas $m = \sum h_i$. Note that the latter sum extends over both the ray and zero classes.

According to these considerations, character tables for ray representations will in general look like Table 5.2. In the top row, $E, \mathscr{C}_2, \ldots, \mathscr{C}_r$ are the ray classes and the rest are the zero classes.

The character table may be constructed in a similar way as for the vector representations (Sect. 4.11). For an irreducible ray representation $D^{(\sigma)}$, we consider the matrices defined by

$$\hat{C}_i = \frac{h_i}{m} \sum_k \hat{D}^{(\sigma)}(k) \hat{D}^{(\sigma)}(i) \hat{D}^{(\sigma)}(k)^{-1} .$$

The sum extends over all the elements of the group. We can also rewrite this as

$$\hat{C}_i = \sum_j p_{ij} \hat{D}^{(\sigma)}(j) ,$$

using p_{ij} given by (5.44). The sum here is taken over the elements j conjugate to i. Now, it can be shown that $\hat{C}_i$ commutes with all the representation matrices $\hat{D}^{(\sigma)}(k)$:

$$\hat{C}_i \hat{D}^{(\sigma)}(k) = \hat{D}^{(\sigma)}(k) \hat{C}_i , \quad k \in \mathscr{L}^1 .$$

Accordingly, $\hat{C}_i$ must be a multiple of the unit matrix, and we have

$$\hat{C}_i = \kappa_i^{(\sigma)} \hat{1}, \quad \kappa_i^{(\sigma)} = \frac{h_i}{d_\sigma} \chi^{(\sigma)}(i) .$$

The product $\hat{C}_i \hat{C}_j$ can also be expressed as a sum of $\hat{C}_k$ as

$$\hat{C}_i \hat{C}_j = \sum_k c_{ij}^k \hat{C}_k$$

with coefficients c_{ij}^k whose absolute values are integers. This then gives

$$\kappa_i^{(\sigma)} \kappa_j^{(\sigma)} = \sum_k c_{ij}^k \kappa_k^{(\sigma)}$$

or

$$h_i h_j \chi^{(\sigma)}(i) \chi^{(\sigma)}(j) = d_\sigma \sum_k c_{ij}^k h_k \chi^{(\sigma)}(k) \ , \tag{5.53}$$

which, together with (5.49), enables us to determine $\chi^{(\sigma)}(i)$ and d_σ.

Example: Let us take double-valued representations of the point group D_6 as an example of ray representations. (For detailed discussion of the double-valued representations, see Sect. 8.5.) Take as an invariant subgroup of the double group D_6' its center $Z_2 = \{E, \bar{E}\}$ and form the factor group $\mathscr{L}^1 = D_6'/Z_2$, which is isomorphic to D_6 (Fig. 5.1). The double-valued representations are ray representations of D_6'/Z_2 with a factor system given by Table 5.3. In the table, the product $k = ij$ of elements i and j is shown in the row i and column j as usual. The bar over k indicates that the corresponding factor is given by $\alpha_{ij}^k = -1$. Otherwise, factors are unity. As noted easily,

$$\alpha_{i'i}^1 = -1 \ , \quad \alpha_{i'j}^k = -\alpha_{j'i}^{k'} \ , \quad j \neq i \ ,$$

if we denote the inverse of i by i'.

The classes $C_2 = \{C_6^3\}$, $3C_2' = \{C_2(1), \ C_2(2), \ C_2(3)\}$, $3C_2'' = \{C_2(1'), C_2(2'), \ C_2(3')\}$ are zero classes. The ray classes are E, $2C_3 = \{C_3, C_3^2\}$ and $2C_6 = \{C_6, C_6^5\}$. If we calculate p_{ij} given by (5.44), we find that $p(i = C_3, j = C_3^2) = -1$, which means the characters for C_3 and C_3^2 differ in sign although they belong to the same class. The same remark applies to C_6 and C_6^5. In the double group D_6', they are regarded as belonging to different classes. The character table for the irreducible ray representations is given by Table 5.4.

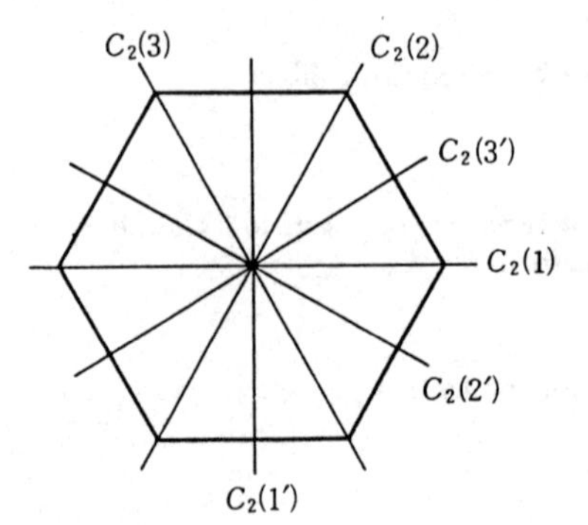

Fig. 5.1. The point group D_6 and its symmetry operations

Table 5.3. Factor system for the ray representations corresponding to the double-valued representations of D_6. To save space, simplified notation is used: $C_6 \to C$, $C_2(m) \to A_m$, $C_2(m') \to B_m$

$i \backslash j$	E	C	C^2	C^3	C^4	C^5	A_1	A_2	A_3	B_1	B_2	B_3
E	E	C	C^2	C^3	C^4	C^5	A_1	A_2	A_3	B_1	B_2	B_3
C^5	C^5	$\bar{E}$	$\bar{C}$	$\bar{C}^2$	$\bar{C}^3$	$\bar{C}^4$	$\bar{B}_2$	$\bar{B}_3$	B_1	A_2	A_3	$\bar{A}_1$
C^4	C^4	C^5	$\bar{E}$	$\bar{C}$	$\bar{C}^2$	$\bar{C}^3$	A_3	$\bar{A}_1$	$\bar{A}_2$	B_3	$\bar{B}_1$	$\bar{B}_2$
C^3	C^3	C^4	C^5	$\bar{E}$	$\bar{C}$	$\bar{C}^2$	$\bar{B}_1$	$\bar{B}_2$	$\bar{B}_3$	A_1	A_2	A_3
C^2	C^2	C^3	C^4	C^5	$\bar{E}$	$\bar{C}$	A_2	A_3	$\bar{A}_1$	B_2	B_3	$\bar{B}_1$
C	C	C^2	C^3	C^4	C^5	$\bar{E}$	B_3	$\bar{B}_1$	$\bar{B}_2$	$\bar{A}_3$	A_1	A_2
A_1	A_1	B_2	$\bar{A}_3$	B_1	$\bar{A}_2$	$\bar{B}_3$	$\bar{E}$	C^4	C^2	$\bar{C}^3$	$\bar{C}$	C^5
A_2	A_2	B_3	A_1	B_2	$\bar{A}_3$	B_1	$\bar{C}^2$	$\bar{E}$	C^4	$\bar{C}^5$	$\bar{C}^3$	$\bar{C}$
A_3	A_3	$\bar{B}_1$	A_2	B_3	A_1	B_2	$\bar{C}^4$	$\bar{C}^2$	$\bar{E}$	C	$\bar{C}^5$	$\bar{C}^3$
B_1	B_1	$\bar{A}_2$	$\bar{B}_3$	$\bar{A}_1$	$\bar{B}_2$	A_3	C^3	C	$\bar{C}^5$	$\bar{E}$	C^4	C^2
B_2	B_2	$\bar{A}_3$	B_1	$\bar{A}_2$	$\bar{B}_3$	$\bar{A}_1$	C^5	C^3	C	$\bar{C}^2$	$\bar{E}$	C^4
B_3	B_3	A_1	B_2	$\bar{A}_3$	B_1	$\bar{A}_2$	$\bar{C}$	C^5	C^3	$\bar{C}^4$	$\bar{C}^2$	$\bar{E}$

Table 5.4. Character table for the irreducible ray representations corresponding to the double-valued irreducible representations of D_6

	E	$2C_3$	$2C_6$	C_2	$3C_2'$	$3C_2''$
Γ_7	2	1	$\sqrt{3}$	0	0	0
Γ_8	2	1	$-\sqrt{3}$	0	0	0
Γ_9	2	-2	0	0	0	0

Finally, a few words on the product representation are in order. As with vector representations, we can also consider a product $D^{(\sigma)} \times D^{(\tau)}$ of two ray representations. With ray representations, however, we must always keep in mind the factor systems. If the factor systems of $D^{(\sigma)}$ and $D^{(\tau)}$ are α and α' respectively, that of the product representation is $\alpha\alpha'$ and reduction must be carried out in terms of irreducible ray representations with this factor system.

*5.5 Construction of Matrices of Irreducible Ray Representations

We discuss how to derive irreducible ray representation matrices in some simple cases.

(I) Suppose $\mathscr{L}^{\mathrm{I}}$ is a cyclic group of order k, so that $\mathscr{L}^{\mathrm{I}} = \{E, R, R^2, \ldots, R^{k-1}\}$.

For a ray representation defined by

$$\hat{D}(R^m)\hat{D}(R^n) = \begin{cases} \alpha_{m,n}^{m+n}\hat{D}(R^{m+n}) & m+n < k \ , \\ \alpha_{m,n}^{m+n-k}\hat{D}(R^{m+n-k}) & m+n \geq k \ , \end{cases}$$

we can show that the following equations hold:

$$\hat{D}(R^m) = \omega_m^{-1}(\hat{D}(R))^m \ , \quad 1 \leqq m \leqq k-1 \ ,$$
$$(\hat{D}(R))^k = \omega \hat{1} \ ,$$

where ω_m and ω are constants with modulus unity (in particular, $\omega_1 = 1$). The factor system α is equivalent to α' given by

$$\alpha'^{m+n}_{m,n} = \frac{\omega_m \omega_n}{\omega_{m+n}} \alpha^{m+n}_{m,n} = 1 \ , \quad m+n < k \ ,$$

$$\alpha'^{m+n-k}_{m,n} = \omega \ , \quad m+n \geqq k \ ,$$

so that we have only to consider ray representations with this factor system:

$$\hat{D}(R^m) = (\hat{D}(R))^m \ , \quad 1 \leqq m \leqq k-1 \ ,$$
$$(\hat{D}(R))^k = \omega \hat{1} \ . \tag{5.54}$$

If we further put $\hat{D}'(R^m) = \omega^{-m/k}\hat{D}(R^m)$, we obtain $(\hat{D}'(R))^k = \hat{1}$ and D' is a vector representation. Therefore, any irreducible ray representation of a cyclic group is equivalent to a vector representation and is one dimensional.

(II) Suppose $\mathscr{L}^1$ has an invariant subgroup $\mathscr{M}^1$ and the factor group $\mathscr{L}^1/\mathscr{M}^1$ is a cyclic group whose order k is prime.

In this case, we can derive irreducible ray representations of $\mathscr{L}^1$ from those of $\mathscr{M}^1$. Let us denote the latter by Δ. We first decompose $\mathscr{L}^1$ into left cosets of $\mathscr{M}^1$ as

$$\mathscr{L}^1 = T_1\mathscr{M}^1 + T_2 M^1 + \cdots + T_k\mathscr{M}^1 \ , \quad T_1 = E \ , \tag{5.55}$$

and form the set of bases

$$\psi_\nu^{(j)} = T_j \psi_\nu$$

from that for Δ. Then, they provide us with the basis set for the irreducible ray representation $\Delta^{(j)}$ of $\mathscr{M}^1$, with

$$\hat{\Delta}^{(j)}(S) = \frac{\alpha(S, T_j)}{\alpha(T_j, S_0)} \hat{\Delta}(S_0) \ , \tag{5.56}$$

$$S_0 = T_j^{-1} S T_j \ , \quad S, S_0 \in \mathscr{M}^1 \ ,$$

where we have used the notation $\alpha(i, j)$ in place of α_{ij}^k. Since k is a prime number, we have two cases: either

(a) the representations $\Delta^{(1)} = \Delta, \Delta^{(2)}, \ldots, \Delta^{(k)}$ are all inequivalent, or
(b) they are all equivalent.

In case (a), $\bar{D} = \Delta \uparrow \mathscr{L}^{\mathrm{I}}$ gives an irreducible ray representation of $\mathscr{L}^{\mathrm{I}}$:

$$\hat{\bar{D}}_{ij}(Q) = \delta_{ij}(Q)\,\frac{\alpha(Q, T_j)}{\alpha(T_i, S_0)}\,\hat{\Delta}(S_0)\;, \tag{5.57}$$

$$\delta_{ij}(Q) = \begin{cases} 1 & \text{when } T_i^{-1} Q T_j = S_0 \in \mathscr{M}^{\mathrm{I}}\;, \\ 0 & \text{otherwise}\;. \end{cases}$$

In case (b), we have

$$\begin{aligned} \hat{\Delta}^{(m+1)}(S) &= \frac{\alpha(S, R^m)}{\alpha(R^m, R^{-m} S R^m)}\,\hat{\Delta}(R^{-m} S R^m) \\ &= \hat{U}^{-m}\hat{\Delta}(S)\hat{U}^m, \quad m = 1, 2, \ldots, k-1\;, \end{aligned} \tag{5.58}$$

putting $T_m = R^{m-1}$. Since $R^k \in \mathscr{M}^{\mathrm{I}}$, we also have

$$\frac{\alpha(S, R^k)}{\alpha(R^k, R^{-k} S R^k)}\,\hat{\Delta}(R^{-k} S R^k) = (\hat{\Delta}(R^k))^{-1}\hat{\Delta}(S)\hat{\Delta}(R^k) = \hat{U}^{-k}\hat{\Delta}(S)\hat{U}^k\;,$$

so that $\hat{U}^k = \omega\hat{\Delta}(R^k)$. If we determine $\hat{U}$ so as to satisfy this with (5.58), we obtain the following irreducible ray representation of $\mathscr{L}^{\mathrm{I}}$:

$$\hat{D}(R^m) = \omega_m \hat{U}^m\;, \qquad m = 1, 2, \ldots, k-1\;,$$

$$\hat{D}(S) = \hat{\Delta}(S)\;, \qquad S \in \mathscr{M}^{\mathrm{I}}\;.$$

The constants ω and ω_m with modulus unity are to be chosen in accordance with the given factor system.

Irreducible ray representations of a direct-product group $\mathscr{L}^{\mathrm{I}} \times \mathscr{L}^{\mathrm{I}'}$, can be derived from irreducible ray representations D of $\mathscr{L}^{\mathrm{I}}$ (with its factor system α) and D' of $\mathscr{L}^{\mathrm{I}'}$ (with factor system α') as $D \times D'$. Note, however, that ray representations obtained in this way are limited naturally to those with the factor system $\alpha \times \alpha'$.

Example 1. As a simple example, let us derive an irreducible ray representation of the group of wavevectors $\mathscr{G}_{\mathrm{X}}$ at the point $\mathrm{X} = (\pi/a, 0)$ of the BZ of the two-dimensional space group Pma. As shown in Fig. 5.2, we have (besides translations) inversion, reflection and glide as symmetry operations:

$$I = \{I|\boldsymbol{\tau}\}\;, \quad m = \{m_x|0\}\;, \quad g = \{m_y|\boldsymbol{\tau}\}\;,$$

where $\boldsymbol{\tau} = (a/2, 0)$. Since none of them change the $\boldsymbol{k}$-vector of the X point, $\mathscr{G}_{\mathrm{X}}$ is decomposed as

$$\mathrm{Pma} = \mathscr{G}_{\mathrm{X}} = \mathscr{T} + I\mathscr{T} + m\mathscr{T} + g\mathscr{T}\;,$$

using the translation group $\mathscr{T}$. The group $\mathscr{T}$ is represented by $\Delta(\{\varepsilon|\boldsymbol{t}_n\}) = \exp(\mathrm{i}\boldsymbol{k}\cdot\boldsymbol{t}_n)$, with $\boldsymbol{k} = (\pi/a, 0)$. When we calculate the factor system for $\mathscr{L}^{\mathrm{I}} = \mathscr{G}_{\mathrm{X}}/\mathscr{T}$ by (5.29), we obtain the result given in Table 5.5.

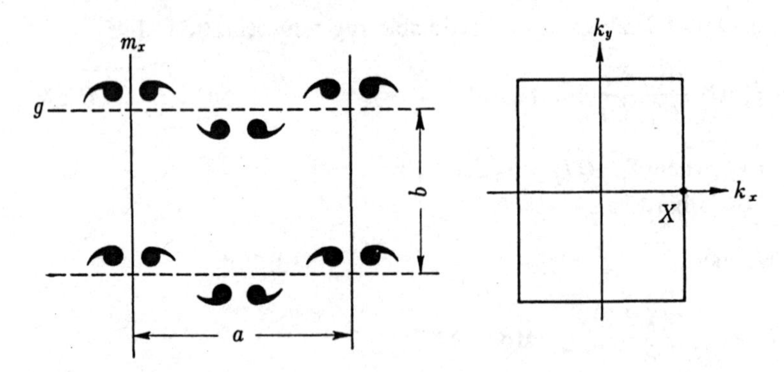

Fig. 5.2. The two-dimensional space group Pma and its Brillouin zone

The group $\mathscr{M}^1 = \{E, I\}$ is an invariant subgroup of $\mathscr{L}^1$ and leads to the decomposition

$$\mathscr{L}^1 = \mathscr{M}^1 + g\mathscr{M}^1 \ .$$

We have only two irreducible representations Δ_1 and Δ_2 of $\mathscr{M}^1$. They are respectively even and odd under inversion. If we construct the representation $\Delta^{(g)}$ from Δ_1 by means of (5.56) as

$$\hat{\Delta}^{(g)}(S) = [\alpha(S, g)/\alpha(g, gSg)]\hat{\Delta}_1(gSg) \ ,$$

the result turns out to be Δ_2. (If we start from Δ_2, we obtain Δ_1.) This is case (a) and $\Delta_1 \uparrow \mathscr{L}^1$ gives the irreducible ray representation of $\mathscr{L}^1$. If we denote the basis function of Δ_1 as ψ, the set $\{\psi, g\psi\}$ provides the basis set for this representation. The representation matrices are given by

$$\hat{E} = \begin{bmatrix} 1 & 0 \\ 0 & 1 \end{bmatrix}, \quad \hat{I} = \begin{bmatrix} 1 & 0 \\ 0 & -1 \end{bmatrix}, \quad \hat{m} = \begin{bmatrix} 0 & -1 \\ -1 & 0 \end{bmatrix}, \quad \hat{g} = \begin{bmatrix} 0 & -1 \\ 1 & 0 \end{bmatrix} .$$

Table 5.5. Factor system for the ray representations of the factor group $\mathscr{G}_X/\mathrm{T}$ associated with the two-dimensional space group Pma. The negative sign (−) indicates that $\alpha_{ij}^k = -1$ for the product $k = ij$

$i \backslash j$	E	I	m	g
E	E	I	m	g
I	I	E	g	m
m	m	$(-)g$	E	$(-)I$
g	g	$(-)m$	I	$(-)E$

As is confirmed easily by evaluating the characters, this is the only irreducible ray representation in the present problem.

Example 2. A more complicated example is furnished by the single- and double-valued representations at the X point of the space group D_{4h}^{14}. This group corresponds to the rutile structure (Fig. 11.3). The representative elements of $\mathscr{G}_X/\mathscr{T}$ are the eight operations given in (11.33, 34) of Sect. 11.8. The factor systems in the present case are given in Table 5.6. The sign $(-)$ before the product gives the factor arising from $\Delta(\{\varepsilon|\boldsymbol{t}_0\}) = \exp(\mathrm{i}\boldsymbol{k}\cdot\boldsymbol{t}_0)$ and the bar over the element indicates that the corresponding angle of rotation is larger by 2π when regarded as an operation of the double group. When we consider double-valued representations, this also brings in a factor -1, so that for $(-)\bar{R}$ the factor becomes equal to $+1$.

The factor group $\mathscr{G}_X/\mathscr{T}$ is isomorphic to the point group D_{2h}. However, the factor systems are different for the single- and double-valued representations, so that ray classes become entirely different. The character table for the irreducible ray representations is given by Table 11.6, where X_1, X_2 are single-valued and X_3, X_4 are double-valued.

Exercise 5.7. Derive Table 11.6 by decomposing D_{2h} as $D_{2h} = D_2 + ID_2$.

Finally, we would like to discuss briefly a method of deriving factor systems for a group $\mathscr{G}$ from that for its invariant subgroup $\mathscr{H}$. As usual, we consider the case where $\mathscr{G}/\mathscr{H}$ is a cyclic group of prime order k, so that the left coset decomposition of $\mathscr{G}$ with respect to $\mathscr{H}$ is given by

$$\mathscr{G} = \mathscr{H} + R\mathscr{H} + R^2\mathscr{H} + \cdots + R^{k-1}\mathscr{H} . \tag{5.59}$$

If we denote h elements of the group $\mathscr{H}$ by $S_1, S_2, \ldots, S_h$, we have

$$R^{-q}S_iR^q = S_{\varphi^q(i)} \tag{5.60}$$

Table 5.6. Factor system for the ray representations corresponding to the single- and double-valued representations at the X point of the BZ for the space group D_{4h}^{14} of rutile structure

	E	C_{2z}	C_{2x}	C_{2y}	I	σ_h	σ_{vx}	σ_{vy}
E	E	C_{2z}	C_{2x}	C_{2y}	I	σ_h	σ_{vx}	σ_{vy}
C_{2z}	C_{2z}	$\bar{E}$	$(-)C_{2y}$	$(-)\bar{C}_{2x}$	σ_h	$\bar{I}$	$(-)\sigma_{vy}$	$(-)\bar{\sigma}_{vx}$
C_{2x}	C_{2x}	$\bar{C}_{2y}$	$\bar{E}$	C_{2z}	σ_{vx}	$\bar{\sigma}_{vy}$	$\bar{I}$	σ_h
C_{2y}	C_{2y}	C_{2x}	$(-)\bar{C}_{2z}$	$(-)\bar{E}$	σ_{vy}	σ_{vx}	$(-)\bar{\sigma}_h$	$(-)\bar{I}$
I	I	σ_h	$(-)\sigma_{vx}$	$(-)\sigma_{vy}$	E	C_{2z}	$(-)C_{2x}$	$(-)C_{2y}$
σ_h	σ_h	$\bar{I}$	σ_{vy}	$\bar{\sigma}_{vx}$	C_{2z}	$\bar{E}$	C_{2y}	$\bar{C}_{2x}$
σ_{vx}	σ_{vx}	$\bar{\sigma}_{vy}$	$(-)\bar{I}$	$(-)\sigma_h$	C_{2x}	$\bar{C}_{2y}$	$(-)\bar{E}$	$(-)C_{2z}$
σ_{vy}	σ_{vy}	σ_{vx}	$\bar{\sigma}_h$	$\bar{I}$	C_{2y}	C_{2x}	$\bar{C}_{2z}$	$\bar{E}$

for $0 \leqq q \leqq k$. We will regard $\varphi^q(i)$ as a function of i determined through this equation. [We put $\varphi^0(i) = i$, $\varphi^1(i) = \varphi(i)$. Note that $\varphi^q(i) = \varphi(\varphi^{q-1}(i))$.] Since R^k must belong to $\mathscr{H}$, we put

$$R^k = S_w. \tag{5.61}$$

Here S_w is not necessarily the identity element E. When $S_w = E$ and $\varphi(i) = i$, the group $\mathscr{G}$ is the direct-product group of $\mathscr{H}$ and C_k, where C_k is the cyclic group: $C_k = \{E, R, R^2, \ldots, R^{k-1}\}$. When $\varphi(i) \neq i$ though $S_w = E$ and $\varphi^k(i) = i$, the group $\mathscr{G}$ is the *semi-direct product* group of $\mathscr{H}$ and C_k, denoted as $\mathscr{H} \odot C_k$. (For example, $D_4 = C_4 \odot C_2'$, $T = D_2 \odot C_3'$ and $O = T \odot C_2'$. Note that the invariant subgroup $\mathscr{H}$ is written first.)

Let us use the notation $\alpha(S_i, S_j)$ for the known factors of $\mathscr{H}$. The factor system for C_k is taken as

$$\alpha(R^p, R^q) = \begin{cases} 1 & \text{for} \quad p + q < k \,, \\ \lambda_0 & \text{for} \quad p + q \geqq k \,, \end{cases} \tag{5.62}$$

for p and q in the interval $0 \leqq p, q \leqq k - 1$. Then we can show that, without loss of generality, factor systems for $\mathscr{G}$ are obtained by putting

$$\alpha(R^p, S_j) = 1 \tag{5.63}$$

and

$$\alpha(S_i, R) = \lambda_i \;. \tag{5.64}$$

Then, the associativity condition gives the following general expression for the factors of the system:

$$\alpha(R^p S_i, R^q S_j) = \alpha(R^p, R^q) \cdot \alpha(R^{p+q}, S_{\varphi^q(i)} S_j) \cdot \Lambda_q(i) \alpha(S_{\varphi^q(i)}, S_j) \;, \tag{5.65}$$

where

$$\alpha(R^{p+q}, S_{\varphi^q(i)} S_j) = \begin{cases} 1 \,, & p + q < k \,, \\ \alpha(S_w, S_{\varphi^q(i)} S_j) \,, & p + q \geqq k \,, \end{cases} \tag{5.66}$$

and $\alpha(S_i, R^q)$, $q \leqq k - 1$, is given by

$$\begin{aligned} \Lambda_q(i) &= \prod_{v=0}^{q-1} \lambda_{\varphi^v(i)} \\ &= \prod_{v=0}^{q-1} \alpha(R^{-v} S_i R^v, R) \,, \quad \text{with} \\ \Lambda_0(i) &= 1 \;. \end{aligned} \tag{5.67}$$

The factors λ_i must satisfy the sets of equations

$$\alpha(S_i, S_j)\lambda_l = \alpha(S_{\varphi(i)}, S_{\varphi(j)})\lambda_i\lambda_j \ , \tag{5.68}$$

where we have assumed $S_iS_j = S_l$. Besides, the equations

$$\alpha(S_i, S_w) = \Lambda_k(i)\alpha(S_w, S_w^{-1}S_iS_w) \tag{5.69}$$

must also hold among them.

Example. It is instructive to derive factor systems for $C_{2nh} = C_{2n} \times C_i$ by the method given above. We choose the group C_{2n} as $\mathscr{H}$ and C_i plays the role of C_k with $k = 2$ and $R = I$ (inversion). If we put

$$\alpha(R, C^i) = 1 \ ,$$

and

$$\alpha(C^i, R) = \lambda_i$$

for $1 \leqq i \leqq 2n - 1$, and note that $\varphi(i) = i$, we find that

$$\lambda_1^2 = 1 \ , \quad \lambda_1 = \pm 1$$

from (5.69) and

$$\lambda_i = \lambda_1^i$$

from (5.68).

We thus obtain two inequivalent factor systems. The one with $\lambda_1 = 1$ is found to be equivalent to that obtained as a direct product of factor systems for C_{2n} and C_i. The other one, that is, the system obtained by choosing $\lambda_1 = -1$, cannot be derived as a direct product. For the number of inequivalent factor systems allowed for point groups, see [5.2].

6. Group Representations in Quantum Mechanics

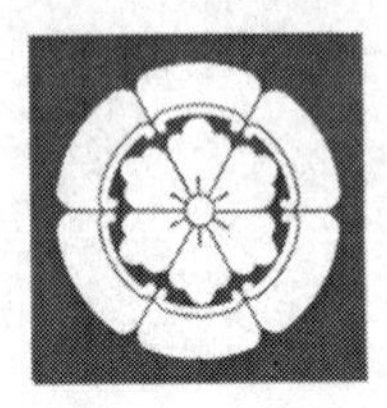

This chapter presents fundamentals required in applications of group theory to quantum-mechanical problems. Physical systems considered in quantum mechanics, in general, possess some symmetry, and the symmetry is reflected in the Hamiltonian of the system. As a result, eigenfunctions of the Hamiltonian will have certain transformation properties. These symmetry considerations are facilitated by explicit use of the concepts of groups and representations. In addition, group theory is powerful in discussing energy level splitting due to perturbation and in deriving various selection rules.

6.1 Symmetry Transformations of Wavefunctions and Quantum-Mechanical Operators

In Sect. 4.3, we defined how a symmetry operation R transforms wavefunctions $\psi(\boldsymbol{r})$. The value of the transformed function $R\psi$ at a point $\boldsymbol{r}$ is equal to the value of the original function ψ at the point $\boldsymbol{r}_0 = R^{-1}\boldsymbol{r}$, which is taken into $\boldsymbol{r}$ by the operation R,

$$R\psi(\boldsymbol{r}) = \psi(R^{-1}\boldsymbol{r}) \ . \tag{6.1}$$

The operator R on the left-hand side may be understood as a quantum-mechanical operator, since it transforms a wavefunction ψ to another wavefunction $R\psi$.

The operator R is unitary in general. In other words, if we define the inner product of two wavefunctions ϕ and ψ by

$$(\phi, \psi) = \int \phi^* \psi \, d\tau \ , \tag{6.2}$$

the equality

$$(R\phi, R\psi) = (\phi, \psi) \tag{6.3}$$

holds. This is because symmetry transformations such as spatial rotations, translations, and permutation of identical particles leave the inner product (6.2) invariant.

Let us now proceed to transformation of quantum mechanical operators T, such as momentum, angular momentum and the Hamiltonian. Let T' denote the quantum-mechanical operator T transformed by a symmetry operation R. It is most natural to define T' by

$$(\phi, T\psi) = (R\phi, T'R\psi) \tag{6.4}$$

for two arbitrary wavefunctions ϕ and ψ. The above definition requires that the matrix element of T be equal to that of T' evaluated for the transformed functions $R\phi$ and $R\psi$. Unitarity (6.3) allows transcription of the left-hand side of (6.4) as

$$(R\phi, RT\psi) = (R\phi, T'R\psi) .$$

This equality holds for any ϕ and ψ only when

$$T' = RTR^{-1} , \tag{6.5}$$

which explicitly defines transformation of the operator T. Note that the right-hand side contains the inverse operation R^{-1}, which terminates the R operation there and does not let R operate on further wavefunctions or operators that may follow on the right-hand side.

Exercise 6.1. Consider a rotation R in two dimensions as discussed in Sect. 4.3. Show that R transforms the coordinate operator $\boldsymbol{r} = (x, y)$ and momentum operator $\boldsymbol{p} = (p_x, p_y)$ as follows:

$$\begin{aligned} R\boldsymbol{r}R^{-1} &= \boldsymbol{r}\hat{R} , \\ R\boldsymbol{p}R^{-1} &= \boldsymbol{p}\hat{R} . \end{aligned} \tag{6.6}$$

Here $\boldsymbol{r}$ and $\boldsymbol{p}$ are understood as row vectors, and the orthogonal matrix $\hat{R}$ is defined by (3.9). The above result (6.6) remains valid for three-dimensional rotations as well, when the 3×3 orthogonal matrix $\hat{R}$ is defined by (7.7) in terms of the Euler angles.

6.2 Eigenstates of the Hamiltonian and Irreducibility

A Hamiltonian which describes some quantum-mechanical system has some symmetry in general. In this section, we wish to consider how this symmetry affects the transformation properties of the eigenfunctions of the Hamiltonian. Before going on to the general case, we begin with an elementary example.

Consider a particle with mass m confined in the one-dimensional potential well

$$V(x) = \begin{cases} \infty, & x < -a , \\ 0, & -a < x < a , \\ \infty, & a < x . \end{cases}$$

The Hamiltonian

$$H = p^2/2m + V(x) \tag{6.7}$$

has eigenvalues $E_n = (n\pi\hbar)^2/8ma^2$ and corresponding eigenfunctions,

$$\psi_n(x) = \begin{cases} \sqrt{\dfrac{1}{a}} \cos\dfrac{n\pi}{2a}x \ , & n = 1, 3, 5, \ldots, \\ \sqrt{\dfrac{1}{a}} \sin\dfrac{n\pi}{2a}x \ , & n = 2, 4, 6, \ldots, \end{cases} \tag{6.8}$$

that satisfy

$$H\psi_n = E_n\psi_n \ . \tag{6.9}$$

Now consider the inversion I. This operator changes the sign of the argument when it operates on a function $\psi(x)$:

$$I\psi(x) = \psi(-x) \ .$$

It also changes the sign of the coordinate and momentum operators:

$$IxI^{-1} = -x \ , \quad IpI^{-1} = -p \ ,$$

while the Hamiltonian (6.7) is invariant under I,

$$IHI^{-1} = H \ , \quad \text{or}$$

$$HI = IH \ , \tag{6.10}$$

i.e., H and I commute.

Let both sides of (6.10) operate on ψ_n and use (6.9). Then we have

$$H(I\psi_n) = E_n(I\psi_n) \ .$$

We find that $I\psi_n$ and ψ_n belong to the same eigenvalue E_n. Since all the energy eigenvalues are nondegenerate in the present problem, $I\psi_n$ and ψ_n should represent the same quantum state, or they should be the same function apart from a constant factor c,

$$I\psi_n = c\psi_n \ .$$

A second operation of I on $I\psi_n$ should bring it back to ψ_n, so that $c^2 = 1$. Therefore we have $c = \pm 1$. The eigenfunctions ψ_n should then be either even or odd under the inversion I, as is actually seen in (6.8). Linear combinations of even and odd functions are forbidden as the eigenfunctions of the Hamiltonian (6.7).

In the above arguments, we have explicitly taken the potential well $V(x)$, but the conclusion remains valid as long as the Hamiltonian (6.7) has the inversion symmetry (6.10). It is not the explicit expression for the Hamiltonian but its symmetry that matters in the above arguments.

The Hamiltonian of the above example was invariant with respect to inversion I and the identity operation E. A general quantum-mechanical system may have more symmetry operations R that leave the Hamiltonian invariant:

$$RHR^{-1} = H \ , \tag{6.11}$$

or

$$RH = HR \ . \tag{6.12}$$

Consider a set $\mathscr{G}$ of these symmetry transformation operations R. This set constitutes a group and is called the *symmetry group* of the Hamiltonian.

To confirm that $\mathscr{G}$ is really a group, one needs to see that the four axioms of Sect. 2.1 are satisfied. For two symmetry operations R and S belonging to $\mathscr{G}$, we have, by definition,

$$RHR^{-1} = H \ , \quad SHS^{-1} = H \ ,$$

which yields

$$(RS)H(RS)^{-1} = RSHS^{-1}R^{-1} = RHR^{-1} = H \ ,$$

so that the product $RS \in \mathscr{G}$. Next, operators in quantum mechanics satisfy the associative law. The identity operation E is the unit element of $\mathscr{G}$. Finally, (6.11) can be written as $R^{-1}H(R^{-1})^{-1} = H$, which implies $R^{-1} \in \mathscr{G}$.

Let us enumerate some familiar examples of the symmetry group.

Rotation Group. The Hamiltonian

$$H = p^2/2m + V(r) \ , \tag{6.13}$$

which describes a particle of mass m in a spherically symmetric potential $V(r)$, is invariant with respect to all rotations about the origin. The set of such rotations is called the rotation group.

Point Group. The potential experienced by an electron in a molecule has the symmetry resulting from the atomic arrangement. The Hamiltonian is then invariant with respect to those symmetry operations that bring the atomic arrangement into coincidence with itself. The set of these operations is called a point group.

Translation Group. Electron wavefunctions (Bloch functions) extended in a perfect crystal should feel a periodic potential. The Hamiltonian is then invariant with respect to all the lattice translations, which form the translation group.

Symmetric Group. Consider a system consisting of n electrons. Since the n electrons are identical particles, the Hamiltonian is invariant to any permutation of the n electron coordinates. The set of these permutations is called the symmetric group.

Having some physical system in mind as exemplified above, consider a d-fold degenerate level of a Hamiltonian H,

$$H\phi_n = E\phi_n \ , \quad n = 1, 2, \ldots, d \ . \tag{6.14}$$

The eigenfunctions ϕ_n are taken orthonormalized,

$$(\phi_m, \phi_n) = \delta_{mn} \ . \tag{6.15}$$

A symmetry operation R belonging to the symmetry group $\mathscr{G}$ satisfies (6.12), so we have

$$HR\phi_n = ER\phi_n \ . \tag{6.16}$$

This equation means that $R\phi_n$ belongs to the same energy eigenvalue E as ϕ_n. Therefore, $R\phi_n$ should be expressible as a linear combination of the d functions. If we write the coefficients of the linear combination as

$$(\phi_m, R\phi_n) = D_{mn}(R) \ , \tag{6.17}$$

we have

$$R\phi_n = \sum_{m=1}^{d} \phi_m D_{mn}(R) \ . \tag{6.18}$$

Equation (6.18) tells us that the matrices $\hat{D}(R)$ form a representation of the group $\mathscr{G}$ and that the d functions $\{\phi_n\}$ form a basis for that representation.

The representation D obtained above is unitary. To see this, evaluate the inner product $(R\phi_m, R\phi_n)$ using (6.18). From unitarity (6.3) of the symmetry operation R and the orthonormality (6.15), we obtain

$$\hat{D}(R)^\dagger \hat{D}(R) = \hat{1} \ , \tag{6.19}$$

which shows unitarity of the matrix $\hat{D}(R)$.

We have seen above that the eigenfunctions belonging to a degenerate level of a Hamiltonian form a basis for a unitary representation D of the symmetry group $\mathscr{G}$. Now, this representation D is irreducible. If it were reducible, it would be possible to regroup the functions ϕ_n into at least two sets such that each set is closed under the symmetry operations. Now, according to the arguments beginning with (6.16), functions connected by the symmetry operations belong to the same eigenvalue. Therefore we can expect in general that the above two sets of functions should have different eigenvalues. It is, of course, possible that the two eigenvalues coincide by accident (*accidental degeneracy*: degeneracy for reasons other than symmetry), but we have no symmetry reasons for the coincidence. Therefore if the representation D is reducible, there should result at least two different eigenvalues from the functions $\{\phi_n\}$. This contradicts the premise made at the outset that the ϕ_n's are degenerate to the same eigenvalue.

Table 6.1 Irreducible representations of the group C_i

	E	I
A_g	1	1
A_u	1	-1

Consequently, the representation D must be irreducible. In short, *the eigenfunctions belonging to a degenerate level form a basis for an irreducible representation of the symmetry group.* The degeneracy d is equal to the dimension of the representation. This irreducibility required by symmetry plays a central role in various physical applications of group theory.[1]

As an example of the irreducibility, let us look back at the potential well in one dimension. The symmetry group of the Hamiltonian (6.7) is $C_i = \{E, I\}$. It has two irreducible representations, shown in Table 6.1. The representations A_g and A_u admit even and odd functions as their basis functions. Both representations are one dimensional and the energy levels are nondegenerate.

As a more realistic example, consider the Hamiltonian (6.13). Its eigenfunctions are written as

$$\psi_{nlm}(r) = P_{nl}(r) Y_{lm}(\theta, \phi)$$

in spherical coordinates. The radial wavefunction $P_{nl}(r)$ depends on the potential $V(r)$. The spherical harmonics $Y_{lm}(\theta, \phi)$ (defined by (7.65)) appear as a result of the spherical symmetry. The symmetry group is here the rotation group and a spatial rotation R transforms Y_{lm} into a linear combination of Y_{lm}'s:

$$RY_{lm} = \sum_{m'=-l}^{l} Y_{lm'} D^{(l)}_{m'm}(R) .$$

The matrices $\hat{D}^{(l)}(R)$ of the coefficients form the irreducible representation $D^{(l)}$ of the rotation group, and Y_{lm} are nothing other than the basis for that representation. The dimension of $D^{(l)}$ is $2l + 1$, which gives the degeneracy.

6.3 Splitting of Energy Levels by a Perturbation

The irreducibility requirement mentioned in the preceding section helps in studying how a degenerate energy level is split by a perturbation.

[1] Reality (time-reversal symmetry) of the Hamiltonian can bring in additional degeneracy. This topic will be discussed in Chap. 13.

Suppose that a perturbation H_1 is applied to an original Hamiltonian H_0. If the symmetry of H_1 is not lower than that of H_0, the perturbed Hamiltonian

$$H = H_0 + H_1$$

has the same symmetry as H_0. In this case, the symmetry group $\mathcal{G}$ remains unchanged, and the eigenfunctions of the Hamiltonian remain to provide the basis for the same irreducible representation. The perturbation causes only a shift of the energy levels and leaves the degeneracy unchanged.

The situation becomes different when H_1 has lower symmetry than H_0. In this case, the symmetry group $\mathcal{G}$ of the perturbed Hamiltonian H is a subgroup of the original symmetry group $\mathcal{G}_0$. Such a perturbation of lower symmetry can lift the degeneracy.

Consider a degenerate unperturbed level that forms an irreducible representation Γ_0 of the group $\mathcal{G}_0$. Under the small perturbation H_1, the wavefunctions belonging to this level form a basis for a representation of the subgroup $\mathcal{G}$. Now the representation Γ_0 may indeed be irreducible as a representation of $\mathcal{G}_0$, but, in general, it is not necessarily irreducible under its subgroup $\mathcal{G}$. The level splitting can be found by regarding Γ_0 as a representation[2] of the subgroup $\mathcal{G}$ and reducing it into irreducible representations of $\mathcal{G}$. If the result of this reduction

$$\Gamma_0 \longrightarrow \Gamma_\alpha + \Gamma_\beta + \cdots .$$

contains n irreducible representations, the irreducibility requirement tells us that the level is split into n levels.

A typical example is a d electron placed in a cubic crystal field. In that case, one needs to reduce the irreducible representation $D^{(2)}$ of the rotation group in the cubic point group O, as will be shown in Sect. 9.6.

6.4 Orthogonality of Basis Functions

Let $D^{(\alpha)}$ be an irreducible unitary representation of a group $\mathcal{G}$ with basis functions $\{\phi_m^{(\alpha)}\}$ and $\{\psi_m^{(\alpha)}\}$, which are transformed as

$$R\phi_m^{(\alpha)} = \sum_{m'} \phi_{m'}^{(\alpha)} D_{m'm}^{(\alpha)}(R) \tag{6.20}$$

by symmetry operations R belonging to $\mathcal{G}$. For any two sets of basis functions that transform like (6.20), the following orthogonality relation holds:

$$(\phi_m^{(\alpha)}, \psi_l^{(\beta)}) = \delta_{\alpha\beta}\delta_{ml} \times \text{constant independent of } m \text{ and } l . \tag{6.21}$$

---

[2] This representation (subduced representation) is sometimes denoted by $\Gamma_0 \downarrow \mathcal{G}$.

The inner product is nonvanishing only when $\alpha = \beta$ (same irreducible representation) and $m = l$ (same partner), and is furthermore independent of m and l.

This orthogonality results from the unitarity (6.3) of symmetry operations and the great orthogonality theorem (4.46). The unitarity (6.3) allows us to replace the inner product of two functions ϕ and ψ by

$$(\phi, \psi) = \frac{1}{g}\sum_R (R\phi, R\psi) , \tag{6.22}$$

where the summation is taken over the g elements that constitute the group $\mathcal{G}$. Thus we have

$$\begin{aligned}(\phi_m^{(\alpha)}, \psi_l^{(\beta)}) &= \frac{1}{g}\sum_R (R\phi_m^{(\alpha)}, R\psi_l^{(\beta)}) \\ &= \sum_{m'l'} (\phi_{m'}^{(\alpha)}, \psi_{l'}^{(\beta)}) \frac{1}{g}\sum_R D_{m'm}^{(\alpha)}(R)^* D_{l'l}^{(\beta)}(R) \\ &= \delta_{\alpha\beta}\delta_{ml} \times \frac{1}{d_\alpha}\sum_{m'} (\phi_{m'}^{(\alpha)}, \psi_{m'}^{(\alpha)}) .\end{aligned} \tag{6.23}$$

Here, d_α denotes the dimension of the representation $D^{(\alpha)}$. The right-hand side does not depend on m and l apart from in the Kronecker delta.

6.5 Selection Rules

When calculating various matrix elements in quantum mechanics, some of them happen to vanish for reasons of symmetry. Representation theory greatly helps in judging whether or not a given matrix element vanishes by symmetry.

To begin with, let us examine a matrix element like

$$(\phi_m^{(\alpha)}, H\psi_l^{(\beta)}) ,$$

where H is the Hamiltonian or any other operator that is invariant under the symmetry group $\mathcal{G}$. Invariance of H means that $H\psi_l^{(\beta)}$ transforms in exactly the same way as $\psi_l^{(\beta)}$. Then the orthogonality relation (6.21) indicates that the above matrix element is nonvanishing only when $\alpha = \beta$ and $m = l$, and that it is independent of m. That is, the Hamiltonian has nonvanishing matrix elements only between the functions that have the same transformation properties.

To consider matrix elements of general operators, it is convenient to define irreducible tensor operators. In parallel with the basis functions $\phi_m^{(\alpha)}$ transforming like (6.20), the operators $T_m^{(\alpha)}$ which transform according to

$$RT_m^{(\alpha)}R^{-1} = \sum_{m'} T_{m'}^{(\alpha)} D_{m'm}^{(\alpha)}(R) \tag{6.24}$$

are called *irreducible tensor operators.* For instance, in the group C_{3v}, components of the coordinate $\boldsymbol{r}$, momentum $\boldsymbol{p}$, and angular momentum $\boldsymbol{l}$ vectors constitute irreducible tensor operators belonging to the following irreducible representations:

$$\begin{aligned} &A_1: \ z, p_z \\ &A_2: \ l_z \\ &E: \ \{x, y\}, \{p_x, p_y\}, \{l_x, l_y\} \ . \end{aligned} \tag{6.25}$$

Whether or not a matrix element of the irreducible tensor operator $T_j^{(\alpha)}$

$$(\phi_m^{(\gamma)}, T_j^{(\alpha)}\psi_l^{(\beta)}) \tag{6.26}$$

vanishes by symmetry is examined in the following way. The functions $T_j^{(\alpha)}\psi_l^{(\beta)}$ transform as basis functions of the product representation $D^{(\alpha)} \times D^{(\beta)}$. So, the function $T_j^{(\alpha)}\psi_l^{(\beta)}$ contains in itself irreducible components that are obtained by reduction of the product representation $D^{(\alpha)} \times D^{(\beta)}$. *In order that the matrix element (6.26) should not vanish by symmetry,* it has to contain the $\phi_m^{(\gamma)}$ component because of the orthogonality relation (6.21). This requires that *the representation $D^{(\gamma)}$ should appear at least once in the reduction of the product representation $D^{(\alpha)} \times D^{(\beta)}$.* The number q_γ of times the representation $D^{(\gamma)}$ appears in $D^{(\alpha)} \times D^{(\beta)}$ is calculated using (4.69). The matrix element (6.26) vanishes by symmetry if $q_\gamma = 0$.

As an example, let us examine the selection rule for photoabsorption by an atom placed in a field of C_{3v} symmetry. Interaction of an electron with light has the form $\boldsymbol{p} \cdot \boldsymbol{\varepsilon}$, where $\boldsymbol{\varepsilon}$ stands for the polarization vector of the electric field. Suppose that the initial state belongs to some irreducible representation Γ_i. For light polarized in the z direction, the optical transition takes place through p_z. Since the operator p_z belongs to the A_1 representation of C_{3v}, the final state must be $A_1 \times \Gamma_i = \Gamma_i$. So, p_z causes transitions between states with the same symmetry.

When the light is polarized in the xy-plane, the transition takes place through p_x and p_y, which belong to the representation E. The symmetry of the final state is obtained by reducing $E \times \Gamma_i$. The result of this reduction turns out to be E, E, $A_1 + A_2 + E$ for $\Gamma_i = A_1, A_2, E$.

Exercise 6.2. Find the photoabsorption selection rule for the group C_{4v}.

The matrix elements (6.26) we have examined concern two different sets of basis functions. The selection rule becomes more stringent for so-called diagonal matrix elements

$$(\phi_m, T_j^{(\alpha)}\phi_l) \ , \tag{6.27}$$

which have higher symmetry than (6.26). In (6.27), we have dropped the superscript that specifies the irreducible representation D to which the ϕ_m's belong. We neglect here electron spin and consider real orbital functions ϕ_m, putting off the discussion of the general case involving electron spin until the end of Sect. 10.4. Then the selection rule for the diagonal matrix elements reads: The matrix element (6.27) is nonvanishing when the following criterion is satisfied.

(a) If $T^{(\alpha)}$ is a real Hermitian operator, such as coordinates or potential, the symmetric product representation $[D \times D]$ contains $D^{(\alpha)}$.
(b) If $T^{(\alpha)}$ is an imaginary Hermitian operator, such as momentum or angular momentum, the antisymmetric product representation $\{D \times D\}$ contains $D^{(\alpha)}$.

As an example, let us consider again C_{3v} and take the two-dimensional representation E. For this representation, we have already seen in Sect. 4.8 that

$$[\mathrm{E} \times \mathrm{E}] = \mathrm{A}_1 + \mathrm{E} \ , \quad \{\mathrm{E} \times \mathrm{E}\} = \mathrm{A}_2 \ . \tag{6.28}$$

Therefore, among the operators listed in (6.25), only z, x, y, and l_z have nonvanishing diagonal matrix elements in the twofold degenerate level with E symmetry.

*6.5.1 Derivation of the Selection Rule for Diagonal Matrix Elements

The above selection rule is derived in the following way. For a real Hermitian operator $T_j^{(\alpha)}$, we have

$$\begin{aligned}(\phi_m, T_j^{(\alpha)}\phi_l) &= (T_j^{(\alpha)}\phi_m, \phi_l)\\ &= (\phi_l, T_j^{(\alpha)}\phi_m)^*\\ &= (\phi_l, T_j^{(\alpha)}\phi_m) \ ,\end{aligned} \tag{6.29}$$

where the first equality rests on the hermiticity of T and the last one on the reality of ϕ and T. If we write $M_{ml}(j)$ for this matrix element for simplicity, it is symmetric with respect to the interchange of m and l,

$$M_{ml}(j) = M_{lm}(j) \ . \tag{6.30}$$

Using again the unitarity (6.22) and this symmetry, we obtain

$$\begin{aligned}M_{ml}(j) &= \frac{1}{g}\sum_R (R\phi_m, RT_j^{(\alpha)}R^{-1}R\phi_l)\\ &= \sum_{m'j'l'} M_{m'l'}(j')\frac{1}{g}\sum_R D_{m'm}(R)D_{j'j}^{(\alpha)}(R)D_{l'l}(R)\\ &= \sum_{m'j'l'} M_{m'l'}(j')\frac{1}{2g}\sum_R D_{j'j}^{(\alpha)}(R)\\ &\quad \times [D_{m'm}(R)D_{l'l}(R) + D_{l'm}(R)D_{m'l}(R)] \ .\end{aligned} \tag{6.31}$$

The quantity in the square brackets is the representation matrix element of the symmetric product representation $[D \times D]$, see (4.74). The summation over R vanishes if $[D \times D]$ does not contain $D^{(\alpha)}$.

For an imaginary Hermitian operator $T_j^{(\alpha)}$, we have similarly

$$M_{ml}(j) = -M_{lm}(j) \ .$$

As a result, we obtain $-$ in place of $+$ in (6.31). Then the matrix element vanishes if $\{D \times D\}$ does not contain $D^{(\alpha)}$.

6.6 Projection Operators

As has been mentioned repeatedly, basis functions for an irreducible representation transform like (6.20), so the representation matrices can be found once some set of basis functions is given for that representation. Readers may be anxious about the converse problem, however. When unitary matrices $\hat{D}^{(\alpha)}(R)$ of an irreducible representation are given, how is it possible to construct basis functions (symmetry-adapted functions) that transform according to those matrices? The method of projection operators provides us with the solution to the problem.

Consider an arbitrary function f. It contains, in general, components of various irreducible representations, so that it may be decomposed as

$$f = \sum_{\alpha} \sum_{m} c_m^{(\alpha)} \phi_m^{(\alpha)} \ , \tag{6.32}$$

where the $c_m^{(\alpha)}$'s are coefficients of the expansion. Now, a projection operator defined by

$$P_{l(m)}^{(\beta)} = \frac{d_\beta}{g} \sum_{R} D_{lm}^{(\beta)}(R)^* R \ , \tag{6.33}$$

when applied to the above function f, picks up the symmetry-adapted function $\phi_l^{(\beta)}$:

$$P_{l(m)}^{(\beta)} f = c_m^{(\beta)} \phi_l^{(\beta)} \ . \tag{6.34}$$

Thus, if the function f contains an irreducible component $\phi_m^{(\beta)}$, operation of the above projection operator on f generates $\phi_l^{(\beta)}$. If, unfortunately, f does not contain the $\phi_m^{(\beta)}$ component, (6.34) vanishes. In that event, one chooses another function f or changes m, to get the desired basis function. In many applications, it suffices to use the diagonal projection operators $P_{l(l)}^{(\beta)}$.

Exercise 6.3. Prove (6.34) using the great orthogonality theorem (4.46).

Exercise 6.4. Show that

$$P^{(\alpha)}_{l(m)} P^{(\beta)}_{l'(m')} = \delta_{\alpha\beta}\delta_{ml'} P^{(\alpha)}_{l(m')} .$$

Another less useful but simpler projection operator is obtained by summing the diagonal projection operators,

$$P^{(\beta)} \equiv \sum_l P^{(\beta)}_{l(l)} = \frac{d_\beta}{g} \sum_R \chi^{(\beta)}(R)^* R , \tag{6.35}$$

which can readily be constructed from the character $\chi^{(\beta)}(R)$. When it operates on (6.32), we have

$$P^{(\beta)} f = \sum_m c^{(\beta)}_m \phi^{(\beta)}_m . \tag{6.36}$$

Thus, $P^{(\beta)}$ projects the function f onto the subspace of the representation $D^{(\beta)}$.

These projection operators are extensively used in constructing explicit basis functions, as will be found in later chapters (e.g., Sects. 9.3, 10.2, 12.4, 12.11).

Exercise 6.5. Show that the projection operators for the three irreducible representations of the point group C_{3v} are

$$P^{(A_1)} = \tfrac{1}{6}(E + C_3 + C_3^{-1} + \sigma_1 + \sigma_2 + \sigma_3) ,$$

$$P^{(A_2)} = \tfrac{1}{6}(E + C_3 + C_3^{-1} - \sigma_1 - \sigma_2 - \sigma_3) ,$$

$$P^{(E)}_{1(1)} = \tfrac{1}{6}(2E - C_3 - C_3^{-1} + 2\sigma_1 - \sigma_2 - \sigma_3) ,$$

$$P^{(E)}_{2(2)} = \tfrac{1}{6}(2E - C_3 - C_3^{-1} - 2\sigma_1 + \sigma_2 + \sigma_3) ,$$

$$P^{(E)}_{1(2)} = \frac{\sqrt{3}}{6}(-C_3 + C_3^{-1} - \sigma_2 + \sigma_3) ,$$

$$P^{(E)}_{2(1)} = \frac{\sqrt{3}}{6}(C_3 - C_3^{-1} - \sigma_2 + \sigma_3) ,$$

$$P^{(E)} = \tfrac{1}{3}(2E - C_3 - C_3^{-1}) .$$

Exercise 6.6. Use the above projection operators to obtain the basis functions given in (4.38, 39) of Sect. 4.4.

Exercise 6.7. In Sect. 4.8, we saw that the product representation E × E in C_{3v} is reduced to $A_1 + A_2 + E$. Use the projection operators to find the basis functions for the A_1, A_2, and E representations given in (4.72).

Exercise 6.8. Show that $\sum_\alpha P^{(\alpha)} = E$ (unit element).

Hint: Use (4.82, 83).

Exercise 6.9. Show that $P^{(\alpha)} P^{(\beta)} = \delta_{\alpha\beta} P^{(\alpha)}$.

Exercise 6.10. Prove that for two sets of functions

$$\phi^{(\alpha)}_l \equiv P^{(\alpha)}_{l(m)} f , \quad \text{and} \quad \psi^{(\alpha)}_l \equiv P^{(\alpha)}_{l(m')} g$$

generated by the projection operators from two functions f and g, the matrix element of the Hamiltonian H is given by

$$\begin{aligned}(\phi_l^{(\alpha)}, H\,\psi_l^{(\alpha)}) &= (P_{m'(m)}^{(\alpha)} f, H\,g) \\ &= (f, H\,P_{m(m')}^{(\alpha)} g)\ .\end{aligned}$$

Hint: Use (6.3) and the result of Exercise 6.4. Prove the second equality first.

7. The Rotation Group

The purpose of the first half of the present chapter is to derive explicit expressions for the representations of the rotation group. For this purpose, the rotations are described in terms of the Euler angles. Then they are given their expressions through infinitesimal rotations or angular momentum operators. As a result, representation matrices can be constructed from the matrices for the angular momentum operators well known in quantum mechanics. By using the representation matrices obtained in this way, we discuss some properties of spherical harmonics as the bases of the representation, besides confirming directly the orthogonality theorems for the representation matrices and characters.

The second half mainly deals with applications. Expressions for the Wigner coefficients or the vector coupling coefficients are derived, which provide us with the unitary matrix to reduce a product representation. By using these coefficients, we prove the Wigner–Eckart theorem for general tensor operators. When additions of more than three angular momenta are being considered, use of the Racah coefficients is indispensable. This applies to the calculation of the matrix elements as well as to the construction of the many-electron wavefunctions with proper symmetry.

It must be said that the subjects treated in the second half have already been fully developed as applications to the electronic structures of atoms. However, they still remain important and are worth learning even now as a successful classical example of applications of the representation theory of groups.

7.1 Rotations

Let us write $R_\theta(\lambda, \mu, \nu)$ for the rotation through an angle θ with the axis of rotation specified by a unit vector whose components are given by the direction cosines (λ, μ, ν). The angle θ is positive for the rotation sense forming a right-handed system with the vector direction. Since we have

$$R_\theta(\lambda, \mu, \nu) = R_{2\pi-\theta}(-\lambda, -\mu, -\nu) , \tag{7.1}$$

we may take $0 \leqq \theta \leqq \pi$.

The result of carrying out a rotation $R_\beta(2)$ through an angle β about the axis in the direction 2 after the rotation $R_\alpha(1)$ through angle α about the axis in the direction 1 is the same as that of another rotation $R_\gamma(3)$ through a certain angle γ

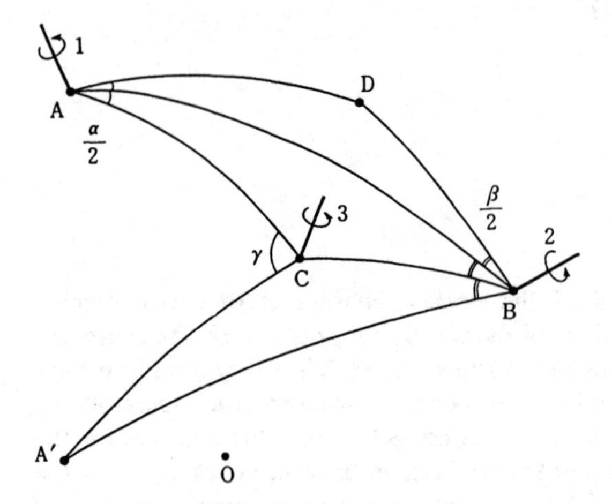

Fig. 7.1. Composition of rotations

about an axis in a certain direction 3. This is expressed in the following way (composition of rotations):

$$R_\beta(2)R_\alpha(1) = R_\gamma(3) \; . \tag{7.2}$$

The direction 3 and the angle γ can be found geometrically from 1, 2, α and β by the following procedure.

Draw a unit sphere centered at the origin O, i.e., the center of rotation, and suppose the axes 1 and 2 pierce the surface of the sphere at points A and B, respectively (Fig. 7.1). Next consider two planes AOC and AOD containing the axis 1, such that the angle α between them is bisected by the plane AOB. Two planes BOC and BOD are fixed in a similar way. The straight line OC is the line of intersection of the two planes AOC and BOC. Fix another plane $A'OB$ so that it makes an angle $\beta/2$ with the plane BOC and take point A' on the arc of the great circle such that the length of the arc $A'B$ is equal to that of the arc AB. Now, the point C is carried to D by $R_\alpha(1)$ and back to the original position by $R_\beta(2)$, which means that C stays unmoved after successive rotations. This implies that OC is the direction of the axis 3. The point A stays unmoved after the first rotation, but is carried into point A' by $R_\beta(2)$. Namely, the rotation about OC brings A to A'. This then means that γ is given by the angle ACA'.

The set of rotations $R_\theta(\lambda, \mu, \nu)$ with the product defined by (7.2) form a group, the *rotation group*. In the rotation group, rotations with an equal angle belong to one and the same class, regardless of the axis. Suppose rotations R and R' are conjugate so that

$$R' = QRQ^{-1} \; . \tag{7.3}$$

In this case Q is the rotation which sends the rotation axis of R into that of R', both R and R' having the same angle of rotation, though with different axes.

7.2 Rotation and Euler Angles

As is seen below, it is more convenient to express a rotation in terms of Euler angles than to use expressions like $R_\theta(\lambda, \mu, \nu)$.

Any rotation can be regarded as a result of the following three successive rotations:

(i) Rotation $R_\alpha(z)$ through an angle α about the z-axis ($0 \leqq \alpha < 2\pi$). This rotation carries the coordinate axes x, y and z into x_1, y_1 and $z_1 = z$ (Fig. 7.2).

(ii) Rotation $R_\beta(y_1)$ through an angle β about the y_1-axis ($0 \leqq \beta \leqq \pi$). This carries x_1, y_1 and z_1 into x_2, $y_2 = y_1$ and z_2. According to (7.3), $R_\beta(y_1)$ can be written as

$$R_\beta(y_1) = R_\alpha(z)R_\beta(y)R_\alpha(z)^{-1} .$$

(iii) Rotation $R_\gamma(z_2) = R_\beta(y_1)R_\gamma(z_1)R_\beta(y_1)^{-1}$ through an angle γ about the z_2-axis ($0 \leqq \gamma < 2\pi$). The axes x_2, y_2 and z_2 are carried into X, Y and $Z = z_2$.

If we note that

$$R_\gamma(z_1) = R_\alpha(z)R_\gamma(z)R_\alpha(z)^{-1} = R_\gamma(z) ,$$

we can put the result in the form

$$\begin{aligned} R(\alpha, \beta, \gamma) &= R_\gamma(z_2)R_\beta(y_1)R_\alpha(z) \\ &= R_\alpha(z)R_\beta(y)R_\gamma(z) . \end{aligned} \tag{7.4}$$

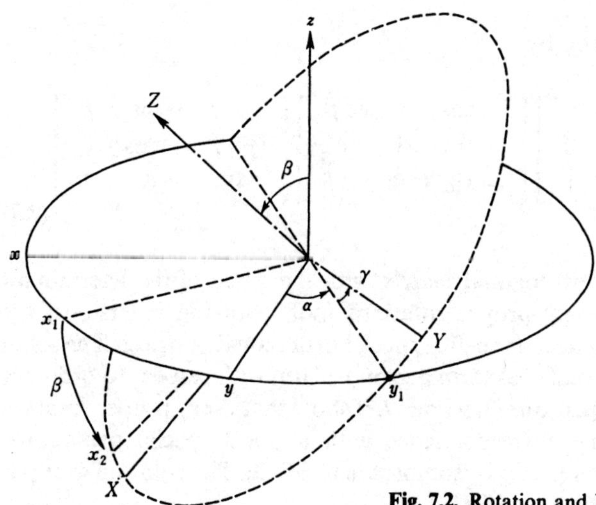

Fig. 7.2. Rotation and Euler angles

In this book, rotations will always be described referring to the original coordinate axes x, y and z fixed in space. If we use a moving coordinate system for each rotation, the order of the product will be inverted, as the first line of (7.4) shows. As will be seen later (Sect. 7.5), the rotation $R(\alpha, \beta, \gamma)$ can be expressed as $R_\theta(\lambda, \mu, \nu)$ with parameters given by

$$\cos\frac{\theta}{2} = \cos\frac{\beta}{2}\cos\frac{\alpha+\gamma}{2} \tag{7.5}$$

and

$$\begin{aligned} \lambda &= -\frac{\sin\dfrac{\beta}{2}\sin\dfrac{\alpha-\gamma}{2}}{\sin\theta/2} , \\ \mu &= \frac{\sin\dfrac{\beta}{2}\cos\dfrac{\alpha-\gamma}{2}}{\sin\theta/2} , \\ \nu &= \frac{\cos\dfrac{\beta}{2}\sin\dfrac{\alpha+\gamma}{2}}{\sin\theta/2} . \end{aligned} \tag{7.6}$$

The rotation $R(\alpha, \beta, \gamma)$ brings the coordinate axes x, y and z to the axes X, Y and Z. The *coordinates* (x, y, z) and (X, Y, Z) of *a point fixed in space* are then related to each other by the equation

$$[X, Y, Z] = [x, y, z]\hat{R} ,$$

where $\hat{R}$ is the matrix given by

$$\hat{R} = \begin{bmatrix} \cos\alpha & -\sin\alpha & 0 \\ \sin\alpha & \cos\alpha & 0 \\ 0 & 0 & 1 \end{bmatrix} \begin{bmatrix} \cos\beta & 0 & \sin\beta \\ 0 & 1 & 0 \\ -\sin\beta & 0 & \cos\beta \end{bmatrix} \begin{bmatrix} \cos\gamma & -\sin\gamma & 0 \\ \sin\gamma & \cos\gamma & 0 \\ 0 & 0 & 1 \end{bmatrix} . \tag{7.7}$$

The matrix $\hat{R}$ is a 3×3 orthogonal matrix with the value of its determinant equal to unity, because only proper rotations (not involving inversions) are considered here. Such matrices are called special orthogonal matrices. The set of $n \times n$ special orthogonal matrices form a group SO(n) with respect to ordinary matrix multiplication. Equations (7.4) and (7.7) show that every proper rotation can be put in one-to-one correspondence with a 3×3 special orthogonal matrix, so that the rotation group is isomorphic to SO(3). The rotation group is therefore sometimes denoted as SO(3).

7.3 Rotations as Operators; Infinitesimal Rotations

Let us consider a function ψ of the position vector $\boldsymbol{r} = (x, y, z)$ and define rotation as an operator acting on ψ in the following manner: When the rotation R takes the point $\boldsymbol{r}$ to $\boldsymbol{r}' = (x', y', z')$, the result of operating R on ψ is defined by

$$\psi' = R\psi \ , \tag{7.8}$$

where

$$\psi'(\boldsymbol{r}') = \psi(\boldsymbol{r}) \ . \tag{7.9}$$

The value of the rotated function $R\psi$ at $\boldsymbol{r}$ is equal to that of ψ at the point which is brought to $\boldsymbol{r}$ by R, that is, at $\hat{R}^{-1}\boldsymbol{r}$ if we use the matrix $\hat{R}$ given by (7.7), see Sect. 4.3.

When $R\psi$ is defined by (7.8, 9), we find that the inner product of the rotated functions is given by

$$(R\phi, R\psi) = \int d\tau' [R\phi(\boldsymbol{r}')]^* R\psi(\boldsymbol{r}') = \int d\tau' \phi(\boldsymbol{r})^* \psi(\boldsymbol{r}) \ .$$

Since the rotation leaves the volume element invariant, the right-hand side turns out to be equal to

$$\int d\tau \phi(\boldsymbol{r})^* \psi(\boldsymbol{r}) \ ,$$

so that

$$(R\phi, R\psi) = (\phi, \psi) \ . \tag{7.10}$$

This means that R is a unitary operator.

Suppose R is the rotation $R_\varepsilon(z)$ through an infinitesimal angle ε about the z-axis. When this rotation carries $\boldsymbol{r}_0$ to $\boldsymbol{r}$, we find that the coordinates (x, y, z) are expressed as

$$\begin{aligned} x_0 &= r\cos(\varphi - \varepsilon) = x + \varepsilon y \ , \\ y_0 &= r\sin(\varphi - \varepsilon) = -\varepsilon x + y \ , \\ z_0 &= z \ , \end{aligned}$$

in terms of (x, y, z) to first order in ε. See Fig. 7.3. The result of operating $R_\varepsilon(z)$ on ψ is thus given by

$$\begin{aligned} R_\varepsilon(z)\psi(x, y, z) &= \psi(x_0, y_0, z_0) \\ &= \psi(x + \varepsilon y, y - \varepsilon x, z) \\ &= (1 - \mathrm{i}\varepsilon l_z)\psi(x, y, z) \ , \end{aligned}$$

where l_z is the z component of the angular momentum operator in quantum

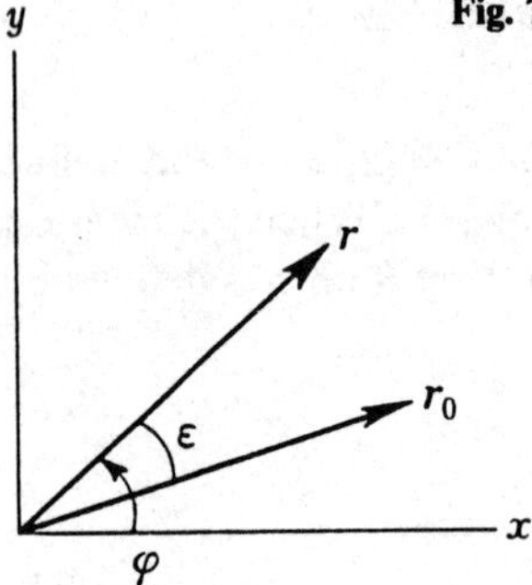

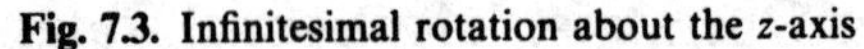

Fig. 7.3. Infinitesimal rotation about the z-axis

mechanics and is given by

$$l_z = -\mathrm{i}\left(x\frac{\partial}{\partial y} - y\frac{\partial}{\partial x}\right) = -\mathrm{i}\frac{\partial}{\partial \varphi} . \tag{7.11}$$

It then follows that the infinitesimal rotation operator is related to the Hermitian operator l_z through the equation

$$R_\varepsilon(z) = 1 - \mathrm{i}\varepsilon l_z . \tag{7.12}$$

For the rotation $R_\theta(z)$ through a finite angle θ, we find

$$R_\theta(z) = \lim_{n\to\infty} [R_{\theta/n}(z)]^n = \lim_{n\to\infty}\left(1 - \mathrm{i}\frac{\theta}{n} l_z\right)^n = \exp(-\mathrm{i}\theta l_z) . \tag{7.13}$$

When the axis of rotation is specified by the direction cosines (λ, μ, ν), it suffices to replace l_z by the component of l in that direction:

$$R_\theta(\lambda, \mu, \nu) = \exp[-\mathrm{i}\theta(\lambda l_x + \mu l_y + \nu l_z)] . \tag{7.14}$$

It is, however, more convenient to use the expression in terms of Euler angles for the purpose of deriving the representation matrices of rotations:

$$R(\alpha, \beta, \gamma) = \mathrm{e}^{-\mathrm{i}\alpha l_z}\mathrm{e}^{-\mathrm{i}\beta l_y}\mathrm{e}^{-\mathrm{i}\gamma l_z} . \tag{7.15}$$

In this form the problem is turned into the derivation of the matrices for l_z and l_y. As is well known, commutation relations hold between l_x, l_y and l_z:

$$[l_x, l_y] = \mathrm{i}l_z , \quad [l_y, l_z] = \mathrm{i}l_x , \quad [l_z, l_x] = \mathrm{i}l_y . \tag{7.16}$$

Representations for l_x, l_y and l_z are set up from these relations as given in the next section.

The commutation relations (7.16) are usually proved by using explicit expressions such as (7.11) for the components of l. They also follow as a natural consequence from a relation among infinitesimal rotations. Let us denote by

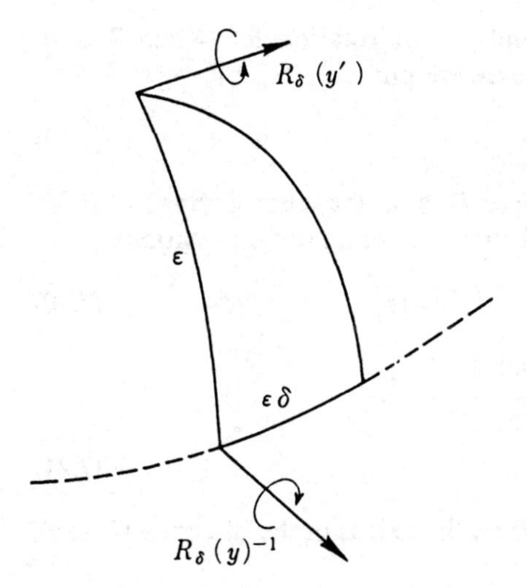

Fig. 7.4. Composition of infinitesimal rotations

$R_\varepsilon(x)$ and $R_\delta(y)$ the rotations through ε and δ about the x- and y-axes. The y-axis is assumed to be carried into y'-axis by $R_\varepsilon(x)$ (Fig. 7.4). Suppose we first perform $R_\delta(y)^{-1}$ (rotation through an angle $-\delta$ about the y-axis) and then $R_\delta(y')$, which is given by

$$R_\delta(y') = R_\varepsilon(x)R_\delta(y)R_\varepsilon(x)^{-1} \ .$$

According to the rule of composition, the result of these successive rotations leaves the z-axis unmoved (for $\varepsilon, \delta \sim 0$) and the point on the unit sphere pierced by the y-axis travels the distance $\varepsilon\delta$ in the xy-plane in the direction of $-x$. In other words, the product is equal to $R_{\varepsilon\delta}(z)$, so that we have

$$R_\varepsilon(x)R_\delta(y)\,R_\varepsilon(x)^{-1}\,R_\delta(y)^{-1} = R_{\varepsilon\delta}(z) \ . \tag{7.17}$$

If we represent $R_\varepsilon(x)$ etc. by l_x etc., following (7.12), and compare terms of order $\varepsilon\delta$ on both sides, we obtain the first equation of (7.16).

7.4 Representation of Infinitesimal Rotations

Let us now consider representation of the rotation group by matrices. We define transformation of the basis ψ_M of representation by

$$R\psi_M = \sum_{M'} \psi_{M'} D_{M'M}(R) \ , \tag{7.18}$$

when the unitary matrix $\hat{D}(R)$ corresponds to the rotation R.[1] When R is an infinitesimal rotation $R_\varepsilon(z)$ about the z-axis, we put

$$D_{MM'}(R_\varepsilon(z)) = \delta_{MM'} - \mathrm{i}\varepsilon(\hat{J}_z)_{MM'} , \tag{7.19}$$

where the matrix $\hat{J}_z$ is Hermitian, because D is unitary. Employing (7.19) for $R_\varepsilon(x)$, $R_\delta(y)$ and $R_{\varepsilon\delta}(z)$ in (7.17), we find that the commutation relations

$$[\hat{J}_x, \hat{J}_y] = \mathrm{i}\hat{J}_z, \quad [\hat{J}_y, \hat{J}_z] = \mathrm{i}\hat{J}_x, \quad [\hat{J}_z, \hat{J}_x] = \mathrm{i}\hat{J}_y \tag{7.20}$$

must hold among the matrices $\hat{J}_x$, $\hat{J}_y$ and $\hat{J}_z$.

From (7.19) we also have

$$D_{MM'}(R_\theta(z)) = [\exp(-\mathrm{i}\theta\hat{J}_z)]_{MM'} \tag{7.21}$$

for rotations through a finite angle θ. When the axis is in the direction (λ, μ, ν), equation (7.21) takes the form

$$D_{MM'}(R_\theta(\lambda, \mu, \nu)) = \{\exp[-\mathrm{i}\theta(\lambda\hat{J}_x + \mu\hat{J}_y + \nu\hat{J}_z)]\}_{MM'} . \tag{7.22}$$

When the rotations are specified in terms of the Euler angles, (7.4) leads to

$$D_{MM'}(R(\alpha, \beta, \gamma)) = [\exp(-\mathrm{i}\alpha\hat{J}_z)\exp(-\mathrm{i}\beta\hat{J}_y)\exp(-\mathrm{i}\gamma\hat{J}_z)]_{MM'} . \tag{7.23}$$

Construction of the representation matrices is now straightforward once we know the matrices for $\exp(-\mathrm{i}\alpha\hat{J}_z)$ and $\exp(-\mathrm{i}\beta\hat{J}_y)$. First we must find those for $\hat{J}_x$, $\hat{J}_y$ and $\hat{J}_z$. Let us put

$$\hat{J}_\pm = \hat{J}_x \pm \mathrm{i}\hat{J}_y . \tag{7.24}$$

Then the commutation relations (7.20) take the forms

$$[\hat{J}_z, \hat{J}_\pm] = \pm\hat{J}_\pm, \quad [\hat{J}_+, \hat{J}_-] = 2\hat{J}_z . \tag{7.25}$$

Choose the matrix $\hat{J}_z$ in diagonal form and put

$$(\hat{J}_z)_{MM'} = M\delta_{MM'} , \tag{7.26}$$

where the eigenvalue M is real. Note that the values of the diagonal elements are being used to label the rows and columns of the matrices of $\boldsymbol{J}$. If we take the MM' element of the first matrix equation of (7.25), we have

$$(\hat{J}_\pm)_{MM'}(M - M' \mp 1) = 0 ,$$

so that the element $(J_\pm)_{MM'}$ will be nonvanishing only when $M = M' \pm 1$. If we take this fact into account when writing down the second equation of (7.25), we

---

[1] Although the definition of $R\psi$ given here is more abstract than the one in the previous section, it enables us to include the case where ψ is a function of the spin coordinate (Sect. 7.4.1).

find

$$(\hat{J}_+)_{M,M-1}(\hat{J}_-)_{M-1,M} - (\hat{J}_-)_{M,M+1}(\hat{J}_+)_{M+1,M} = 2M \ .$$

When we put

$$f(M) = (\hat{J}_+)_{M,M-1}(\hat{J}_-)_{M-1,M} = |(\hat{J}_-)_{M-1,M}|^2 \ , \tag{7.27}$$

this implies that

$$f(M) - f(M+1) = 2M \ ,$$

and

$$f(M+k) = f(M) - 2Mk - k(k-1) \ , \tag{7.28}$$

where k is a nonnegative integer.

Now it is also well known that the following commutation relations follow from (7.20) or (7.25):

$$[\hat{J}^2, \hat{J}_z] = 0 \ , \quad [\hat{J}^2, \hat{J}_\pm] = 0 \tag{7.29}$$

for

$$\hat{J}^2 = \hat{J}_x^2 + \hat{J}_y^2 + \hat{J}_z^2 \ .$$

According to (7.29), the matrix $\hat{J}^2$ is diagonal and its eigenvalues are independent of M. Since $\hat{J}^2 - \hat{J}_z^2$ is equal to $\hat{J}_x^2 + \hat{J}_y^2$, the eigenvalues of $\hat{J}_z^2$ can never exceed those of $\hat{J}^2$.[2] Let us write J for the largest eigenvalue of $\hat{J}_z$. Then it follows that

$$f(J+1) = 0 \ ,$$

which means that the values of $J - M$ have to be nonnegative integers. Starting from $M = J$, we can determine $f(M)$ from (7.28) as

$$f(M) = (J+M)(J-M+1) \ . \tag{7.30}$$

This shows that there is a minimum in the values of M which is $-J$, and $\hat{J}^2$ is given by

$$\hat{J}^2 = \hat{J}_+\hat{J}_- + \hat{J}_z^2 - \hat{J}_z = J(J+1)\hat{1} \ , \tag{7.31}$$

where $\hat{1}$ is the unit matrix.

The difference between the maximum and minimum values of M has to be an integer, so that possible values of J are given by

$$2J = 0, 1, 2, \ldots . \tag{7.32}$$

[2] $(\hat{J}^2)_{MM} - M^2 = \frac{1}{2}\{|(\hat{J}_-)_{M-1,M}|^2 + |(\hat{J}_+)_{M+1,M}|^2\} \geqq 0$

To summarize, we have obtained all Hermitian representation matrices for J_x, J_y and J_z. They are characterized by the value of J ($= 0, 1/2, 1, 3/2, \ldots$) and of dimensionality $2J + 1$. Explicitly, the only nonvanishing matrix elements in this representation are given by

$$(\hat{J}_+)_{M+1,M} = \sqrt{(J-M)(J+M+1)} \;,$$
$$(\hat{J}_-)_{M-1,M} = \sqrt{(J+M)(J-M+1)} \;, \qquad M = -J, \; -J+1, \ldots, J \;,$$
$$(\hat{J}_z)_{M,M} = M \;, \tag{7.33}$$

where we have chosen the phase factor in the elements of $\hat{J}_+$ and $\hat{J}_-$ to be unity (without loss of generality).

When $J = 1/2$, It is customary to use s for J. The matrices of s, the spin angular momentum operator, are given by the Pauli matrices

$$\hat{\sigma}_x = \begin{bmatrix} 0 & 1 \\ 1 & 0 \end{bmatrix}, \quad \hat{\sigma}_y = \begin{bmatrix} 0 & -\mathrm{i} \\ \mathrm{i} & 0 \end{bmatrix}, \quad \hat{\sigma}_z = \begin{bmatrix} 1 & 0 \\ 0 & -1 \end{bmatrix}, \tag{7.34}$$

as

$$\hat{s} = \tfrac{1}{2}\hat{\sigma} \;. \tag{7.35}$$

7.4.1 Rotation of Spin Functions

It is not so easy to see the meaning of applying rotations to the spin functions χ_m ($m = 1/2, \; -1/2$), contrary to the case of rotating the functions of coordinates in real space (7.8, 9). In Sect. 7.3, we rotated the position vector $\boldsymbol{r}$ to $\boldsymbol{r}'(= \hat{R}\boldsymbol{r})$ by the operation R. Let us assume that the coordinate axes are also rotated by R as in Fig. 7.5, and denote the original system of axes by K and the rotated system of axes (X, Y, Z) by K'. Application of the rotation R on the up-spin state $\chi_{1/2}$ in the K system yields the up-spin state $\chi'_{1/2}$ in the K' system. In fact, using the

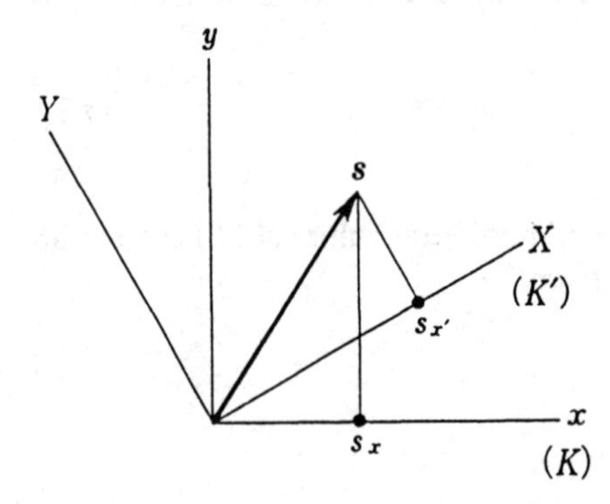

Fig. 7.5. Rotation of spin functions

definition (7.18) of $R\chi_m$, that is,

$$R\chi_m = \sum_{m'} \chi_{m'} D_{m'm}(R) \tag{7.36}$$

with $D = \hat{D}^{(1/2)}(R(\alpha, \beta, \gamma))$, we can show that the following relations hold:

$$(R\chi_m, s_{j'} R\chi_n) = (\chi_m, s_j \chi_n) \tag{7.37}$$

for the components of the spin vector s in the K' system given by

$$s_{j'} = \sum_i s_i R_{ij} \ , \quad i, j = x, y, z \ . \tag{7.38}$$

For simplicity, we have used the notation $s_{x'}, \ldots,$ in place of $s_X, \ldots$. With (7.36 and 38), equation (7.37) implies that the following (matrix) equation must be valid:

$$\hat{D}^{-1} \sum_i \hat{s}_i R_{ij} \hat{D} = \hat{s}_j \ , \tag{7.39}$$

which can indeed be verified by using (7.7) and the expression for the matrix $\hat{D}$

$$\hat{D} = e^{-i\alpha\hat{s}_z} e^{-i\beta\hat{s}_y} e^{-i\gamma\hat{s}_z}$$

to be given in (7.52).

7.5 Representations of the Rotation Group

According to (7.23, 33), we have

$$D_{MK}(R(\alpha, \beta, \gamma)) = e^{-iM\alpha} (e^{-i\beta\hat{J}_y})_{MK} e^{-iK\gamma} \ , \tag{7.40}$$

so our next task is to obtain

$$D^{(J)}_{MK}(\beta) = (e^{-i\beta\hat{J}_y})_{MK} \tag{7.41}$$

Since we know that

$$D^{(J)}_{MK}(R)^* = D^{(J)}_{KM}(R^{-1})$$

for $R = R(\alpha, \beta, \gamma)$, we may equally attempt to find the representation matrix for R^{-1}. Now, for

$$\hat{D}(R^{-1}) = e^{i\gamma\hat{J}_z} e^{i\beta\hat{J}_y} e^{i\alpha\hat{J}_z} \ ,$$

we see that the following equations hold:

$$
\begin{aligned}
&-\mathrm{i}\frac{\partial}{\partial\alpha}\hat{D}(R^{-1})=\hat{D}(R^{-1})\hat{J}_z\ ,\\
&-\mathrm{i}\frac{\partial}{\partial\beta}\hat{D}(R^{-1})=\hat{D}(R^{-1})\mathrm{e}^{-\mathrm{i}\alpha\hat{J}_z}\hat{J}_y\mathrm{e}^{\mathrm{i}\alpha\hat{J}_z}=\hat{D}(R^{-1})\hat{J}_{y_1}\ ,\\
&-\mathrm{i}\frac{\partial}{\partial\gamma}\hat{D}(R^{-1})=\hat{J}_z\hat{D}(R^{-1})=\hat{D}(R^{-1})\hat{J}_Z\ ,
\end{aligned}
\tag{7.42}
$$

where y_1 and Z are the coordinate axes shown in Fig. 7.2 and J_{y_1} and J_Z are the components of $\boldsymbol{J}$ in those directions. Operating both sides of the following equation on $\hat{D}(R^{-1})$ from the left

$$
\begin{aligned}
J_\pm &= \mathrm{e}^{\pm\mathrm{i}\alpha}(J_{x_1}\pm\mathrm{i}J_{y_1})\\
&= \mathrm{e}^{\pm\mathrm{i}\alpha}\left(\pm\mathrm{i}J_{y_1}+\frac{J_Z-J_z\cos\beta}{\sin\beta}\right)
\end{aligned}
$$

and using (7.42), we obtain

$$
\mathrm{e}^{\pm\mathrm{i}\alpha}\left\{\pm\frac{\partial}{\partial\beta}+\frac{\mathrm{i}}{\sin\beta}\left(-\frac{\partial}{\partial\gamma}+\cos\beta\frac{\partial}{\partial\alpha}\right)\right\}\hat{D}(R^{-1})=\hat{D}(R^{-1})\hat{J}_\pm \tag{7.43}
$$

Taking the KM element of both sides and remembering (7.40, 33), we arrive at

$$
\begin{aligned}
&\left\{\pm\frac{\partial}{\partial\beta}+\frac{1}{\sin\beta}(K-M\cos\beta)\right\}D^{(J)}_{MK}(\beta)^*\\
&\quad = D^{(J)}_{M\pm1,K}(\beta)^*\sqrt{(J\mp M)(J\pm M+1)}\ .
\end{aligned}
\tag{7.44}
$$

This is the recurrence equation determining $D^{(J)}_{MK}(\beta)^*$. Let us put $M=J$ here. Equation (7.44) then reads

$$
\left\{\frac{\partial}{\partial\beta}+\frac{1}{\sin\beta}(K-J\cos\beta)\right\}D^{(J)}_{JK}(\beta)^*=0
$$

whose solution is given by

$$
D^{(J)}_{JK}(\beta)=(-1)^{J-K}\sqrt{\frac{(2J)!}{(J+K)!(J-K)!}}\left(\sin\frac{\beta}{2}\right)^{J-K}\left(\cos\frac{\beta}{2}\right)^{J+K}\ . \tag{7.45a}
$$

The constant factor in the above equation was determined by expanding the right-hand side of

$$
D^{(J)}_{JK}(\beta)^*=[\exp(\mathrm{i}\beta\hat{J}_y)]_{KJ}=\left[\exp\left(\frac{\beta}{2}(\hat{J}_+-\hat{J}_-)\right)\right]_{KJ}
$$

in a power series of β and identifying the coefficient of β^{J-K} with it. Note that the right-hand side of (7.45a) is real, so that the symbol * (signifying the complex conjugate) has been dropped. For later use, we give also the expression for $D^{(J)}_{MK}(\beta)$ for $M = -J$ here:

$$D^{(J)}_{-JK}(\beta) = \sqrt{\frac{(2J)!}{(J+K)!(J-K)!}} \left(\sin\frac{\beta}{2}\right)^{J+K} \left(\cos\frac{\beta}{2}\right)^{J-K} . \tag{7.45b}$$

In terms of the new variable ζ defined by

$$\zeta = \sin^2\frac{\beta}{2} , \tag{7.46}$$

the recurrence equation (7.44) can be written as

$$\begin{aligned} &\pm \zeta^{\pm(M-K\pm 1)/2}(1-\zeta)^{\pm(M+K\pm 1)/2} \frac{d}{d\zeta}[\zeta^{\mp(M-K)/2}(1-\zeta)^{\mp(M+K)/2} D^{(J)}_{MK}(\beta)^*] \\ &\quad = D^{(J)}_{M\pm 1,K}(\beta)^* \sqrt{(J\mp M)(J\pm M+1)} . \end{aligned} \tag{7.47}$$

Repeated application of the above recurrence relation with the lower signs to (7.45a) leads to

$$\begin{aligned} D^{(J)}_{MK}(\beta) &= (-1)^{M-K} \sqrt{\frac{(J+M)!}{(J-M)!(J+K)!(J-K)!}} \\ &\quad \times \zeta^{-(M-K)/2}(1-\zeta)^{-(M+K)/2}\left(\frac{d}{d\zeta}\right)^{J-M} \zeta^{J-K}(1-\zeta)^{J+K} . \end{aligned} \tag{7.48}$$

Using the Jacobi polynomial

$$\begin{aligned} &F(-p, a+p; b|x) \\ &\quad = \frac{(b-1)!}{(b+p-1)!} x^{1-b}(1-x)^{b-a}\left(\frac{d}{dx}\right)^p x^{b-1+p}(1-x)^{a-b+p} \\ &\quad = \sum_{r=0}^{p} \frac{(b-1)!(a-b+p)!p!(-x)^{p-r}(1-x)^r}{(p-r)!r!(b-1+p-r)!(a-b+r)!} \end{aligned}$$

this result may also be expressed as

$$\begin{aligned} D^{(J)}_{MK}(\beta) &= \frac{(-1)^{M-K}}{(M-K)!} \sqrt{\frac{(J+M)!(J-K)!}{(J-M)!(J+K)!}} \zeta^{(M-K)/2}(1-\zeta)^{(M+K)/2} \\ &\quad \times F(-J+M, J+M+1; M-K+1|\zeta) . \end{aligned} \tag{7.49}$$

The matrix element has the following symmetry properties:

$$\begin{aligned} D^{(J)}_{MK}(\beta) &= (-1)^{M-K} D^{(J)}_{KM}(\beta) \\ &= (-1)^{M-K} D^{(J)}_{-M,-K}(\beta) \\ &= (-1)^{M-K} D^{(J)}_{MK}(-\beta) \\ &= (-1)^{J+M} D^{(J)}_{M,-K}(\pi - \beta) \; . \end{aligned} \tag{7.50}$$

Especially, when $J = 1/2$, we may derive the expression for $\hat{D}^{(1/2)}(\beta)$ directly, because the matrix exponential of $\hat{\sigma}_y$ can be evaluated easily as

$$D^{(1/2)}_{MK}(\beta) = \left[\exp\left(-\mathrm{i}\frac{\beta}{2}\hat{\sigma}_y\right)\right]_{MK}$$

with

$$\exp\left(-\mathrm{i}\frac{\beta}{2}\hat{\sigma}_y\right) = \hat{1}\cos\frac{\beta}{2} - \mathrm{i}\hat{\sigma}_y\sin\frac{\beta}{2} = \begin{bmatrix} \cos\frac{\beta}{2} & -\sin\frac{\beta}{2} \\ \sin\frac{\beta}{2} & \cos\frac{\beta}{2} \end{bmatrix} \tag{7.51}$$

if we note that[3] $\hat{\sigma}_y^2 = \hat{1}$. This then gives us

$$\hat{D}^{(1/2)}(R(\alpha, \beta, \gamma)) = \begin{bmatrix} \mathrm{e}^{-\mathrm{i}(\alpha+\gamma)/2}\cos\frac{\beta}{2} & -\mathrm{e}^{-\mathrm{i}(\alpha-\gamma)/2}\sin\frac{\beta}{2} \\ \mathrm{e}^{\mathrm{i}(\alpha-\gamma)/2}\sin\frac{\beta}{2} & \mathrm{e}^{\mathrm{i}(\alpha+\gamma)/2}\cos\frac{\beta}{2} \end{bmatrix} \tag{7.52}$$

for $J = 1/2$. Equation (7.51) indicates that we should have

$$\begin{aligned} \hat{D}^{(1/2)}(R_\theta(\lambda, \mu, \nu)) &= \hat{1}\cos\frac{\theta}{2} - \mathrm{i}(\lambda\hat{\sigma}_x + \mu\hat{\sigma}_y + \nu\hat{\sigma}_z)\sin\frac{\theta}{2} \\ &= \begin{bmatrix} \cos\frac{\theta}{2} - \mathrm{i}\nu\sin\frac{\theta}{2} & -\mathrm{i}(\lambda - \mathrm{i}\mu)\sin\frac{\theta}{2} \\ -\mathrm{i}(\lambda + \mathrm{i}\mu)\sin\frac{\theta}{2} & \cos\frac{\theta}{2} + \mathrm{i}\nu\sin\frac{\theta}{2} \end{bmatrix} . \end{aligned} \tag{7.53}$$

Comparison of (7.53) with (7.52) then leads to (7.5, 6).

---

[3] In the same way, we can show that

$$\hat{D}^{(1)}(\beta) = \exp(-\mathrm{i}\beta\hat{J}_y) = \hat{1} - \mathrm{i}\hat{J}_y\sin\beta + \hat{J}_y^2(\cos\beta - 1) \; ,$$

using $\hat{J}_y^3 = \hat{J}_y$ in the case of $J = 1$.

7.6 SU(2), SO(3) and O(3)

The value of the determinant of the matrix in (7.52) is unity, so that the matrix is a 2×2 special unitary matrix. Conversely, every 2×2 special unitary matrix can be expressed in the form of (7.52), provided the range of parameters is extended to

$$0 \leqq \alpha < 2\pi \ , \quad 0 \leqq \gamma < 4\pi \ , \quad 0 \leqq \beta \leqq \pi \ . \tag{7.54}$$

The set of such matrices form the group SU(2), the special unitary group in two dimensions. In fact, it would be impossible to represent the element of SU(2) which corresponds to $\gamma = 2\pi$, $\alpha = \beta = 0$,

$$\bar{E} = \begin{bmatrix} -1 & 0 \\ 0 & -1 \end{bmatrix} \longleftrightarrow R(0, 0, 2\pi) \ ,$$

if we stayed within the range chosen so far, i.e.,

$$0 \leqq \alpha, \gamma < 2\pi \ , \quad 0 \leqq \beta \leqq \pi \ . \tag{7.55}$$

We note here that $R(0, 0, 2\pi)$ and $R(0, 0, 0)$ which corresponds to E (2×2 unit matrix) are one and the same operation as spatial rotations. This implies therefore that $D^{(1/2)}$, the representation by SU(2) of the rotation group SO(3), is double valued. In fact, we observe that

$$\hat{D}^{(1/2)}(R_\theta(\lambda, \mu, \nu)) = -\hat{D}^{(1/2)}(R_{2\pi-\theta}(-\lambda, -\mu, -\nu)) \ ,$$

so that the correspondence is 1:2 in the sense that both matrices $\hat{D}^{(1/2)}(R)$ and $-\hat{D}^{(1/2)}(R)$ correspond to the same rotation R. This 1:2 correspondence is also found in the representations $D^{(j)}$ with j half of an odd integer[4]. As with the case of $j = 1/2$, here, too, two different representation matrices with parameters (α, β, γ) and $(\alpha, \beta, \gamma + 2\pi)$ correspond to the same rotation $R(\alpha, \beta, \gamma)$ when we extend the range of parameters as in (7.54).

There is a way of turning this 1:2 correspondence into 1:1. This is to introduce the double rotation group SO(3)′ ($\cong$ SU(2)). In SO(3)′, $R(0, 0, 2\pi)$ is regarded as a rotation entirely different from the identity $E = R(0, 0, 0)$ and is denoted as $\bar{E}$. The product $\bar{E}R$ with $R = R_\theta(\lambda, \mu, \nu)$, or

$$\bar{E}R = R_{2\pi-\theta}(-\lambda, -\mu, -\nu)$$

is an element $\bar{R}$ different from R. By this convention, the correspondence becomes 1:1 even for j half of an odd integer, because we now have $R \leftrightarrow \hat{D}^{(j)}(R)$ and $\bar{R} \leftrightarrow -\hat{D}^{(j)}(R)$.

---

[4] Note that the invariant subgroup $Z_2 = \{E, \bar{E}\}$ is the center of SU(2) and the factor group $SU(2)/Z_2$ is isomorphic to SO(3). The group SU(2) is thus the covering group of SO(3). The double-valued representations $D^{(j)}$ of SO(3) with j half of an odd integer constitute the ray representations of $SU(2)/Z_2$ (Sects. 5.3, 5.4).

Table 7.1. Irreducible representations of O(3) derived from the single-valued representation $D^{(l)}$ of SO(3)

	R	IR
$D_g^{(l)}$	$D^{(l)}(R)$	$D^{(l)}(R)$
$D_u^{(l)}$	$D^{(l)}(R)$	$-D^{(l)}(R)$

Table 7.2. Irreducible representations of O′(3) derived from the double-valued representation $D^{(J)}$ of SO(3)

	R	$\bar{R}$	IR	$I\bar{R}$
$D_g^{(J)}$	$D^{(J)}(R)$	$-D^{(J)}(R)$	$D^{(J)}(R)$	$-D^{(J)}(R)$
$D_u^{(J)}$	$D^{(J)}(R)$	$-D^{(J)}(R)$	$-D^{(J)}(R)$	$D^{(J)}(R)$

So far we have considered SO(3), which consists only of pure (proper) rotations. If we include the operation I, namely, inversion with respect to the origin, we obtain the full rotation group or the rotation-reflection group consisting of all proper and improper rotations. This is the orthogonal group in three dimensions $O(3) = SO(3) \times C_i$ with $C_i = \{E, I\}$. The group O(3) is the group of all 3×3 (real) orthogonal matrices whose determinants can be equal to either 1 or -1. Since the group C_i has only two irreducible representations, A_g (even parity) and A_u (odd parity), we can derive all the single-valued irreducible representations of O(3) as given in Table 7.1 from $D^{(l)}$ of SO(3) with nonnegative integers l.

In contrast, for j half of an odd integer, we obtain only $D^{(j)}$ as the irreducible representation of O(3), because it is double-valued and IR as well as R corresponds to $\hat{D}^{(j)}(R)$ and $-\hat{D}^{(j)}(R)$, so that no new representation is introduced.

We may, however, also consider here the double rotation-reflection group $O(3)' = SO(3)' \times C_i$ with $I^2 = E$. For this double group, representations $D_{g,u}^{(J)}$ are conceivable as shown in Table 7.2, with the proviso that operation of I upon the spin functions $\alpha(=\chi_{1/2})$ and $\beta(=\chi_{-1/2})$ are defined by $I\alpha = \alpha$ and $I\beta = \beta$.

7.7 Basis of Representations

Let us define the operation of the rotation Q on a function of Euler angles ψ (α, β, γ) in the following way. Suppose rotations R and R_0 which correspond to the angles α, β, γ and $\alpha_0, \beta_0, \gamma_0$, respectively, are connected with each other by Q as

$$Q^{-1} R = R_0 \ . \tag{7.56}$$

In other words, Q rotates the coordinate axes described by the angles $\alpha_0, \beta_0, \gamma_0$ to bring them to the axes described by α, β, γ. We then put

$$Q\psi(\alpha, \beta, \gamma) = \psi(\alpha_0, \beta_0, \gamma_0) \ . \tag{7.57}$$

If we set

$$\psi_M(\alpha, \beta, \gamma) = D^{(J)}_{MK}(R(\alpha, \beta, \gamma))^* \tag{7.58}$$

here, we find that the ψ_M's form the basis for the representation $D^{(J)}$, because we have from the definition (7.57)

$$Q\psi_M(\alpha, \beta, \gamma) = \psi_M(\alpha_0, \beta_0, \gamma_0) = D^{(J)}_{MK}(Q^{-1}R)^* = \sum_N D^{(J)}_{NK}(R)^* D^{(J)}_{NM}(Q) \ , \tag{7.59a}$$

or

$$Q\psi_M(\alpha, \beta, \gamma) = \sum_N \psi_N(\alpha, \beta, \gamma) D^{(J)}_{NM}(Q) \ , \tag{7.59b}$$

where we have made use of the unitarity of the representation $D^{(J)}$.

Normalizing the functions (7.58), we put

$$\chi^{(J)}_{MK}(\alpha, \beta, \gamma) = \sqrt{\frac{2J+1}{16\pi^2}}\, D^{(J)}_{MK}(R(\alpha, \beta, \gamma))^* \ . \tag{7.60}$$

They are normalized in the range (7.54)[5] for the variables α, β and γ besides satisfying the orthogonality relations

$$\int_0^{2\pi} d\alpha \int_0^{\pi} \sin\beta\, d\beta \int_0^{4\pi} d\gamma\, \chi^{(J)}_{MK}(\alpha, \beta, \gamma)^* \chi^{(J')}_{M'K'}(\alpha, \beta, \gamma) = \delta_{JJ'}\delta_{MM'}\delta_{KK'} \ . \tag{7.61}$$

The orthogonality follows from the fact that the functions $\chi^{(J)}_{MK}$ are the simultaneous eigenfunctions of the operators J_z, J_Z, and $\boldsymbol{J}^2$:

$$J_z \chi^{(J)}_{MK} \equiv -\mathrm{i}\frac{\partial}{\partial\alpha}\chi^{(J)}_{MK} = M\chi^{(J)}_{MK} \ , \tag{7.62a}$$

$$J_Z \chi^{(J)}_{MK} \equiv -\mathrm{i}\frac{\partial}{\partial\gamma}\chi^{(J)}_{MK} = K\chi^{(J)}_{MK} \ , \tag{7.62b}$$

$$\begin{aligned} \boldsymbol{J}^2 \chi^{(J)}_{MK} &\equiv -\left\{\frac{\partial^2}{\partial\beta^2} + \cot\beta\frac{\partial}{\partial\beta} + \frac{1}{\sin^2\beta}\left(\frac{\partial^2}{\partial\gamma^2} - 2\cos\beta\frac{\partial^2}{\partial\gamma\partial\alpha} + \frac{\partial^2}{\partial\alpha^2}\right)\right\}\chi^{(J)}_{MK} \\ &= J(J+1)\chi^{(J)}_{MK} \ , \end{aligned} \tag{7.62c}$$

[5] $\chi^{(J)}_{MK}(\alpha, \beta, \gamma + 2\pi) = (-1)^{2J}\chi^{(J)}_{MK}(\alpha, \beta, \gamma)$

where $J_z = -\mathrm{i}\partial/\partial\alpha$ etc. on the left-hand side of these equations are the expressions for J_z, J_Z and $\boldsymbol{J}^2 = J_+J_- + J_z^2 - J_z$ as differential operators acting upon $\chi_{MK}^{(J)}$, see (7.42, 43).

The normalization may be proved in the following way. We first note that the left-hand side of (7.61) becomes

$$\frac{2J+1}{2}\int_0^{\pi} \sin\beta\, d\beta\, D_{MK}^{(J)}(\beta)\, D_{MK}^{(J)}(\beta)$$

when $J = J'$, $M = M'$ and $K = K'$, which is independent of M, as shown by the use of the recurrence relation (7.44) or (7.47) followed by partial integration. For $M = J$, the above integral becomes equal to

$$\frac{(2J+1)!}{(J+K)!(J-K)!}\int_0^1 d\zeta\, \zeta^{J-K}(1-\zeta)^{J+K}$$

by (7.45a) and this is certainly equal to unity.

7.8 Spherical Harmonics

When J is a nonnegative integer L, the function $\chi_{MK}^{(L)}(\alpha, \beta, \gamma)$ becomes single valued and the functions $\sqrt{(2L+1)/8\pi^2}\, D_{MK}^{(L)}(R(\alpha, \beta, \gamma))^*$ with the range of variables given by (7.55) form an orthonormal set. They are nothing other than the wavefunctions of the symmetric top [7.1]. When we further put $K = 0$, we obtain the spherical harmonics:

$$\begin{aligned} Y_{LM}(\beta, \alpha) &= \sqrt{\frac{2L+1}{4\pi}}\, D_{M0}^{(L)}(R(\alpha, \beta, \gamma))^* \\ &= \sqrt{\frac{2L+1}{2}}\, D_{M0}^{(L)}(\beta)\, \frac{1}{\sqrt{2\pi}}\, \mathrm{e}^{\mathrm{i}M\alpha} \\ &\equiv \sqrt{\frac{2L+1}{4\pi}}\, C_M^{(L)}(\beta, \alpha)\ . \end{aligned} \tag{7.63}$$

Equation (7.59) now takes the form

$$QY_{LM}(\beta, \alpha) = Y_{LM}(\beta_0, \alpha_0)\ , \tag{7.64a}$$

$$QY_{LM} = \sum_N Y_{LN} D_{NM}^{(L)}(Q)\ . \tag{7.64b}$$

If we regard β and α respectively as the polar angle θ and azimuthal angle φ of the position vector $\boldsymbol{r}$, (7.64a) agrees with the usual definition of rotation as an operator acting on a function of $\boldsymbol{r}$. Equation (7.64b) of course asserts that the spherical harmonics Y_{LM} form the basis for the representation $D^{(L)}$ of the rotation group SO(3). As the basis for the irreducible representation of the rotation-reflection group O(3), the Y_{LM}'s give rise to $D_{\mathrm{g}}^{(L)}$ (L even) or $D_{\mathrm{u}}^{(L)}$ (L odd) according to the parity of L. This is because these functions behave like

$$IY_{LM}(\theta, \varphi) \equiv Y_{LM}(\pi - \theta, \varphi + \pi) = (-1)^L Y_{LM}(\theta, \varphi)$$

under the operation of inversion I.

Explicitly, the spherical harmonics are given by

$$Y_{LM}(\theta, \varphi) = (-1)^{(M+|M|)/2} \sqrt{\frac{2L+1}{2}\frac{(L-|M|)!}{(L+|M|)!}}\, P_L^{|M|}(\cos\theta)\,\frac{\mathrm{e}^{\mathrm{i}M\varphi}}{\sqrt{2\pi}}\,, \tag{7.65}$$

$$P_L^M(t) = (1-t^2)^{M/2}\frac{d^M}{dt^M}P_L(t)\,, \quad M \geqq 0\,, \tag{7.66}$$

$$P_L(t) = \frac{1}{2^L L!}\frac{d^L}{dt^L}(t^2-1)^L\,, \tag{7.67}$$

where $P_L^M(t)$ and $P_L(t)$ are respectively the associated Legendre function and Legendre polynomial.

Important properties of the spherical harmonics are reiterated below:

1) Complex conjugation

$$Y_{LM}(\theta, \varphi)^* = (-1)^M Y_{L,-M}(\theta, \varphi)\,. \tag{7.68}$$

2) L and Y_{LM}

$$L_z Y_{LM} \equiv -\mathrm{i}\frac{\partial}{\partial\varphi}Y_{LM} = MY_{LM}\,,$$

$$L_\pm Y_{LM} \equiv \mathrm{e}^{\pm\mathrm{i}\varphi}\left(\pm\frac{\partial}{\partial\theta} + \mathrm{i}\cot\theta\frac{\partial}{\partial\varphi}\right)Y_{LM}$$

$$= \sqrt{(L\mp M)(L\pm M+1)}\,Y_{L,M\pm 1}\,,$$

$$L^2 Y_{LM} \equiv -\left(\frac{1}{\sin\theta}\frac{\partial}{\partial\theta}\sin\theta\frac{\partial}{\partial\theta} + \frac{1}{\sin^2\theta}\frac{\partial^2}{\partial\varphi^2}\right)Y_{LM}$$

$$= L(L+1)Y_{LM}\,. \tag{7.69}$$

3) Addition theorem

When Θ is the angle between the directions specified by angles (θ_0, φ_0) and (θ_1, φ_1), the addition theorem for the spherical harmonics reads

$$\sum_M C_M^{(L)}(\theta_0, \varphi_0)\,C_M^{(L)}(\theta_1, \varphi_1)^* = P_L(\cos\Theta)\,. \tag{7.70}$$

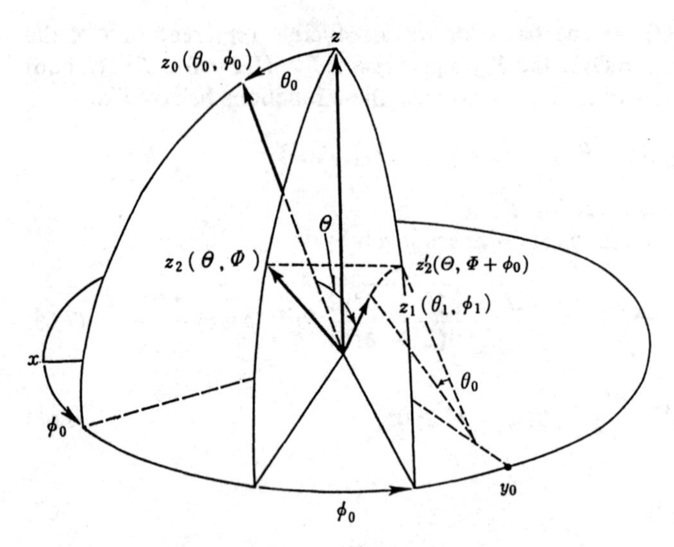

Fig. 7.6. Addition theorem for the spherical harmonics

This can be shown as follows: Suppose that R_0 and R_1 are rotations with Euler angles $(\varphi_0, \theta_0, 0)$ and $(\varphi_1, \theta_1, 0)$ respectively and that the rotation R_2 given by

$$R_0^{-1} R_1 = R_2 \tag{7.71}$$

corresponds to Euler angles (Φ, Θ, γ). Note that the rotation R_2 brings the z-axis to the z_2-axis with the direction (Θ, Φ) (Fig. 7.6). Now, by the rotation R_0 the z-axis is carried to the z_0-axis with (θ_0, φ_0) and the z_2-axis to the z_1-axis with (θ_1, φ_1), because $R_1 = R_0 R_2$. The angle between the z_0-axis and z_1-axis must therefore be equal to Θ. Writing down (7.71) in terms of the representation matrices, we find

$$\sum_M D_{M0}^{(L)}(R(\varphi_0, \theta_0, 0))^* D_{M0}^{(L)}(R(\varphi_1, \theta_1, 0)) = D_{00}^{(L)}(R(\Phi, \Theta, \gamma)) \ ,$$

which is precisely (7.70).

7.9 Orthogonality of Representation Matrices and Characters

Needless to say, (7.61) stands for the orthogonality relation of the representation matrices. Let us write

$$\omega = R(\alpha, \beta, \gamma) \ , \qquad d\omega = d\alpha \sin\beta d\beta d\gamma \ , \tag{7.72}$$

for simplicity. Then (7.61) may be expressed as

$$\int d\omega\, D^{(J)}_{MK}(\omega)^* D^{(J')}_{M'K'}(\omega) = \frac{\Omega}{2J+1}\, \delta_{JJ'}\delta_{MM'}\delta_{KK'} \tag{7.73}$$

with

$$\Omega = \int d\omega = 16\pi^2$$

in the form of the orthogonality theorem[6] for the irreducible representations.

If we put $M = K$, $M' = K'$ here and sum over M and M', we obtain an orthogonality relation (of the first kind) for the irreducible characters

$$\chi^{(J)}(\omega) = \sum_M D^{(J)}_{MM}(\omega) \tag{7.74}$$

as

$$\int d\omega\, \chi^{(J)}(\omega)^* \chi^{(J')}(\omega) = \Omega\delta_{JJ'} \ . \tag{7.75}$$

As mentioned before, rotations through equal angles are conjugate to each other. The rotation ω through an angle θ is conjugate to the rotation $R_\theta(z)$ about the z-axis, so that the corresponding character is given by

$$\chi^{(J)}(\omega) \equiv \chi^{(J)}(\theta) = \sum_M \mathrm{e}^{-\mathrm{i}M\theta} = \frac{\sin\dfrac{2J+1}{2}\theta}{\sin\theta/2} \ . \tag{7.76}$$

After some calculations using (7.5, 6), $d\omega$ of (7.72) can be expressed in terms of the angle of rotation θ and the direction of the axis given by (ϑ, ψ) as

$$d\omega = 4\sin^2\frac{\theta}{2}\, d\theta \sin\vartheta\, d\vartheta\, d\psi \ ,$$

$$0 \leqq \theta \leqq 2\pi \ , \quad 0 \leqq \vartheta \leqq \pi \ , \quad 0 \leqq \psi < 2\pi \ ,$$

where we have restricted the range of θ to $0 \leqq \theta \leqq 2\pi$,[7] bearing in mind that

$$\hat{D}^{(J)}(R_\theta(\lambda, \mu, \nu)) = \hat{D}^{(J)}(R_{4\pi-\theta}(-\lambda, -\mu, -\nu)) \ .$$

Since

$$\int d\omega \ldots = 4\int_0^{2\pi} \sin^2\frac{\theta}{2}\, d\theta \int_0^{\pi} \sin\vartheta\, d\vartheta \int_0^{2\pi} d\psi \ldots \ , \tag{7.77}$$

---

[6] When both J and J' are nonnegative integers, we may restrict the range of integration as given in (7.55) and put $\Omega = 8\pi^2$ in (7.73).

[7] When both J and J' are nonnegative integers, we may assume $0 \leqq \theta \leqq \pi$. In this case the right-hand side of (7.77) becomes $8\int_0^\pi \sin^2(\theta/2)d\theta \ldots$.

Let us put $\theta' = 0$ in (7.79). This leads to

$$\sum_J (2J+1)\chi^{(J)}(\omega) = \frac{\Omega\delta(\theta)}{16\pi \sin^2\theta/2} \ . \tag{7.80}$$

On the other hand, we have

$$\sum_{MM'} D^{(J)}_{MM'}(\omega) D^{(J)}_{MM'}(\omega')^* = \sum_M D^{(J)}_{MM}(\omega\omega'^{-1}) = \chi^{(J)}(\omega\omega'^{-1}) \ ,$$

because the representation is unitary. If we multiply both sides of this equation by $(2J+1)$, sum over J and use (7.80), we arrive at the completeness relation for the representation matrices

$$\sum_{JMM'} (2J+1) D^{(J)}_{MM'}(\omega) D^{(J)}_{MM'}(\omega')^* = \Omega\delta(\omega, \omega') \ , \tag{7.81}$$

where $\delta(\omega, \omega')$ is equal to zero when $\omega \neq \omega'$ and satisfies

$$\int d\omega \delta(\omega, \omega') = 1 \ .$$

If we further integrate both sides of (7.81) over γ from 0 to 4π, the terms with $M' = 0$ and J a nonnegative integer remain in the summation on the left-hand side and (7.81) reduces to the completeness relation for the spherical harmonics.

7.10 Wigner Coefficients

It is easy to see that the product representation $D^{(J_1)} \times D^{(J_2)}$ is completely reducible in general and when reduced becomes a direct sum of $D^{(J)}$ with J satisfying the triangular condition $|J_1 - J_2| \leqq J \leqq J_1 + J_2$ (each $D^{(J)}$ appearing once in the decomposition), because the character of the product representation can be expressed as

$$\chi^{(J_1)}(\theta)\chi^{(J_2)}(\theta) = \sum_{J=|J_1-J_2|}^{J_1+J_2} \chi^{(J)}(\theta) \ . \tag{7.82}$$

Taking an appropriate linear combination of the basis $\psi_{J_1M_1}\phi_{J_2M_2}$ of the product representation, we can construct the basis Ψ_{JM} for $D^{(J)}$:

$$\Psi_{JM} = \sum_{M_1M_2} \psi_{J_1M_1}\phi_{J_2M_2} \langle J_1 M_1 J_2 M_2 | JM \rangle \ . \tag{7.83}$$

This problem is known as the addition of angular momenta $\boldsymbol{J}_1$ and $\boldsymbol{J}_2$, $\boldsymbol{J} = \boldsymbol{J}_1 + \boldsymbol{J}_2$, in quantum mechanics. The coefficients $\langle J_1 M_1 J_2 M_2 | JM \rangle$ on the right-hand side are called *Wigner coefficients* (sometimes Clebsch–Gordan or vector coupling coefficients). Since the basis functions are usually normalized, unitarity of the transformation (7.82) leads to

$$\sum_{JM} \langle J_1 M_1 J_2 M_2 | JM \rangle \langle JM | J_1 M_1' J_2 M_2' \rangle = \delta_{M_1 M_1'} \delta_{M_2 M_2'} \,, \tag{7.84a}$$

$$\sum_{M_1 M_2} \langle JM | J_1 M_1 J_2 M_2 \rangle \langle J_1 M_1 J_2 M_2 | J' M' \rangle = \delta_{JJ'} \delta_{MM'} \,, \tag{7.84b}$$

with

$$\langle JM | J_1 M_1 J_2 M_2 \rangle \equiv \langle J_1 M_1 J_2 M_2 | JM \rangle^* \,.$$

These relations of course imply

$$\psi_{J_1 M_1} \phi_{J_2 M_2} = \sum_{JM} \Psi_{JM} \langle JM | J_1 M_1 J_2 M_2 \rangle \,. \tag{7.85}$$

The Wigner coefficients may be obtained in an elementary way. We begin with the state with $J = M = J_1 + J_2$, which is given by

$$\Psi_{J_1+J_2,\,J_1+J_2} = \psi_{J_1 J_1} \phi_{J_2 J_2}$$

and operate $J_- = J_{1-} + J_{2-}$ upon this function:

$$\begin{aligned} J_- \Psi_{J_1+J_2,\,J_1+J_2} &= \sqrt{2(J_1+J_2)}\,\Psi_{J_1+J_2,\,J_1+J_2-1} \\ &= \sqrt{2J_2}\,\psi_{J_1 J_1} \phi_{J_2,\,J_2-1} + \sqrt{2J_1}\,\psi_{J_1,\,J_1-1} \phi_{J_2 J_2} \,. \end{aligned}$$

This then gives us Ψ_{JM} with $J = J_1 + J_2$, $M = J_1 + J_2 - 1$. If we choose a linear combination of $\psi_{J_1 J_1} \phi_{J_2,\,J_2-1}$ and $\psi_{J_1,\,J_1-1} \phi_{J_2 J_2}$ orthogonal to it, we obtain the basis Ψ_{JM} with $J = M = J_1 + J_2 - 1$:

$$\begin{aligned} \Psi_{J_1+J_2-1,\,J_1+J_2-1} &= \sqrt{\frac{J_1}{J_1+J_2}}\,\psi_{J_1 J_1} \phi_{J_2,\,J_2-1} \\ &\quad - \sqrt{\frac{J_2}{J_1+J_2}}\,\psi_{J_1,\,J_1-1} \phi_{J_2 J_2} \,. \end{aligned}$$

If we further operate J_- on these functions to obtain $J_- \Psi_{J_1+J_2,\,J_1+J_2-1}$ and $J_- \Psi_{J_1+J_2-1,\,J_1+J_2-1}$ and construct a linear combination of $\psi_{J_1 M_1} \phi_{J_2 M_2}$ with $M_1 + M_2 = J_1 + J_2 - 2$ orthogonal to them, the result provides us with Ψ_{JM} with $J = M = J_1 + J_2 - 2$. This process can be repeated until we reach $J = -M = |J_1 - J_2|$.[8]

We can, however, deduce a general formula for the coefficients by making use of the results obtained so far. As a matter of fact, Wigner proceeded as

[8] The Wigner coefficients obtained by this procedure agree with those given by the general formula (7.89), when we keep to the phase convention $\langle J_1 M_1 = J_1, J_2 M_2 = J_2 - k | J_1 + J_2 - k, J_1 + J_2 - k \rangle > 0$ (in the case of $J_1 \geqq J_2$).

follows to derive his formula. We calculate the matrix element

$$(\psi_{J_1M_1}\phi_{J_2M_2}, R(\psi_{J_1M'_1}\phi_{J_2M'_2}))$$

directly as well as using (7.85). We then have

$$D^{(J_1)}_{M_1M'_1}(R)D^{(J_2)}_{M_2M'_2}(R) = \sum_{JMM'} \langle J_1M_1J_2M_2|JM\rangle D^{(J)}_{MM'}(R) \times \langle JM'|J_1M'_1J_2M'_2\rangle \ . \tag{7.86}$$

Use of the orthogonality relation for the representation matrices then leads to

$$\int d\omega\, D^{(J_1)}_{M_1M'_1}(\omega)D^{(J_2)}_{M_2M'_2}(\omega)D^{(J)}_{MM'}(\omega)^* = \frac{\Omega}{2J+1}\langle J_1M_1J_2M_2|JM\rangle\langle JM'|J_1M'_1J_2M'_2\rangle \ , \tag{7.87}$$

where ω has been used for R. If we put $M_1 = M'_1 = J_1$, $M_2 = M'_2 = -J_2$ and use (7.45a, b, 48 and 72) to evaluate the integral on the left-hand side, we obtain the square of the absolute value of $\langle J_1M_1{=}J_1, J_2M_2{=}-J_2|JM{=}J_1-J_2\rangle$. Choosing its value to be real and positive, we find

$$\langle J_1J_1J_2, -J_2|J, J_1-J_2\rangle = \sqrt{\frac{(2J+1)(2J_1)!(2J_2)!}{(J_1+J_2+J+1)!(J_1+J_2-J)!}} \ . \tag{7.88}$$

We next put $M_1 = J_1$, $M_2 = -J_2$, $M = J_1 - J_2$ in (7.87) and calculate the left-hand side. Since the right-hand side is the product of (7.88) with $\langle JM'|J_1M'_1J_2M'_2\rangle$, we obtain

$$\langle J_1M_1J_2M_2|JM\rangle = \delta(M_1+M_2, M)\sqrt{2J+1}\,\Delta(J_1J_2J) \times \sqrt{\frac{(J+M)!(J-M)!}{(J_1+M_1)!(J_1-M_1)!(J_2+M_2)!(J_2-M_2)!}} \times \sum_z (-1)^{J_2+M_2+z}\frac{(J+J_2+M_1-z)!(J_1-M_1+z)!}{(J-J_1+J_2-z)!(J+M-z)!z!(z+J_1-J_2-M)!} \ , \tag{7.89}$$

where

$$\Delta(J_1J_2J) = \sqrt{\frac{(J_1+J_2-J)!(J_1+J-J_2)!(J_2+J-J_1)!}{(J_1+J_2+J+1)!}} \ . \tag{7.90}$$

The summation over z on the right-hand side of (7.89) is to be taken over nonnegative integers for which the factorials occurring there are meaningful, with $0! = 1$. Equation (7.89) can also be written as

$$\langle J_1 M_1 J_2 M_2 | JM \rangle = (-1)^{J_1 - J_2 + M} \sqrt{2J+1} \begin{pmatrix} J_1 & J_2 & J \\ M_1 & M_2 & -M \end{pmatrix}. \tag{7.91}$$

The quantity on the right-hand side is called the Wigner *3j symbol* and is given by the formula

$$\begin{pmatrix} J_1 & J_2 & J_3 \\ M_1 & M_2 & M_3 \end{pmatrix} = \delta(M_1 + M_2 + M_3, 0)\Delta(J_1 J_2 J_3)$$
$$\times \sum_z \frac{(-1)^{J_1 - J_2 - M_3 + z} \sqrt{\begin{array}{r}(J + M_1)!(J_1 - M_1)!(J_2 + M_2)!(J_2 - M_2)! \\ \times (J_3 + M_3)!(J_3 - M_3)!\end{array}}}{\begin{array}{c} z!(J_1 + J_2 - J_3 - z)!(J_1 - M_1 - z)!(J_2 + M_2 - z)! \\ \times (J_3 - J_2 + M_1 + z)!(J_3 - J_1 - M_2 + z)! \end{array}}. \tag{7.92}$$

Although the summations in (7.89) and (7.92) take different forms, each of them can be transformed into the other.

Table 7.3. Wigner coefficients for $J_2 = 1/2$ and $J_2 = 1$

$[J_2 = \frac{1}{2}]$	$\langle J_1 M - M_2 \frac{1}{2} M_2 \vert JM \rangle$	
	$M_2 = \frac{1}{2}$	$M_2 = -\frac{1}{2}$
$J = J_1 + \frac{1}{2}$	$\sqrt{\frac{J_1 + M + 1/2}{2J_1 + 1}}$	$\sqrt{\frac{J_1 - M + 1/2}{2J_1 + 1}}$
$J = J_1 - \frac{1}{2}$	$-\sqrt{\frac{J_1 - M + 1/2}{2J_1 + 1}}$	$\sqrt{\frac{J_1 + M + 1/2}{2J_1 + 1}}$

$[J_2 = 1]$	$\langle J_1 M - M_2 1 M_2 \vert JM \rangle$		
	$M_2 = 1$	$M_2 = 0$	$M_2 = -1$
$J = J_1 + 1$	$\sqrt{\frac{(J_1 + M)(J_1 + M + 1)}{(2J_1 + 1)(2J_1 + 2)}}$	$\sqrt{\frac{(J_1 - M + 1)(J_1 + M + 1)}{(2J_1 + 1)(2J_1 + 1)}}$	$\sqrt{\frac{(J_1 - M)(J_1 - M + 1)}{(2J_1 + 1)(2J_1 + 2)}}$
$J = J_1$	$-\sqrt{\frac{(J_1 + M)(J_1 - M + 1)}{2J_1(J_1 + 1)}}$	$\frac{M}{\sqrt{J_1(J_1 + 1)}}$	$\sqrt{\frac{(J_1 - M)(J_1 + M + 1)}{2J_1(J_1 + 1)}}$
$J = J_1 - 1$	$\sqrt{\frac{(J_1 - M)(J_1 - M + 1)}{2J_1(2J_1 + 1)}}$	$-\sqrt{\frac{(J_1 - M)(J_1 + M)}{J_1(2J_1 + 1)}}$	$\sqrt{\frac{(J_1 + M)(J_1 + M + 1)}{2J_1(2J_1 + 1)}}$

The $3j$ symbol has the following symmetry:

(i)
$$\begin{pmatrix} J_1 & J_2 & J_3 \\ M_1 & M_2 & M_3 \end{pmatrix} = \begin{pmatrix} J_2 & J_3 & J_1 \\ M_2 & M_3 & M_1 \end{pmatrix} = \begin{pmatrix} J_3 & J_1 & J_2 \\ M_3 & M_1 & M_2 \end{pmatrix}, \tag{7.93a}$$

(ii)
$$\begin{pmatrix} J_2 & J_1 & J_3 \\ M_2 & M_1 & M_3 \end{pmatrix} = (-1)^{J_1+J_2+J_3} \begin{pmatrix} J_1 & J_2 & J_3 \\ M_1 & M_2 & M_3 \end{pmatrix}, \tag{7.93b}$$

(iii)
$$\begin{pmatrix} J_1 & J_2 & J_3 \\ -M_1 & -M_2 & -M_3 \end{pmatrix} = (-1)^{J_1+J_2+J_3} \begin{pmatrix} J_1 & J_2 & J_3 \\ M_1 & M_2 & M_3 \end{pmatrix}. \tag{7.93c}$$

They correspond to the following properties of the Wigner coefficients:

(i)
$$\langle J_1 M_1 J_2 M_2 | JM \rangle = (-1)^{J_2+M_2} \sqrt{\frac{2J+1}{2J_1+1}} \langle J - M J_2 M_2 | J_1 - M_1 \rangle, \tag{7.94a}$$

(ii)
$$\langle J_1 M_1 J_2 M_2 | JM \rangle = (-1)^{J_1+J_2-J} \langle J_2 M_2 J_1 M_1 | JM \rangle, \tag{7.94b}$$

(iii)
$$\langle J_1 M_1 J_2 M_2 | JM \rangle = (-1)^{J_1+J_2-J} \langle J_1 - M_1 J_2 - M_2 | J - M \rangle. \tag{7.94c}$$

Expressions for the $3j$ symbol for special values of the parameters are given below. In Table 7.3, formulae for the Wigner coefficients for the cases $J_2 = 1/2$ and $J_2 = 1$ are reproduced.

$$\begin{pmatrix} J & J' & 0 \\ M & M' & 0 \end{pmatrix} = (-1)^{J-M} \frac{\delta_{JJ'}\delta_{M,-M'}}{\sqrt{2J+1}}, \tag{7.95a}$$

$$\begin{pmatrix} l & l' & k \\ 0 & 0 & 0 \end{pmatrix} = (-1)^g \Delta(ll'k) \frac{g!}{(g-l)!(g-l')!(g-k)!}, \tag{7.95b}$$

(nonvanishing only when $l + l' + k$ is even $(=2g)$),

$$\sqrt{\left(j+\frac{1}{2}\right)\left(j'+\frac{1}{2}\right)} \begin{pmatrix} j & j' & k \\ 1/2 & -1/2 & 0 \end{pmatrix}$$
$$= \begin{cases} (-1)^{g-1} \Delta(jj'k) \dfrac{g!}{\left(g-j-\frac{1}{2}\right)!\left(g-j'-\frac{1}{2}\right)!(g-k)!}, \\ \qquad (j+j'+k \text{ is even } (=2g)), \\ (-1)^g \Delta(jj'k) \dfrac{(g+1)!}{\left(g-j+\frac{1}{2}\right)!\left(g-j'+\frac{1}{2}\right)!(g-k)!}, \\ \qquad (j+j'+k \text{ is odd } (=2g+1)). \end{cases} \tag{7.95c}$$

7.11 Tensor Operators

Following the definition of operating a rotation R upon a wavefunction ψ, we define rotating an operator T as follows. Since the rotated operator T' should satisfy

$$T' R\psi = RT\psi \ ,$$

we require

$$T' = RTR^{-1} \ . \tag{7.96}$$

When the following transformation property holds for an operator $\mathbf{T}^{(k)} = \{T_q^{(k)}\}$ with $2k+1$ components ($k = 0, 1, 2, \ldots : q = -k, -k+1, \ldots, k$),

$$RT_q^{(k)}R^{-1} = \sum_{q'} T_{q'}^{(k)} D_{q'q}^{(k)}(R) \ , \tag{7.97}$$

we call $\mathbf{T}^{(k)}$ an irreducible tensor operator of degree k. The spin and orbital angular momentum operators $\boldsymbol{S}$, $\boldsymbol{L}$, the position vector $\boldsymbol{r}$ and the momentum vector $\boldsymbol{p}$ are examples of tensor operators of degree 1, or vector operators. The meaning of the rotated operator $RT_q^{(k)}R^{-1}$ is obvious. It represents the q-component of the tensor $\mathbf{T}^{(k)}$ in the rotated coordinate system K' as in the discussion on the rotation of spin functions given at the end of Sect. 7.4.

Equation (7.97) means that $T_q^{(k)}$ transforms like the qth partner of the representation $D^{(k)}$ under rotation. Therefore, as a tensor operator, the components of $\boldsymbol{L}$ are defined as

$$\begin{aligned} L_{+1}^{(1)} &= -\frac{1}{\sqrt{2}} L_+ = -\frac{1}{\sqrt{2}}(L_x + \mathrm{i}L_y) \ , \\ L_0^{(1)} &= L_z \ , \\ L_{-1}^{(1)} &= \frac{1}{\sqrt{2}} L_- = \frac{1}{\sqrt{2}}(L_x - \mathrm{i}L_y) \ . \end{aligned} \tag{7.98}$$

Another example is the quadrupole moment operator

$$Q_{\alpha\beta} = e(x_\alpha x_\beta - \tfrac{1}{3}r^2\delta_{\alpha\beta}) \ , \quad \alpha, \beta = x, y, z \ ,$$

which is related to the irreducible tensor operator of degree 2

$$\mathbf{Q}^{(2)} = er^2\mathbf{C}^{(2)}(\theta, \varphi)$$

by

$$Q_0^{(2)} = \tfrac{3}{2} Q_{zz} \ , \tag{7.99}$$

where the component $C_q^{(k)}(\theta, \varphi)$ of the tensor $\mathbf{C}^{(k)}(\theta, \varphi)$ is defined by the spherical harmonics $Y_{kq}(\theta, \varphi)$ times $\sqrt{4\pi/(2k+1)}$, see (7.63).

If we choose an infinitesimal rotation $1 - i\varepsilon J_\alpha$ about the α-axis for R in (7.97), we obtain commutation relations between J_α and $T_q^{(k)}$:

$$[J_\alpha, T_q^{(k)}] = \sum_{q'} T_{q'}^{(k)} (\hat{J}_\alpha)_{q'q} ,$$

or

$$[J_z, T_q^{(k)}] = qT_q^{(k)} ,$$

$$[J_\pm, T_q^{(k)}] = \sqrt{(k \mp q)(k \pm q + 1)}\, T_{q\pm 1}^{(k)} . \tag{7.100}$$

Irreducible tensor operators may also be defined through these commutation relations.

If we operate $T_q^{(k)}$ upon $\psi_{\alpha' J' M'}$, the result will transform under rotation like the basis of the product representation $D^{(J')} \times D^{(k)}$. This means that the functions

$$\Psi(k\alpha' J' JM) = \sum_{qM'} T_q^{(k)} \psi_{\alpha' J' M'} \langle J' M' kq | JM \rangle$$

will transform like the basis of the irreducible representation $D^{(J)}$. Solving for $T_q^{(k)} \psi$, we find

$$T_q^{(k)} \psi_{\alpha' J' M'} = \sum_{J_0 M_0} \Psi(k\alpha' J' J_0 M_0) \langle J_0 M_0 | J' M' kq \rangle .$$

The inner product of this function with $\psi_{\alpha JM}$, or the matrix element of $T_q^{(k)}$, is then written as

$$\langle \alpha JM | T_q^{(k)} | \alpha' J' M' \rangle \equiv (\psi_{\alpha JM}, T_q^{(k)} \psi_{\alpha' J' M'})$$

$$= \sum_{J_0 M_0} (\psi_{\alpha JM}, \Psi(k\alpha' \hat{J}' J_0 M_0)) \langle J_0 M_0 | J' M' kq \rangle . \tag{7.101}$$

The inner product $(\psi_{\alpha JM}, \Psi(k\alpha' J' J_0 M_0))$ appearing on the right-hand side vanishes unless $J = J_0$ and $M = M_0$, and the value for $M = M_0$ is independent of M. These follow from the fact that both ψ and Ψ are eigenfunctions of $\boldsymbol{J}^2$ and J_z and satisfy

$$\psi_{\alpha JM} = \frac{J_+ \psi_{\alpha J, M-1}}{\sqrt{(J+M)(J-M+1)}} ,$$

$$J_- \Psi(\ldots JM) = \sqrt{(J+M)(J-M+1)}\, \Psi(\ldots J, M-1) .$$

We may therefore put

$$(\psi_{\alpha JM}, \Psi(k\alpha' J' J_0 M_0)) = \delta_{JJ_0} \delta_{MM_0} \frac{\langle \alpha J \| \mathbf{T}^{(k)} \| \alpha' J' \rangle}{\sqrt{2J+1}} .$$

Then (7.101) gives the *Wigner–Eckart theorem*:

$$\langle \alpha JM|T_q^{(k)}|\alpha' J'M'\rangle = \langle \alpha J\|\mathbf{T}^{(k)}\|\alpha' J'\rangle (2J+1)^{-1/2}\langle JM|J'M'kq\rangle$$

$$= \langle \alpha J\|\mathbf{T}^{(k)}\|\alpha' J'\rangle (-1)^{J'-k+M}\begin{pmatrix} J & J' & k \\ -M & M' & q \end{pmatrix} . \qquad (7.102)$$

The calculation of the matrix elements of the tensor operators thus reduces to the evaluation of the geometrical factor (namely, the Wigner coefficient) that depends on M, M' and q and the physical factor that is independent of them. The latter factor $\langle \alpha J\|\mathbf{T}^{(k)}\|\alpha' J'\rangle$ is called the reduced matrix element. In practice, it is convenient to evaluate the matrix element on the left-hand side of (7.102) for some particular combination of (M, q, M') and divide it by the corresponding Wigner coefficient on the right-hand side to obtain the value of the reduced matrix element. Then the values of the matrix elements for any other combinations of (M, q, M') can be found from those of the Wigner coefficients.

We find that the reduced matrix elements have the following property. We first examine the commutation relation like (7.100) that $T_q^{(k)\dagger}$, the Hermitian conjugate of $T_q^{(k)}$, obeys and find that it transforms like $(-1)^q T_{-q}^{(k)}$. Let us therefore define the tensor $\mathbf{T}^{(k)\dagger}$, the Hermitian conjugate of $\mathbf{T}^{(k)}$, by

$$(\mathbf{T}^{(k)\dagger})_q = (-1)^q T_{-q}^{(k)\dagger} . \qquad (7.103)$$

It then follows that

$$\langle \alpha J\|\mathbf{T}^{(k)}\|\alpha' J'\rangle = (-1)^{J-J'}\langle \alpha' J'\|\mathbf{T}^{(k)\dagger}\|\alpha J\rangle^* . \qquad (7.104)$$

In particular, for a Hermitian tensor $\mathbf{T}^{(k)\dagger} = \mathbf{T}^{(k)}$, we have

$$T_q^{(k)\dagger} = (-1)^q T_{-q}^{(k)} \qquad (7.105)$$

and

$$\langle \alpha J\|\mathbf{T}^{(k)}\|\alpha' J'\rangle = (-1)^{J-J'}\langle \alpha' J'\|\mathbf{T}^{(k)}\|\alpha J\rangle^* . \qquad (7.106)$$

We give below expressions for reduced matrix elements for several important operators.

(i) Scalar operator $T_0^{(0)} = V(r)$:
Putting $\psi_{\alpha lm} = R_{\alpha l}(r) Y_{lm}(\theta, \varphi)$,

$$\langle \alpha l\|V\|\alpha' l'\rangle = \delta_{ll'}\sqrt{2l+1}\,(V)_{\alpha l, \alpha' l'} , \qquad (7.107)$$

$$(V)_{\alpha l, \alpha' l'} = \int_0^\infty r^2 dr R_{\alpha l}(r)^* V(r) R_{\alpha' l'}(r) .$$

(ii) Position and momentum vectors:

$$\langle \alpha l \| \boldsymbol{r} \| \alpha' l' \rangle = \langle l \| \mathbf{C}^{(1)} \| l' \rangle (r)_{\alpha l, \alpha' l'} , \tag{7.108a}$$

$$\langle \alpha l \| \boldsymbol{p} \| \alpha' l' \rangle = \langle l \| \mathbf{C}^{(1)} \| l' \rangle (p)_{\alpha l, \alpha' l'} ,$$

$$(p)_{\alpha l, \alpha' l+1} = -\mathrm{i}\hbar \left(\frac{d}{dr} + \frac{l+2}{r} \right)_{\alpha l, \alpha' l+1} , \tag{7.108b}$$

$$(p)_{\alpha l, \alpha' l-1} = -\mathrm{i}\hbar \left(\frac{d}{dr} - \frac{l-1}{r} \right)_{\alpha l, \alpha' l-1} .$$

(iii) Angular momentum:

$$\langle \alpha l \| \boldsymbol{l} \| \alpha' l' \rangle = \delta_{\alpha\alpha'} \delta_{ll'} \sqrt{l(l+1)(2l+1)} . \tag{7.109}$$

Table 7.4. $c^k(lm, l'm')$ for $l, l' \leqq 3$, see (7.111)

l	l'	m	m'	c^1	c^3	c^5
s	p	0	± 1	$-1/\sqrt{3}$	0	
		0	0	$+1/\sqrt{3}$	0	
p	d	± 1	± 2	$-\sqrt{6}/\sqrt{15}$	$+\sqrt{3}/7\sqrt{5}$	
		± 1	± 1	$+\sqrt{3}/\sqrt{15}$	$-3/7\sqrt{5}$	
		± 1	0	$-1/\sqrt{15}$	$+3\sqrt{2}/7\sqrt{5}$	
		0	± 2	0	$+\sqrt{15}/7\sqrt{5}$	
		0	± 1	$-\sqrt{3}/\sqrt{15}$	$-2\sqrt{6}/7\sqrt{5}$	
		0	0	$+2/\sqrt{15}$	$+3\sqrt{3}/7\sqrt{5}$	
		± 1	∓ 2	0	$+3\sqrt{5}/7\sqrt{5}$	
		± 1	∓ 1	0	$-\sqrt{30}/7\sqrt{5}$	
s	f	0	± 3	0	$-1/\sqrt{7}$	
		0	± 2	0	$+1/\sqrt{7}$	
		0	± 1	0	$-1/\sqrt{7}$	
		0	0	0	$+1/\sqrt{7}$	

Table 7.4 (continued)

l	l'	m	m'	c^1	c^3	c^5
d	f	± 2	± 3	$-\sqrt{15}/\sqrt{35}$	$+\sqrt{10}/3\sqrt{35}$	$-5/33\sqrt{35}$
		± 2	± 2	$+\sqrt{5}/\sqrt{35}$	$-2\sqrt{5}/3\sqrt{35}$	$+5\sqrt{5}/33\sqrt{35}$
		± 2	± 1	$-1/\sqrt{35}$	$+2\sqrt{6}/3\sqrt{35}$	$-5\sqrt{15}/33\sqrt{35}$
		± 2	0	0	$-2\sqrt{5}/3\sqrt{35}$	$+5\sqrt{35}/33\sqrt{35}$
		± 1	± 3	0	$+5/3\sqrt{35}$	$-5\sqrt{7}/33\sqrt{35}$
		± 1	± 2	$-\sqrt{10}/\sqrt{35}$	$-\sqrt{15}/3\sqrt{35}$	$+10\sqrt{6}/33\sqrt{35}$
		± 1	± 1	$+2\sqrt{2}/\sqrt{35}$	$+\sqrt{2}/3\sqrt{35}$	$-25\sqrt{2}/33\sqrt{35}$
		± 1	0	$-\sqrt{3}/\sqrt{35}$	$+\sqrt{2}/3\sqrt{35}$	$+20\sqrt{5}/33\sqrt{35}$
		0	± 3	0	$+5/3\sqrt{35}$	$-10\sqrt{7}/33\sqrt{35}$
		0	± 2	0	0	$+15\sqrt{7}/33\sqrt{35}$
		0	± 1	$-\sqrt{6}/\sqrt{35}$	$-3/3\sqrt{35}$	$-15\sqrt{10}/33\sqrt{35}$
		0	0	$+3/\sqrt{35}$	$+4/3\sqrt{35}$	$+50/33\sqrt{35}$
		± 2	∓ 3	0	0	$-5\sqrt{210}/33\sqrt{35}$
		± 2	∓ 2	0	0	$+15\sqrt{14}/33\sqrt{35}$
		± 2	∓ 1	0	$+\sqrt{10}/3\sqrt{35}$	$-5\sqrt{70}/33\sqrt{35}$
		± 1	∓ 3	0	0	$-10\sqrt{21}/33\sqrt{35}$
		± 1	∓ 2	0	$+5/3\sqrt{35}$	$+20\sqrt{7}/33\sqrt{35}$
		± 1	∓ 1	0	$-\sqrt{15}/3\sqrt{35}$	$-5\sqrt{105}/33\sqrt{35}$

l	l'	m	m'	c^0	c^2	c^4	c^6
s	s	0	0	$+1$			
p	p	± 1	± 1	$+1$	$-1/5$		
		± 1	0	0	$+\sqrt{3}/5$		
		0	0	$+1$	$+2/5$		
		± 1	∓ 1	0	$-\sqrt{6}/5$		

Table 7.4 (continued)

l	l'	m	m'	c^0	c^2	c^4	c^6
s	d	0	± 2	0	$1/\sqrt{5}$		
		0	± 1	0	$-1/\sqrt{5}$		
		0	0	0	$1/\sqrt{5}$		
d	d	± 2	± 2	1	$-2/7$	$+1/21$	
		± 2	± 1	0	$+\sqrt{6}/7$	$-\sqrt{5}/21$	
		± 2	0	0	$-2/7$	$+\sqrt{15}/21$	
		± 1	± 1	1	$+1/7$	$-4/21$	
		± 1	0	0	$+1/7$	$+\sqrt{30}/21$	
		0	0	1	$+2/7$	$+6/21$	
		± 2	∓ 2	0	0	$+\sqrt{70}/21$	
		± 2	∓ 1	0	0	$-\sqrt{35}/21$	
		± 1	∓ 1	0	$-\sqrt{6}/7$	$-2\sqrt{10}/21$	
p	f	± 1	± 3	0	$+3\sqrt{5}/5\sqrt{7}$	$-1/3\sqrt{21}$	
		± 1	± 2	0	$-\sqrt{30}/5\sqrt{7}$	$+\sqrt{3}/3\sqrt{21}$	
		± 1	± 1	0	$+3\sqrt{2}/5\sqrt{7}$	$-\sqrt{6}/3\sqrt{21}$	
		± 1	0	0	$-3/5\sqrt{7}$	$+\sqrt{10}/3\sqrt{21}$	
		0	± 3	0	0	$-\sqrt{7}/3\sqrt{21}$	
		0	± 2	0	$+\sqrt{15}/5\sqrt{7}$	$+2\sqrt{3}/3\sqrt{21}$	
		0	± 1	0	$-2\sqrt{6}/5\sqrt{7}$	$-\sqrt{15}/3\sqrt{21}$	
		0	0	0	$+3\sqrt{3}/5\sqrt{7}$	$+4/3\sqrt{21}$	
		± 1	∓ 3	0	0	$-2\sqrt{7}/3\sqrt{21}$	
		± 1	∓ 2	0	0	$+\sqrt{21}/3\sqrt{21}$	
		± 1	∓ 1	0	$+\sqrt{3}/5\sqrt{7}$	$-\sqrt{15}/3\sqrt{21}$	

Table 7.4 (continued)

l	l'	m	m'	c^0	c^2	c^4	c^6
f	f	± 3	± 3	1	$-5/15$	$+3/33$	$-5/429$
		± 3	± 2	0	$+5/15$	$-\sqrt{30}/33$	$+5\sqrt{7}/429$
		± 3	± 1	0	$-\sqrt{10}/15$	$+3\sqrt{6}/33$	$-10\sqrt{7}/429$
		± 3	0	0	0	$-3\sqrt{7}/33$	$+10\sqrt{21}/429$
		± 2	± 2	1	0	$-7/33$	$+30/429$
		± 2	± 1	0	$+\sqrt{15}/15$	$+4\sqrt{2}/33$	$-5\sqrt{105}/429$
		± 2	0	0	$-2\sqrt{5}/15$	$-\sqrt{3}/33$	$+20\sqrt{14}/429$
		± 1	± 1	1	$+3/15$	$+1/33$	$-75/429$
		± 1	0	0	$+\sqrt{2}/15$	$+\sqrt{15}/33$	$+25\sqrt{14}/429$
		0	0	1	$+4/15$	$+6/33$	$+100/429$
		± 3	∓ 3	0	0	0	$-10\sqrt{231}/429$
		± 3	∓ 2	0	0	0	$+5\sqrt{462}/429$
		± 3	∓ 1	0	0	$+\sqrt{42}/33$	$-5\sqrt{210}/429$
		± 2	∓ 2	0	0	$+\sqrt{70}/33$	$+30\sqrt{14}/429$
		± 2	∓ 1	0	0	$-\sqrt{14}/33$	$-15\sqrt{42}/429$
		± 1	∓ 1	0	$-2\sqrt{6}/15$	$-2\sqrt{10}/33$	$-10\sqrt{105}/429$

(iv) Spherical harmonics:

$$\begin{aligned}\langle l \| \mathbf{C}^{(k)} \| l' \rangle &= \sqrt{2l'+1}\,\langle l0|l'0k0\rangle \\ &= (-1)^l \sqrt{(2l+1)(2l'+1)} \begin{pmatrix} l & l' & k \\ 0 & 0 & 0 \end{pmatrix}, \\ &\quad (l + l' + k \text{ even})\end{aligned} \tag{7.110}$$

with

$$\begin{aligned}
&\langle lm|C^{(k)}_{m-m'}|l'm'\rangle \\
&\quad = \iint \sin\theta d\theta d\varphi\, Y_{lm}(\theta,\varphi)^* C^{(k)}_{m-m'}(\theta,\varphi)\, Y_{l'm'}(\theta,\varphi) \\
&\quad = \langle l\|\mathbf{C}^{(k)}\|l'\rangle(2l+1)^{-1/2}\langle lm|l'm'k, m-m'\rangle \\
&\quad = c^k(lm, l'm') = (-1)^{m-m'}c^k(l'm', lm)\ . \qquad (7.111)
\end{aligned}$$

Numerical tables of $c^k(lm, l'm')$ are given in Table 7.4.

According to (iv), the following formula holds for spherical harmonics:

$$\langle l_1 0 l_2 0|l0\rangle C^{(l)}_m(\theta,\varphi) = \sum_{m_1 m_2} C^{(l_1)}_{m_1}(\theta,\varphi)C^{(l_2)}_{m_2}(\theta,\varphi)\langle l_1 m_1 l_2 m_2|lm\rangle\ . \qquad (7.112)$$

This can be proved by setting $R = R(\varphi, \theta, 0)$ in (7.86), taking the complex conjugate of both sides and making use of (7.63), together with (7.84).

7.12 Operator Equivalents

According to the Wigner–Eckart theorem (7.102), within the manifold of definite J any vector operator $\boldsymbol{T}$ may be replaced by $c\boldsymbol{J}$, the angular momentum operator times a constant c, as

$$\langle \alpha JM|T_\xi|\alpha' JM'\rangle = c\langle JM|J_\xi|JM'\rangle\ . \qquad (7.113)$$

The constant c depends on α, α', J and $\boldsymbol{T}$ but not on M, M' and $\xi(=x, y, z)$, and may be expressed as

$$c = \frac{\langle \alpha J\|\boldsymbol{T}\|\alpha' J\rangle}{\sqrt{J(J+1)(2J+1)}} \qquad (7.114)$$

in terms of the reduced matrix element. The operator $c\boldsymbol{J}$ is called an *operator equivalent* to $\boldsymbol{T}$ (in the manifold of definite J).

The idea of operator equivalents finds applications in many branches of physics. As a most elementary example, consider the effect of the spin-orbit interaction on the multiplet term $\alpha^{2S+1}L$. First of all, the spin-orbit interaction

$$H_{\mathrm{SO}} = \sum_{i=1}^{x} \xi(r_i)\boldsymbol{l}_i\cdot\boldsymbol{s}_i \qquad (7.115)$$

can be replaced by

$$H_{\mathrm{SO}} = \zeta_{nl}\sum_{i=1}^{x} \boldsymbol{l}_i\cdot\boldsymbol{s}_i\ , \quad \zeta_{nl} = \int_0^\infty r^2 dr R_{nl}(r)^2\xi(r) \qquad (7.116)$$

within the electron configuration $(nl)^x$. This may further be replaced by

$$H_{\text{SO}} = \lambda \boldsymbol{S} \cdot \boldsymbol{L} \tag{7.117}$$

within the manifold with definite S and L. The splitting of the multiplet term due to the spin-orbit interaction and the corresponding eigenfunctions can be obtained by diagonalizing this simple Hamiltonian. The parameter λ is given by

$$\lambda = \frac{\langle l^x \alpha SL \| \zeta_{nl} \sum_{i=1}^{x} \boldsymbol{l}_i \cdot \boldsymbol{s}_i \| l^x \alpha' SL \rangle}{\sqrt{S(S+1)(2S+1)L(L+1)(2L+1)}} .$$

As mentioned before, the value of the reduced matrix element can be obtained by evaluating the matrix element of (7.116) for an appropriate combination of M_S, M_L, and M'_S, M'_L.

Naturally, we can consider operator equivalents also for tensor operators of higher degrees. For example,

$$\langle \alpha LM | \sum_i r_i^k C_q^{(k)}(\theta_i, \varphi_i) | \alpha' LM' \rangle = c \langle LM | C_q^{(k)}(\boldsymbol{L}) | LM' \rangle , \tag{7.118}$$

where $C_q^{(k)}(\boldsymbol{L})$ appearing on the right-hand side is the q component of the irreducible tensor operator $\mathbf{C}^{(k)}(\boldsymbol{L})$ of degree k defined as follows. The component with $q = k$ is given by[9]

$$C_k^{(k)}(\boldsymbol{L}) = (-1)^k \sqrt{\frac{(2k-1)!!}{(2k)!!}} \, (L_+)^k . \tag{7.119a}$$

For $q < k$, the components are obtained from

$$[L_-, C_q^{(k)}(\boldsymbol{L})] = \sqrt{(k+q)(k-q+1)} \, C_{q-1}^{(k)}(\boldsymbol{L}) . \tag{7.119b}$$

This tensor may also be obtained from the tensor $r^k \mathbf{C}^{(k)}(\theta, \varphi)$ by expressing it in terms of x, y and z, replacing x, y and z by L_x, L_y and L_z, respectively, and then symmetrizing the resulting expression, so that, for example, the product xy is replaced by $(L_x L_y + L_y L_x)/2$. The reduced matrix element of this tensor is given by

$$\langle L \| \mathbf{C}^{(k)}(\boldsymbol{L}) \| L \rangle = \frac{k!}{2^k \Delta(LLk)} = 2^{-k} \sqrt{\frac{(2L+k+1)!}{(2L-k)!}} . \tag{7.120}$$

As a simple example, let us construct an operator equivalent in the state with nuclear spin $I (\geqq 1)$ for the nuclear quadrupole moment operator

$$Q_{\alpha\beta} = |e| \sum_n \left(x_{n\alpha} x_{n\beta} - \frac{r_n^2 \delta_{\alpha\beta}}{3} \right) , \quad \alpha, \beta = x, y, z , \tag{7.121}$$

---

[9] $(2k)!! = 2k(2k-2) \cdots 2$, $(2k-1)!! = (2k-1)(2k-3) \cdots 3 \cdot 1$.

where $x_{n\alpha}$ is the α coordinate of the nth proton. The operator equivalent corresponding to

$$Q_{zz} = \tfrac{2}{3} Q_0^{(2)}$$

is given by

$$= \tfrac{2}{3} c C_0^{(2)}(I) = c(I_z^2 - \tfrac{1}{3} I^2)$$

with

$$c = \frac{\langle I \| \mathbf{Q}^{(2)} \| I \rangle}{\langle I \| \mathbf{C}^{(2)}(I) \| I \rangle} = \frac{Q}{I(2I-1)} . \qquad (7.122a)$$

The constant Q is defined by

$$\langle I, M = I | Q_{zz} | I, M = I \rangle = \frac{1}{3} Q = \int dr \left(z^2 - \frac{1}{3} r^2 \right) [\varrho_n(r)]_{M=I} , \qquad (7.122b)$$

using the nuclear charge distribution $[\varrho_n(r)]_{M=I}$ in the state M (the eigenvalue of I_z) $= I$. The operator equivalent to $Q_{\alpha\beta}$ is thus found to be

$$Q_{\alpha\beta} = \frac{Q}{I(2I-1)} \left\{ \frac{1}{2} (I_\alpha I_\beta + I_\beta I_\alpha) - \delta_{\alpha\beta} \frac{1}{3} I(I+1) \right\} . \qquad (7.123)$$

7.13 Addition of Three Angular Momenta; Racah Coefficients

The addition of three angular momenta J_1, J_2 and J_3 is carried out successively:

$$J_1 + J_2 = J_{12} , \quad J_{12} + J_3 = J .$$

This corresponds to the reduction of the product representation $D^{(J_1)} \times D^{(J_2)} \times D^{(J_3)}$ in two steps:

$$D^{(J_1)} \times D^{(J_2)} = \sum_{J_{12}} D^{(J_{12})} , \quad D^{(J_{12})} \times D^{(J_3)} = \sum_{J} D^{(J)} .$$

In terms of the basis of the representations,

$$\psi(J_1 J_2 J_{12} M_{12}) = \sum_{M_1 M_2} \psi_{J_1 M_1} \psi_{J_2 M_2} \langle J_1 M_1 J_2 M_2 | J_{12} M_{12} \rangle ,$$

$$\Psi(J_1 J_2 (J_{12}) J_3 J M) = \sum_{M_{12} M_3} \psi(J_1 J_2 J_{12} M_{12}) \psi_{J_3 M_3} \times \langle J_{12} M_{12} J_3 M_3 | J M \rangle , \qquad (7.124)$$

where J_1, J_2, J_{12} and J_{12}, J_3, J have to satisfy the triangular conditions

$$|J_1 - J_2| \leqq J_{12} \leqq J_1 + J_2 \ , \quad |J_{12} - J_3| \leqq J \leqq J_{12} + J_3 \ .$$

Now the addition may be carried out in another order as

$$\boldsymbol{J}_2 + \boldsymbol{J}_3 = \boldsymbol{J}_{23} \ , \quad \boldsymbol{J}_1 + \boldsymbol{J}_{23} = \boldsymbol{J} \ .$$

The basis functions corresponding to this process are given by

$$\psi(J_2 J_3 J_{23} M_{23}) = \sum_{M_2 M_3} \psi_{J_2 M_2} \psi_{J_3 M_3} \langle J_2 M_2 J_3 M_3 | J_{23} M_{23} \rangle \ ,$$

$$\begin{aligned} &\Psi(J_1, J_2 J_3 (J_{23}) JM) \\ &\quad = \sum_{M_1 M_{23}} \psi_{J_1 M_1} \psi(J_2 J_3 J_{23} M_{23}) \langle J_1 M_1 J_{23} M_{23} | JM \rangle \ , \end{aligned} \tag{7.125}$$

where

$$|J_2 - J_3| \leqq J_{23} \leqq J_2 + J_3 \ , \quad |J_1 - J_{23}| \leqq J \leqq J_1 + J_{23} \ .$$

The two sets of bases (7.124) and (7.125) are both eigenfunctions of $\boldsymbol{J}^2$ and J_z, and those with the same J and M are related to each other by a unitary transformation:

$$\begin{aligned} \Psi(J_1, J_2 J_3 (J_{23}) JM) = &\sum_{J_{12}} \Psi(J_1 J_2 (J_{12}) J_3 JM) \\ &\times \langle J_1 J_2 (J_{12}) J_3 J | J_1, J_2 J_3 (J_{23}) J \rangle \ . \end{aligned}$$

The unitary transformation matrix is independent of M and is given by

$$\begin{aligned} &(\Psi(J_1 J_2 (J_{12}) J_3 JM), \Psi(J_1, J_2 J_3 (J_{23}) J'M')) \\ &\quad = \delta_{JJ'} \delta_{MM'} \langle J_1 J_2 (J_{12}) J_3 J | J_1, J_2 J_3 (J_{23}) J \rangle \\ &\quad = \sum_{\substack{M_1 M_2 M_3 \\ M_{12} M_{23}}} \langle JM | J_{12} M_{12} J_3 M_3 \rangle \langle J_{12} M_{12} | J_1 M_1 J_2 M_2 \rangle \\ &\qquad \times \langle J_2 M_2 J_3 M_3 | J_{23} M_{23} \rangle \langle J_1 M_1 J_{23} M_{23} | J'M' \rangle \ . \end{aligned} \tag{7.126}$$

We put this transformation matrix in the form

$$\begin{aligned} \langle J_1 J_2 (J_{12}) J_3 J | J_1, J_2 J_3 (J_{23}) J \rangle = &\sqrt{(2J_{12} + 1)(2J_{23} + 1)} \\ &\times W(J_1 J_2 J J_3; J_{12} J_{23}) \end{aligned} \tag{7.127}$$

and call $W(abcd; ef)$ on the right-hand side the *Racah coefficient* (or W coefficient). Although the derivation is rather tedious, the following expression for the Racah coefficient can be obtained from the definition (7.126) through

Wigner coefficients:

$$W(abcd;ef) = \Delta(abe)\Delta(cde)\Delta(acf)\Delta(bdf) \times \sum_z \frac{(-1)^z(a+b+c+d+1-z)!}{(a+b-e-z)!(c+d-e-z)!(a+c-f-z)!(b+d-f-z)!(e+f-a-d+z)!(e+f-b-c+z)!z!}\,, \tag{7.128}$$

where the summation over z is to be taken over all non-negative integers for which the factorials make sense. Wigner's *6j symbol* defined by

$$\begin{Bmatrix} a & b & e \\ d & c & f \end{Bmatrix} = (-1)^{a+b+c+d}\, W(abcd;ef)$$

is also sometimes used instead of the Racah coefficient [7.2].

The W coefficients have the following symmetry for the permutations of their arguments:

$$\begin{aligned} W(abcd;ef) &= W(badc;ef) = W(cdab;ef) = W(acbd;fe) \\ &= (-1)^{a+d-e-f}\, W(ebcf;ad) = (-1)^{b+c-e-f}\, W(aefd;bc)\,. \end{aligned} \tag{7.129}$$

From the unitarity of the transformation matrix, we have

$$\begin{aligned} \sum_e (2e+1)\, W(abcd;ef)\, W(abcd;ef') &= \frac{\delta_{ff'}}{2f+1}\,, \\ \sum_f (2f+1)\, W(abcd;ef)\, W(abcd;e'f) &= \frac{\delta_{ee'}}{2e+1}\,. \end{aligned} \tag{7.130}$$

As in (7.126), the transformation between the functions (7.124) and $\Psi(J_1J_3(J_{13})J_2JM)$ can also be expressed by W:

$$\begin{aligned} &\langle J_1J_2(J_{12})J_3J|J_1J_3(J_{13})J_2J\rangle \\ &= \sqrt{(2J_{12}+1)(2J_{13}+1)}\, W(J_2J_{12}J_{13}J_3;J_1J)\,. \end{aligned} \tag{7.131}$$

Similarly, we have

$$\begin{aligned} &\langle J_1J_3(J_{13})J_2J|J_1,J_2J_3(J_{23})J\rangle \\ &= (-1)^{J_2+J_3-J_{23}}\sqrt{(2J_{13}+1)(2J_{23}+1)}\, W(J_1J_3JJ_2;J_{13}J_{23})\,. \end{aligned} \tag{7.132}$$

The matrix multiplication of (7.131) with (7.132) must be equal to (7.127). This leads to

$$\sum_e (-1)^{a+b-e}(2e+1)\, W(abcd;ef)\, W(bacd;ef') = W(aff'b;cd) \tag{7.133}$$

in terms of the W coefficients.

Another identity between the W coefficients is

$$\sum_{\lambda}(2\lambda+1)W(a'\lambda be;ae')W(a'\lambda fc;ac')W(c\lambda de';c'e)$$

$$= W(abcd;ef)W(a'bc'd;e'f) \ . \tag{7.134}$$

This can be derived by considering transformations between the bases obtained by coupling four angular momenta in different ways. The details are, however, left to the reader as an exercise.

The Racah coefficients play important roles also in the calculation of the matrix elements of tensor operators.

Suppose the system under consideration consists of two parts 1 and 2 and the tensor operators $\mathbf{T}^{(k)}$ and $\mathbf{U}^{(k)}$ act on 1 and 2, respectively. We define the scalar product of $\mathbf{T}^{(k)}$ and $\mathbf{U}^{(k)}$ by

$$\mathbf{T}^{(k)}\cdot\mathbf{U}^{(k)} = \sum_{q=-k}^{k}(-1)^q T_q^{(k)}U_{-q}^{(k)} \ , \tag{7.135}$$

which reduces to the ordinary scalar product of vectors when $k=1$. A familiar example of the scalar product of irreducible tensors is provided by

$$\mathbf{C}^{(k)}(\theta_1\varphi_1)\cdot\mathbf{C}^{(k)}(\theta_2\varphi_2) = P_k(\cos\Theta) \ . \tag{7.136}$$

By the addition theorem, the left-hand side is equal to the Legendre polynomial of degree k whose argument is equal to the cosine of the angle Θ between the two directions (θ_1, φ_1) and (θ_2, φ_2).

The matrix elements of such a scalar product connecting the states $\psi(\gamma_1\gamma_2J_1J_2JM)$ are calculated as

$$\langle\gamma_1\gamma_2J_1J_2JM|\mathbf{T}^{(k)}\cdot\mathbf{U}^{(k)}|\gamma_1'\gamma_2'J_1'J_2'J'M'\rangle$$

$$= \delta_{JJ'}\delta_{MM'}\sum_{\substack{qM_1M_2\\M_1'M_2'}}(-)^q\langle JM|J_1M_1J_2M_2\rangle\langle\gamma_1J_1M_1|T_q^{(k)}|\gamma_1'J_1'M_1'\rangle$$

$$\times\langle\gamma_2J_2M_2|U_{-q}^{(k)}|\gamma_2'J_2'M_2'\rangle\langle J_1'M_1'J_2'M_2'|JM\rangle \ .$$

We apply here the Wigner–Eckart theorem (7.102), make use of the symmetry (7.94a, b, c) of the Wigner coefficients, and compare the result with the definition (7.126, 7) of the Racah coefficients. We then find that the expression above reduces to

$$= \delta_{JJ'}\delta_{MM'}(-1)^k W(J_1'kJJ_2;J_1J_2')\langle\gamma_1J_1\|\mathbf{T}^{(k)}\|\gamma_1'J_1'\rangle$$

$$\times\langle\gamma_2J_2\|\mathbf{U}^{(k)}\|\gamma_2'J_2'\rangle \ . \tag{7.137}$$

It is interesting to apply (7.137) in the case of operator equivalents, where we put

$$\mathbf{T}^{(k)} = \mathbf{C}^{(k)}(J_1) , \quad \mathbf{U}^{(k)} = \mathbf{C}^{(k)}(J_2)$$

and define

$$P_k(J_1 \cdot J_2) \equiv \mathbf{C}^{(k)}(J_1) \cdot \mathbf{C}^{(k)}(J_2) \tag{7.138}$$

after (7.136). The matrix element of this operator Legendre polynomial reads

$$\langle J_1 J_2 JM | P_k(J_1 \cdot J_2) | J_1 J_2 J' M' \rangle$$
$$= \delta_{JJ'} \delta_{MM'} (-1)^k W(J_1 k J J_2; J_1 J_2) \langle J_1 \| \mathbf{C}^{(k)}(J_1) \| J_1 \rangle \langle J_2 \| \mathbf{C}^{(k)}(J_2) \| J_2 \rangle$$

according to (7.137). The operator Legendre polynomials are defined step by step by the recurrence relation

$$(2k+1)\left(J_1 \cdot J_2 + \frac{k(k+1)}{4} \right) P_k(J_1 \cdot J_2)$$
$$= (k+1) P_{k+1}(J_1 \cdot J_2) + k\left(J_1(J_1+1) - \frac{k^2-1}{4} \right)\left(J_2(J_2+1) - \frac{k^2-1}{4} \right) P_{k-1}(J_1 \cdot J_2) .$$

For example, for small values of k,

$$P_0(J_1 \cdot J_2) = 1 , \quad P_1(J_1 \cdot J_2) = J_1 \cdot J_2 ,$$
$$P_2(J_1 \cdot J_2) = \tfrac{3}{2}(J_1 \cdot J_2)(J_1 \cdot J_2 + \tfrac{1}{2}) - \tfrac{1}{2} J_1(J_1+1) J_2(J_2+1) .$$

When J_1, J_2 tend to infinity, we have

$$P_k(J_1 \cdot J_2) \sim P_k(\cos\theta_{12}) \frac{\langle J_1 \| \mathbf{C}^{(k)}(J_1) \| J_1 \rangle \langle J_2 \| \mathbf{C}^{(k)}(J_2) \| J_2 \rangle}{\sqrt{(2J_1+1)(2J_2+1)}} ,$$

θ_{12} being the angle between J_1 and J_2, so that

$$(-1)^k \sqrt{(2J_1+1)(2J_2+1)}\, W(J_1 k J J_2; J_1 J_2) \sim P_k(\cos\theta_{12})$$

in this limit.

In a similar fashion, we can evaluate reduced matrix elements of the operator $\mathbf{T}^{(k)}$, which commutes with J_2, and of $\mathbf{U}^{(k)}$, which commutes with J_1:

$$\langle \gamma J_1 J_2 J \| \mathbf{T}^{(k)} \| \gamma' J_1' J_2 J' \rangle$$
$$= \langle \gamma J_1 \| \mathbf{T}^{(k)} \| \gamma' J_1' \rangle \sqrt{(2J+1)(2J'+1)}\, W(J_1 J_2 k J'; J J_1') , \tag{7.139a}$$

$$\langle \gamma J_1 J_2 J \| \mathbf{U}^{(k)} \| \gamma' J_1 J_2' J' \rangle$$
$$= (-1)^{J+J_2'-J'-J_2} \langle \gamma J_2 \| \mathbf{U}^{(k)} \| \gamma' J_2' \rangle \sqrt{(2J+1)(2J'+1)}\, W(J_2 J_1 k J'; J J_2') . \tag{7.139b}$$

We will attempt later (Sect. 7.16) to calculate the energy of the Coulomb interaction between electrons by applying (7.137). However, we consider here a simpler application, namely, calculation of the interaction between the nuclear quadrupole moment and the electric field gradient at the nucleus due to electrons. The interaction between protons and electrons is given by

$$H_{\text{Coul}} = \sum_{nj} \frac{-e^2}{|\boldsymbol{r}_n - \boldsymbol{r}_j|}$$

$$= -e^2 \sum_{nj} \sum_{k} \frac{r_n^k}{r_j^{k+1}} P_k(\cos \Theta_{nj})$$

$$= -e^2 \sum_{nj} \left\{ \frac{1}{r_j} + \frac{r_n}{r_j^2} \mathbf{C}^{(1)}(\theta_n \varphi_n) \cdot \mathbf{C}^{(1)}(\theta_j \varphi_j) \right.$$

$$\left. + \frac{r_n^2}{r_j^3} \mathbf{C}^{(2)}(\theta_n \varphi_n) \cdot \mathbf{C}^{(2)}(\theta_j \varphi_j) + \cdots \right\}, \tag{7.140}$$

where $(r_n, \theta_n, \varphi_n)$ and $(r_j, \theta_j, \varphi_j)$ are the polar coordinates of the proton n and electron j, respectively, and Θ_{nj} represents the angle between $\boldsymbol{r}_n$ and $\boldsymbol{r}_j$. The second line of (7.140) can be derived from the first by making use of the series expansion of the generating function for the Legendre polynomials assuming $r_n < r_j$. The third line follows from the addition theorem (7.136). The first term on the third line of (7.140) represents the potential energy of the electrons in the Coulomb field due to the nucleus and is usually taken into account in the Schrödinger equation for the electrons. The second term need not be considered, because nuclei do not have a permanent dipole moment. The third term is the interaction between the nuclear quadrupole moment and the electric field gradient:

$$H_{\text{Q}} = -e^2 \sum_{n} r_n^2 \mathbf{C}^{(2)}(\theta_n \varphi_n) \cdot \sum_{j} r_j^2 \mathbf{C}^{(2)}(\theta_j \varphi_j)/r_j^5 = -\frac{1}{2} \sum_{\alpha\beta} Q_{\alpha\beta} (\nabla \boldsymbol{E})_{\alpha\beta}$$

$$= -\mathbf{Q}^{(2)} \cdot (\nabla \boldsymbol{E})^{(2)} . \tag{7.141}$$

The definition of $Q_{\alpha\beta}$ has been given in (7.121). The electric field gradient can be expressed as

$$(\nabla \boldsymbol{E})_{\alpha\beta} = |e| \sum_{j} \frac{1}{r_j^5} (3 x_{j\alpha} x_{j\beta} - r_j^2 \delta_{\alpha\beta}) . \tag{7.142}$$

In the state with nuclear spin I ($\geqq 1$), $Q_{\alpha\beta}$ may be replaced by an operator equivalent as in (7.123). In a similar way, $\nabla \boldsymbol{E}$ can be expressed by an operator equivalent in the state with total electronic angular momentum J (> 1). This will then enable us to diagonalize H_{Q}. We will, however, calculate the eigenvalues of H_{Q} more straightforwardly by working in the states where I and J are coupled to give the total angular momentum $\boldsymbol{F} = \boldsymbol{I} + \boldsymbol{J}$ of the whole system, which is conserved. Keeping in mind that $\mathbf{Q}$ and $\nabla \boldsymbol{E}$ are tensors of degree 2, we find

$$E_Q \equiv \langle IJFM_F | H_Q | IJFM_F \rangle$$
$$= -\langle I \| \mathbf{Q}^{(2)} \| I \rangle \langle J \| (\nabla \boldsymbol{E})^{(2)} \| J \rangle W(I2FJ; IJ) ,$$

where F is the magnitude of $\boldsymbol{F}$. The reduced matrix element of $\mathbf{Q}^{(2)}$ has been given in (7.122a, b). For $\nabla \boldsymbol{E}$ we obtain

$$\langle J \| (\nabla \boldsymbol{E})^{(2)} \| J \rangle = \frac{-3|e| q_J \langle J \| \mathbf{C}^{(2)}(\boldsymbol{J}) \| J \rangle}{J(2J-1)} ,$$

$$-|e| q_J/3 = \int d\boldsymbol{r} \frac{1}{r^5} \left(z^2 - \frac{r^2}{3} \right) [\rho_e(\boldsymbol{r})]_{M=J} ,$$

where $[\rho_e(\boldsymbol{r})]_{M=J}$ is the electron charge density in the state with $M = J$. Putting these together and writing out the expression of $W(I2FJ; IJ)$ explicitly, we have

$$E_Q = \frac{-3|e| q_J Q}{8I(2I-1)J(2J-1)} \left[K(K+1) - \frac{4}{3} I(I+1) J(J+1) \right]$$

with

$$K = F(F+1) - I(I+1) - J(J+1) = 2(\boldsymbol{I} \cdot \boldsymbol{J}) .$$

Let us next consider an example of (7.139a, b). Our purpose is to show that in the state with total angular momentum J denoted as $\alpha^{2S+1} L_J$ (a split component of multiplet term $\alpha^{2S+1} L$ under the influence of the spin-orbit interaction), the total magnetic moment of the electrons given by

$$\mu = -\beta(\boldsymbol{L} + 2\boldsymbol{S}) , \quad \beta = |e| \frac{\hbar}{2mc} ,$$

can be expressed as

$$\mu = -g\beta \boldsymbol{J} , \quad g = \frac{3}{2} + \frac{S(S+1) - L(L+1)}{2J(J+1)} . \tag{7.143}$$

The constant g is called the Landé g-factor. This result can be derived as follows. We note from (7.139a, b) that

$$\langle \alpha SLJ \| \boldsymbol{L} \| \alpha SLJ \rangle = \langle L \| \boldsymbol{L} \| L \rangle (2J+1) W(LS1J; JL) ,$$
$$\langle \alpha SLJ \| \boldsymbol{S} \| \alpha SLJ \rangle = \langle S \| \boldsymbol{S} \| S \rangle (2J+1) W(SL1J; JS) .$$

The reduced matrix elements of $\boldsymbol{L}$ and $\boldsymbol{S}$ are obtained from (7.109). The Racah coefficients appearing here are found to be given as

$$W(LS1J; JL) = \frac{J(J+1) + L(L+1) - S(S+1)}{2\langle J \| \boldsymbol{J} \| J \rangle \langle L \| \boldsymbol{L} \| L \rangle} .$$

We put them into the expression for g,

$$g = \frac{\langle \alpha SLJ \| \boldsymbol{L} + 2\boldsymbol{S} \| \alpha SLJ \rangle}{\langle J \| \boldsymbol{J} \| J \rangle} ,$$

and obtain (7.143). This value of g may be obtained in a more elementary way by evaluating the component of $\boldsymbol{L} + 2\boldsymbol{S}$ parallel to $\boldsymbol{J}$:

$$g = \frac{(\boldsymbol{L} + 2\boldsymbol{S})\cdot(\boldsymbol{L} + \boldsymbol{S})}{\boldsymbol{J}^2} = \frac{\boldsymbol{L}^2 + 2\boldsymbol{S}^2 + \frac{3}{2}(\boldsymbol{J}^2 - \boldsymbol{L}^2 - \boldsymbol{S}^2)}{\boldsymbol{J}^2} .$$

However, it is not so easy to derive the (off-diagonal) matrix elements of the component perpendicular to $\boldsymbol{J}$ by such a semiclassical treatment, though not impossible.[10] They are given by

$$\begin{aligned}\langle \alpha SLJ \| \boldsymbol{L} \| \alpha SLJ' \rangle &= -\langle \alpha SLJ \| \boldsymbol{S} \| \alpha SLJ' \rangle \\ &= -\langle L \| \boldsymbol{L} \| L \rangle \sqrt{(2J+1)(2J'+1)}\, W(LS1J';JL)\end{aligned}$$

with $J' = J \pm 1$ so that

$$\langle \alpha SLJ \| \boldsymbol{L} \| \alpha SLJ' \rangle$$

$$= \begin{cases} -\sqrt{\dfrac{[(S+L+1)^2 - J^2][J^2 - (S-L)^2]}{4J}} , & J' = J-1 , \\ \sqrt{\dfrac{[(S+L+1)^2 - (J+1)^2][(J+1)^2 - (S-L)^2]}{4(J+1)}} & J' = J+1 . \end{cases}$$

7.14 Electronic Wavefunctions for the Configuration $(nl)^x$

The wavefunction of a two-electron system in the configuration $nln'l'$ can be obtained by putting the electrons in each shell, adding their spin and orbital angular momenta to give total S and L and antisymmetrizing the resulting function with regard to the exchange of both spin and orbital coordinates of the two electrons:

$$\Psi(nln'l'SLM_SM_L) = A_2 \Phi(nln'l'SLM_SM_L) , \tag{7.144}$$

$$\begin{aligned}\Phi(nln'l'SLM_SM_L) = \sum_{\substack{m_s m_s' \\ m_l m_l'}} &\phi_1(nlm_sm_l)\phi_2(n'l'm_s'm_l') \\ &\times \langle \tfrac{1}{2}m_s \tfrac{1}{2}m_s' | SM_S \rangle \langle lm_l l'm_l' | LM_L \rangle , \end{aligned} \tag{7.145}$$

[10] Semiclassically, the component rotates in the plane perpendicular to $\boldsymbol{J}$ because of precession. See, for example, [7.3].

where A_2 is the antisymmetrizer

$$A_2 = \frac{1}{\sqrt{2}}(E - (12)) .$$

The notation (12) stands for the permutation operator that exchanges the coordinates 1 and 2 of the electrons. Note that we have used, for example, subscript 1 to ϕ to indicate the coordinates (including spin coordinate) of the electron 1. Since the permutation (12) and A_2 commute with $\boldsymbol{S} = \boldsymbol{s}_1 + \boldsymbol{s}_2$ and with $\boldsymbol{L} = \boldsymbol{l}_1 + \boldsymbol{l}_2$, the function Ψ constructed above is certainly antisymmetric and at the same time is a normalized eigenfunction of $\boldsymbol{S}^2$, $\boldsymbol{L}^2$, S_z and L_z. In this case all possible combinations of the values of $S = 0, 1$; $L = |l - l'|, |l - l'| + 1, \ldots, l + l'$ are allowed. If we write k_1 and k_2 in place of $nlm_s m_l$ and $n'l'm_s'm_l'$ respectively, (7.144) can be written as

$$\Psi(nln'l'SLM_SM_L) = \sum_{k_1 k_2} |\phi_1(k_1), \phi_2(k_2)| C_{k_1 k_2} \tag{7.146}$$

in terms of the (normalized) Slater determinant

$$|\phi_1(k_1), \phi_2(k_2)| \equiv \frac{1}{\sqrt{2}} \begin{vmatrix} \phi_1(k_1) & \phi_1(k_2) \\ \phi_2(k_1) & \phi_2(k_2) \end{vmatrix} , \tag{7.147}$$

where we have written, for simplicity, $C_{k_1 k_2}$ for the product of Wigner coefficients on the right-hand side of (7.145).

When the two electrons are accommodated in the same shell, i.e. $(nl)^2$, the method described above leads in some cases to a function vanishing identically and in other cases to a function not correctly normalized. This may be seen as follows. According to (7.94b), we find

$$\begin{aligned}
&(12)\Phi(nlnlSLM_SM_L) \\
&\quad = \sum \phi_2(nlm_s m_l)\phi_1(nlm_s'm_l')\langle \tfrac{1}{2}m_s \tfrac{1}{2}m_s'|SM_S\rangle \langle lm_l lm_l'|LM_L\rangle \\
&\quad = \sum \phi_1(nlm_s'm_l')\phi_2(nlm_s m_l) \\
&\qquad \times (-1)^{1-S}(-1)^{2l-L}\langle \tfrac{1}{2}m_s' \tfrac{1}{2}m_s|SM_S\rangle \langle lm_l' lm_l|LM_L\rangle \\
&\quad = -(-1)^{S+L}\Phi(nlnlSLM_SM_L) .
\end{aligned} \tag{7.148}$$

This means that for $S + L$ odd, Φ vanishes identically when antisymmetrized. For $S + L$ even, Φ is already antisymmetric and normalized, so that further operation of A_2 will result in incorrect normalization. The correct wavefunction for the $(nl)^2$ configuration is thus given by

$$\Psi((nl)^2 SLM_SM_L) = \Phi(nlnlSLM_SM_L) , \quad S + L \text{ even} . \tag{7.149}$$

The allowed multiplet terms ${}^{2S+1}L$ are ${}^1S$, ${}^3P$, ${}^1D, \ldots, {}^1L$ $(L = 2l)$.

Let us next consider the case where we have more than two electrons in the same nl-shell. To simplify notation, we denote by l^n the configuration with n electrons accommodated in this shell. (There will be no danger of confusing this n with the principal quantum number.) In the case of configuration l^n, it can happen that multiplet terms with the same S and L occur more than once. For example, ^2D appears twice in d^3. Let us put the label α before S and L to distinguish between them. Suppose we know all the wavefunctions for the electron configuration l^{n-1}:

$$\Psi(l^{n-1}\alpha_1 S_1 L_1 M_{S1} M_{L1}) \ .$$

Then wavefunctions of $l^n\ SL$ will be obtained from (7.144, 145) by operating the antisymmetrizer

$$A_n = \frac{1}{\sqrt{n}}\left(E - \sum_{j=1}^{n-1} (jn)\right)$$

on the functions (see footnote 11 on p. 163)

$$\begin{aligned}\Phi(l^{n-1}(\alpha_1 S_1 L_1) lSLM_S M_L) &= \sum_{\substack{m_s m_l \\ M_{S_1} M_{L_1}}} \Psi(l^{n-1}\alpha_1 S_1 L_1 M_{S1} M_{L1})\phi_n(lm_s m_l) \\ &\times \langle S_1 M_{S1} \tfrac{1}{2} m_s | SM_S\rangle \langle L_1 M_{L1} l m_l | LM_L\rangle \ .\end{aligned} \quad (7.150)$$

The result will in general be a linear combination of the correct orthonormal wavefunctions $\Psi(l^n\alpha SLM_S M_L)$ of l^n.

$$A_n\Phi(l^{n-1}(\alpha_1 S_1 L_1) lSLM_S M_L) = \sum_{\alpha} \Psi(l^n\alpha SLM_S M_L) C(\alpha SL, \alpha_1 S_1 L_1) \ . \quad (7.151)$$

This result shows that, when the multiplet term SL appears only once in l^n, the function $\Psi(l^n SLM_S M_L)$ can be obtained by starting from an appropriate $\alpha_1 S_1 L_1$ of l^{n-1} and antisymmetrizing (7.150) by A_n. When the same SL appears more than once, it is necessary to choose several sets of $\alpha_1 S_1 L_1$, construct the left-hand sides of (7.151) and seek the required number of linearly independent wavefunctions.

Note that we have

$$\begin{aligned} C(\alpha SL, \alpha_1 S_1 L_1) &= (\Psi(l^n\alpha SLM_S M_L), A_n\Phi(l^{n-1}(\alpha_1 S_1 L_1) lSLM_S M_L)) \\ &= (A_n\Psi, \Phi) \\ &= \sqrt{n}(\Psi(l^n\alpha SLM_S M_L), \Phi(l^{n-1}(\alpha_1 S_1 L_1) lSLM_S M_L)) \end{aligned} \quad (7.152)$$

from (7.151).

Instead of antisymmetrizing the functions (7.150), we could just as well attempt to construct a linear combination of them such that the resulting expression has the correct symmetry:

$$\Psi(l^n \alpha SLM_S M_L) = \sum_{\alpha_1 S_1 L_1} \Phi(l^{n-1}(\alpha_1 S_1 L_1) lSLM_S M_L) \times \langle l^{n-1}(\alpha_1 S_1 L_1) lSL | \} l^n \alpha SL \rangle \, , \tag{7.153}$$

where the coefficients (called *coefficients of fractional parentage*, abbreviated c.f.p) are to be determined so that the right-hand side is totally antisymmetric. According to (7.152), the coefficients C and the c.f.p. are related to each other by

$$C(\alpha SL, \alpha_1 S_1 L_1) = \sqrt{n} \langle l^n \alpha SL \{ | l^{n-1}(\alpha_1 S_1 L_1) lSL \rangle \, ,$$
$$\langle l^n \alpha SL \{ | l^{n-1}(\alpha_1 S_1 L_1) lSL \rangle \equiv \langle l^{n-1}(\alpha_1 S_1 L_1) lSL | \} l^n \alpha SL \rangle^* \, . \tag{7.154}$$

One practical way to determine these coefficients is as follows. We first apply (7.153), replacing n by $n-1$ to express $\Psi(l^{n-1}\alpha_1 S_1 L_1)$ in terms of $\Psi(l^{n-2}\alpha_2 S_2 L_2)$ (dropping M_S and M_L, for simplicity):

$$\Psi(l^n \alpha SL) = \sum_{\substack{\alpha_1 S_1 L_1 \\ \alpha_2 S_2 L_2}} \Phi(l^{n-2}(\alpha_2 S_2 L_2) l(S_1 L_1) lSL) \times \langle l^{n-2}(\alpha_2 S_2 L_2) lS_1 L_1 | \} l^{n-1} \alpha_1 S_1 L_1 \rangle \times \langle l^{n-1}(\alpha_1 S_1 L_1) lSL | \} l^n \alpha SL \rangle \, . \tag{7.155}$$

From the definition of the Racah coefficient, we have

$$\Phi(l^{n-2}(\alpha_2 S_2 L_2) l(S_1 L_1) lSL) = \sum_{S'L'} \Phi(l^{n-2}(\alpha_2 S_2 L_2) l^2(S'L') SL) \times \langle S_2, \tfrac{1}{2}\tfrac{1}{2}(S') S | S_2 \tfrac{1}{2}(S_1) \tfrac{1}{2} S \rangle \langle L_2, ll(L') L | L_2 l(L_1) lL \rangle \, . \tag{7.156}$$

On the right-hand side of this equation, the angular momenta of the electrons $n-1$ and n are coupled to give the total spin and orbital angular momenta S' and L'. In order that the wave function (7.155) be antisymmetric with respect to the interchange of the coordinates of $n-1$ and n, terms with $S' + L'$ an odd integer must not appear when (7.156) is substituted into (7.155). This leads to

$$\sum_{\alpha_1 S_1 L_1} \langle S_2, \tfrac{1}{2}\tfrac{1}{2}(S') S | S_2 \tfrac{1}{2}(S_1) \tfrac{1}{2} S \rangle \langle L_2, ll(L') L | L_2 l(L_1) lL \rangle \times \langle l^{n-2}(\alpha_2 S_2 L_2) lS_1 L_1 | \} l^{n-1} \alpha_1 S_1 L_1 \rangle \langle l^{n-1}(\alpha_1 S_1 L_1) lSL | \} l^n \alpha SL \rangle = 0 \tag{7.157}$$

for $S' + L'$ an odd integer. If we know the values of the c.f.p for l^{n-1}, we can determine the values of $\langle l^{n-1} \ldots | \} l^n \ldots \rangle$ from (7.157) together with the normalization condition

$$\sum_{\alpha_1 S_1 L_1} \langle l^n \alpha SL\{|l^{n-1}(\alpha_1 S_1 L_1) lSL\rangle\langle l^{n-1}(\alpha_1 S_1 L_1) lSL|\}l^n \alpha' SL\rangle = \delta_{\alpha\alpha'} \,. \tag{7.158}$$

The condition (7.157) was derived from the requirement that the right-hand side of (7.155) be antisymmetric with respect to the interchange of $n-1$ and n. This guarantees that (7.155) is totally antisymmetric, because the right-hand side is already antisymmetric under any permutation of the electron coordinates $1, 2, 3, \ldots, n-1$.

It should be remarked that the matrix of c.f.p. whose rows and columns are specified by $\alpha_1 S_1 L_1$ and αSL is not an ordinary unitary matrix, but only a rectangular matrix which is a part of a unitary one, because its columns do not exhaust all states of $l^{n-1}l$ but only those that are allowed in l^n. The matrix multiplication (7.158) is meaningful, but multiplication in the opposite order has no sense.

For $n = 2$, we find from (7.149) that the coefficient of fractional parentage is equal to unity. For n greater than 2, use of (7.157) provides us with the necessary c.f.p. and thus the wavefunctions of the multiplet terms of l^n. In Table 7.5, c.f.p. in the simplest case $p^2 \to p^3$ are given [7.4].

In the electron configuration l^n, the multiplet term with the largest S, and if such a term appears more than once, the one with the largest value of L among them, lies lowest in energy (Hund's rule), so that the wavefunction of the ground state can be easily written down in terms of the Slater determinant without using complicated procedures as given above. Namely, we may put for the ground state

$$\begin{aligned}\Psi(l^n SL, M_S = S, M_L = L) \\ = |\phi_1(\tfrac{1}{2}, l), \phi_2(\tfrac{1}{2}, l-1), \ldots, \phi_n(\tfrac{1}{2}, l-n+1)|\end{aligned} \tag{7.159}$$

with $S = n/2$ and $L = l + (l-1) + \ldots + (l-n+1)$. This is, however, for the case of $n \leqq 2l+1$. When $n > 2l+1$, we put $n - 2l - 1$ electrons in the down-spin states $\phi(-1/2, l), \phi(-1/2, l-1), \ldots$ after filling up the $2l+1$ up-spin states.

Table 7.5. $\langle p^2(S_1 L_1)pSL|p^3 SL\rangle$

S_1L_1 \ SL	^4S	^2P	^2D
^1S	0	$\sqrt{2/3}$	0
^3P	1	$-1/\sqrt{2}$	$1/\sqrt{2}$
^1D	0	$-\sqrt{5/3}\sqrt{2}$	$-1/\sqrt{2}$

7.15 Electrons and Holes

In the previous section we described how to construct the wavefunctions for l^n. It is unnecessary to perform this procedure for all n up to $4l + 2$ for which the l shell is closed. This is because the wavefunctions of the configuration l^{4l+2-n} with n "holes" can be correlated with the wavefunctions of the n-electron configuration l^n, as seen below, and matrix elements of the operators within one configuration can be derived from those in the other by simple rules.

The wavefunction of the closed shell $l^{4l+2}\ {}^1S$ is expressed as

$$\Psi(l^{4l+2}\ {}^1S) = |\phi_1(\tfrac{1}{2}, l), \phi_2(-\tfrac{1}{2}, l), \phi_3(\tfrac{1}{2}, l-1), \ldots, \phi_{4l+2}(-\tfrac{1}{2}, -l)| \ , \tag{7.160}$$

where we write $\phi(m_s m_l)$ (or simply $\phi(k)$ below) in place of $\phi(l m_s m_l)$ in the previous section. Now according to the Laplace expansion of the Slater determinant we may write the right-hand side as

$$\begin{aligned} \Psi(l^{4l+2}\ {}^1S) &= \binom{4l+2}{n}^{-1/2} \sum_k |\phi_1(k_1), \phi_2(k_2), \ldots, \phi_n(k_n)| \\ &\quad \times (-1)^{n(n+1)/2 - \Sigma k_j} |\phi_{n+1}(k'_1), \ldots, \phi_{4l+2}(k'_{4l+2-n})| \\ &= \binom{4l+2}{n}^{-1/2} \sum_k D_k \bar{D}_k \ , \end{aligned} \tag{7.161}$$

where $\phi(k_j)$ is the k_jth spin-orbital when $\phi(m_s m_l)$ are arranged as in (7.160) (e.g. $\phi(3) = \phi(1/2, l-1)$). In the first line of (7.161),

$$D_k = |\phi_1(k_1), \phi_2(k_2), \ldots, \phi_n(k_n)| \tag{7.162}$$

stands for the minor (divided by $\sqrt{n!}$) obtained from the determinant (7.160) by taking the first n rows and the columns $k_1, k_2, \ldots, k_n$, where $k_1 < k_2 < \ldots < k_n$, and

$$|\phi_{n+1}(k'_1), \ldots, \phi_{4l+2}(k'_{4l+2-n})|$$

is the minor complementary to it (divided by $\sqrt{(4l+2-n)!}$). On the second line, k stands for the set of numbers $k_1, k_2, \ldots, k_n$ and the summation runs over all possible choices of n columns.

Let us define ϕ^+ with respect to ϕ as[11]

$$\phi^+(m_s m_l) = (-1)^{(1/2)+l-m_s-m_l} \phi(-m_s, -m_l) \ . \tag{7.163}$$

With this definition, it can be shown that the functions $L_k = D_k$ and

$$R_k = \bar{D}_k^+ = (-1)^{n(n+1)/2 - \Sigma k_j} |\phi^+_{n+1}(k'_1), \ldots, \phi^+_{4l+2}(k'_{4l+2-n})|$$

[11] This + operation is essentially the time reversal, θ (Sect. 13.1). It turns out in fact that $\phi^+ = (-1)^l \theta\phi$, so that for a complex number c, $c^+ = c^*$.

both have the same matrix elements for operators L and S. This may be seen in the following way. For one-electron operators

$$F = \sum_{j=1}^{n} f_j \quad (\text{acting on } L_k) \ , \qquad F = \sum_{j=n+1}^{4l+2} f_j \quad (\text{acting on } R_k) \ , \tag{7.164}$$

we find that the diagonal elements are given by

$$\langle L_k|F|L_k\rangle = \sum_{j=1}^{n} \langle k_j|f|k_j\rangle \ , \tag{7.165a}$$

$$\langle R_k|F|R_k\rangle = \sum_{\nu=1}^{4l+2} \langle \nu|f|\nu\rangle - \sum_{j=1}^{n} \langle k_j^+|f|k_j^+\rangle \ . \tag{7.165b}$$

The summation of the first term of the right-hand side of (7.165b) is to be taken over all the spin orbitals appearing in (7.160). For $\boldsymbol{F} = \boldsymbol{L}, \boldsymbol{S}$ this sum vanishes. For off-diagonal elements between L_k and $L_{k'}$ where L_k differs from $L_{k'}$ only in the occupancy $\phi(k_\lambda)$ in place of $\phi(k_\mu)$, we have

$$\langle L_k|F|L_{k'}\rangle = \pm \langle k_\lambda|f|k_\mu\rangle \ , \tag{7.166a}$$

$$\langle R_k|F|R_{k'}\rangle = \mp \langle k_\mu^+|f|k_\lambda^+\rangle \ , \tag{7.166b}$$

where the sign on the right-hand side of (7.166b) is to be chosen opposite to that of (7.166a), as indicated.

When f is a Hermitian operator that does not change sign under time reversal (Sect. 13.1), we have

$$\langle k_\mu^+|f|k_\lambda^+\rangle = \langle k_\lambda|f|k_\mu\rangle \ , \tag{7.167}$$

which means that (7.166a) and (7.166b) have opposite signs. In constrast, when f changes sign under time reversal like l and s, we find

$$\langle k_\mu^+|f|k_\lambda^+\rangle = -\langle k_\lambda|f|k_\mu\rangle \ , \tag{7.168}$$

so that the matrix element between R states is exactly equal to the one between corresponding L states.

A correspondence like (7.165a, b) and (7.166a, b) holds also for two-electron operators

$$G = \sum_{i<j} g_{ij} \tag{7.169}$$

in the following form. For diagonal elements, we obtain

$$\langle L_k|G|L_k\rangle = \sum_{j<l} \{\langle k_j k_l|g|k_j k_l\rangle - \langle k_j k_l|g|k_l k_j\rangle\} \ , \tag{7.170a}$$

$$\begin{aligned}\langle R_k|G|R_k\rangle &= (2l+1-n) \sum_{\nu=1}^{4l+2} \{\langle k\nu|g|k\nu\rangle - \langle k\nu|g|\nu k\rangle\} \\ &\quad + \sum_{j<l} \{\langle k_j^+ k_l^+|g|k_j^+ k_l^+\rangle - \langle k_j^+ k_l^+|g|k_l^+ k_j^+\rangle\} \ . \end{aligned} \tag{7.170b}$$

When g is a scalar operator, the first sum on the right-hand side of (7.170b) takes the same value irrespective of the spin-orbital $\phi(k)$ and is independent of k. When L_k and $L_{k'}$ differ in the occupancy of only one pair of spin orbitals, we have

$$\langle L_k|G|L_{k'}\rangle = \pm \sum_j \{\langle k_\lambda k_j|g|k_\mu k_j\rangle - \langle k_\lambda k_j|g|k_j k_\mu\rangle\} , \tag{7.171a}$$

$$\langle R_k|G|R_{k'}\rangle = \pm \sum_j \{\langle k_\mu^+ k_j^+|g|k_\lambda^+ k_j^+\rangle - \langle k_\mu^+ k_j^+|g|k_j^+ k_\lambda^+\rangle\} . \tag{7.171b}$$

Further, we find

$$\langle L_k|G|L_{k''}\rangle = \pm \{\langle k_\lambda k_\rho|g|k_\mu k_\sigma\rangle - \langle k_\lambda k_\rho|g|k_\sigma k_\mu\rangle\} , \tag{7.172a}$$

$$\langle R_k|G|R_{k''}\rangle = \pm \langle k_\mu^+ k_\sigma^+|g|k_\lambda^+ k_\rho^+\rangle - \langle k_\mu^+ k_\sigma^+|g|k_\rho^+ k_\lambda^+\rangle\} , \tag{7.172b}$$

when L_k and $L_{k'}$ differ in two pairs. The matrix elements appearing on the right-hand side of these equations satisfy the relations similar to (7.167) or (7.168) according as the Hermitian operator g remains invariant under time reversal or changes its sign.

Since L_k and R_k have exactly the same matrix elements of $\boldsymbol{L}$ as well as $\boldsymbol{S}$, the (normalized) eigenfunction of a multiplet term αSL of l^n constructed as a linear combination of D_k

$$\Psi(l^n\alpha SLM_SM_L) = \Psi_L(l^n\alpha SLM_SM_L) \equiv \sum_k D_k u_{k,\alpha SLM_sM_L} \tag{7.173}$$

provides us with

$$\Psi_R(l^{4l+2-n}\alpha SLM_SM_L) \equiv (-1)^n \sum_k \bar{D}_k^+ u_{k,\alpha SLM_sM_L} \tag{7.174}$$

with the same coefficients u, and this is precisely the eigenfunction of the multiplet term αSL of l^{4l+2-n}:

$$\Psi(l^{4l+2-n}\alpha SLM_SM_L) .$$

The n-electron states and n-hole states put into one-to-one correspondence in this way are said to be "conjugated" to each other.

If we observe that (7.163) can be generalized to[12]

$$\Psi^+(l^n\alpha SLM_SM_L) = (-1)^{S+L-M_S-M_L}\Psi(l^n\alpha SL-M_S-M_L) \tag{7.175}$$

and take the unitarity of the coefficients u into account, we may rewrite (7.161) as

$$\begin{aligned}\Psi(l^{4l+2}\,{}^1S) = \binom{4l+2}{n}^{-1/2} &\sum_{\alpha SLM_SM_L} \Psi_L(l^n\alpha SLM_SM_L) \\ &\times (-1)^{S+L-M_S-M_L}\Psi_R(l^{4l+2-n}\alpha SL-M_S-M_L) . \end{aligned}\tag{7.176}$$

[12] The coefficient u on the right-hand side of (7.173) may be expressed through products of several Wigner coefficients. Equation (7.175) follows from the symmetry of the Wigner coefficient (7.94c) and the definition (7.163), yielding the behavior of eigenfunctions of angular momenta under time reversal.

Let us follow the convention of using Ψ_{L} for l^n $(n \leqq 2l+1)$ and Ψ_{R} for l^{4l+2-n}. Then we have, for the one-electron operator (7.164),

$$\langle l^{4l+2-n}\gamma | F | l^{4l+2-n}\gamma' \rangle_{\mathrm{R}} = \delta_{\gamma\gamma'} \sum_{\nu=1}^{4l+2} \langle \nu | f | \nu \rangle - \langle l^n \gamma'^+ | F | l^n \gamma^+ \rangle_{\mathrm{L}} \tag{7.177}$$

according to (7.165a–166b). Note that we have used the notation γ in place of $\alpha SLM_S M_L$, for simplicity. When F is Hermitian, the second term of the right-hand side may be rewritten as in (7.167) or (7.168) according as F is invariant under time reversal or not. The first term is nonvanishing only when F is scalar with respect to rotations both in spin and ordinary space. For a double tensor $\mathbf{F}^{(\kappa k)}$ of degree κ $(\kappa = 0, 1)$ with regard to spin and degree k $(k = 0, 1, \ldots, 2l)$ with regard to orbit, we obtain the following relation between the reduced matrix elements from (7.177) when $(\kappa k) \neq (00)$:

$$\langle l^{4l+2-n}\alpha SL \| \mathbf{F}^{(\kappa k)} \| l^{4l+2-n}\alpha' S' L' \rangle_{\mathrm{R}} = -(-1)^{\kappa+k} \langle l^n \alpha SL \| \mathbf{F}^{(\kappa k)\dagger} \| l^n \alpha' S' L' \rangle_{\mathrm{L}}^*$$

As a result, for example, the splitting pattern due to the spin-orbit interaction in the configuration l^n becomes inverted in l^{4l+2-n}.

In the case of two-electron operators, or more specifically when $g_{ij} = e^2/r_{ij}$, we can show that the energy of the state $\Psi(l^n\alpha SL)$ coincides with that of $\Psi(l^{4l+2-n}\alpha SL)$ apart from a constant energy shift independent of αSL. This means that the relative arrangements of the multiplet terms are the same in both l^n and l^{4l+2-n}.

It is therefore unnecessary to repeat the calculation of matrix elements of any operator for l^{4l+2-n} once they have been obtained for l^n $(n \leqq 2l+1)$. This is also true with respect to c.f.p., because we have

$$\begin{aligned} &\langle l^{4l+1-n}(\alpha_1 S_1 L_1) lSL | \} l^{4l+2-n}\alpha SL \rangle_{\mathrm{R}} \\ &= (-1)^{S+S_1+L+L_1-1/2-l} \sqrt{\frac{(n+1)(2S_1+1)(2L_1+1)}{(4l+2-n)(2S+1)(2L+1)}} \\ &\quad \times \langle l^n(\alpha SL) lS_1 L_1 | \} l^{n+1}\alpha_1 S_1 L_1 \rangle_{\mathrm{L}} \end{aligned} \tag{7.178}$$

from (7.176).

7.16 Evaluation of the Matrix Elements of Operators

As discussed in Sect. 7.14, the calculation of the matrix elements of opeators is straightforward using (7.165a, 166a, 170a, 171a, 172a), once the wave functions of l^n are expressed as linear combinations of Slater determinants.

It should be remarked, however, that we can evaluate the matrix elements for l^n if we know them for l^{n-1} and the c.f.p. for $l^n\alpha SL$, as seen below.

In the case of a one-electron operator $F = \sum f_j$, we note that the matrix element of F is equal to n times that of f_n, so that

$$\begin{aligned}
&\langle l^n \alpha SLM_S M_L | F | l^n \alpha' S' L' M'_S M'_L \rangle \\
&= n \sum_{\alpha_1 S_1 L_1} \langle l^n \alpha SL\{| l^{n-1}(\alpha_1 S_1 L_1) lSL \rangle \langle l^{n-1}(\alpha_1 S_1 L_1) lS' L' |\} l^n \alpha' S' L' \rangle \\
&\times \sum_{\substack{M_{S_1} m_s m'_s \\ M_{L_1} m_l m'_l}} \langle lm_s m_l | f | lm'_s m'_l \rangle \langle SM_S | S_1 M_{S1} \tfrac{1}{2} m_s \rangle \langle LM | L_1 M_{L1} lm_l \rangle \\
&\times \langle S_1 M_{S1} \tfrac{1}{2} m'_s | S' M'_S \rangle \langle L_1 M_{L1} lm'_l | L' M'_L \rangle . \qquad (7.179)
\end{aligned}$$

When f is an irreducible tensor operator, the second summation on the right-hand side may be carried out using the Racah coefficient and can be put into a simpler form.

A similar consideration applies to the calculation of the matrix elements of the two-electron scalar operator $G = \sum g_{ij}$. In this case we multiply the matrix element of G for l^{n-1} by the factor $(1/2)n(n-1)/(1/2)(n-1)(n-2) = n/(n-2)$:

$$\begin{aligned}
&\langle l^n \alpha SL | G | l^n \alpha' SL \rangle \\
&= \frac{n}{n-2} \sum_{\alpha_1 \alpha'_1 S_1 L_1} \langle l^n \alpha SL\{| l^{n-1}(\alpha_1 S_1 L_1) lSL \rangle \langle l^{n-1} \alpha_1 S_1 L_1 | G | l^{n-1} \alpha'_1 S_1 L_1 \rangle \\
&\times \langle l^{n-1}(\alpha'_1 S_1 L_1) lSL |\} l^n \alpha' SL \rangle , \qquad (7.180)
\end{aligned}$$

where we have omitted $M_S M_L$ after SL, because the matrix elements diagonal with respect to them do not depend upon them. Use of (7.180) enables us to obtain the term values of l^n starting from those of l^2 [7.5].

We will therefore only briefly treat the calculation of the energies of the Coulomb interaction for a two-electron system. The interaction operator can be expanded in terms of the products of spherical harmonics as in (7.140):

$$g = e^2 \sum_k \frac{r_<^k}{r_>^{k+1}} \mathbf{C}^{(k)}(\theta_1 \varphi_1) \cdot \mathbf{C}^{(k)}(\theta_2 \varphi_2) , \qquad (7.181)$$

where $r_<$ and $r_>$ stand for the smaller and larger of r_1 and r_2, respectively. According to (7.137), the matrix elements of g are given by

$$\langle l^2 SL | g | l^2 SL \rangle = \sum_k f_k(llL) F^k(nl, nl) , \qquad (7.182)$$

$$\begin{aligned}
f_k(llL) &= \langle llL | \mathbf{C}^{(k)}(\theta_1 \varphi_1) \cdot \mathbf{C}^{(k)}(\theta_2 \varphi_2) | llL \rangle \\
&= (-1)^L W(llll; kL) \langle l \| \mathbf{C}^{(k)}(\theta\varphi) \| l \rangle^2 ,
\end{aligned}$$

$$F^k(nl, nl) = \int_0^\infty r_1^2 \, dr_1 \int_0^\infty r_2^2 \, dr_2 R_{nl}(r_1)^2 R_{nl}(r_2)^2 e^2 \frac{r_<^k}{r_>^{k+1}} .$$

The quantity F^k is called the Slater integral. Possible values of k here are $0, 2, \ldots, 2l$. As simple examples, we give below the term values (the right-hand side of (7.182)) for p^2, p^3 and d^2.

$$\begin{aligned}
p^2:\ & {}^1S = F_0 + 10F_2\ , \qquad {}^1D = F_0 + F_2\ , \\
& {}^3P = F_0 - 5F_2\ , \\
p^3:\ & {}^4S = 3F_0 - 15F_2\ , \qquad {}^2D = 3F_0 - 6F_2\ , \\
& {}^2P = 3F_0\ ,
\end{aligned}$$

where

$$F_0 = F^0(pp)\ , \quad F_2 = \frac{F^2(pp)}{25}\ ,$$

and

$$\begin{aligned}
d^2:\ & {}^1S = A + 14B + 7C\ , \\
& {}^3P = A + 7B\ , \\
& {}^1D = A - 3B + 2C\ , \\
& {}^3F = A - 8B\ , \\
& {}^1G = A + 4B + 2C\ .
\end{aligned}$$

where

$$A = F^0(dd) - \frac{1}{9} F^4(dd)\ ,$$

$$B = \frac{1}{49} F^2(dd) - \frac{5}{441} F^4(dd)\ ,$$

$$C = \frac{5}{63} F^4(dd)\ .$$

8. Point Groups

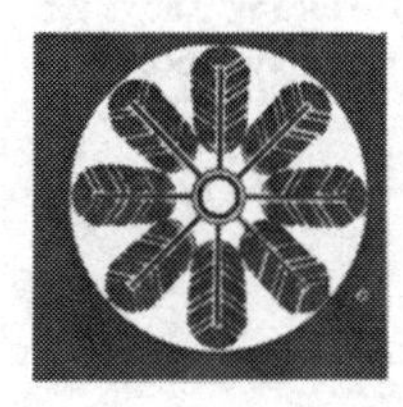

In Sect. 2.2, we considered the symmetry groups of an equilateral triangle and a square as elementary examples of a group. These groups, which consist of rotations and reflections, are called, in general, point groups. Point groups describe the microscopic symmetry of molecules and the macroscopic symmetry of crystals. They are therefore frequently used in studying electronic states and vibrations of molecules as well as the symmetry of the macroscopic properties of crystals. In this chapter, we describe point groups, their notation, and their irreducible representations. Applications to molecules are given in Chaps. 9 and 10.

8.1 Symmetry Operations in Point Groups

A *point group* is defined as a group of symmetry operations that leave invariant a finite object like a molecule. The symmetry operations are confined to rotations (and their combinations with space inversion) about a fixed point of that object.

As an example, consider the configuration depicted in Fig. 8.1. The following eight operations bring this configuration into coincidence with itself:

- E: This is the identity operation, which leaves the configuration as it stands. Recognition of this operation as a group element is necessary in order to satisfy the group axiom.
- C_4^2: Rotation through π about the vertical axis.
- IC_4: Rotation through $\pi/2$ about the vertical axis followed by inversion.
- IC_4^{-1}: Rotation through $3\pi/2$ about the vertical axis followed by inversion.
- $2C_2'$: Rotations through π about the horizontal x- and y-axes.
- $2\sigma_d$: Mirror reflections in the diagonal planes.

Note that there exist two perpendicular C_2' axes and two perpendicular σ_d mirror planes. These eight operations are closed if we understand their products as successive operations. This may be verified by tracing the movement of the circles in Fig. 8.1 under the symmetry operations. The eight operations therefore consitute a group, which is designated as D_{2d} or $\bar{4}2m$.

In general, geometrical symmetry operations are denoted by the following symbols:

- E: Identity operation.
- C_n: Rotation through an angle $2\pi/n$. The rotation axis is called an n-fold axis.

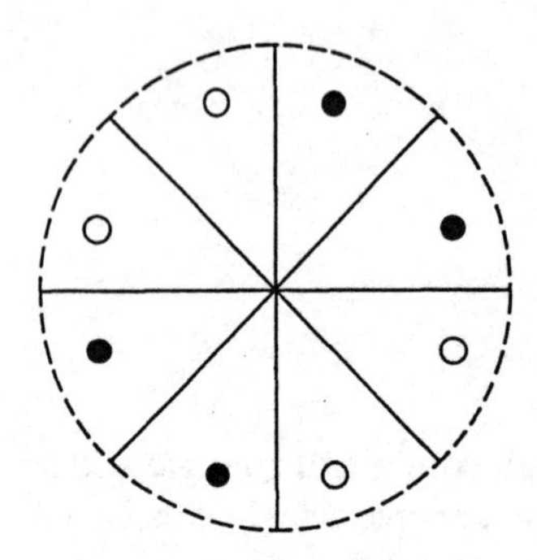

Fig. 8.1. Point group D_{2d}. White circles are located above the sheet, and black ones the same distance below the sheet

Rotations through π perpendicular to the principal rotation axis are denoted by C_2' and are called *Umklappung*.

I: Space inversion. It takes $\boldsymbol{r}$ into $-\boldsymbol{r}$.

σ: Mirror reflection. It carries three kinds of suffixes according to the property of the mirror plane.

σ_h: Mirror reflection in the horizontal plane.

σ_v: Mirror reflection in the vertical plane.

σ_d: Mirror reflection in the vertical diagonal plane.

IC_n: Rotatory inversion. Rotation through the angle $2\pi/n$ followed by inversion.

As may be seen from Fig. 8.2, the rotatory inversion IC_2 is nothing but a mirror reflection σ, the mirror plane being perpendicular to the twofold axis. In general, a rotatory inversion may also be understood as a rotatory reflection (rotation followed by reflection). If we denote the rotatory reflection $\sigma_h C_n$ by S_n, and use $\sigma_h = IC_2$, we find

$$S_n = IC_2 C_n \ . \tag{8.1}$$

In particular, when $n = 4$, we have

$$S_4 = IC_4^3 \quad (= IC_4^{-1}) \ ,$$

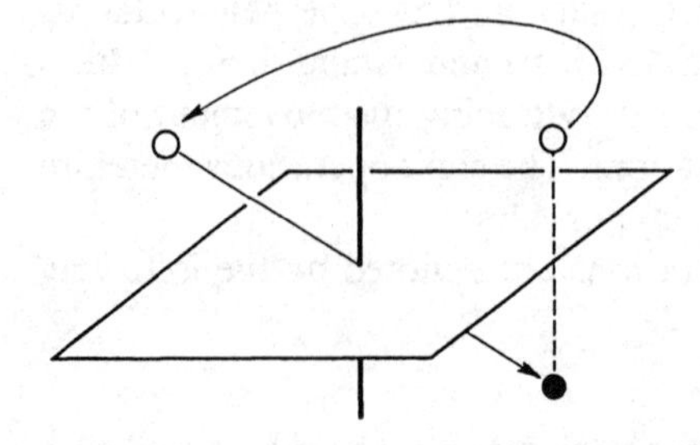

Fig. 8.2. Rotatory inversion IC_2 is equivalent to mirror reflection σ

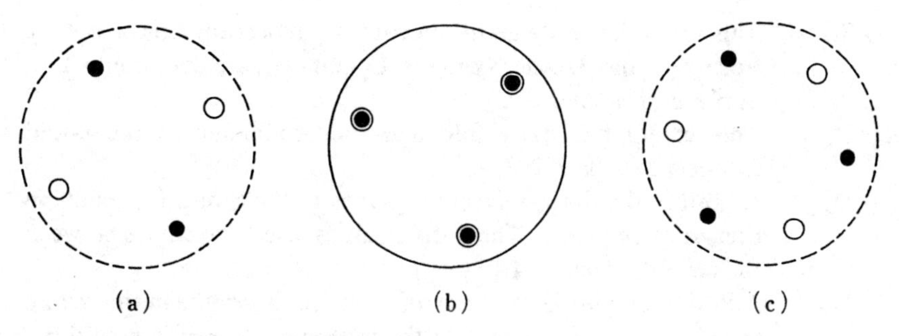

Fig. 8.3. Point groups **(a)** S_4, **(b)** C_{3h}, and **(c)** S_6 (C_{3i})

and when $n = 3$,

$$S_3 = IC_2C_3 = IC_6^5 \quad (= IC_6^{-1}) \ . \tag{8.2}$$

Successive S_3 operations produce the configuration shown in Fig. 8.3b, which tells us that the S_3 axis is accompanied by C_3 and σ_h operations. When $n = 6$, we have

$$S_6 = IC_2C_6 = IC_3^2 \quad (= IC_3^{-1}) \ . \tag{8.3}$$

In this case, Fig. 8.3c demonstrates the C_3 and I symmetries. Note that the suffixes 3 and 6 are interchanged in (8.2) and (8.3).

Rotatory inversions and rotatory reflections are sometimes called *improper rotations.* Simple rotations are then called *proper rotations.*

8.2 Point Groups and Their Notation

As far as rotations C_n about the origin are concerned, n can take on any integer values. In crystals, however, we have to take account of translational symmetry as well. Rotational symmetry compatible with translational symmetry restricts the value of n to 1, 2, 3, 4, and 6 (Sect. 11.4). Point groups composed of these restricted kinds of rotations and inversion are called *crystallographic point groups.* Let us enumerate the 32 point groups below in terms of the *Schönflies symbols.*

Group C_n: This group has only an n-fold rotation axis. It is a cyclic group of order n consisting of $E, C_n, C_n^2, \ldots, C_n^{n-1}$ ($n = 1, 2, 3, 4, 6$).

Group C_i: This group is composed of the inversion I and identity E.

Group C_{nv}: This group has n vertical mirror planes and an n-fold axis ($n = 2, 3, 4, 6$).

Group C_{nh}: This group has a horizontal mirror plane and an n-fold axis ($n = 1, 2, 3, 4, 6$). It contains the inversion I for $n = 2, 4, 6$.

Group S_n: This group has only an n-fold rotatory reflection axis ($n = 4, 6$). For $n = 2$ and 3, other symbols, C_i and C_{3h}, are commonly used in place of S_2 and S_3.

Group D_n: This group has n twofold axes perpendicular to the n-fold rotation axis ($n = 2, 3, 4, 6$).

Group D_{nd}: Addition of n diagonal mirror planes to the group D_n results in this group ($n = 2, 3$). The mirror planes bisect the angles between the twofold axes.

Group D_{nh}: Addition of a horizontal mirror plane to D_n results in this group ($n = 2, 3, 4, 6$). D_{nh} contains the inversion for $n = 2, 4$, and 6.

Besides the above 27 groups, we have cubic point groups O_h, O, T_d, T_h, and T. Group O is the symmetry group of a cube and a regular octahedron consisting of 24 proper rotations, as shown in Fig. 8.4. Inclusion of inversion leads to the *octahedral group* $O_h = O \times C_i$, which describes the full symmetry of a cube.

The group T is composed of the 12 proper rotations that leave a regular tetrahedron invariant, as shown in Fig. 8.5. Its direct product with C_i is $T_h = T \times C_i$. (Note that a tetrahedron is not invariant under inversion, so T_h is not the symmetry group of a tetrahedron.) It is the *tetrahedral group* T_d that describes the full symmetry of a tetrahedron. The group T_d is obtained by adding 6 IC_4 and 6 σ_d operations to T. Of these five cubic point groups, O_h and T_d often appear in physical applications. They are related by $O_h = T_d \times C_i$.

In addition to the above 32 crystallographic point groups, the following two groups describe the symmetry of linear molecules:

Group $C_{\infty v}$: This group consists of rotations of arbitrary angles about the molecular axis and vertical mirror reflections.

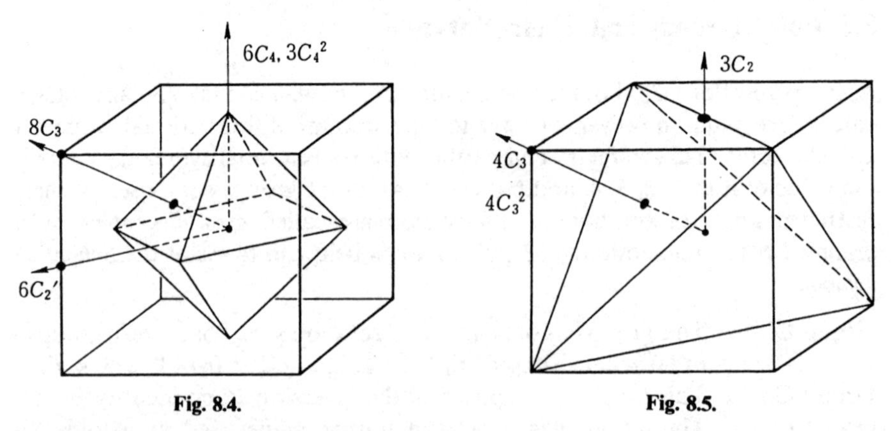

Fig. 8.4. **Fig. 8.5.**

Fig. 8.4. Rotational operations of the group O. The numbers of similar operations are written in front of the rotation symbols

Fig. 8.5. Rotational operations of the group T

Group $D_{\infty h}$: Homonuclear diatomic molecules have this symmetry. Addition of the horizontal mirror plane to $C_{\infty v}$ leads to this group.

We have so far adopted the Schönflies symbol for naming the point groups. Another system, called the *international notation*, or Hermann–Mauguin notation, is also commonly used. For a comparison of the symbols of the two systems, see Appendix A. The international notation was established from crystallographic motives. Various symmetry elements are expressed by means of the following symbols:

rotation axis. 1, 2, 3, 4, 6

rotatory inversion axis . . . . $\bar{1}, \bar{3}, \bar{4}, \bar{6}$

mirror reflection. . . . . m

Rotatory reflections are given no particular symbols. The mirror symbol m does not distinguish between σ_h, σ_v, and σ_d by itself. Instead, a horizontal mirror plane is indicated by writing n/m, which is understood to mean that the mirror plane is perpendicular to the n-fold axis. Furthermore, nm means that the mirror plane contains the n-fold axis. The international symbol for the point groups is constructed from the above rules.

For instance, the group D_{2d}, or $\bar{4}2$m, shown in Fig. 8.1 has a fourfold rotatory inversion axis and two twofold axes. The mirror planes contain the $\bar{4}$ axis.

For some lengthy international symbols, shorthand symbols are preferred (see Appendix A).

8.3 Class Structure in Point Groups

Classification of group elements into classes can be done by examining conjugate relations between group elements, as was mentioned in Sect. 2.8. In the case of geometrical operations as in point groups, one can achieve this classification by means of intuitive geometrical considerations.

Suppose that A and B denote two rotations of the same angle and that a rotation R brings the A axis to the B axis (Fig. 8.6). We then have

$$B = RAR^{-1} \,. \tag{8.4}$$

An example of this relation has already appeared in Sect. 2.8. Equation (8.4) means that B is conjugate to A. In (8.4), the operations A, B and R may be improper rotations, because the inversion I commutes with any operations. We can then conclude that any two symmetry operations that can be brought into each other by another operation in the group are conjugate to each other.

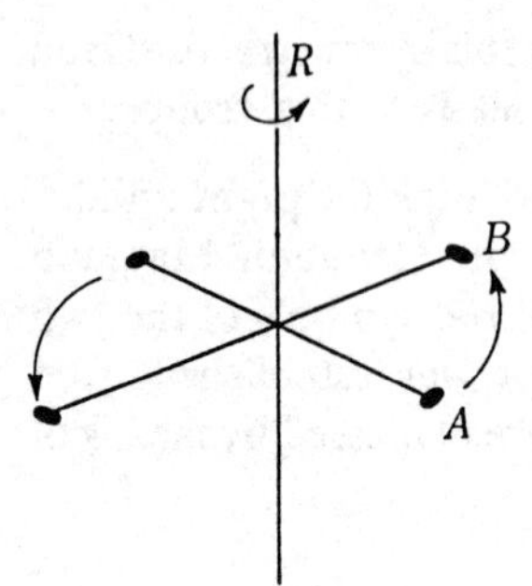

Fig. 8.6. Conjugate relation of two similar operations A and B connected by another operation R

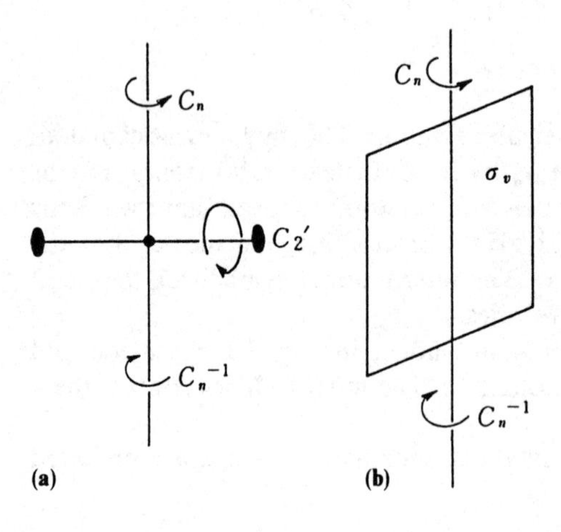

Fig. 8.7. (**a**) An n-fold axis with a perpendicular twofold axis. (**b**) An n-fold axis with a σ_v mirror plane

Let us next consider two different operations about the same axis. As Fig. 8.7a shows, the *Umklappung* C_2' reverses the n-fold axis, so we have

$$C_2' C_n C_2'^{-1} = C_n^{-1} . \tag{8.5}$$

A vertical mirror plane σ_v also makes C_n and C_n^{-1} conjugate,

$$\sigma_v C_n \sigma_v^{-1} = C_n^{-1} , \tag{8.6}$$

because σ_v reverses the sense of the rotation (Fig. 8.7b). In either case, C_n and C_n^{-1} belong to the same class. Note that the horizontal mirror plane σ_h has no such power. We have simply $\sigma_h C_n \sigma_h^{-1} = C_n$, because σ_h reverses both the n-fold axis and the sense of rotation.

Using the above rule, the eight elements of group D_{2d} are classified into five classes:

$$E , \quad C_4^2 , \quad 2IC_4 , \quad 2C_2' , \quad 2\sigma_d .$$

Groups containing inversion symmetry can be expressed as a direct-product group $G \times C_i$. Once the elements of G have been classified, the elements of the direct-product group are readily classified, since inversion commutes with any operation. For example, the 16 elements of $D_{4h} = D_{2d} \times C_i$ are divided into ten classes: the above five classes plus the following five classes:

$$I \ , \quad \sigma_h \ , \quad 2C_4 \ , \quad 2\sigma_v \ , \quad 2C_2'' \ .$$

8.4 Irreducible Representations of Point Groups

Irreducible representations of a group are determined from the class structure of the group, see Sect. 4.11. Let us again consider the point group D_{2d}. Since we have $g = 8$ elements in $n_c = 5$ classes, the group should have $n_r = 5$ irreducible representations. The dimensions of those representations are determined by (4.83), or $1^2 + 1^2 + 1^2 + 1^2 + 2^2 = 8$. Then we have four 1-dimensional representations and one 2-dimensional representation. Their characters are obtained by using the first and second orthogonality relations. The irreducible characters obtained in this way are compiled in Appendix B at the end of this book for the 32 crystallographic point groups. The representations below the broken line are double-valued representations, which will be discussed in the next section.

Two conventions are currently used in labelling these irreducible representations. For instance, the representations of D_{2d} are called either A_1, A_2, B_1, B_2, E or Γ_1, Γ_2, Γ_3, Γ_4, Γ_5. In the latter notation, which is due to Bethe, the representations are simply numbered by the suffixes to Γ.

The former notation, initiated by Mulliken for molecular vibrations, is systematic. The dimensions of the irreducible representations of point groups are at most 3. Symbols E and T are assigned to two- and three-dimensional representations, respectively.

The Mulliken notation for one-dimensional representations calls for a little caution. When the character of a one-dimensional representation is complex, it is degenerate with its complex conjugate representation on account of time-reversal symmetry. In the Bethe notation, such a pair of representations is counted as two independent one-dimensional representations, whereas the Mulliken notation regards it as a two-dimensional representation in the spirit that it is physically irreducible. When the character is real, it is $+1$ or -1. According as the character for the n-fold rotation is $+1$ or -1, symbol A or B is assigned. Furthermore, they are distinguished by subscripts 1, 2 according as the character for rotation about the y-axis is $+1$ or -1; and by ($'$), ($''$) for horizontal reflection. The above rule is abandoned as an exception in the dihedral group D_2 (and D_{2h}). Since the z-, y-, and x-axes are equivalent in this group, the four 1-dimensional representations are designated as A, B_1, B_2, and B_3.

As a final rule of the Mulliken notation, if the group has inversion, the symbol carries an extra suffix "g" or "u" to distinguish between even and odd parity under inversion. Apart from the group C_i itself, ten point groups contain inversion as their group element and can be expressed as a direct product of C_i with a point group of proper rotations:

$$\begin{aligned} &O_h = O \times C_i \quad && D_{6h} = D_6 \times C_i \quad && D_{2h} = D_2 \times C_i \\ &T_h = T \times C_i && C_{6h} = C_6 \times C_i && C_{2h} = C_2 \times C_i \\ &D_{4h} = D_4 \times C_i && D_{3d} = D_3 \times C_i && \\ &C_{4h} = C_4 \times C_i && C_{3i} = C_3 \times C_i \ . && \end{aligned} \tag{8.7}$$

These groups have twice as many irreducible representations as those on the right-hand sides. Half of them are "g" representations and the other half "u" ones.

8.5 Double-Valued Representations and Double Groups

We have so far neglected spin degrees of freedom. If we consider electron spin, spatial rotations will influence the spin functions as well as orbital functions. Rotation operator R about an axis $\boldsymbol{n} = (\lambda, \mu, \nu)$ through an angle θ is represented by

$$\begin{aligned} R &= \exp(-\mathrm{i}\theta \boldsymbol{n} \cdot \boldsymbol{s}) \\ &= \exp[-\mathrm{i}\theta(\lambda s_x + \mu s_y + \nu s_z)] \end{aligned} \tag{8.8}$$

in terms of the spin angular momentum operator $\boldsymbol{s}$. The explicit matrix representation for this rotation operator is given by (7.53), which is nothing other than the irreducible representation $D^{(1/2)}$ of the rotation group (Sects. 7.5, 7.6). The spin functions α and β, which are eigenstates of s_z,

$$s_z \alpha = \tfrac{1}{2}\alpha \ , \quad s_z \beta = -\tfrac{1}{2}\beta \ , \tag{8.9}$$

are transformed by the above rotation R as

$$[R\alpha, R\beta] = [\alpha, \beta]\hat{D}^{(1/2)}(R) \ . \tag{8.10}$$

Now a peculiar feature arises from the matrix (7.53). Suppose $\bar{R}$ denotes a rotation in the same direction $\boldsymbol{n}$ but with a rotation angle $\theta + 2\pi$ in place of θ. R and $\bar{R}$ should be physically the same operation. Nevertheless, we have

$$\hat{D}^{(1/2)}(\bar{R}) = -\hat{D}^{(1/2)}(R)$$

from (7.53). Such a representation is said to be *double-valued*, because it has two representation matrices for a physically identical rotation.

The usual representation theory assumes only single-valued representations, which associate a single matrix with a single group element. Admission of double-valued representations requires some generalization of the group representation theory. We have two possible approaches to this: In one approach, the group is left as it stands and the representation theory is generalized to cover ray representations (see Sect. 5.4 and the worked example for D_6). The other approach is to enlarge the group (double the number of group elements) to make the correspondence one-to-one, and remain in the conventional representation theory. We choose here the latter.

Since the 2π rotation changes the sign of the spin functions α and β, we have to distinguish this rotation $\bar{E}$ from the unit element E. Similarly, we distinguish between the elements R of the original group (called the single group) and its product with $\bar{E}$,

$$\bar{R} = R\bar{E} \ .$$

The group obtained in this way, which has twice as many elements, is called the *double group.* It is designated as G' by attaching a prime to the Schönflies symbol G.

It is important to note that products of group elements in a double group do not parallel those in the single group. The C_n rotation yields the unit element E only after $2n$ successive operations:

$$C_n^{2n} = E \ , \quad C_n^n = \bar{E} \ .$$

Its inverse element is

$$C_n^{-1} = C_n^{2n-1} = \bar{E}C_n^{n-1} \ .$$

In particular, if $n = 2$, we have $C_2^{-1} = \bar{C}_2$. As to the inversion I, we have nothing to worry about because it commutes with any rotation (including $\bar{E}$). So, the relation $I^2 = E$ holds as in the single group.

Mirror reflection σ, being defined by $\sigma = IC_2$, satisfies

$$\sigma^{-1} = \bar{\sigma} \ , \quad \sigma^2 = \bar{E} \ .$$

From these relations, rotatory reflections and rotatory inversions are connected through

$$S_4 = I\bar{C}_4^{-1} \ , \quad S_3 = I\bar{C}_6^{-1} \ , \quad S_6 = I\bar{C}_3^{-1}$$

in double groups. Note the difference from (8.2, 3).

The conjugate relations (8.4–6) can be used in double groups as well. Let us take as an example the D'_{2d} double group. The 16 group elements are divided into 7 classes as shown in the top row of Table 8.1. Doubling the group elements does not necessarily mean doubling the number of classes, because R and $\bar{R}$ can belong to the same class. The group should have as many irreducible representations as the number of classes. From $1^2 + 1^2 + 1^2 + 1^2 + 2^2 + 2^2 + 2^2 = 16$, we obtain four 1-dimensional representations and three 2-dimensional

Table 8.1. Classification of the elements of the double point group D'_{2d} and its irreducible representations

		E	$\bar{E}$	IC_4 IC_4^{-1}	$I\bar{C}_4$ $I\bar{C}_4^{-1}$	C_4^2 $\bar{E}C_4^2$	$2C'_2$ $2\bar{C}'_2$	$2\sigma_d$ $2\bar{\sigma}_d$
A_1	Γ_1	1	1	1	1	1	1	1
A_2	Γ_2	1	1	1	1	1	-1	-1
B_1	Γ_3	1	1	-1	-1	1	1	-1
B_2	Γ_4	1	1	-1	-1	1	-1	1
E	Γ_5	2	2	0	0	-2	0	0
$E_{1/2}$	Γ_6	2	-2	$\sqrt{2}$	$-\sqrt{2}$	0	0	0
$E_{3/2}$	Γ_7	2	-2	$-\sqrt{2}$	$\sqrt{2}$	0	0	0

ones, whose characters are given in Table 8.1. Of these seven representations, Γ_1–Γ_5 are single-valued representations, which also appeared in the single group. Only Γ_6 and Γ_7 are new members and their characters are really double-valued,

$$\chi(\bar{R}) = -\chi(R) . \tag{8.11}$$

In this sense, the double-valued representations are sometimes called additional, extra, or specific representations of the double group. The first and second orthogonalities of the characters hold among these seven representations.

In double-valued representations, the character $\chi(R)$ vanishes if R and $\bar{R}$ belong to the same class. This can happen in (8.5, 6) when $n = 2$. Therefore, the character vanishes for all double-valued representations when C_2 and $\bar{C}_2$ (σ and $\bar{\sigma}$) belong to the same class.

A few words are in order here on the notation for double-valued representations. In the Bethe notation, the double-valued representations are labelled by simply increasing the number of the suffix to Γ. On the other hand, the Mulliken notation for double-valued representations is not well established. The notation adopted in Appendix B obeys the following rule [8.1]. Double-valued irreducible representations of the point groups are either four, two, or one dimensional. The four-dimensional representation is designated as $G_{3/2}$, which derives from the representation $D^{(3/2)}$ of the rotation group. Two-dimensional representations corresponding to atomic $M = \pm 1/2$, $\pm 3/2$, $\pm 5/2$ levels are labelled $E_{1/2}$, $E_{3/2}$, and $E_{5/2}$ respectively. Real one-dimensional representations are designated as B.

So far we have studied double-valued representations in terms of the double group. When viewed from the theory of ray representations, the above result may be understood as follows (see Sect. 5.4 for more details): The double group G' has an invariant subgroup $H = \{E, \bar{E}\}$, and the factor group G'/H is isomorphic to the single group G. The double-valued representation of G is a ray representation of G'/H. In addition, if R and $\bar{R}$ belong to the same class in the double group, R belongs to the zero class of G'/H. The character of ray representations vanishes for zero classes, while it is nonvanishing for ray classes. As may be seen from the example of Table 8.1, the following relations hold in general in ray representations:

number of irreducible double-valued representations = number of ray classes = 2 in Table 8.1 ,

square sum of the dimensions of irreducible double-valued representations = order of $G = 8$ in Table 8.1.

As an example of the application of double-valued representations, let us consider how one-electron states are varied by introduction of electron spin. Suppose that orbital functions ψ_ν form a basis for a representation Γ of the single group. The dimension d of Γ stands for the orbital degeneracy. Now, inclusion of electron spin gives $2d$ spin orbitals $\psi_\nu\alpha$, $\psi_\nu\beta$. Since the spin functions $\{\alpha, \beta\}$ form a basis for the representation $E_{1/2}$ of the double group (which corresponds to the $D^{(1/2)}$ representation of the rotation group), the above spin orbitals form the basis for the direct-product representation $\Gamma \times E_{1/2}$. Reduction of this product tells us the symmetry of resulting electron states as well as possible lifting of the degeneracy.

For example, choose $\Gamma = E$ of D_{2d}. This two-dimensional representation has the basis transforming like $\{x, y\}$. Reduction of $\Gamma \times E_{1/2}$ becomes here

$$E \times E_{1/2} = E_{1/2} + E_{3/2} .$$

Appearance of two representations on the right-hand side means that inclusion of the spin-orbit interaction leads to energy level splitting into 2 twofold-degenerate levels.

8.6 Transformation of Spin and Orbital Functions

When electron wavefunctions contain both orbital and spin functions, the rotational operation

$$R = \exp(-\mathrm{i}\theta \boldsymbol{n} \cdot \boldsymbol{l}) \exp(-\mathrm{i}\theta \boldsymbol{n} \cdot \boldsymbol{s}) \tag{8.12}$$

has to be applied correctly to both parts. Although the operation is well defined by itself, incorrect interpretation sometimes leads to erroneous results. Table 8.2 lists the result of some representative rotations. For improper rotations, operate inversion on the orbital part after the rotation. Inversion leaves the spin functions unchanged.

*8.7 Constructive Derivation of Point Groups Consisting of Proper Rotations

In this section, we wish to show that Table 8.3 exhausts the point groups consisting of proper rotations.

Table 8.2. Effects of typical rotations on the orbital functions $f(x, y, z)$ and the spin functions α, β

Rotation R axis	angle	$Rf(x, y, z)$	$R\alpha$	$R\beta$
z-axis	θ	$f(x\cos\theta + y\sin\theta,$ $y\cos\theta - x\sin\theta, z)$	$\alpha e^{-i\theta/2}$	$\beta e^{i\theta/2}$
y-axis	θ	$f(x\cos\theta - z\sin\theta, y,$ $z\cos\theta + x\sin\theta)$	$\alpha\cos\frac{\theta}{2} + \beta\sin\frac{\theta}{2}$	$\beta\cos\frac{\theta}{2} - \alpha\sin\frac{\theta}{2}$
x-axis	θ	$f(x, y\cos\theta + z\sin\theta,$ $z\cos\theta - y\sin\theta)$	$\alpha\cos\frac{\theta}{2} - i\beta\sin\frac{\theta}{2}$	$\beta\cos\frac{\theta}{2} - i\alpha\sin\frac{\theta}{2}$
[110]	π	$f(y, x, -z)$	$(1-i)\beta/\sqrt{2}$	$(-1-i)\alpha/\sqrt{2}$
$[1\bar{1}0]$	π	$f(-y, -x, -z)$	$(-1-i)\beta/\sqrt{2}$	$(1-i)\alpha/\sqrt{2}$
[111]	$2\pi/3$	$f(y, z, x)$	$(1-i)(\alpha+\beta)/2$	$(1+i)(\beta-\alpha)/2$

Table 8.3. Point groups consisting of proper rotations. For the meaning of the superscript a, see the text

Number of stars λ	Group	Order g		Star (m_i, v_i) $i=1$	$i=2$	$i=3$	Rotations
2	C_n	n		$(n, 1)$	$(n, 1)^a$		$1C_n$
3	D_n	$2n$	$n=2k$	$(2, n)$	$(2, n)$	$(n, 2)$	$1C_n, kC'_2, kC''_2$
			$n=2k+1$	$(2, n)$	$(2, n)^a$	$(n, 2)$	$1C_n, nC'_2$
	T	12		(2, 6)	(3, 4)	$(3, 4)^a$	$3C_2, 4C_3$
	O	24		(2, 12)	(3, 8)	(4, 6)	$6C'_2, 4C_3, 3C_4$
	P	60		(2,30)	(3,20)	(5,12)	$15C_2, 10C_3, 6C_5$

A rotational operation R about the origin O sends the points on a unit sphere to other points on the sphere. Only the poles p and $\bar{p}$ at which the rotation axis pierces the sphere remain unmoved (Fig. 8.8). When R is a $2\pi/m$ rotation, we will call both p and $\bar{p}$ poles of order m. The corresponding cyclic group C_m is a subgroup of the point group G.

Let us decompose the group G into left cosets with respect to the subgroup C_m,

$$G = C_m + R_2 C_m + R_3 C_m + \cdots + R_v C_m ,$$

$$R_1 = E , \quad g = vm .$$

The first coset C_m leaves the pole p unmoved. Other cosets, say $R_j C_m$, move the pole p to other poles p_j. The v poles $p, p_2, p_3, \ldots, p_v$ are all different, and have the same order m. (The group $R_j C_m R_j^{-1}$, which is conjugate and isomorphic to

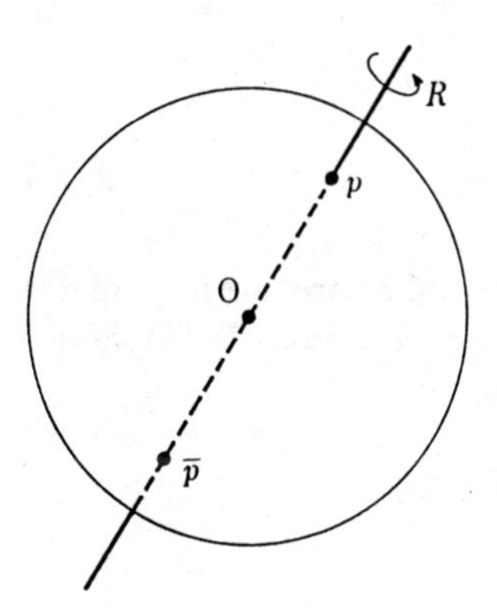

Fig. 8.8. Poles p and $\bar{p}$ for a rotation R

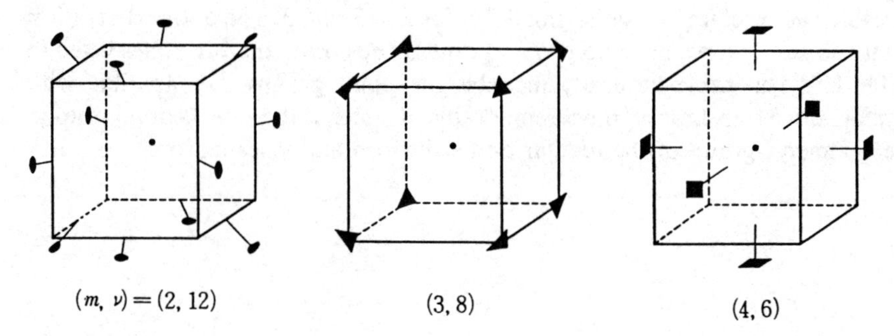

Fig. 8.9. Poles of the octahedral group O

the group C_m, leaves the pole p_j unmoved. It is a cyclic group of order m as C_m.) Let us call the set of these ν poles the *star of p*. A star is characterized by a pair (m, ν). For example, the octahedral group O has three kinds of stars shown in Fig. 8.9. In this example, p and $\bar{p}$ always belong to the same star, which holds only when the group contains an operation that takes p to $\bar{p}$. In general, p and $\bar{p}$ can belong to different stars. (Such stars are given a superscript a in Table 8.3.)

Now, apart from the identity operation, each of the other $(g-1)$ rotations has a couple of poles. So the number of poles amounts to $2(g-1)$ in total, where a pole of order m is counted $(m-1)$ times.

Suppose the group G has λ stars and number them by an index i. Then we have

$$\sum_{i=1}^{\lambda} \nu_i(m_i - 1) = 2(g-1) \ , \tag{8.13}$$

$$\nu_i m_i = g \ . \tag{8.14}$$

Elimination of v_i gives

$$\sum_{i=1}^{\lambda}\left(1-\frac{1}{m_i}\right)=2\left(1-\frac{1}{g}\right). \tag{8.15}$$

Except for the trivial group C_1 ($g = 1, \lambda = 0$), both g and m_i are integers greater than or equal to 2. Therefore, the left-hand and right-hand sides of (8.15) should fall in the intervals

$$\frac{\lambda}{2} \leqq \sum_{i=1}^{\lambda}\left(1-\frac{1}{m_i}\right) < \lambda ,$$

$$1 \leqq 2\left(1-\frac{1}{g}\right) < 2 .$$

Equality of the both sides restricts λ to 2 and 3. Table 8.3 summarizes the sets of possible values of (m_i, v_i) satisfying (8.15) for $\lambda = 2$ and 3. The above derivation demonstrates that we have no proper point groups other than the ones listed in Table 8.3. This table naturally includes the point groups incompatible with crystal lattice translations, most remarkable of which is the icosahedral group P, the symmetry group of the regular dodecahedron and icosahedron.

9. Electronic States of Molecules

We treat in this chapter the symmetry of electronic states of molecules as an application of the representation theory of point groups. Our first problem is to characterize the orbitals of a molecule as the basis functions for the irreducible representations of the molecular symmetry group and to construct such orbitals in the form of linear combinations of atomic orbitals (LCAO). This is equivalent to reducing a representation based on the atomic orbitals into its irreducible components. Its inverse problem is to derive directed hybridized orbitals from constituent atomic orbitals. As another application, we consider the splitting of atomic levels in ligand fields (subduction to subgroups). When we go over to many-electron problems in the theory of molecules, we have to set up many-electron wavefunctions having correct symmetry and satisfying the Pauli exclusion principle. This can be solved in the same way as in the case of atomic systems.

9.1 Molecular Orbitals

The method of molecular orbitals starts from the view that each electron in the molecule moves independently of the others in the field of an average potential $V(\boldsymbol{r})$ due to the nuclei and the other electrons. In this approximation, the electronic structure of the molecule is described as the state in which each electron occupies a *molecular orbital* (abbreviated MO) $\psi_j(\boldsymbol{r})$ extended over the whole system. Each orbital ψ_j is assumed to satisfy the Schrödinger equation

$$H\psi_j \equiv \left(-\frac{\hbar^2}{2m}\nabla^2 + V(\boldsymbol{r})\right)\psi_j(\boldsymbol{r}) = \varepsilon_j\psi_j(\boldsymbol{r}) \ . \tag{9.1}$$

When the molecule possesses a certain symmetry, it will be reflected in the potential energy $V(\boldsymbol{r})$. The molecular orbital method assumes that $V(\boldsymbol{r})$ has the same symmetry as the whole molecule.

Let us write G for the group of symmetry operations R of the molecule. We define R as an operator acting upon an orbital ψ by

$$\psi'(\boldsymbol{r}) = R\psi(\boldsymbol{r}) \ , \quad \psi'(\boldsymbol{r}') = \psi(\boldsymbol{r}) \ , \tag{9.2}$$

when R carries the point $\boldsymbol{r}$ in space to $\boldsymbol{r}'$ (Sect. 4.3). The electron at $\boldsymbol{r}$ will be carried by R to $\boldsymbol{r}'$, but it will then feel the same field as at $\boldsymbol{r}$. We thus have

$$V(\boldsymbol{r}) = V(\boldsymbol{r}') \ , \quad \text{or} \tag{9.3}$$

$$V(\boldsymbol{r})\psi(\boldsymbol{r}) = V(\boldsymbol{r}')\psi'(\boldsymbol{r}') \ . \tag{9.4}$$

The left-hand side must be equal to the value of $RV\psi$ at $\boldsymbol{r}'$, while the right-hand side is that of $VR\psi$ at the same point so that

$$RV\psi = VR\psi \ . \tag{9.5}$$

The same equation holds if we replace V by the Laplacian operator:

$$R\nabla^2\psi = \nabla^2 R\psi \ . \tag{9.6}$$

In the case of the Laplacian, we have only to show that $\nabla_{\boldsymbol{r}}^2 = \nabla_{\boldsymbol{r}'}^2$, corresponding to (9.3). It is not so difficult to confirm it directly by a change of variables, taking, for example, the rotation about the z-axis through an angle α as R.

According to (9.5, 6), the Hamiltonian H in (9.1) commutes with any symmetry operation R of the molecule:

$$RH = HR \ , \qquad R \in G \ , \tag{9.7}$$

or H is invariant under R.

It then follows from (9.7) that

$$HR\psi_j = RH\psi_j = \varepsilon_j R\psi_j \ ,$$

which means $R\psi_j$ is also an eigenfunction of H with the same eigenvalue ε_j as ψ_j. Suppose the level is d-fold degenerate and let us use α and ν instead of j. That is, we denote by $\psi_\nu^{(\alpha)}$ $(\nu = 1, 2, \ldots, d)$ the eigenfunctions with the same eigenvalue $\varepsilon^{(\alpha)}$. Then $R\psi_\nu^{(\alpha)}$ will be expressed as

$$R\psi_\nu^{(\alpha)} = \sum_{\mu=1}^{d} \psi_\mu^{(\alpha)} D_{\mu\nu}^{(\alpha)}(R) \tag{9.8}$$

in terms of $\psi_\nu^{(\alpha)}$. We observe thus that the matrices on the right-hand side of (9.8) provide us with a representation of the group G and the eigenfunctions form a basis for it. Unless there is some particular reason to consider otherwise, we may assume the representation to be irreducible.

This leads to the following consequences.

(1) Each level and the associated eigenfunctions may be labelled by the symmetry of the orbital functions or by the species Γ of the irreducible representation for which they form a basis. To label one-electron levels and orbitals, we use small letters such as a, b, e, t instead of Mulliken's notation in capital letters A, B, E, T for the irreducible representations of point groups. Of course, there are an infinite number of levels and associated functions so that symmetry alone will be insufficient to classify them. In the case of one-electron levels and orbitals, for example, figures are attached to indicate the order of appearance, starting from the lowest (as 1a, 2a, 1b, etc.).

(2) Matrix elements of the Hamiltonian between the wavefunctions of different symmetry Γ and Γ' will necessarily vanish. This can be shown directly by the use of orthogonality relations for the matrices of two inequivalent irreducible representations (Sect. 6.5). On the other hand, there is in general no reason to suppose that they vanish when Γ and Γ' are the same.

When we want to diagonalize the Hamiltonian H within the manifold of a finite number of atomic orbitals, say, $\varphi_1, \varphi_2, \ldots, \varphi_N$, our first problem will be somehow, for example by symmetry considerations, to set up linear combinations of them such that they form basis sets for the irreducible representations Γ of the group G:

$$\phi_\gamma^{\nu(\Gamma)} = \sum_{i=1}^{N} c_i(\nu\Gamma\gamma)\varphi_i \ , \tag{9.9}$$

where we have introduced the superscript ν to ϕ to distinguish between independent sets within the same symmetry species Γ. Once we succeed in preparing the wavefunctions (9.9), we shall be left only with the diagonalization of the matrix of H with respect to ν.

9.2 Diatomic Molecules: LCAO Method

We begin with a very simple molecule, i.e., H_2 or H_2^+. Since the symmetry group of a homonuclear diatomic molecule is $D_{\infty h}$, one-electron wavefunctions as well as orbital levels are characterized by its irreducible representations a_{1g}, a_{2u}, $e_{1g}, \ldots$ (see Appendix B at the end of the book). Instead of this Mulliken notation, however, the symbols σ_g^+, σ_u^+, $\pi_g, \ldots$ derived from those for a heteronuclear molecule (with symmetry $C_{\infty v}$) are used more commonly. The Greek letters $\sigma, \pi, \delta, \ldots$ represent the eigenvalues $\lambda = 0, \pm 1, \pm 2, \ldots$ of angular momentum about the molecular axis (the z-axis). All states except σ have double degeneracy corresponding to the clockwise and counterclockwise rotation.

The basis function for the representation σ (called the σ orbital) is axially symmetric about the z-axis. When the position $\boldsymbol{r}$ of the electron is expressed in terms of the cylindrical coordinates (ϱ, φ, z), the function does not depend on the angle φ. The φ dependence of the basis functions ψ_λ and $\psi_{-\lambda}$ with $\lambda \neq 0$ are given by

$$\psi_{\pm\lambda}(\boldsymbol{r}) = f(\varrho, z)\mathrm{e}^{\pm \mathrm{i}\lambda\varphi} \ , \tag{9.10}$$

so that these functions transform according to the rule

$$R_\alpha \psi_{\pm\lambda} = \mathrm{e}^{\mp \mathrm{i}\lambda\alpha}\psi_{\pm\lambda}$$

under the rotation R_α about the z-axis through an angle α. For the basis functions for the representations π, δ (or π, δ orbitals), real forms are sometimes preferred to the complex ones like (9.10). For example, orbitals

$$\psi_x = \frac{x}{\varrho} f(\varrho, z) , \quad \psi_y = \frac{y}{\varrho} f(\varrho, z)$$

are often used instead of π orbitals ψ_{+1} and ψ_{-1}.

The + and − superscripts to σ indicate whether the function remains invariant or changes its sign under the mirror reflection σ_v in the plane containing the molecular axis. The indices g and u stand for the parity (g: even, u: odd) of the function under inversion. In the case of heteronuclear molecules, there is no distinction between g and u. Note that $D_{\infty h} = C_{\infty v} \times C_i$.

These considerations merely tell us what kind of symmetry is possible for a level, or more precisely, for a molecular orbital of the homonuclear diatomic molecule. They tell us nothing about which kinds of levels have lower energies nor how orbitals with the same symmetry differ as functions of their arguments. We can, however, obtain some insight into these matters through the following analysis for H_2^+.

Suppose we separate the two protons A and B of the H_2^+ ion on the z-axis by a very large distance, keeping its center fixed at the origin. During this process, the symmetry of the molecule does not change, although the internuclear distance R does. The shape of each molecular orbital ψ will vary so that it coincides at an infinite distance, apart from a constant factor, with the atomic orbitals $\phi_A(nlm)$ and $\phi_B(nlm)$ around the A and B atoms, respectively. These atomic orbitals must have the same symmetry as the molecular orbital with respect to the rotation C_∞ about the z-axis and the reflection σ_v in the xz-plane. If we note that we have $I\phi_A(nlm) = (-1)^l \phi_B(nlm)$ under the inversion I, we can write down explicit forms of ψ at $R = \infty$ from symmetry considerations as follows:

$$R = \infty: \psi(\sigma^+_{g,u}) = \frac{1}{\sqrt{2}} \{\phi_A(nl0) \pm (-1)^l \phi_B(nl0)\} , \tag{9.11a}$$

$$\psi(\pi_{g,u}) = \frac{1}{\sqrt{2}} \{\phi_A(nl, \pm 1) \pm (-1)^l \phi_B(nl, \pm 1)\} , \tag{9.11b}$$

where the $\pm$ signs in front of $(-1)^l$ correspond to the parity g and u.

Conversely, if we let the distance R decrease so that A and B merge into a united atom C at the origin, the orbital ψ will turn into an atomic orbital $\phi_C(n'l'm')$ of C at $R = 0$. We thus have

$$R = 0: \quad \psi(\sigma^+_{g,u}) = \phi_C(n'l'0) , \qquad \text{g:} \quad l'\text{even} , \tag{9.12a}$$

$$\psi(\pi_{g,u}) = \phi_C(n'l', \pm 1) , \qquad \text{u:} \quad l'\text{odd} . \tag{9.12b}$$

How the nlm states of separated atoms turn into the $n'l'm'$ state of the united atom via the molecular state with decreasing R may be seen by applying the *non-*

crossing rule, starting from the levels with lowest energies. The non-crossing rule demands that the levels with the same symmetry should never cross when R is changed. This rule is based on the fact that the nondiagonal elements between wavefunctions of the same symmetry are nonvanishing in general, as remarked in Sect. 9.1, leading to repulsion between the levels. The state with the lowest energy is the σ_g^+ state which becomes $\phi_C(1s)$ when united and turns into $\{\phi_A(1s) + \phi_B(1s)\}/\sqrt{2}$ when separated infinitely. Let us denote this orbital as $\sigma_g^+\,1s$, the notation $1s$ standing for the $1s$ state of separated atoms. In contrast, the σ_u^+ or $\sigma_u^+\,1s^*$ state related to $\{\phi_A(1s) - \phi_B(1s)\}/\sqrt{2}$ at $R = \infty$ will have higher energy, because the wavefunction has a nodal plane at $z = 0$, although it is degenerate with σ_g^+ at $R = \infty$. As a matter of fact, the orbital turns into $\phi_C(2p0)$ of the united atom. In this way, we can draw the correlation diagram given in Fig. 9.1. The diagram is only schematic: the order of levels at both ends ($R = 0$ and $R = \infty$) is correct, whereas the arrangement of molecular levels in between is by no means definite.

If we want to calculate orbital energies for an actual molecule, we need explicit expressions for the molecular orbitals. As a practical means, we have the LCAO method (Linear Combination of Atomic Orbitals). As the name suggests, this approximates the molecular orbital by a linear combination of atomic orbitals. In the case of σ_g^+ of H_2, this amounts to using (9.11), which is valid for $R = \infty$, also for finite R. For example, we put

$$\psi(\sigma_g^+\,1s) = \frac{\phi_A(1s) + \phi_B(1s)}{\sqrt{2(1+S)}} \tag{9.13}$$

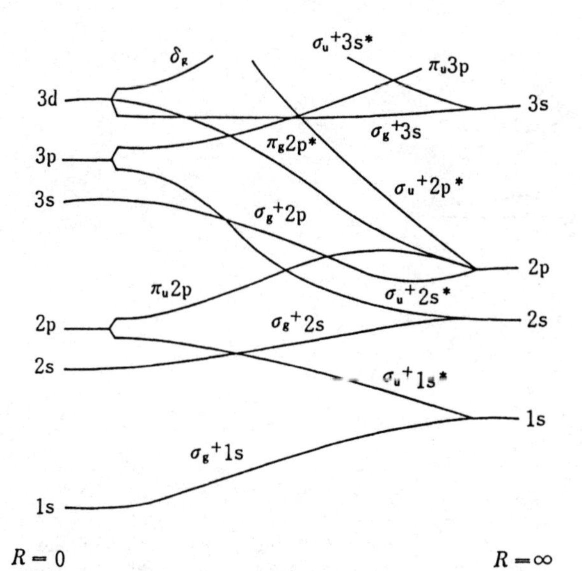

Fig. 9.1. Correlation diagram for homonuclear diatomic molecules

and

$$\psi(\sigma_u^+ 1s^*) = \frac{\phi_A(1s) - \phi_B(1s)}{\sqrt{2(1-S)}} , \tag{9.14}$$

where

$$S = \int d\tau \phi_A(1s)\phi_B(1s)$$

is the overlap integral. At a finite distance R, ϕ_A and ϕ_B need not necessarily be the $1s$ function of the hydrogen atom. The radial wavefunction should rather be determined by a variational means so as to minimize the total energy of the system.

In Fig. 9.2, we plot the values of the functions (9.13, 14) along the molecular axis. When the two hydrogen atoms get closer and come into the region where molecular binding occurs, the electrons lower their (mainly kinetic) energies by moving into the molecular orbital $\psi(\sigma_g^+ 1s)$ spread over the whole molecule with opposite spins rather than remaining in the localized atomic orbitals ϕ_A and ϕ_B. This is the molecular-orbital point of view for the binding. In this sense $\psi(\sigma_g^+ 1s)$ is called the *bonding orbital*. Contrary to this, the binding will become unstable if we put an electron in the orbital $\psi(\sigma_u^+ 1s^*)$ given by (9.14), so that the latter is called the *antibonding orbital*. The LCAO-MO denotes antibonding orbitals with an asterisk *.

We give below the electron configuration for the ground state of some simple homonuclear diatomic molecules in terms of the molecular orbitals given in

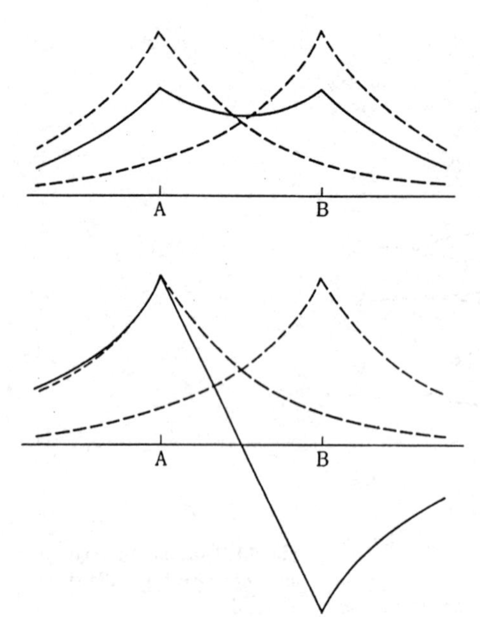

Fig. 9.2. Bonding (*upper*) and antibonding (*lower*) LCAO molecular orbitals in homonuclear diatomic molecules. The atomic orbitals are assumed to be $1s$

Fig. 9.1:

$H_2 \quad (\sigma_g^+ 1s)^2$

$Li_2 \quad (\sigma_g^+ 1s)^2(\sigma_u^+ 1s^*)^2(\sigma_g^+ 2s)^2$

$N_2 \quad (\quad)^2(\quad)^2(\sigma_g^+ 2s)^2(\sigma_u^+ 2s^*)^2(\sigma_g^+ 2p)^2(\pi_u 2p)^4$

$O_2 \quad (\quad)^2(\quad)^2(\quad)^2(\quad)^2(\sigma_g^+ 2p)^2(\pi_u 2p)^4(\pi_g 2p^*)^2$

Note that the oxygen molecule has an incomplete shell in the ground configuration.

9.3 Construction of LCAO-MO: The π-Electron Approximation for the Benzene Molecule

In the language of LCAO-MOs, the ground and several low excited states of the benzene molecule (C_6H_6) are described as follows: Each of the six carbon atoms has a π orbital ($2p_z = \phi(2p0)$) directed perpendicular to the benzene ring. Appropriate linear combinations of these six π orbitals will give rise to six MOs (three bonding and three antibonding orbitals), among which six electrons from carbon atoms are distributed.

The $1s$ electrons of C are tightly bound to the nucleus and will not contribute much to the binding. From the orbitals $2s$, $2p_x$ [$=\{-\phi(2p1)+\phi(2p-1)\}/\sqrt{2}$], $2p_y$ [$=\mathrm{i}\{\phi(2p1)+\phi(2p-1)\}/\sqrt{2}$] of six C and $1s$ of six H, 24 molecular orbitals may be formed, half of them being bonding orbitals. Among the valence electrons totalling 30 from six C and six H ($30 = 4\times 6 + 1\times 6$), 24 are accommodated in the 12 bonding orbitals. In the π-electron approximation, they are simply assumed to give rise to an average potential field for the six π electrons.

By choosing appropriate linear combinations of $2s$, $2p_x$, $2p_y$ of each C atom, we can form three hybridized orbitals extending toward the neighboring H and two C atoms. As with diatomic molecules, such orbitals are called σ orbitals. We thus call the $2p_z$ orbitals here π orbitals, because they are extended toward the direction perpendicular to the σ bond similarly to those in diatomic molecules.

We now examine what kinds of molecular orbitals can be formed from linear combinations of the six π orbitals of C. Let us number the six C atoms as shown in Fig. 9.3 and label the π orbitals as $\phi_1, \phi_2, \ldots, \phi_6$ corresponding to Fig. 9.3.

Under the symmetry transformations of the group D_{6h} for the benzene molecule, the six functions are transformed into each other as

$$\begin{aligned}
&C_6\phi_k = \phi_{k+1}\,,\\
&C_{2y}\phi_1 = -\phi_1, C_{2y}\phi_4 = -\phi_4, C_{2y}\phi_2 = -\phi_6, C_{2y}\phi_3 = -\phi_5, \ldots,\\
&C_{2x}\phi_1 = -\phi_4, C_{2x}\phi_4 = -\phi_1, C_{2x}\phi_2 = -\phi_3, C_{2x}\phi_3 = -\phi_2, \ldots,\\
&I\phi_k = -\phi_{k+3}\,,
\end{aligned} \tag{9.15}$$

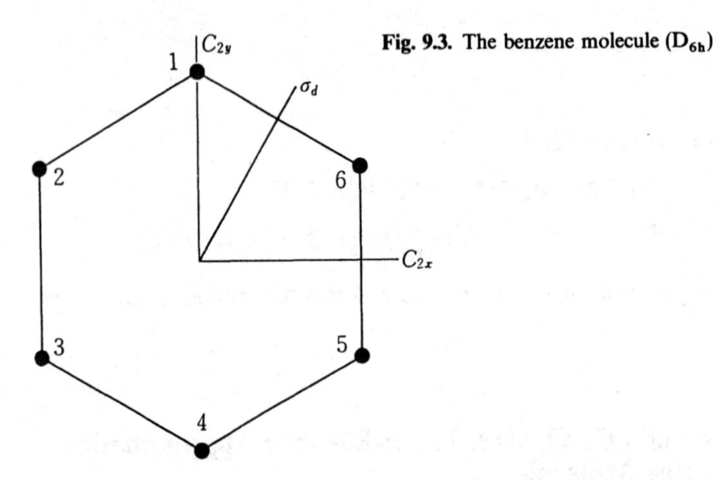

Fig. 9.3. The benzene molecule (D_{6h})

where we put $\phi_{k+6} = \phi_k$. They thus form the basis set for a six-dimensional representation Γ of the group. The irreducible representations that arise from the decomposition of it may be found by calculating its character $\chi^{(\Gamma)}(R)$ and expressing it as a sum of irreducible characters. The values of the character may be obtained easily by picking up only those orbitals that do not change position under R. The result is shown in Table 9.1 and leads to the decomposition

$$\Gamma = A_{2u} + B_{2g} + E_{1g} + E_{2u} \ . \tag{9.16}$$

The molecular orbitals made up of the π orbitals are thus $\psi(a_{2u})$, $\psi(b_{2g})$, $\{\psi(e_{1g}1), \psi(e_{1g}2)\}$ and $\{\psi(e_{2u}1), \psi(e_{2u}2)\}$.

Exercise 9.1. Show that molecular orbitals with symmetry A_{1g}, B_{1u}, E_{1u}, E_{2g} are obtained from 2*s* orbitals of C and 1*s* orbitals of H. Show also that molecular orbitals of symmetry A_{2g}, B_{2u}, E_{1u}, E_{2g} as well as A_{1g}, B_{1u}, E_{1u}, E_{2g} result from $2p_x$ and $2p_y$ orbitals of C.

Several methods are available to determine explicit expressions for the linear combination (the basis sets for the irreducible representations, or *symmetry-adapted functions*).

(1) In simple cases, they may be written down almost intuitively. For example, we have

$$\psi(a_{2u}) = \frac{1}{\sqrt{6}}(\phi_1 + \phi_2 + \phi_3 + \phi_4 + \phi_5 + \phi_6) \ ,$$

Table 9.1. Character of the representation based on the six π orbitals of the benzene molecule

D_{6h}	E	$2C_6$	$2C_3$	C_2	$3C_{2y}$	$3C_{2x}$	I	$2S_3$	$2S_6$	σ_h	$3\sigma_d$	$3\sigma_v$
$\chi^{(\Gamma)}(R)$	6	0	0	0	−2	0	0	0	0	−6	0	2

where we have neglected in its normalization the overlap integral between orbitals on different sites.

(2) In the case of a one-dimensional representation, the projection-operator method of Sect. 6.6 is convenient. Here the character itself plays the role of the representation matrices and, for example,

$$\psi_1 = \sum_R \chi^{(B_{2g})}(R)^* R\phi_1 \tag{9.17}$$

is the molecular orbital with symmetry B_{2g}. The summation on the right-hand side is to be taken over all symmetry operations belonging to D_{6h}. Normalization then yields

$$\psi(b_{2g}) = \frac{1}{\sqrt{6}}(\phi_1 - \phi_2 + \phi_3 - \phi_4 + \phi_5 - \phi_6) \ .$$

Equation (9.17) may also be employed when the dimension of the representation is two or more. Take E_{1g}, for example. First we obtain from ϕ_1 the function

$$\psi_1 = \sum_R \chi^{(E_{1g})}(R)^* R\phi_1 = 2(2\phi_1 + \phi_2 - \phi_3 - 2\phi_4 - \phi_5 + \phi_6) \ .$$

This must be a linear combination of the basis functions for the irreducible representation E_{1g}. Since E_{1g} is two dimensional, another linearly independent combination is required. By applying the same procedure to ϕ_2, we obtain

$$\psi_2 = 2(\phi_1 + 2\phi_2 + \phi_3 - \phi_4 - 2\phi_5 - \phi_6) \ .$$

The function ψ_2 is linearly independent of ψ_1, but not orthogonal to it. By taking the combination $\psi_2 - \psi_1/2$, which is orthogonal to ψ_1, and normalizing it, we obtain, say, $\psi(e_{1g}1)$:

$$\psi(e_{1g}1) = \tfrac{1}{2}(-\phi_2 - \phi_3 + \phi_5 + \phi_6) \ .$$

As the second partner, $\psi(e_{1g}2)$, we choose a normalized form of ψ_1:

$$\psi(e_{1g}2) = \frac{1}{\sqrt{12}}(2\phi_1 + \phi_2 - \phi_3 - 2\phi_4 - \phi_5 + \phi_6) \ .$$

In a similar way, we obtain for E_{2u} the set

$$\psi(e_{2u}1) = \tfrac{1}{2}(-\phi_2 + \phi_3 - \phi_5 + \phi_6) \ ,$$

$$\psi(e_{2u}2) = \frac{1}{\sqrt{12}}(-2\phi_1 + \phi_2 + \phi_3 - 2\phi_4 + \phi_5 + \phi_6) \ .$$

Derivation of the basis sets for two-dimensional representations may look very laborious compared to that for one-dimensional representations. However, there is a quicker way to determine the coefficients of ϕ_k when character tables with typical sets of bases are available, as in Appendix B of this book. If we compare the form of LCAO-MOs given above with the basis functions in the right column of the table for D_6, we note that the coefficients of ϕ_k may be determined by a simple rule. For example, the coefficients of ϕ_k in $\psi(e_{1g}1)$ and $\psi(e_{2g}2)$ are proportional to the values of x and y at the site k. We can usually take advantage of this fact to write down explicit expressions for symmetry-adapted LCAO-MOs.

*9.3.1 Further Methods for Determining the Basis Sets

There are several other methods of determining the basis sets. They are less intuitive but surer, though quite laborious in some cases.

(1) When we know all the irreducible representation matrices, the method of projection operators will give the required basis sets. The representation matrices may be constructed explicitly, if we have some simple analytical expressions for the basis sets as in Appendix B.

(2) From the six-dimensional representation matrices $\hat{D}^{(\Gamma)}(R)$ given by (9.15), we set up the matrices given by

$$\hat{D}(\mathscr{C}_j) = \sum_{R \in \mathscr{C}_j} \hat{D}^{(\Gamma)}(R) \ ,$$

where we sum over R belonging to the jth class $\mathscr{C}_j$ on the right-hand side. It can be shown that the matrices $\hat{D}(\mathscr{C}_j)$ commute with each other (and with any one of $\hat{D}^{(\Gamma)}(R)$) so that all of them are diagonalized simultaneously if we move over to the basis sets for irreducible representations. We obtain explicit forms of the symmetry-adapted functions from the transformations of basis sets ϕ_k to the latter sets. When diagonalized, $\hat{D}(\mathscr{C}_j)$ will have on its diagonal d_α times the eigenvalues $\lambda_j^{(\alpha)} = h_j \chi^{(\alpha)}(\mathscr{C}_j)/d_\alpha$ for each allowed irreducible representation α, where d_α is the dimension of α and h_j is the number of group elements belonging to the class $\mathscr{C}_j$, see (4.107).

(3) When we have explicit forms of the irreducible representation matrices $\hat{D}^{(\alpha)}(R)$, we can transform $\hat{D}^{(\Gamma)}(R)$ so that the reduced forms for the representation α coincide with the known $\hat{D}^{(\alpha)}(R)$. We obtain explicit expressions for the basis set from the transformation matrix. This process need not be carried out for all R. We may choose only the generating elements of the group ($R = C_6, C_{2y}$, and I in the case of D_{6h}).

(4) Method of descent in symmetry. This makes use of the representations of the subgroups. In the present example, we note that

$$D_{6h} = D_6 \times C_i \ , \quad D_6 = C_6 + C_{2y}C_6$$

with the compatibility relations (Table 5.1) between the irreducible representations of the group D_{6h} and its subgroups. Suppose six H atoms change their positions slightly so that the D_{6h} symmetry of the molecule is lowered to C_6. The molecular orbitals, which were basis functions for the irreducible representations of D_{6h}, must now be the ones of C_6. Since the irreducible representations of C_6 are all one dimensional, the basis functions may all be obtained as in (9.17):

$$\psi^{(m)} = \frac{1}{\sqrt{6}} \sum_{k=0}^{5} \chi^{(m)}(C_6^k)^* C_6^k \phi_1 \ , \quad m = 0, \pm 1, \pm 2, 3 \ .$$

They are the basis functions for the mth irreducible representation of C_6, namely,

$$C_6 \psi^{(m)} = \omega^m \psi^{(m)} \ ,$$

$$\chi^{(m)}(C_6) = \omega^m \equiv \exp\left(\frac{2\pi i m}{6}\right) .$$

On the other hand, we know that we have A_{2u}, B_{2g}, E_{1g}, and E_{2u} as irreducible representations of D_{6h} and that in C_6 they become $A(m = 0)$, $B(m = 3)$, $E_1(m = 1, -1)$, and $E_2(m = 2, -2)$, respectively. This immediately tells us that

$$\psi(a_{2u}) = \psi^{(0)} \ , \quad \psi(b_{2g}) = \psi^{(3)} \ .$$

For E_{1g} and E_{2u}, we find that the basis sets derived earlier are expressed as

$$\psi(e_{1g}1) = \frac{1}{\sqrt{2}i}(\psi^{(1)} - \psi^{(-1)}) \ , \quad \psi(e_{1g}2) = \frac{1}{\sqrt{2}}(\psi^{(1)} + \psi^{(-1)}) \ ,$$

$$\psi(e_{2u}1) = \frac{1}{\sqrt{2}i}(\psi^{(2)} - \psi^{(-2)}) \ , \quad \psi(e_{2u}2) = \frac{1}{\sqrt{2}}(-\psi^{(2)} - \psi^{(-2)}) \ .$$

9.4 The Benzene Molecule (Continued)

We have derived explicit expressions for all the π orbitals of the benzene molecule within the framework of LCAO. However, this does not tell us much about the level structure. Information on the orbital energies is to be obtained by solving the Schrödinger equation or by diagonalizing the Hamiltonian matrix within the manifold of six ϕ_k orbitals. We put

$$\psi_j = \sum_{k=1}^{6} c_k \phi_k \tag{9.18}$$

and substitute it into (9.1). If we take inner products of ϕ_k with both sides of the resulting equation, we obtain

$$\sum_{k'} (\phi_k, H\phi_{k'})c_{k'} = \varepsilon_j \sum_{k'} (\phi_k, \phi_{k'})c_{k'} \ . \tag{9.19}$$

Let us assume, for simplicity, that different ϕ_k's are orthogonal and the resonance integrals $(\phi_k, H\phi_{k'})$ are nonvanishing only when k and k' are nearest neighbors. Since Coulomb integrals $(\phi_k, H\phi_k)$ will take the same value for any k, we put

$$(\phi_k, H\phi_k) = \alpha \ .$$

For the same reason, we may set

$$(\phi_k, H\phi_{k\pm 1}) = \beta \ .$$

Equation (9.19) then takes the form

$$(\alpha - \varepsilon_j)c_k + \beta(c_{k+1} + c_{k-1}) = 0 \ , \quad c_{k+6} \equiv c_k \ ,$$

and ε_j are the roots of the secular equation[1]

$$\begin{bmatrix} \alpha-\varepsilon & \beta & & & & \beta \\ \beta & \alpha-\varepsilon & \beta & & & \\ & \beta & \alpha-\varepsilon & \beta & & \\ & & \beta & \alpha-\varepsilon & \beta & \\ & & & \beta & \alpha-\varepsilon & \beta \\ \beta & & & & \beta & \alpha-\varepsilon \end{bmatrix} = 0 \ .$$

The eigenvalues and the corresponding coefficients c_k are easily determined. (As a matter of fact, we already know the coefficients c_k, so that the process of diagonalization is actually unnecessary here.) The result is

$$\begin{aligned} \varepsilon = \alpha - 2\beta \ , &\quad c_k = -c_{k+1} \ , &&: \mathrm{B_{2g}} \ , \\ \alpha - \beta \ , &\quad c_k = -c_{k+1} - c_{k-1} \ , &&: \mathrm{E_{2u}} \ , \\ \alpha + \beta \ , &\quad c_k = c_{k+1} + c_{k-1} \ , &&: \mathrm{E_{1g}} \ , \\ \alpha + 2\beta \ , &\quad c_k = c_{k+1} \ , &&: \mathrm{A_{2u}} \ . \end{aligned}$$

In the right-most column we have given the symmetry of the relevant level by comparing the coefficients c_k with those obtained earlier.

[1] The secular equation may be expressed as $\det|\alpha\hat{1} + \beta\hat{D}(2C_6) - \varepsilon\hat{1}| = 0$ with $\hat{D}(2C_6) \equiv \hat{D}(C_6) + \hat{D}(C_6^{-1})$, so that diagonalization of the Hamiltonian is equivalent to diagonalizing $\hat{D}(2C_6)$ (see (2) in Sect. 9.3.1).

Since $\beta < 0$, corresponding to the fact that $\sigma_u^+ 1s^*$ of H_2 has higher energy than $\sigma_g^+ 1s$, the level a_{2u} lies lowest. Then come the levels e_{1g}, e_{2u} and b_{2g}, in this order. In the ground configuration, four of six π electrons are accommodated in e_{1g} with opposite spins and two in a_{2u}. See Fig. 9.4.

9.5 Hybridized Orbitals

In the Heitler–London method (or the valence-bond method), we adopt the view that a molecule is formed as a result of interactions between the constituent atoms. Specifically, in the electron-pair bond approximation, we consider for each atom and bond an orbital extending toward the bonding partner, put an electron in it and let a pair of them make a covalent bond (electron-pair bond) as in the case of the hydrogen molecule. The more the orbitals are extended toward the partner, the larger the overlap between them, leading to a stronger chemical bond. Such orbitals may be formed by choosing several atomic orbitals having nearly equal energies (like 2*s* and 2*p* orbitals of the C atom) and superposing them. The extended orbitals constructed in this way are called the *hybridized orbitals.* A description of the electronic structure of molecules in terms of such hybridized orbitals is not suited for quantitative calculations of molecular energies because of the difficulty caused by nonorthogonality of orbitals. Nevertheless, it appeals to our intuitive idea of chemical bonds and provides us with a natural explanation for directed valence as seen below. It is probably for this reason that the valence-bond theory has not yet lost its importance for a qualitative understanding of the nature of chemical bonds. It should also be remarked that the hybridized orbitals are also called into play in molecular orbital theory as building materials for the MOs in place of the raw atomic orbitals.

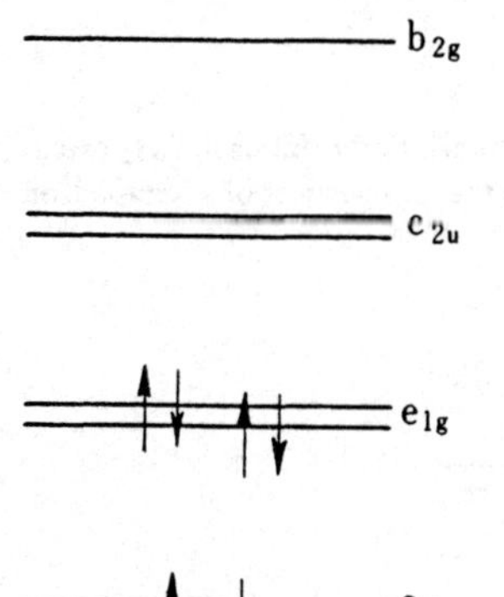

Fig. 9.4. Ground configuration of the benzene molecule

9.5.1 Methane and sp^3-Hybridization

If we use the 2*s* (or *s*, for simplicity,) orbital together with the three 2*p* orbitals (p_x, p_y, p_z) of C, we can construct four (equivalent) orbitals extending toward the H atoms on the four vertices of a tetrahedron as in the methane molecule CH_4 (Fig. 9.5).

Suppose we have obtained such orbitals, φ_1, φ_2, φ_3 and φ_4. Since they are transformed into each other by the symmetry operations of T_d, which is the symmetry group of CH_4, they form a basis set for a (reducible) representation Γ of T_d. By evaluating the character of Γ, we can easily see what kind of irreducible representations of T_d appear when it is reduced. The character may be obtained immediately (Table 9.2), if we pay attention to φ_i's that remain invariant under symmetry operations R. The result is

$$\Gamma = A_1 + T_2 \ . \tag{9.20}$$

Functions transforming according to the irreducible representation A_1 of T_d are s, $f_{xyz}, \ldots$, whereas those transforming according to T_2 are (p_x, p_y, p_z), $(d_{yz}, d_{zx}, d_{xy}), \ldots$. Since we can obtain functions of such symmetry by taking linear combinations of φ_1 through φ_4, we can derive them conversely from, for example, s and p_x, p_y, p_z in the following way. Let us put

$$\varphi_1 = as + bp_x + cp_y + dp_z \ .$$

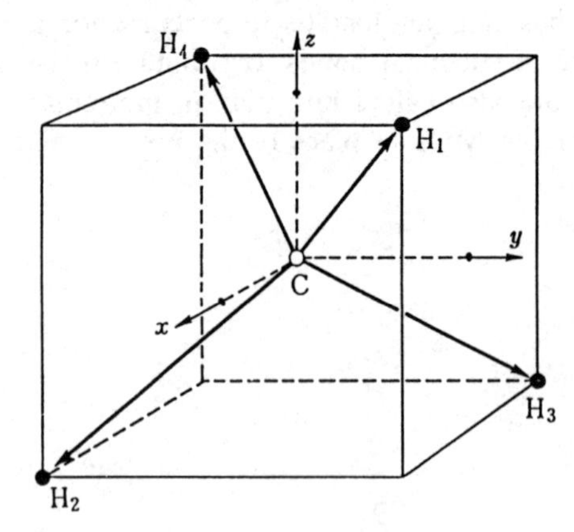

Fig. 9.5. sp^3 hybridized orbitals in CH_4 extending towards the four vertices of a tetrahedron

Table 9.2. Character of the representation based on the four hybridized orbitals

T_d	E	$6IC_4$	$3C_2$	$6\sigma_d$	$8C_3$
$\chi^{(\Gamma)}(R)$	4	0	0	2	1

If we require this to be invariant under the rotations C_3 and C_3^2 about the [111]-axis, we must have $b = c = d$, so that

$$\varphi_1 = as + b(p_x + p_y + p_z) . \tag{9.21}$$

Operation of IC_4 on this function turns it into φ_2:

$$\varphi_2 = as + b(p_x - p_y - p_z) .$$

Then operation of C_2 leads to the expression for φ_3:

$$\varphi_3 = as + b(- p_x + p_y - p_z) .$$

If we operate C_2 on φ_1, we obtain φ_4:

$$\varphi_4 = as + b(- p_x - p_y + p_z) .$$

In the above, we could have used admixtures of p_x and d_{yz} etc. in place of p_x etc. The d components are neglected usually because of their high orbital energy.

The magnitudes of the coefficients a and b are not determined by the symmetry considerations alone. If we impose orthogonality between φ_1 and φ_2, we have $a^2 = b^2$. With the normalization condition $a^2 + 3b^2 = 1$, we find $a = b = 1/2$.

If we assume the same radial function for $2s$ and $2p$, the orbital φ_1 will take the form

$$\varphi_1 = \{a + \sqrt{3}b(x + y + z)/r\} f(r) .$$

If we want to maximize the value of φ_1 in the [111] direction, which is given by $(a + 3b) f(r)$ under $a^2 + 3b^2 = 1$, we are also led to $a = b = 1/2$.

If we consider four $1s$ orbitals of H in place of φ_1, φ_2, φ_3 and φ_4 and solve (9.21) and the following three equations for s and three p functions, we will obtain molecular orbitals $\psi(\mathrm{a_1})$ and three $\psi(\mathrm{t_2})$. Note that the procedure of constructing directed hybridized orbitals is the reverse of building molecular orbitals from atomic orbitals, that is, deriving a basis set for a reducible representation from those for irreducible representations. Some typical examples of hybridized orbitals are given in Table 9.3.

Table 9.3. Important types of hybridization

Hybridized orbitals	Symmetry	Resulting hybrids
sp	$D_{\infty h}$	Linear
sp^2	C_{3h}	Trigonal plane
sp^3	T_d	Tetrahedral
dsp^2	D_{4h}	Tetragonal plane
d^2sp^3	O_h	Octahedral

9.6 Ligand Field Theory

Ions M of iron group elements in solutions or crystals are usually found in the form of octahedral complex ions $[MX_6]$ with six ligands X. As the ligands X, negative ions such as halogen ions, NO_2^-, CN^-, OH^- and neutral polar molecules H_2O, NH_3 are common. The system $[MX_6]$ is naturally to be treated as a whole as a molecular ion. The ligand (or crystal) field theory, however, assumes simply that the 3d elections of the central ion M are subject to the electric field V_c of cubic symmetry exerted by the six ligands X.

Suppose six negative ions X with charge $-Ze$ are located on the x-, y-, z-axes at a distance a from the origin (Fig. 9.6). When $r < a$, the electrostatic potential energy due to the ligand field is given by

$$V_c(\boldsymbol{r}) = V_0 + D(x^4 + y^4 + z^4 - \tfrac{3}{5}r^4) + \dots \tag{9.22}$$

with

$$V_0 = \frac{6Ze^2}{a}, \quad D = \frac{35Ze^2}{4a^5},$$

which is an explicit expression for the cubic ligand field in the point charge approximation.

The symmetry of the Hamiltonian for 3d electrons is now lowered from spherical to cubic because of the presence of V_c. The group that left invariant the Hamiltonian H_0 for an isolated ion M was the (improper) rotation group and the states with definite orbital angular momentum l formed the basis set for its irreducible representation $D^{(l)}$. The symmetry group of the Hamiltonian $H_0 + V_c$ for MX_6 is the octahedral group O_h, a subgroup of the rotation group,

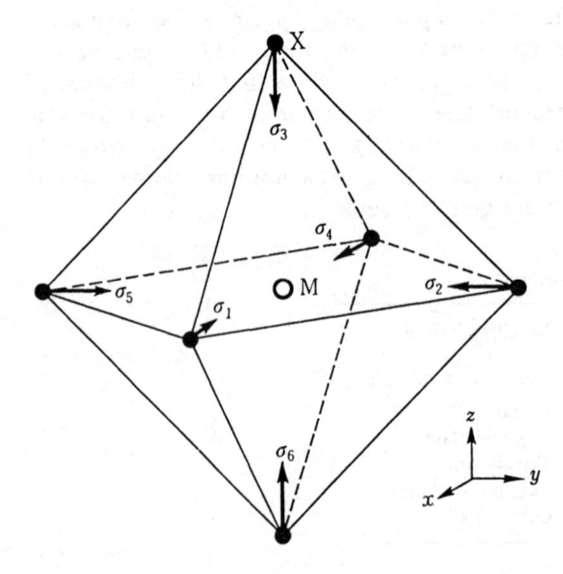

Fig. 9.6. σ orbitals of the six ligands in an MX_6 complex ion

and the levels are classified according to the irreducible representations of O_h. Accordingly, the levels with a definite value of l will be split into sublevels E_α, $E_\beta, \ldots$ corresponding to the irreducible representations $\Gamma_\alpha, \Gamma_\beta, \ldots$ of O_h.

In the language of the first-order perturbation theory for degenerate states, the situation may be described as follows: The eigenvalues $E_\alpha, E_\beta, \ldots$ under the perturbation V_c can be obtained by diagonalizing the matrix of V_c constructed within the manifold of $2l+1$ degenerate states. The eigenfunctions corresponding to $E_\alpha, E_\beta, \ldots$ may be derived as linear combinations of the $2l+1$ wavefunctions using the unitary matrix that diagonalizes V_c. The results must give the basis sets for the irreducible representations $\Gamma_\alpha, \Gamma_\beta, \ldots$. This procedure is just the reduction of $D^{(l)}$ to obtain irreducible representations $\Gamma_\alpha, \Gamma_\beta, \ldots$ of O_h:

$$D^{(l)} \downarrow O_h = \Gamma_\alpha + \Gamma_\beta + \ldots \quad (9.23)$$

Let us consider, as an example, splitting of the $3d$ level. We use

$$\phi(3dm) = R_{3d}(r) Y_{2m}(\theta, \varphi) , \qquad m = \pm 2, \pm 1, 0 ,$$

as the $3d$ orbitals. To obtain the matrix elements of V_c given by (9.22) with respect to these functions, we first rewrite (9.22) in terms of the spherical harmonics as

$$V_c = V_0 + \frac{2}{5} D r^4 \left\{ C_0^{(4)}(\theta, \varphi) + \sqrt{\frac{5}{14}} \left(C_4^{(4)}(\theta, \varphi) + C_{-4}^{(4)}(\theta, \varphi) \right) \right\} , \quad (9.24)$$

where we have employed $C_m^{(l)}(\theta, \varphi)$ given by (7.63). Then it is straightforward to derive the matrix of $H = H_0 + V_c$ with the help of (7.111) and Table 7.4:

$$\begin{array}{rc} m = & 2 \\ & 1 \\ & 0 \\ & -1 \\ & -2 \end{array} \begin{bmatrix} \varepsilon' + Dq & & & & 5Dq \\ & \varepsilon' - 4Dq & & & \\ & & \varepsilon' + 6Dq & & \\ & & & \varepsilon' - 4Dq & \\ 5Dq & & & & \varepsilon' + Dq \end{bmatrix} , \quad (9.25)$$

where we have put $\varepsilon' = \varepsilon_{3d} + V_0$, ε_{3d} being the eigenvalue of H_0. The role of the term V_0 is simply to shift the $3d$ levels as a whole so that it will be discarded hereafter. The quantity q is given by

$$q = \frac{2}{105} \langle r^4 \rangle , \qquad \langle r^4 \rangle = \int_0^\infty r^4 R_{3d}(r)^2 r^2 dr .$$

We need not consider terms of degree higher than four, since we are using pure d functions to calculate matrix elements.

The eigenvalues of the matrix (9.25) and the corresponding eigenfunctions are given by

$$\varepsilon = \varepsilon' - 4Dq: \quad \phi(3d1), \phi(3d-1), \frac{1}{\sqrt{2}}\{\phi(3d2) - \phi(3d-2)\} , \tag{9.26}$$

and

$$\varepsilon = \varepsilon' + 6Dq: \quad \phi(3d0), \frac{1}{\sqrt{2}}\{\phi(3d2) + \phi(3d-2)\} . \tag{9.27}$$

The three functions of (9.26) are the basis for the irreducible representation T_{2g} and the two functions of (9.27) are the basis for E_g. Actually, the following real forms for the basis are preferred, because they manifestly comply with the cubic symmetry:

$$\mathrm{t}_{2g}\left\{\begin{aligned} d_{yz} &= \phi_\xi = \frac{\mathrm{i}}{\sqrt{2}}\{\phi(3d1) + \phi(3d-1)\} &&= \sqrt{3}yz \times \sqrt{\frac{5}{4\pi}}R_{3d}(r)/r^2 ,\\ d_{zx} &= \phi_\eta = \frac{1}{\sqrt{2}}\{-\phi(3d1) + \phi(3d-1)\} &&= \sqrt{3}zx \times \ldots ,\\ d_{xy} &= \phi_\zeta = \frac{\mathrm{i}}{\sqrt{2}}\{-\phi(3d2) + \phi(3d-2)\} &&= \sqrt{3}xy \times \ldots , \end{aligned}\right. \tag{9.28}$$

$$\mathrm{e}_g\left\{\begin{aligned} d_{z^2} &= \phi_u = \phi(3d0) &&= \frac{1}{2}(3z^2 - r^2) \times \ldots ,\\ d_{x^2-y^2} &= \phi_v = \frac{1}{\sqrt{2}}\{\phi(3d2) + \phi(3d-2)\} &&= \frac{\sqrt{3}}{2}(x^2 - y^2) \times \ldots . \end{aligned}\right. \tag{9.29}$$

Let us generalize the problem a little and consider the reduction of $D^{(l)} \downarrow \mathrm{O}$ for later reference. Since the character of $D^{(l)}$ for the rotation of angle θ is given by (7.76), we can write down the values immediately for the operations belonging to the group O:

$$\chi^{(l)}(C_2) = (-1)^l , \qquad \chi^{(l)}(C_4) = \begin{cases} 1 & l = 0, 1 (\mathrm{mod}\ 4), \\ -1 & \quad 2, 3 (\mathrm{mod}\ 4), \end{cases} \qquad \chi^{(l)}(C_3) = \begin{cases} 1 & l = 0 (\mathrm{mod}\ 3), \\ 0 & \quad 1 (\mathrm{mod}\ 3), \\ -1 & \quad 2 (\mathrm{mod}\ 3). \end{cases}$$

For example, we have the character table for $l = 2$ as in Table 9.4 and

$$D^{(2)} \downarrow \mathrm{O} = \mathrm{T}_2 + \mathrm{E}$$

as a result. Results of the reduction of $D^{(l)} \downarrow \mathrm{O}$ for some other values of l are given in Table 9.5.

Exercise 9.2. Using Table 9.6, construct the T_{2g} and E_g orbitals from $\phi(3dm)$ by means of the method of descent in symmetry.

Table 9.4. Character of the representation $D^{(2)} \downarrow O$

O	E	$6C_4$	$3C_4^2$	$6C_2'$	$8C_3$
$\chi^{(2)}(R)$	5	−1	1	1	−1

Table 9.5. Results of the decomposition of $D^{(l)} \downarrow O$

l	$D^{(l)} \downarrow O$
0	A_1
1	T_1
2	$E + T_2$
3	$A_2 + T_1 + T_2$
4	$A_1 + E + T_1 + T_2$
5	$E + 2T_1 + T_2$
6	$A_1 + A_2 + E + T_1 + 2T_2$

Table 9.6. Compatibility relations between O_h, D_{4h}, C_{4v}, and C_4

O_h	D_{4h}	D_{4h}	C_{4v}	C_4	(m)
A_{1g}	A_{1g}	A_{1g}	A_1	A	(0)
A_{1u}	A_{1u}	A_{1u}	A_2		
A_{2g}	B_{1g}	B_{1g}	B_1	B	(2)
A_{2u}	B_{1u}	B_{1u}	B_2		
E_g	$A_{1g}B_{1g}$	A_{2g}	A_2	A	
E_u	$A_{1u}B_{1u}$	A_{2u}	A_1		
T_{1g}	$A_{2g}E_g$	B_{2g}	B_2	B	
T_{1u}	$A_{2u}E_u$	B_{2u}	B_1		
T_{2g}	$B_{2g}E_g$	E_g	E	E	(± 1)
T_{2u}	$B_{2u}E_u$	E_u	E		

Since $D > 0$ in (9.22), t_{2g} will have lower energy than e_g. This is confirmed in practice by the analysis of optical spectra of complex ions; $\Delta = 10Dq$ is found in the range 2–3 eV in most complexes.

So far we have taken advantage of the explicit form of the cubic field given by (9.22) to obtain the wavefunctions (9.28, 29) for the split levels. It need hardly be emphasized that they may be derived solely from the facts that the potential energy or the Hamiltonian has cubic symmetry and that the d level splits into sublevels t_{2g} and e_g in this environment. In fact, if we take the point of view of molecular orbital theory, there is no reason to suppose that t_{2g} and e_g orbitals should be made up only of the pure $3d$ orbitals of the metal ion as in (9.28, 29). Mixing may take place with the linear combinations of the ligand orbitals with symmetry T_{2g} or E_g. Suppose the ligand X is the F^- ion. We can construct orbitals with T_{2g} or E_g symmetry from $2s$ and $2p$ of F^- as shown below. From the six $2s$ orbitals $s_1, s_2, \ldots, s_6$, we obtain orbitals with E_g symmetry:

$$\begin{cases} \chi_{su} = \dfrac{1}{\sqrt{12}}(2s_3 + 2s_6 - s_1 - s_4 - s_2 - s_5) \ , \\ \chi_{sv} = \dfrac{1}{2}(s_1 + s_4 - s_2 - s_5) \ . \end{cases}$$

It is also possible to build up $\chi_{\sigma u}$ and $\chi_{\sigma v}$ using six $2p\ \sigma$ orbitals $\sigma_1, \sigma_2, \ldots, \sigma_6$ by replacing s_i in the above by σ_i, where σ_i is the $2p$ orbital of the ith ion extended towards the central metal ion, e.g., $\sigma_1 = -p_{1x}, \sigma_2 = -p_{2y}$ (Fig. 9.6).

Exercise 9.3. Besides E_g, we can also derive orbitals with A_{1g} and T_{1u} symmetry. Write down the corresponding expressions.

Similarly, twelve $2p\pi$-orbitals $p_{1y}, p_{1z}, p_{2x}, p_{2z}, \ldots$, give rise to the basis sets for irreducible representations T_{1g}, T_{2g}, T_{1u}, and T_{2u}. Only the result for T_{2g} will be given here:

$$\begin{cases} \chi_{\pi\xi} = \frac{1}{2}(p_{3y} - p_{6y} + p_{2z} - p_{5z}) \ , \\ \chi_{\pi\eta} = \frac{1}{2}(p_{1z} - p_{4z} + p_{3x} - p_{6x}) \ , \\ \chi_{\pi\zeta} = \frac{1}{2}(p_{1y} - p_{4y} + p_{2x} - p_{5x}) \ . \end{cases}$$

When the ligand field theory is interpreted within the framework of the molecular orbital theory, the orbitals (9.28, 29) are replaced by (antibonding) molecular orbitals such as

$$\psi_\xi = \frac{1}{\sqrt{N_t}}(\phi_\xi - \lambda_\pi \chi_{\pi\xi}) \ , \quad (\xi, \eta, \zeta) \ ,$$

$$\psi_u = \frac{1}{\sqrt{N_e}}(\phi_u - \lambda_\sigma \chi_{\sigma u} - \lambda_s \chi_{su}) \ , \quad (u, v) \ ,$$

where $\lambda_\pi, \lambda_\sigma, \lambda_s$ are the mixing coefficients and N_t and N_e are normalization constants. (For ψ_η replace ξ by η, etc.) From this point of view, $\Delta = 10Dq$ is equal to the energy difference between the e_g and t_{2g} orbitals and contains contributions from origins other than V_c, the simple electrostatic potential.

So far we have dealt with the case of a single d electron in a cubic field. For a system d^N of more than one d electron, two different approaches are possible to take account of the Coulomb interaction between electrons. One is to start from the configurations $t_2^m e^n (m + n = N)$ and take the Coulomb interaction afterwards. This treatment is called the strong field scheme. In this scheme, the first problem is to construct the wavefunctions with correct symmetry, that is, $\Psi(t_2^m e^n S\Gamma)$ with a definite value of S and belonging to the irreducible representation Γ of the group O. This will be discussed in the next section. The other is called the weak field scheme, which, in the first approximation, regards the ligand field as the cause of splitting of the multiplet terms $3d^N \alpha^{2S+1}L$ of the central metal ion. In the higher approximations, we have to include the effect of off-diagonal elements of V_c between the multiplet terms with the same S but different L. In short, the ligand field is first taken into account and the interactions between electrons come in next in the strong field scheme, while, in the weak field scheme, the Coulomb interactions are considered first and then the ligand field is brought into consideration as a perturbation.

Leaving the treatment of the strong field case to the next section, we will confine ourselves here to discussion of the first-order splitting of multiplet terms ^{2S+1}L due to V_c in the weak field scheme. We already know from Table 9.5 what kind of multiplet terms $^{2S+1}\Gamma$ follow from ^{2S+1}L. The method of operator equivalents (or the effective Hamiltonian) of Sect. 7.12 tells us how ^{2S+1}L is split. According to this, the potential energy operator for the ligand field

$$V_c = \sum_{i=1}^{N} D\left(x_i^4 + y_i^4 + z_i^4 - \frac{3}{5}r_i^4\right) \tag{9.30}$$

can be replaced within the manifold of a definite orbital angular momentum L by the operator equivalent

$$\begin{aligned} V_c &= \beta\frac{2}{5}D\langle r^4\rangle\left\{C_0^{(4)}(L) + \sqrt{\frac{5}{14}}(C_4^{(4)}(L) + C_{-4}^{(4)}(L))\right\} \\ &= \beta\frac{1}{20}D\langle r^4\rangle\Big[35L_z^4 - 30L(L+1)L_z^2 + 25L_z^2 + 3L^2(L+1)^2 \\ &\quad - 6L(L+1) + \frac{5}{2}\{L_+^4 + L_-^4\}\Big], \end{aligned} \tag{9.31}$$

where the definition of the operator $C_q^{(k)}(L)$ has been given in (7.119). The proportionality constant $\beta = \beta(3d^{N\,2S+1}L)$ may be determined by putting the result of direct evaluation of (9.30) with respect to one of the states specified by M_S and M_L of ^{2S+1}L equal to the expectation value of (9.31) in that state. We saw in Sect. 7.14 that the state with $M_S = S$ and $M_L = L$ of the ground multiplet ^{2S+1}L of the $3d^N$ configuration was expressible in a very simple fashion by a single Slater determinant, see (7.159). This circumstance makes the evaluation of (9.30) fairly simple. We give below the values of β for the ground multiplet terms of $3d^N$ and the splitting patterns for them (Fig. 9.7) which are obtained by diagonalizing (9.31) within the manifold of ^{2S+1}L:

$$\left.\begin{aligned} \beta(d^1\ {}^2D) &= -\beta(d^9\ {}^2D) = \frac{2}{63}, \\ \beta(d^2\ {}^3F) &= -\beta(d^8\ {}^3F) = -\frac{2}{315}, \\ \beta(d^3\ {}^4F) &= -\beta(d^7\ {}^4F) = \frac{2}{315}, \\ \beta(d^4\ {}^5D) &= -\beta(d^6\ {}^5D) = -\frac{2}{63}. \end{aligned}\right\} \tag{9.32}$$

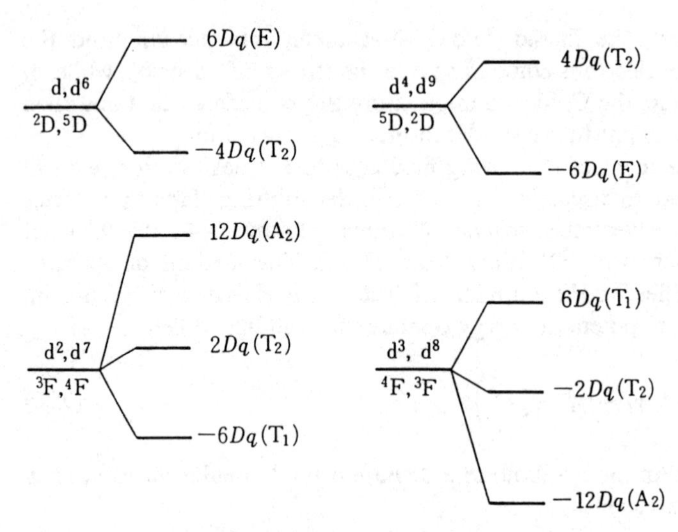

Fig. 9.7. Splitting of the ground states of d^N due to a cubic field

The symmetry labels for the split levels can be determined from transformation properties of the corresponding eigenfunctions under symmetry operations of the point group O, or by knowing the character of the representation based on these eigenfunctions. For example, in the case of the multiplet ^{2S+1}F, $^{2S+1}\mathrm{T}_1$ can be distinguished from $^{2S+1}\mathrm{T}_2$ by the transformation properties of the obtained eigenfunctions under the rotation C_4.

When both the ligand field and the Coulomb interaction are fully taken into account, we should reach the same result by either the weak or strong field schemes. In fact, if we vary the parameter $\Delta = 10Dq$ from 0 to $+\infty$, the levels will change continuously from the multiplet terms of $3d^N$ of a free ion to the sets of multiplet terms of $\mathrm{t}_2^m\mathrm{e}^n$ $(m + n = N)$. This may be seen in the energy level diagram (Fig. 9.8) for the d^2 system in a cubic field. The abscissa is Δ and the ordinate E is the excitation energy from the ground state. The levels drawn parallel in the strong field limit $\Delta \to \infty$ belong to the same configuration in the strong field scheme.

9.7 Multiplet Terms in Molecules

Let G denote the symmetry group of a molecule. Since the Hamiltonian for electrons of the molecule

$$H = \sum_i \left(-\frac{\hbar^2}{2m} \nabla_i^2 + V(\boldsymbol{r}_i) \right) + \sum_{i>j} \frac{e^2}{r_{ij}} \tag{9.33}$$

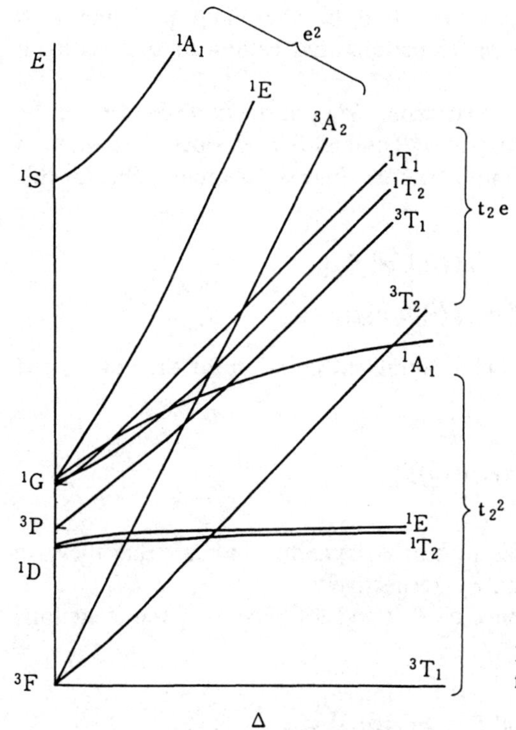

Fig. 9.8. Energy diagram for the d^2 configuration in a cubic field

is invariant under any symmetry operation R of G, its eigenfunctions $\Psi_\nu^{(\mu)}$ will transform according to one of the irreducible representations $D^{(\mu)}$ of G. Here we define the operation of R on the many-electron wavefunction Ψ by

$$R\Psi(\boldsymbol{r}_1', \boldsymbol{r}_2', \ldots) = \Psi(\boldsymbol{r}_1, \boldsymbol{r}_2, \ldots) , \tag{9.34}$$

when the position vector $\boldsymbol{r}_i$ of the ith electron is carried to $\boldsymbol{r}_i'$ by R. Note that all the electrons are subjected to the same and simultaneous transformation of coordinates.

The wavefunction $\Psi_\nu^{(\mu)}$ must be, at the same time, an eigenfunction of S^2 and S_z, as the Hamiltonian (9.33) does not contain spins of electrons explicitly. If we write Γ in place of $D^{(\mu)}$, the states of the many-electron system are characterized by S and Γ. The corresponding multiplet term is denoted as $\alpha^{2S+1}\Gamma$, where α is introduced to distinguish between terms with the same sets of S and Γ. In accordance with it, the wavefunctions are expressed as $\Psi(\alpha S\Gamma M\gamma)$ with M, the eigenvalue of S_z, and γ specifying the partners in the representation Γ.

As we saw in the discussion of the π-electron approximation for a benzene molecule, the molecular orbital theory starts from the electron configuration where each electron is accommodated in one-electron orbitals with various symmetry. The ground configuration for benzene is $(a_{2u})^2(e_{1g})^4$, while the first excited configuration is given by $(a_{2u})^2(e_{1g})^3e_{2u}$. This section studies how to construct correct wavefunctions permitted by the Pauli principle with given symmetry $S\Gamma$ by using the wavefunctions that belong to such electron configurations.

Let us begin with the case of two electrons. We put the two electrons in the orbitals $\phi(i)$ and $\varphi(j)$. The symmetry of $\phi(i)$ and $\varphi(j)$ is left open for a while. If we take the spin degree of freedom into account, four states are possible for this system:

$$\Psi_{ij}^{(1)} = |\phi(i)\alpha, \varphi(j)\alpha| \ , \quad \Psi_{ij}^{(3)} = |\phi(i)\beta, \varphi(j)\alpha| \ ,$$
$$\Psi_{ij}^{(2)} = |\phi(i)\alpha, \varphi(j)\beta| \ , \quad \Psi_{ij}^{(4)} = |\phi(i)\beta, \varphi(j)\beta| \ ,$$

where we have employed the following simplified notation for the normalized Slater determinant:

$$\frac{1}{\sqrt{2}}\begin{vmatrix} \phi_1(i)\alpha_1 & \varphi_1(j)\beta_1 \\ \phi_2(i)\alpha_2 & \varphi_2(j)\beta_2 \end{vmatrix} = |\phi(i)\alpha, \varphi(j)\beta| \ .$$

The suffixes 1 and 2 in the symbols such as $\phi_1(i)$ and α_2 indicate that they are functions of the coordinates $\boldsymbol{r}_1$ and σ_2, respectively.

If we take the sum and difference of $\Psi_{ij}^{(2)}$ and $\Psi_{ij}^{(3)}$, we have four functions given by

$$\begin{aligned} \Psi_{ij}^{(1)} &= \frac{1}{\sqrt{2}}\{\phi(i)\varphi(j) - \varphi(j)\phi(i)\}\alpha\alpha \ , \\ \frac{1}{\sqrt{2}}(\Psi_{ij}^{(2)} + \Psi_{ij}^{(3)}) &= \frac{1}{\sqrt{2}}\{\phi(i)\varphi(j) - \varphi(j)\phi(i)\}\frac{1}{\sqrt{2}}\{\alpha\beta + \beta\alpha\} \ , \\ \Psi_{ij}^{(4)} &= \frac{1}{\sqrt{2}}\{\phi(i)\varphi(j) - \varphi(j)\phi(i)\}\beta\beta \ , \end{aligned} \tag{9.35}$$

$$\frac{1}{\sqrt{2}}(\Psi_{ij}^{(2)} - \Psi_{ij}^{(3)}) = \frac{1}{\sqrt{2}}\{\phi(i)\varphi(j) + \varphi(j)\phi(i)\}\frac{1}{\sqrt{2}}\{\alpha\beta - \beta\alpha\} \ , \tag{9.36}$$

where it is to be understood that the coordinates in the products of orbital and spin functions are arranged in the order 1 and 2. Equation (9.36) represents the state with $S = 0$, whereas (9.35) provides the states $M = 1, 0, -1$ of $S = 1$ in this order.

We may draw the following conclusions from these considerations.

(1) When two electrons are put in a single non-degenerate orbital ϕ belonging to a real one-dimensional (irreducible) representation, the resultant spin S is equal to zero and the normalized wavefunction is given by

$$\Psi = \phi_1\phi_2\frac{1}{\sqrt{2}}(\alpha_1\beta_2 - \beta_1\alpha_2) = |\phi\alpha, \phi\beta| \ .$$

Since the symmetry of $\phi_1\phi_2$ is A_1, we have ${}^{2S+1}\Gamma = {}^1A_1$.

(2) When the sets of functions $\{\phi(i)\}$ $(i = 1, 2, \ldots, d_1)$ and $\{\varphi(j)\}$ $(j = 1, 2, \ldots, d_2)$ represent basis sets for different irreducible representations Γ_1 and Γ_2, respectively, both values $S = 0$ and 1 are allowed and each of the functions (9.35) and (9.36) forms the basis for the product representation $\Gamma_1 \times \Gamma_2$. (Note that we have $4d_1 d_2$ states altogether including spin.) If the product representation is decomposed as $\Gamma_\alpha + \Gamma_\beta + \ldots$, we will have multiplet terms like ${}^1\Gamma_\alpha, {}^1\Gamma_\beta, \ldots, {}^3\Gamma_\alpha, {}^3\Gamma_\beta, \ldots$. Corresponding eigenfunctions can be derived from (9.35 and 36) by applying to them the unitary transformation to reduce the product representation, or by using the Clebsch–Gordan coefficients. Even when $\{\phi(i)\}$ and $\{\varphi(j)\}$ belong to the same irreducible representation Γ, we reach the same conclusion as long as they are linearly independent.

(3) However, when $\phi = \varphi$ and $\{\phi(i)\}$ $(i = 1, 2, \ldots, d)$ form the basis set for an irreducible representation Γ with dimension greater than 1, not all the combinations of S and Γ_α appearing in the reduction of $\Gamma \times \Gamma$ are allowed. For example, we will never have ${}^1\Gamma_\alpha$ and ${}^3\Gamma_\alpha$ at the same time. This is expected if we count the number of states permitted by the Pauli principle. We are putting two electrons in $2d$ spin-orbitals, so that the total number of states is given by ${}_{2d}C_2 = d(2d - 1)$, which is certainly smaller than the value $4d^2$ that we would obtain without the exclusion principle. To distinguish this case from case (2), the present one is called that of (two) equivalent electrons and the electron configuration is denoted as $(\Gamma)^2$.

When we put $\phi = \varphi$, the orbital parts of (9.35) and (9.36) turn respectively into

$$\frac{\phi_1(i)\phi_2(j) - \phi_1(j)\phi_2(i)}{\sqrt{2}}, \quad \frac{\phi_1(i)\phi_2(j) + \phi_1(j)\phi_2(i)}{\sqrt{2}}.$$

They are antisymmetric and symmetric with respect to the interchange of the orbital coordinates 1 and 2 of the electrons, which means they are respectively the basis functions of the antisymmetric representation $\{\Gamma \times \Gamma\}$ and those of the symmetric representation $[\Gamma \times \Gamma]$. Accordingly, for Γ_α obtained by reducing $\{\Gamma \times \Gamma\}$, only ${}^3\Gamma_\alpha$ is allowed, while only ${}^1\Gamma_\beta$ is possible for Γ_β contained in $[\Gamma \times \Gamma]$. The reduction of $\{\Gamma \times \Gamma\}$ and $[\Gamma \times \Gamma]$ may be carried out by decomposing the characters (4.76) and (4.75), respectively.

The method of descent in symmetry works also for examining the kind of allowed multiplet terms. Take, for example, the configuration $(e_{1g})^2$ of benzene. The irreducible representation E_{1g} of the group D_{6h} becomes E_1 $(m = \pm 1)$ of the group C_6. Let us denote the basis functions corresponding to $m = \pm 1$ by $\phi(+1)$ and $\phi(-1)$. In the configuration $(e_1)^2$ of C_6, we obtain (i) two substates $(m = +2)$ or $(m = -2)$ of ${}^1E_2$ by putting two electrons with opposite spins in either $\phi(+1)$ or $\phi(-1)$ and (ii) ${}^1A(m = 0)$ and ${}^3A(m = 0)$ by putting one in $\phi(+1)$ and one in $\phi(-1)$. The orbital part of the former wavefunction is symmetric and given by $\{\phi_1(+1)\phi_2(-1) + \phi_1(-1)\phi_2(+1)\}/\sqrt{2}$, while that of the latter wavefunction is given by the corresponding antisymmetric linear combination. Thus in C_6 symmetry

$$(e_1)^2 = {}^1A + {}^3A + {}^1E_2 \quad (C_6).$$

On the other hand, we have

$$E_{1g} \times E_{1g} = A_{1g} + A_{2g} + E_{2g} \quad (D_{6h}).$$

This together with the compatibility relation given by Table 5.1 and the transformation property

$$C_{2x}\phi_1(+1)\phi_2(-1) = \phi_1(-1)\phi_2(+1)$$

leads to the following result in D_{6h}:

$$(e_{1g})^2 = {}^1A_{1g} + {}^3A_{2g} + {}^1E_{2g} \quad (D_{6h}) \ .$$

The result is of course in accordance with the decomposition

$$[E_{1g} \times E_{1g}] = A_{1g} + E_{2g} \ , \qquad \{E_{1g} \times E_{1g}\} = A_{2g} \ .$$

Exercise 9.4. Confirm that multiplet terms of $(t_{2g})^2$ in O_h are ${}^1A_{1g}$, ${}^1E_g$, ${}^3T_{1g}$, and ${}^1T_{2g}$. Determine allowed terms of $(e_g)^2$ in O_h.

Wavefunctions for two equivalent electrons may be obtained in the way discussed above. However, it is more convenient to derive them by taking into account the requirement of antisymmetry from the beginning, or by using the Slater determinants, because use of them facilitates the extension of the method to more than two equivalent electrons. Let us denote $\phi(i)$ by $\phi(\gamma)$ as the partner γ in the basis for the irreducible representation Γ. For simplicity, we assume that no irreducible representation appears more than once in the reduction of $\Gamma \times \Gamma$, see Sect. 9.8. The wavefunctions for the multiplet terms of the configuration $(\Gamma)^2$ are given by

$$\Psi((\Gamma)^2 S = 1, M = 1, \Gamma_\alpha\gamma_\alpha) = \frac{1}{\sqrt{2}} \sum_{\gamma_1\gamma_2} |\phi(\gamma_1)\alpha, \phi(\gamma_2)\alpha| \langle \Gamma\gamma_1\Gamma\gamma_2|\Gamma_\alpha\gamma_\alpha\rangle \ ,$$

$$\Psi((\Gamma)^2 S = 1, M = 0, \Gamma_\alpha\gamma_\alpha) = \frac{1}{2}\sum \{|\phi(\gamma_1)\alpha, \phi(\gamma_2)\beta| + |\phi(\gamma_1)\beta, \phi(\gamma_2)\alpha|\}\langle \Gamma\gamma_1\Gamma\gamma_2|\Gamma_\alpha\gamma_\alpha\rangle \ ,$$

$$\Psi((\Gamma)^2 S = 1, M = -1, \Gamma_\alpha\gamma_\alpha) = \frac{1}{\sqrt{2}} \sum |\phi(\gamma_1)\beta, \phi(\gamma_2)\beta| \langle \Gamma\gamma_1\Gamma\gamma_2|\Gamma_\alpha\gamma_\alpha\rangle \ ,$$

$$\Psi((\Gamma)^2 S = 0, \Gamma_\beta\gamma_\beta) = \frac{1}{2}\sum \{|\phi(\gamma_1)\alpha, \phi(\gamma_2)\beta| - |\phi(\gamma_1)\beta, \phi(\gamma_2)\alpha|\}\langle \Gamma\gamma_1\Gamma\gamma_2|\Gamma_\beta\gamma_\beta\rangle \ ,$$

where we have set $\{\Gamma \times \Gamma\} = \Gamma_\alpha + \cdots$, $[\Gamma \times \Gamma] = \Gamma_\beta + \cdots$. Note that we have to introduce an extra normalization factor $1/\sqrt{2}$ compared to the case of inequivalent electrons. These results may be put into a single formula

$$\Psi((\Gamma)^2 SM\Gamma_\nu\gamma_\nu) = \frac{1}{\sqrt{2}} \sum_{\substack{m_1m_2 \\ \gamma_1\gamma_2}} |\psi(m_1\gamma_1), \psi(m_2\gamma_2)| \left\langle \frac{1}{2}m_1\frac{1}{2}m_2 \middle| SM \right\rangle \times \langle \Gamma\gamma_1\Gamma\gamma_2|\Gamma_\nu\gamma_\nu\rangle \ , \tag{9.37}$$

if we use notation such as $\psi(m = 1/2, \gamma) \equiv \phi(\gamma)\alpha$ for the spin-orbitals. The expression $\langle \frac{1}{2}m_1 \frac{1}{2}m_2 | SM \rangle$ stands for the Wigner coefficient. Wavefunctions for the multiplet terms of t_2^2, e^2, t_2e configurations in O_h symmetry are given in Table 9.7. They can be obtained by the use of (9.37) and Clebsch–Gordan coefficients given in Table 9.9. In Table 9.7, spin-orbitals such as $\phi_\xi\alpha$, $\phi_v\beta$ are abbreviated as ξ, $\bar{v}$.

In the case of three equivalent electrons we proceed as follows. Let us assume for the time being that the third electron is put in a spin-orbital ψ' with orbital function ϕ' having the same symmetry as ϕ though linearly independent of it, which means that the third electron is inequivalent. Combining ψ'_3 with ψ_{12} given by (9.37), we can write down the wavefunction with total spin S and orbital symmetry Γ_v as

$$\Phi'((\Gamma)^2(S_1\Gamma_1)(\Gamma)SM\Gamma_v\gamma_v)$$

$$= \sum_{\substack{M_1\gamma_1 \\ m\gamma}} \Psi_{12}((\Gamma)^2 S_1 M_1 \Gamma_1 \gamma_1)\psi'_3(m\gamma)\left\langle S_1 M_1 \frac{1}{2} m | SM \right\rangle \langle \Gamma_1\gamma_1\Gamma\gamma|\Gamma_v\gamma_v\rangle \,. \tag{9.38}$$

Table 9.7. Wavefunctions for the multiplet terms of t_2^2, e^2 and t_2e

$$\Psi(t_2^2\ {}^1A_1) = \{|\xi\bar{\xi}| + |\eta\bar{\eta}| + |\zeta\bar{\zeta}|\}/\sqrt{3}$$

$$\left\{\begin{array}{l} \Psi(t_2^2\ {}^1Eu) = \{2|\zeta\bar{\zeta}| - |\xi\bar{\xi}| - |\eta\bar{\eta}|\}/\sqrt{6} \\ \Psi(t_2^2\ {}^1Ev) = \{|\xi\bar{\xi}| - |\eta\bar{\eta}|\}/\sqrt{2} \end{array}\right.$$

$$\left\{\begin{array}{l} \Psi(t_2^2\ {}^3T_1\gamma, M=1) = |\xi\eta| \\ \Psi(t_2^2\ {}^3T_1\gamma, M=0) = \{|\xi\bar{\eta}| - |\eta\bar{\xi}|\}/\sqrt{2} \\ \Psi(t_2^2\ {}^3T_1\gamma, M=-1) = |\bar{\xi}\bar{\eta}| \end{array}\right.$$

$$\Psi(t_2^2\ {}^1T_2\zeta) = \{|\xi\bar{\eta}| + |\eta\bar{\xi}|\}/\sqrt{2}$$

$$\Psi(e^2\ {}^1A_1) = \{|u\bar{u}| + |v\bar{v}|\}/\sqrt{2}$$

$$\left\{\begin{array}{l} \Psi(e^2\ {}^3A_2, M=1) = |uv| \\ \Psi(e^2\ {}^3A_2, M=0) = \{|u\bar{v}| - |v\bar{u}|\}/\sqrt{2} \\ \Psi(e^2\ {}^3A_2, M=-1) = |\bar{u}\bar{v}| \end{array}\right.$$

$$\left\{\begin{array}{l} \Psi(e^2\ {}^1Eu) = \{|v\bar{v}| - |u\bar{u}|\}/\sqrt{2} \\ \Psi(e^2\ {}^1Ev) = \{|u\bar{v}| + |v\bar{u}|\}/\sqrt{2} \end{array}\right.$$

$$\left\{\begin{array}{l} \Psi(t_2e\ {}^3T_1\gamma, M=1) = |\zeta v| \\ \Psi(t_2e\ {}^3T_1\gamma, M=0) = \{|\zeta\bar{v}| + |\bar{\zeta}v|\}/\sqrt{2} \\ \Psi(t_2e\ {}^3T_1\gamma, M=-1) = |\bar{\zeta}\bar{v}| \end{array}\right.$$

$$\left\{\begin{array}{l} \Psi(t_2e\ {}^3T_2\zeta, M=1) = |\zeta u| \\ \Psi(t_2e\ {}^3T_2\zeta, M=0) = \{|\zeta\bar{u}| + |\bar{\zeta}u|\}/\sqrt{2} \\ \Psi(t_2e\ {}^3T_2\zeta, M=-1) = \{|\bar{\zeta}\bar{u}| \end{array}\right.$$

$$\Psi(t_2e\ {}^1T_1\gamma) = \{|\zeta\bar{v}| - |\bar{\zeta}v|\}/\sqrt{2}$$

$$\Psi(t_2e\ {}^1T_2\zeta) = \{|\zeta\bar{u}| - |\bar{\zeta}u|\}/\sqrt{2}$$

If we write simply $\psi'(k)$ for $\psi'(m\gamma)$ on the right-hand side and express Ψ_{12} as a sum of Slater determinants, (9.38) will take the form

$$\Phi' = \sum_{k_1 k_2 k_3} |\psi(k_1), \psi(k_2)|_{12}\psi'_3(k_3) C_{k_1 k_2 k_3} . \tag{9.39}$$

Since this has yet to be antisymmetrized, we apply the operator

$$A_3 = \frac{1}{\sqrt{3}}(E - (13) - (23))$$

to it, where (13), for example, denotes the permutation operator for the coordinates (including spin coordinates) of electrons 1 and 3. The wavefunction fully antisymmetrized now reads

$$A_3 \Phi' = \sum_{k_1 k_2 k_3} |\psi(k_1), \psi(k_2), \psi'(k_3)| C_{k_1 k_2 k_3} . \tag{9.40}$$

When the third electron is inequivalent, the Pauli principle is readily taken care of by replacing the products in (9.39) by the corresponding Slater determinants for three electrons as in (9.40).

If we put $\psi' = \psi$ in (9.40) to obtain the wavefunction for three equivalent electrons, we have

$$A_3 \Phi = \sum |\psi(k_1), \psi(k_2), \psi(k_3)| C_{k_1 k_2 k_3} . \tag{9.41}$$

It then happens that not all the Slater determinants appearing on the right-hand side of (9.41) are linearly independent of each other. As a consequence, it is even possible that for some $S\Gamma_v$ the right-hand side vanishes identically. This means that such combinations of S and Γ_v are forbidden by the Pauli principle. When the right-hand side is nonvanishing the corresponding multiplet term ${}^{2S+1}\Gamma_v$ becomes allowed. Now, if we look back to (9.38), we realize that several different choices of S_1 and Γ_1 may lead to the same set $S\Gamma_v$. When we reach always one and the same function (9.41) (apart from a constant factor) even though we start from different sets of $S_1\Gamma_1$, the matter is simple and we may conclude that there is only one such multiplet term ${}^{2S+1}\Gamma_v$ in the configuration $(\Gamma)^3$. The function $A_3\Phi$ is not necessarily normalized, so the correct wavefunction for the multiplet term in question will be obtained by dividing it by its norm $\|A_3\Phi\|$:

$$\Psi((\Gamma)^3 SM\Gamma_v\gamma_v) = A_3\Phi((\Gamma)^2(S_1\Gamma_1)(\Gamma)SM\Gamma_v\gamma_v)/\|A_3\Phi\| .$$

Wavefunctions of the t_2^3 configuration of O are given in Table 9.8.

(i) When we obtain more than one, say, f linearly independent functions $A_3\Phi$ in the procedure described above, this means that we have f multiplet terms with the same S and Γ_v in $(\Gamma)^3$. In such a case, further diagonalization of the Coulomb interaction matrix within the manifold of these f multiplet terms will be necessary.

(ii) It is also possible here, as in the case of the two-electron problem, to examine separately the permutation symmetry of spin and orbital functions for three electrons and determine allowed

Table 9.8. $\Psi(t_2^3 SM\Gamma\gamma)$

$^{2S+1}\Gamma$	M	γ	Ψ
$^4A_2$	3/2	e_2	$-\lvert\xi\eta\zeta\rvert$
^2E	1/2	u	$\{\lvert\xi\bar{\eta}\zeta\rvert-\lvert\bar{\xi}\eta\zeta\rvert\}/\sqrt{2}$
		v	$\{2\lvert\xi\eta\bar{\zeta}\rvert-\lvert\xi\bar{\eta}\zeta\rvert-\lvert\bar{\xi}\eta\zeta\rvert\}/\sqrt{6}$
$^2T_1$	1/2	α	$\{\lvert\xi\eta\bar{\eta}\rvert-\lvert\xi\zeta\bar{\zeta}\rvert\}/\sqrt{2}$
		β	$\{\lvert\eta\zeta\bar{\zeta}\rvert-\lvert\eta\xi\bar{\xi}\rvert\}/\sqrt{2}$
		γ	$\{\lvert\zeta\xi\bar{\xi}\rvert-\lvert\zeta\eta\bar{\eta}\rvert\}/\sqrt{2}$
$^2T_2$	1/2	ξ	$\{\lvert\xi\eta\bar{\eta}\rvert+\lvert\xi\zeta\bar{\zeta}\rvert\}/\sqrt{2}$
		η	$\{\lvert\eta\zeta\bar{\zeta}\rvert+\lvert\eta\xi\bar{\xi}\rvert\}/\sqrt{2}$
		ζ	$\{\lvert\zeta\xi\bar{\xi}\rvert+\lvert\zeta\eta\bar{\eta}\rvert\}/\sqrt{2}$

combinations of S and Γ_v. For example, the state with $S = 3/2$ and $M = 3/2$ is given by

$$\Psi_{ijk} = \lvert\phi(i)\alpha, \phi(j)\alpha, \phi(k)\alpha\rvert = \lvert\phi(i), \phi(j), \phi(k)\rvert\alpha\alpha\alpha \ ,$$

which shows that the orbital function must be a component of a totally antisymmetric tensor. This fact holds also for other states of $S = 3/2$, i.e. for $M = \pm 1/2, -3/2$. The symmetry Γ_v of the orbital function to be associated with $S = 3/2$ is thus restricted to irreducible representations belonging to the totally antisymmetric representation (which is denoted as $\Gamma^{[1111]}$ in Sects. 15.5–7) contained in the product representation $\Gamma \times \Gamma \times \Gamma$. Although the symmetry with respect to permutation becomes more complicated for the orbital function associated with $S = 1/2$, we can also show that Γ_v is to be restricted to the irreducible representations appearing in the decomposition of the representation denoted as $\Gamma^{[21]}$ of $\Gamma \times \Gamma \times \Gamma$. The formulae to calculate the characters of the representations $\Gamma^{[111]}$ and $\Gamma^{[21]}$ are obtained from those given in Table 15.4 by replacing s_p by $\chi^{(\Gamma)}(R^p)$ [9.1].

Exercise 9.5. Confirm that the multiplet terms of t_2^3 are $^4A_2$, ^2E, $^2T_1$, $^2T_2$. What kind of multiplet terms are expected for the configuration $t_2^2t_2'$?

The method described above for three equivalent electrons can easily be extended to the derivation of wavefunctions for n equivalent electrons from those for $n-1$ electrons. This completes the discussion of allowed multiplet terms $^{2S+1}\Gamma_v$ of the electron configuration $(\Gamma)^n$ and the construction of the corresponding many-electron wavefunctions $\Psi((\Gamma)^n SM\Gamma_v\gamma_v)$.

When the dimension of Γ is d, the shell Γ becomes full for $n = 2d$ and $(\Gamma)^{2d}$ $^1A_1$ is the only allowed state, where each of the d orbitals are occupied by two electrons with opposite spins. In accordance with this, we find that multiplet terms $^{2S+1}\Gamma_v$ allowed in $(\Gamma)^n$ are also allowed in $(\Gamma)^{2d-n}$ and that a simple correlation can be established between the wavefunctions $\Psi((\Gamma)^n SM\Gamma_v\gamma_v)$ and $\Psi((\Gamma)^{2d-n} SM\Gamma_v\gamma_v)$ for the same term. For details, the reader is referred to the discussions given in Sect. 7.15 for the atomic problem.

If we have two incomplete shells such as $(\Gamma)^n$ and $(\Gamma')^m$, we first write down wavefunctions for multiplet terms of each shell and couple them as follows:

$$\begin{aligned}\Phi_1((\Gamma)^n(S\Gamma_\nu)(\Gamma')^m(S'\Gamma_\mu)S_0M_0\Gamma_0\gamma_0) &= \sum_{\substack{MM'\\ \gamma_\mu\gamma_\mu}} \Psi((\Gamma)^n SM\Gamma_\nu\gamma_\nu)\Psi((\Gamma')^m S'M'\Gamma_\mu\gamma_\mu) \\ &\quad \times \langle SMS'M'|S_0M_0\rangle\langle\Gamma_\nu\gamma_\nu\Gamma_\mu\gamma_\mu|\Gamma_0\gamma_0\rangle \\ &= \sum_{\substack{k_1\ldots k_n\\ k_{1'}\ldots k_{m'}}} |\psi(k_1),\ldots,\psi(k_n)||\psi(k_1'),\ldots,\psi(k_m')|C_{k_1\ldots k_n k_{1'}\ldots k_{m'}} \,.\end{aligned}$$

The wavefunction for the total system is obtained by antisymmetrization:

$$\Psi = A\Phi_1 = \sum |\psi(k_1),\ldots,\psi(k_n),\psi(k_1'),\ldots,\psi(k_m')|C_{k_1\ldots k_n k_{1'}\ldots k_{m'}} \,.$$

*9.8 Clebsch–Gordan Coefficients for Simply Reducible Groups and the Wigner–Eckart Theorem

Suppose operator $T_j^{(\alpha)}$ transforms like the jth partner in the irreducible representation $D^{(\alpha)}$ so that

$$RT_j^{(\alpha)}R^{-1} = \sum_i T_i^{(\alpha)}D_{ij}^{(\alpha)}(R) \,.$$

Let us operate $T_j^{(\alpha)}$ on the lth partner of the basis set $\phi(\mu\beta l)$ for the irreducible representation $D^{(\beta)}$, where μ denotes a set of quantum numbers other than βl. Since the result will transform like a basis function of the product representation $D^{(\alpha)} \times D^{(\beta)}$, the function defined by

$$\Psi(\alpha\mu\beta\gamma m) \equiv \sum_{jl} T_j^{(\alpha)}\phi(\mu\beta l)\langle\beta l\alpha j|\gamma m\rangle \tag{9.42}$$

in terms of the *Clebsch–Gordan* (CG) *coefficient* $\langle\beta l\alpha j|\gamma m\rangle$ transforms like the mth partner in the representation $D^{(\gamma)}$. Note that we have assumed that none of the irreducible representations $D^{(\gamma)}$ appear more than once in the decomposition of $D^{(\alpha)} \times D^{(\beta)}$. When this holds, equation (9.42) may be solved for $T_j^{(\alpha)}\phi(\mu\beta l)$ as

$$T_j^{(\alpha)}\phi(\mu\beta l) = \sum_{\gamma m} \Psi(\alpha\mu\beta\gamma m)\langle\gamma m|\beta l\alpha j\rangle \,.$$

It then follows that

$$\begin{aligned}\langle\nu\gamma m|T_j^{(\alpha)}|\mu\beta l\rangle &\equiv (\phi(\nu\gamma m), T_j^{(\alpha)}\phi(\mu\beta l)) \\ &= (\phi(\nu\gamma m), \Psi(\alpha\mu\beta\gamma m))\langle\gamma m|\beta l\alpha j\rangle \,.\end{aligned}$$

Since the first factor on the second line does not depend on m, we may put this result in the following form:

$$\langle \nu\gamma m | T_j^{(\alpha)} | \mu\beta l \rangle = \langle \nu\gamma \| \mathbf{T}^{(\alpha)} \| \mu\beta \rangle (\gamma)^{-1/2} \langle \gamma m | \beta l \alpha j \rangle \ , \tag{9.43}$$

where (γ) stands for the dimension of the representation $D^{(\gamma)}$.

Equation (9.43) is known as the *Wigner–Eckart theorem*. This theorem is well known in the case of the rotation group (Sect. 7.11) and, when it is referred to, it usually means the theorem as given by (7.102). However, the theorem is valid also for other groups as seen above.

In the derivation of (9.43), we have assumed that no irreducible representation occurs more than once in the reduction of the products of two irreducible representations. If we further impose the condition that every element of the group belongs to the same class as its inverse, as with the rotation group, the CG coefficients will become endowed with essentially the same high symmetry as the Wigner coefficients. The groups that satisfy the two requirements are called *simply reducible groups*. An example other than the rotation group is provided by the octahedral group (the point group O). In this section, we will discuss briefly the properties of their irreducible representations and associated CG coefficients.

In simply reducible groups, the characters are all real, because any element belongs to the same class as its inverse. Accordingly, every irreducible representation $D^{(\alpha)}$ is equivalent to its complex conjugate $D^{(\alpha)*}$, so that the representation matrices are transformed into each other by a unitary matrix $\hat{U}$:

$$\hat{D}^{(\alpha)}(R) = \hat{U}\hat{D}^{(\alpha)}(R)^* \hat{U}^{-1} \ , \quad R \in G \ . \tag{9.44}$$

For the matrix $\hat{U}$, we have two possibilities, see Sect. 4.12,

$$\hat{U}\hat{U}^* = \lambda_\alpha \hat{1} \ , \quad \lambda_\alpha = \pm 1 \ .$$

If $\lambda_\alpha = 1$, all the representation matrices $\hat{D}^{(\alpha)}(R)$ may be made real by a suitable equivalence transformation. Representations $D^{(l)}$ of the rotation group with integers l belong to this case. Let us call this kind of representations "integral" representations, keeping this in mind. The five representations A_1, A_2, E, T_1, T_2 of O are all integral. If $\lambda_\alpha = -1$, we know that it is impossible to make all the matrices $\hat{D}^{(\alpha)}(R)$ real simultaneously by any equivalence transformation. An example of this case is given by the representations $D^{(j)}$ of the rotation group with j half of an odd integer. Let us therefore call such representations "half-odd". Irreducible representations of a simply reducible group are either integral or half-odd. This can be determined by calculating the sum

$$\lambda_\alpha = \frac{1}{g} \sum_R \chi^{(\alpha)}(R^2) \ ,$$

where g is the order of the group.

It is then not difficult to show that we have only integral or half-odd representations in the reduction of $D^{(\alpha)} \times D^{(\beta)}$ according as $\lambda_\alpha \lambda_\beta = 1$ or -1. Especially, when $D^{(\alpha)} \times D^{(\alpha)}$ is reduced, we have only integral representations. On the other hand, we know that the product representation may be reduced into a direct sum of the symmetric and antisymmetric representations:

$$D^{(\alpha)} \times D^{(\alpha)} = [D^{(\alpha)} \times D^{(\alpha)}] + \{D^{(\alpha)} \times D^{(\alpha)}\} . \tag{9.45}$$

When $D^{(\alpha)}$ is integral, we call (integral) representations contained in the symmetric product even representations and those in the antisymmetric product odd representations. Conversely, when $D^{(\alpha)}$ is half-odd, representations in the symmetric product are odd and those in the antisymmetric product are even. Irreducible integral representations that appear in the reduction of the left-hand side of (9.45) are thus either even or odd. Integral representations that never appear in the reduction of the product (9.45) are neither even nor odd. These nomenclatures follow the ones used for the rotation group. If we put $D^{(j)} \times D^{(j)} = \sum_J D^{(J)}$ for the rotation group, the parity of J is precisely that of the representation defined here. In the point group O, the representations A_1, E, and T_2 are even, while A_2 and T_1 are odd.

To keep the notation in parallel with that for the rotation group, we denote irreducible representations of a simply reducible group as $D^{(j)}$ with dimension (j) and distinguish the partners of its basis by m. Let us put $\lambda_j = (-1)^{2j} = 1$ for integral representations. For half-odd representations, $\lambda_j = (-1)^{2j} = -1$. We further make the convention for integral representations that $(-1)^j = 1$ and -1 for even and odd ones, respectively. For integral representations that are neither even nor odd, we put $(-1)^j = 1$. For half-odd representations, we choose either i or $-$i as the value of $(-1)^j$ for each $D^{(j)}$ and stick to the choice throughout.

Let us put the elements of the unitary matrix (i.e., CG coefficients) which decomposes the product representation $D^{(j_1)} \times D^{(j_2)}$ into the sum of $D^{(j_3)*}$ in the form

$$\langle j_1 m_1 j_2 m_2 | (j_3 m_3)^* \rangle = \omega(j_1 j_2 j_3) \sqrt{(j_3)} \begin{pmatrix} j_1 & j_2 & j_3 \\ m_1 & m_2 & m_3 \end{pmatrix} , \tag{9.46}$$

where $\omega(j_1 j_2 j_3)$ is a complex factor of modulus unity depending upon j_1, j_2, and j_3. The last factor $\begin{pmatrix} j_1 & j_2 & j_3 \\ m_1 & m_2 & m_3 \end{pmatrix}$ of (9.46) is called the 3j-symbol. Note that (9.46) may be rewritten as

$$\langle j_1 m_1 j_2 m_2 | j_3 m_3 \rangle = \omega(j_1 j_2 j_3) \sqrt{(j_3)} \sum_{\bar{m}} \begin{pmatrix} j_1 & j_2 & j_3 \\ m_1 & m_2 & \bar{m} \end{pmatrix} (\hat{U}^{-1})_{\bar{m} m_3} , \tag{9.47}$$

with the aid of the unitary matrix $\hat{U}$ on the right-hand side of (9.44).

The 3j-symbols satisfy the following identities corresponding to the unitarity of the transformation:

$$\sum_{m_1 m_2} (j) \begin{pmatrix} j_1 & j_2 & j \\ m_1 & m_2 & m \end{pmatrix}^* \begin{pmatrix} j_1 & j_2 & j' \\ m_1 & m_2 & m' \end{pmatrix} = \delta_{jj'} \delta_{mm'} ,$$

$$\sum_{jm} (j) \begin{pmatrix} j_1 & j_2 & j \\ m_1 & m_2 & m \end{pmatrix}^* \begin{pmatrix} j_1 & j_2 & j \\ m'_1 & m'_2 & m \end{pmatrix} = \delta_{m_1 m'_1} \delta_{m_2 m'_2} .$$

We also have from the definition of the unitary matrix

$$D^{(j_1)}_{m_1 m'_1}(R) D^{(j_2)}_{m_2 m'_2}(R) = \sum_{j_3 m_3 m'_3} (j_3) \begin{pmatrix} j_1 & j_2 & j_3 \\ m_1 & m_2 & m_3 \end{pmatrix} D^{(j_3)}_{m_3 m'_3}(R)^* \begin{pmatrix} j_1 & j_2 & j_3 \\ m'_1 & m'_2 & m'_3 \end{pmatrix}^* , \tag{9.48}$$

which corresponds to (7.86) for the rotation group. This leads to the identity

$$\sum_R D^{(j_1)}_{m_1 m'_1}(R) D^{(j_2)}_{m_2 m'_2}(R) D^{(j_3)}_{m_3 m'_3}(R) = g \begin{pmatrix} j_1 & j_2 & j_3 \\ m_1 & m_2 & m_3 \end{pmatrix} \begin{pmatrix} j_1 & j_2 & j_3 \\ m'_1 & m'_2 & m'_3 \end{pmatrix}^* , \tag{9.49}$$

if we take into account the orthogonality theorem for the representation matrices. Keeping in mind the symmetry of the basis sets of irreducible representations in the decomposition of $D^{(j)} \times D^{(j)}$ and (9.49), we assume the following symmetry for the symbols [see (7.93b)]:

$$\begin{pmatrix} j_1 & j_2 & j_3 \\ m_1 & m_2 & m_3 \end{pmatrix} = (-1)^{j_1+j_2+j_3} \begin{pmatrix} j_2 & j_1 & j_3 \\ m_2 & m_1 & m_3 \end{pmatrix} = (-1)^{j_1+j_2+j_3} \begin{pmatrix} j_1 & j_3 & j_2 \\ m_1 & m_3 & m_2 \end{pmatrix} . \tag{9.50}$$

It then follows that

$$\begin{pmatrix} j_1 & j_2 & j_3 \\ m_1 & m_2 & m_3 \end{pmatrix} = \begin{pmatrix} j_2 & j_3 & j_1 \\ m_2 & m_3 & m_1 \end{pmatrix} = \begin{pmatrix} j_3 & j_1 & j_2 \\ m_3 & m_1 & m_2 \end{pmatrix} ,$$

because the 3j-symbols are nonvanishing only when $(-1)^{2j_1+2j_2+2j_3} = 1$, see (7.93a).

When $D^{(j_2)}$ in (9.48) is the identity representation, we have

$$D^{(j)}_{m_1 m'_1}(R) = \sum_{mm'} (j) \begin{pmatrix} j & 0 & j \\ m_1 & 0 & m \end{pmatrix} D^{(j)}_{mm'}(R)^* \begin{pmatrix} j & 0 & j \\ m'_1 & 0 & m' \end{pmatrix}^* ,$$

writing $j_2 = m_2 = 0$ for the identity representation. This means that the unitary matrix $\hat{U}$ in (9.44) is given by

$$U_{mm'} = \sqrt{(j)} \begin{pmatrix} j & 0 & j \\ m & 0 & m' \end{pmatrix} = (-1)^{2j} \sqrt{(j)} \begin{pmatrix} j & 0 & j \\ m' & 0 & m \end{pmatrix} . \tag{9.51}$$

For a simply reducible group, the Wigner–Eckart theorem thus takes the following form:

$$\langle vjm|T_q^{(k)}|\mu j'm'\rangle = \langle vj\|\mathbf{T}^{(k)}\|\mu j'\rangle \omega(j'kj)^* \sqrt{(j)} \sum_{\bar{m}} \begin{pmatrix} j & 0 & j \\ m & 0 & \bar{m} \end{pmatrix} \begin{pmatrix} j & j' & k \\ \bar{m} & m' & q \end{pmatrix}^* .$$

Table 9.9. CG coefficients $\langle \Gamma_1\gamma_1\Gamma_2\gamma_2|\Gamma\gamma\rangle$ for the point group O

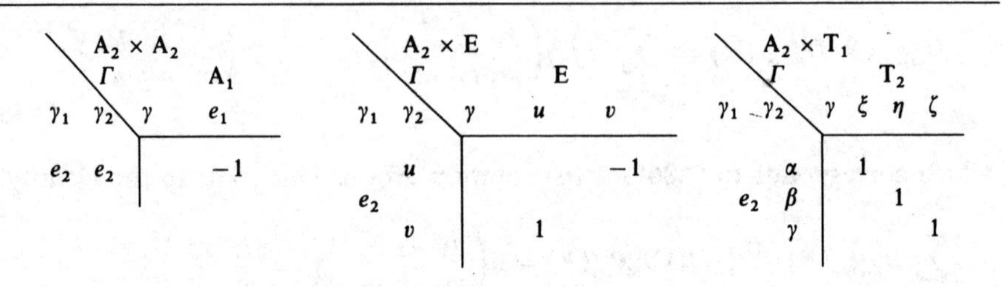

$A_2 \times A_2$: γ_1	γ_2	A_1: e_1
e_2	e_2	-1

$A_2 \times E$: γ_1	γ_2	E: u	v
e_2	u		-1
	v	1	

$A_2 \times T_1$: γ_1	γ_2	T_2: ξ	η	ζ
	α	1		
e_2	β		1	
	γ			1

$A_2 \times T_2$: γ_1	γ_2	T_1: α	β	γ
	ξ	-1		
e_2	η		-1	
	ζ			-1

$E \times E$: γ_1	γ_2	A_1: e_1	A_2: e_2	E: u	v
u	u	$1/\sqrt{2}$		$-1/\sqrt{2}$	
	v		$1/\sqrt{2}$		$1/\sqrt{2}$
v	u		$-1/\sqrt{2}$		$1/\sqrt{2}$
	v	$1/\sqrt{2}$		$1/\sqrt{2}$	

$E \times T_1$: γ_1	γ_2	T_1: α	β	γ	T_2: ξ	η	ζ
	α	$-1/2$			$\sqrt{3}/2$		
u	β		$-1/2$			$-\sqrt{3}/2$	
	γ			1			
	α	$\sqrt{3}/2$			$1/2$		
v	β		$-\sqrt{3}/2$			$1/2$	
	γ						-1

Table 9.9 (continued)

$E \times T_2$		T_1			T_2		
γ_1	γ_2 \ γ	α	β	γ	ξ	η	ζ
u	ξ	$-\sqrt{3}/2$			$-1/2$		
	η		$\sqrt{3}/2$			$-1/2$	
	ζ						1
v	α	$-1/2$			$\sqrt{3}/2$		
	η		$-1/2$			$-\sqrt{3}/2$	
	ζ			1			

$T_1 \times T_1$		A_1	E		T_1			T_2		
γ_1	γ_2 \ γ	e_1	u	v	α	β	γ	ξ	η	ζ
α	α	$-1/\sqrt{3}$	$1/\sqrt{6}$	$-1/\sqrt{2}$						
	β						$-1/\sqrt{2}$			$-1/\sqrt{2}$
	γ					$1/\sqrt{2}$			$-1/\sqrt{2}$	
β	α						$1/\sqrt{2}$			$-1/\sqrt{2}$
	β	$-1/\sqrt{3}$	$1/\sqrt{6}$	$1/\sqrt{2}$						
	γ				$-1/\sqrt{2}$			$-1/\sqrt{2}$		
γ	α					$-1/\sqrt{2}$			$-1/\sqrt{2}$	
	β				$1/\sqrt{2}$			$-1/\sqrt{2}$		
	γ	$-1/\sqrt{3}$	$-2/\sqrt{6}$							

$T_1 \times T_2$		A_2	E		T_1			T_2		
γ_1	γ_2 \ γ	e_2	u	v	α	β	γ	ξ	η	ζ
α	ξ	$-1/\sqrt{3}$	$-1/\sqrt{2}$	$-1/\sqrt{6}$						
	η						$1/\sqrt{2}$			$-1/\sqrt{2}$
	ζ					$1/\sqrt{2}$			$1/\sqrt{2}$	
β	ξ						$1/\sqrt{2}$			$1/\sqrt{2}$
	η	$-1/\sqrt{3}$	$1/\sqrt{2}$	$-1/\sqrt{6}$						
	ζ				$1/\sqrt{2}$			$-1/\sqrt{2}$		
γ	ξ					$1/\sqrt{2}$			$-1/\sqrt{2}$	
	η				$1/\sqrt{2}$			$1/\sqrt{2}$		
	ζ	$-1/\sqrt{3}$		$2/\sqrt{6}$						

Table 9.9 (continued)

$T_2 \times T_2$		Γ: A_1	E		T_1			T_2		
γ_1	γ_2 \ γ	e_1	u	v	α	β	γ	ξ	η	ζ
ξ	ξ	$1/\sqrt{3}$	$-1/\sqrt{6}$	$1/\sqrt{2}$						
	η						$1/\sqrt{2}$			$1/\sqrt{2}$
	ζ					$-1/\sqrt{2}$			$1/\sqrt{2}$	
η	ξ						$-1/\sqrt{2}$			$1/\sqrt{2}$
	η	$1/\sqrt{3}$	$-1/\sqrt{6}$	$-1/\sqrt{2}$						
	ζ				$1/\sqrt{2}$			$1/\sqrt{2}$		
ζ	ξ					$1/\sqrt{2}$			$1/\sqrt{2}$	
	η				$-1/\sqrt{2}$			$1/\sqrt{2}$		
	ζ	$1/\sqrt{3}$	$2/\sqrt{6}$							

In the case of the rotation group, (9.51) and (9.46) are respectively given by

$$U_{mm'} = (-1)^{j+m}\delta_{m,-m'} ,$$

$$\langle j_1 m_1 j_2 m_2 | (j_3 m_3)^* \rangle = \sum_{\bar{m}} \langle j_1 m_1 j_2 m_2 | j_3 \bar{m} \rangle U_{\bar{m}m_3}$$

$$= (-1)^{j_1 - j_2 - j_3} \sqrt{2j_3 + 1} \begin{pmatrix} j_1 & j_2 & j_3 \\ m_1 & m_2 & m_3 \end{pmatrix} ,$$

see (7.95a) and (7.91).

The symmetry property (9.50) of the $3j$-symbol insures that the states with resultant spin $S = 0$ and 1 obtained by coupling two spins $s_1 = s_2 = 1/2$ are antisymmetric and symmetric with respect to the interchange of 1 and 2, respectively. In the addition of two orbital angular momenta $l_1 = l_2 = l$, the states with $L = 0, 2, \ldots, 2l$ are symmetric, whereas those with $L = 1, 3, \ldots, 2l - 1$ are antisymmetric. Putting these results together, we see that only the multiplet terms with $S + L$ even (${}^1S$, ${}^3P$, ${}^1D$, ...) satisfy the Pauli principle and are allowed in the configuration $(nl)^2$. In a similar way, for the point group O, we have only ${}^1A_1$, ${}^1E$, ${}^1T_2$, ${}^3T_1$ in t_1^2 or t_2^2 and ${}^1A_1$, ${}^1E$, ${}^3A_2$ in e^2.

Another consequence of the symmetry of the CG coefficients is that, within the manifold of *real* wavefunctions $\Psi(\mu\Gamma\gamma)$ (γ distinguishes the basis functions for the irreducible representation Γ), non-vanishing *real* operators are restricted to those of type A_1, E, or T_2 and *pure imaginary* ones to type A_2 or T_1.

In Table 9.9, CG coefficients for the group O are given with the symbols for the basis e_1, e_2 for A_1, A_2; $\{u, v\} = \{3z^2 - r^2, \sqrt{3}(x^2 - y^2)\}$ for E; $\{\alpha, \beta, \gamma\} = \{x, y, z\}$ for T_1; $\{\xi, \eta, \zeta\} = \{yz, zx, xy\}$ for T_2. Note that they satisfy the following symmetry relations:

$$\langle \Gamma_1\gamma_1\Gamma_2\gamma_2|\Gamma\gamma\rangle = \varepsilon(\Gamma_1\Gamma_2\Gamma)\langle \Gamma_2\gamma_2\Gamma_1\gamma_1|\Gamma\gamma\rangle \ ,$$

$$\begin{aligned} \varepsilon(\Gamma_1\Gamma_2\Gamma) = -1 \quad & \text{for} \quad \Gamma_1 = \Gamma_2 = \mathrm{E}\ , \quad \Gamma = \mathrm{A}_2\ , \\ & \text{for} \quad \Gamma_1 = \Gamma_2 = \mathrm{T}_1\ , \quad \Gamma = \mathrm{T}_1\ , \\ & \text{for} \quad \Gamma_1 = \Gamma_2 = \mathrm{T}_2\ , \quad \Gamma = \mathrm{T}_1\ , \\ = 1 \quad & \text{otherwise,} \end{aligned}$$

and

$$\langle \Gamma\gamma|\Gamma'\gamma'\bar{\Gamma}\bar{\gamma}\rangle/\sqrt{(\Gamma)} = \delta(\Gamma\bar{\Gamma}\Gamma')\langle \Gamma'\gamma'|\Gamma\gamma\bar{\Gamma}\bar{\gamma}\rangle/\sqrt{(\Gamma')}\ ,$$

$$\delta(\Gamma\mathrm{A}_1\Gamma') = \delta(\Gamma\mathrm{T}_2\Gamma') = 1\ ,$$

$$\delta(\Gamma\mathrm{A}_2\Gamma') = \delta(\Gamma\mathrm{T}_1\Gamma') = -1\ ,$$

$$\delta(\Gamma\mathrm{E}\Gamma') = 1 \quad \text{except for } \delta(\mathrm{T}_1\mathrm{E}\mathrm{T}_2) = \delta(\mathrm{T}_2\mathrm{E}\mathrm{T}_1) = -1.$$

For further details about the $3j$-symbols of O, see [9.2].

10. Molecular Vibrations

Elastic vibration of a molecule is another area where point groups and their representations find good applications. The symmetric structure of molecules allows classification of normal vibration modes according to irreducible representations of the symmetry group of the molecule without explicit knowledge of the force constants. In this chapter, we begin with the elementary theory of normal vibration modes, then study how group representation theory can be applied to the normal modes, and finally consider selection rules associated with infrared and Raman processes as well as interaction of electrons with atomic displacements.

10.1 Normal Modes and Normal Coordinates

Consider a molecule composed of N atoms and number those atoms by an index k. When the atoms perform small vibrations around their equilibrium positions, the total energy E of the system can be expressed as a sum of the kinetic energy and the elastic deformation energy:

$$E = \frac{1}{2}\sum_{k\alpha} m_k \dot{u}_{k\alpha}^2 + \frac{1}{2}\sum_{kl\alpha\beta} \hat{\Phi}\begin{pmatrix} kl \\ \alpha\beta \end{pmatrix} u_{k\alpha} u_{l\beta} \ , \tag{10.1}$$

where m_k stands for the mass of the kth atom, and $u_{k\alpha}$ for its displacement in the α direction (x, y, or z direction). The force constant $\hat{\Phi}\begin{pmatrix} kl \\ \alpha\beta \end{pmatrix}$ is symmetric with respect to the interchange of the indices $(k\alpha) \leftrightarrow (l\beta)$. The displacements then obey the equation of motion

$$m_k \ddot{u}_{k\alpha} = -\sum_{l\beta} \hat{\Phi}\begin{pmatrix} kl \\ \alpha\beta \end{pmatrix} u_{l\beta} \ . \tag{10.2}$$

To find the normal vibration modes, we assume that all the atoms move in phase and put

$$u_{k\alpha}(t) = U_{k\alpha}\cos(\omega t + \phi) \ ,$$

then we obtain

$$m_k \omega^2 U_{k\alpha} = \sum_{l\beta} \hat{\Phi}\begin{pmatrix} kl \\ \alpha\beta \end{pmatrix} U_{l\beta} \ ,$$

which can be brought into a symmetric form,

$$\omega^2\sqrt{m_k}\,U_{k\alpha} = \sum_{l\beta} \hat{D}\begin{pmatrix} kl \\ \alpha\beta \end{pmatrix} \sqrt{m_l}\,U_{l\beta} \,, \tag{10.3}$$

$$\hat{D}\begin{pmatrix} kl \\ \alpha\beta \end{pmatrix} = \frac{\hat{\Phi}\begin{pmatrix} kl \\ \alpha\beta \end{pmatrix}}{\sqrt{m_k m_l}} \,. \tag{10.4}$$

The above matrix $\hat{D}$ of dimension $3N$, called the *dynamical matrix*, is real and symmetric. Equation (10.3) is an eigenvalue problem for the $\hat{D}$ matrix with eigenvalue ω^2 and eigenvector $\{\sqrt{m_k}\,U_{k\alpha}\}$. Let ω_s^2 and $\{e_{k\alpha}^{(s)}\}$ denote $3N$ such eigenvalues and eigenvectors ($s = 1, 2, \ldots, 3N$). Then we have

$$\omega_s^2 e_{k\alpha}^{(s)} = \sum_{l\beta} \hat{D}\begin{pmatrix} kl \\ \alpha\beta \end{pmatrix} e_{l\beta}^{(s)} \,. \tag{10.5}$$

The eigenvectors can be chosen real, since the $\hat{D}$ matrix is real. The vibrations obtained in this way, namely, the vibrations in which all the constituent atoms vibrate with the same phase, are called *normal modes* or simply modes. ω_s is the frequency of the sth mode, and $e_{k\alpha}^{(s)}/\sqrt{m_k}$ is proportional to the amplitude $U_{k\alpha}$.

If we normalize the length of the eigenvectors $e^{(s)}$ to unity, we find

$$\sum_{k\alpha} e_{k\alpha}^{(s)} e_{k\alpha}^{(s')} = \delta_{ss'} \,, \tag{10.6a}$$

because eigenvectors belonging to different eigenvalues are orthogonal to each other. Furthermore, the set of $3N$ eigenvectors is complete,

$$\sum_{s} e_{k\alpha}^{(s)} e_{l\beta}^{(s)} = \delta_{kl}\delta_{\alpha\beta} \,. \tag{10.6b}$$

Mathematically, (10.6a, b) derive from the fact that the set of the eigenvectors forms an orthogonal matrix that diagonalizes the real symmetric $\hat{D}$ matrix. Physically, (10.6b) means that any molecular displacement pattern can be expressed as a linear combination of the $3N$ mode patterns.

Consider now some deformation of the molecule specified by a set of displacements $\{u_{k\alpha}\}$. The above-mentioned completeness allows expansion of this deformation,

$$u_{k\alpha} = m_k^{-1/2} \sum_{s} Q_s e_{k\alpha}^{(s)} \,, \quad k = 1, 2, \ldots, N;\ \alpha = x, y, z \,, \tag{10.7}$$

in terms of the eigenvectors of the normal modes. The coefficient Q_s of the expansion, representing the amplitude of the sth mode, is called the *normal coordinate*. In the left- and right-hand sides of (10.7), $u_{k\alpha}$ and Q_s are regarded as dynamical variables (and hence depend on time). The basis of the expansion $e_{k\alpha}^{(s)}$ does not depend on time.

Equation (10.7) may be solved for Q_s using (10.6a),

$$Q_s = \sum_{k\alpha} \sqrt{m_k}\, e_{k\alpha}^{(s)} u_{k\alpha} \ , \qquad s = 1, 2, \ldots, 3N \ . \tag{10.8}$$

The normal coordinate Q_s is a linear combination of the $3N$ displacements $u_{k\alpha}$. Use of (10.4–8) in (10.2) gives the equation of motion for Q_s,

$$\ddot{Q}_s + \omega_s^2 Q_s = 0 \ , \tag{10.9}$$

which shows that the $3N$ normal coordinates oscillate independently. The energy (10.1) is then written as

$$E = \frac{1}{2} \sum_s (\dot{Q}_s^2 + \omega_s^2 Q_s^2) \ . \tag{10.10}$$

Here the summation over s is taken from 1 to $3N$, among which three correspond to the translational motion and another three to the rotational motion of the molecule as a whole. For these six modes, the intramolecular forces yield no restoring force, and the corresponding ω_s values vanish, leaving only the kinetic energy terms.

10.2 Group Theory and Normal Modes

When a molecule has some geometrical symmetry, its normal modes will have some transformation properties reflecting the molecule's symmetry. Before going on to the general theory, let us consider normal modes for the specific example of the square molecule shown in Fig. 10.1. (The skeleton of carbon atoms in cyclobutane C_4H_8 actually has this structure.) The symmetry group of this molecule is the point group D_{4h}. Now, three degrees of freedom for each of the four atoms add up to 12 for the molecule as a whole. Subtracting three translation and three rotation modes, we are left with 6 intramolecular degrees of freedom, which give rise to the 6 vibration modes shown in Fig. 10.1. The mode patterns are given irreducible representation labels of the symmetry group D_{4h} by comparison with the irreducible characters. The A_{1g} mode maintains the D_{4h} symmetry of the molecule. It is called a totally symmetric mode or a breathing mode. The B_{1g} and B_{2u} modes keep the bond lengths unchanged for small vibration amplitudes, changing only the bond angles. Such modes in general have low frequencies by virtue of the weak restoring force. The E_u mode is twofold degenerate.

Why normal modes belong to irreducible representations may be understood in the following way. As we have seen in (10.9), a normal coordinate Q_s is

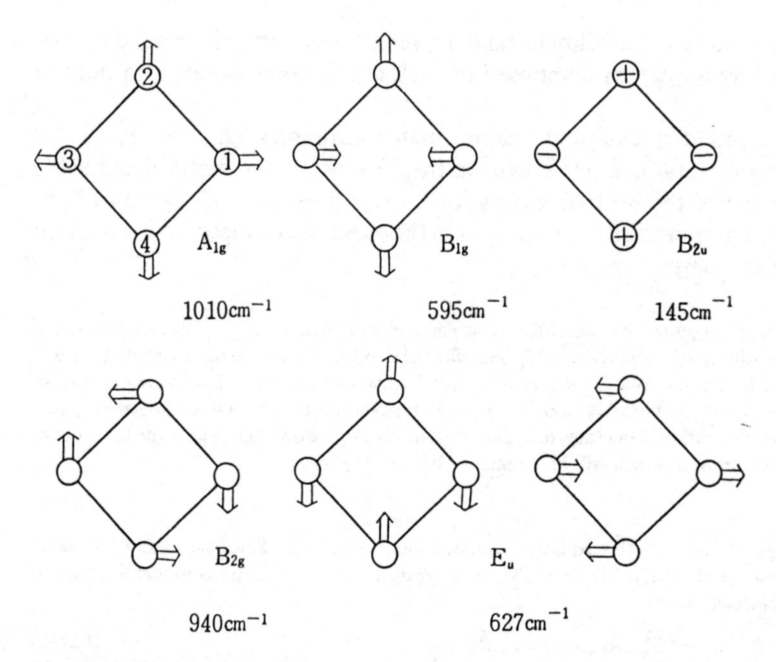

Fig. 10.1. Normal modes of the square molecule X_4. The arrows indicate displacements of the nuclei. The displacements of the B_{2u} mode are perpendicular to the sheet. The frequencies are for cyclobutane, C_4H_8 [10.1]

decoupled from the other normal coordinates. Now, apply some operation R of the symmetry group to Q_s. The displacement pattern described by RQ_s is geometrically the same as the original one described by Q_s, and hence should oscillate with the same frequency ω_s as Q_s does. Thus, if the frequency ω_s is nondegenerate, RQ_s should be equal to Q_s apart from a phase factor. Then Q_s will form a basis for a one-dimensional representation of the symmetry group. If ω_s is degenerate, RQ_s should be a linear combination of Q_s's:

$$RQ_s = \sum_{s'} Q_{s'} D_{s's}(R) \ .$$

The coefficient matrices $\hat{D}(R)$ then form a representation and the set of Q_s's belonging to the same frequency ω_s form a basis for that representation. The argument here parallels the electron problem (Sect. 6.2), and the representation should be irreducible in general. Any two modes belonging to different irreducible representations should have different frequencies, apart from accidental degeneracy due to some coincidence in the force constants.

An important exception to this is the complex conjugate representations. Some of the one-dimensional representations of point groups appear in a pair of complex conjugate ones. Normal coordinates Q_s, being real quantities, cannot be a basis for such a one-dimensional representation. In this case, the two-dimensional representation composed of such a pair corresponds to a normal mode.

We have found above that the normal coordinates Q_s form bases for irreducible representations. How can we then know the symmetry (irreducible representations) of the normal modes for a given molecule? To go into such considerations, it is necessary to define here the effect of symmetry operations on atomic displacements.

Place the origin at the center of mass of the molecule, and let $\boldsymbol{r}_k$ denote the equilibrium position of the kth atom relative to the origin (Fig. 10.2). The effect of a rotation R can be represented by a 3×3 orthogonal matrix $\hat{R}$. If the rotation R takes the atom k to an identical atom l, we have $\hat{R}\boldsymbol{r}_k = \boldsymbol{r}_l$. In such a case, we choose to write $l = R(k)$. Then the above relation may be written as $\hat{R}\boldsymbol{r}_k = \boldsymbol{r}_{R(k)}$.

When the same rotation R operates on a deformed molecule, the displacement of the kth atom in the rotated configuration, which will be designated as $R\boldsymbol{u}_k$, is given by

$$R\boldsymbol{u}_k = \hat{R}\boldsymbol{u}_{R^{-1}(k)} \ . \tag{10.11a}$$

In the language of Sect. 3.2, this equation corresponds to (3.11, 12). Since the symmetry transformation operators are defined by (4.32, 33) in general, the above rotation transforms functions $f(\boldsymbol{u}_k)$ of the displacements as

$$Rf(\boldsymbol{u}_k) = f(R^{-1}\boldsymbol{u}_k) = f(\hat{R}^{-1}\boldsymbol{u}_{R(k)}) = f(\boldsymbol{u}_{R(k)}\hat{R}) \ . \tag{10.11b}$$

In the last expression on the right-hand side, the displacement $\boldsymbol{u}_{R(k)}$, regarded as a row vector, is postmultiplied by the orthogonal matrix $\hat{R}$.

Equation (10.8) shows that the normal coordinates Q_s are linear combinations of the $3N$ displacements $u_{k\alpha}$. Therefore, irreducible decomposition of the $3N$-dimensional representation based on the $3N$ atomic displacements $u_{k\alpha}$ allows us to find the irreducible representations of the normal modes. The decomposition needs only a knowledge of the character of the $3N$-dimensional representation.

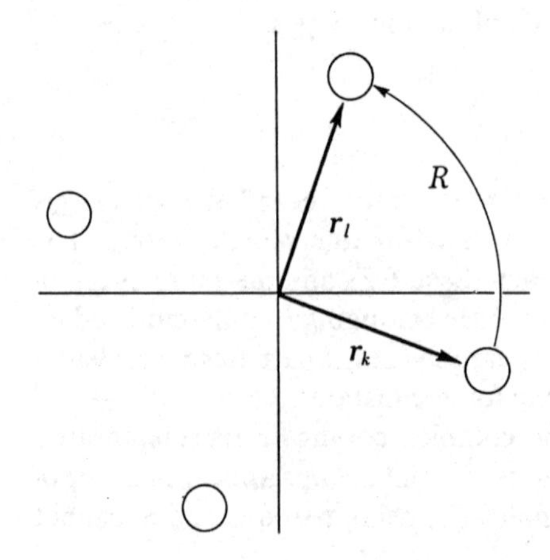

Fig. 10.2. The rotation R brings the atom k to an equivalent atom l

The character $\chi(R)$, in turn, can be obtained in the following way: since nonvanishing contributions to the character $\chi(R)$ come from diagonal matrix elements, we have only to consider those atoms which are not moved by R, i.e., $R(k) = k$.

Consider next how the displacements u_x, u_y, and u_z are transformed by the operation R. When R is a proper rotation about the z-axis through an angle θ, it will transform the displacements according to

$$\begin{aligned} u_x &\to u_x \cos\theta + u_y \sin\theta \ , \\ u_y &\to - u_x \sin\theta + u_y \cos\theta \ , \\ u_z &\to u_z \ . \end{aligned} \tag{10.12}$$

The trace of this transformation matrix is

$$\chi^{(1)}(R) = 1 + 2\cos\theta \quad \text{for proper rotations} \ . \tag{10.13}$$

In crystallographic point groups, the rotation angle θ is limited to the five values given in Table 10.1. In the above, we have assumed the rotation axis to lie in the z direction. For rotations in other directions, the transformation does not obey (10.12), but the trace $\chi^{(1)}(R)$ is still given by (10.13). Therefore, the value of $\chi^{(1)}(R)$ is common to all the atoms that are left unmoved by R. The character $\chi(R)$ of the $3N$-dimensional representation is then given by

$$\chi(R) = N_R \chi^{(1)}(R) \ , \tag{10.14}$$

where N_R stands for the number of atoms that are not moved by R. For rotatory inversions R, the right-hand sides of (10.12) change sign, and we have

$$\chi^{(1)}(R) = -1 - 2\cos\theta \quad \text{for rotatory inversions} \ . \tag{10.15}$$

The character $\chi(R)$ thus obtained will contain three translation and three rotation modes. Since we are interested only in the molecular vibrations, we have to discard such modes. Translational motion of a molecule as a whole is described by the center-of-mass coordinates, which will form a basis for a representation with the character $\chi^{(1)}(R)$. Rotation is described by an axial vector. The corresponding character is given by (10.13) for both proper and improper rotations, because axial vectors do not change sign on inversion.

Table 10.1. Contribution to the character from an unmoved atom. For rotatory inversions, change the signs

Rotation angle θ	0	$\pi/3$	$\pi/2$	$2\pi/3$	π
Character $\chi^{(1)}(R)$	3	2	1	0	-1

Table 10.2. Character of the representation based on the displacement vectors of the square X_4 molecule

D_{4h}	E	$2C_4$	C_4^2	$2C_2'$	$2C_2''$	I	$2IC_4$	σ_h	$2\sigma_v$	$2\sigma_d$
N_R	4	0	0	2	0	0	0	4	2	0
$\chi^{(1)}(R)$	3			-1				1	1	
$\chi(R)$	12	0	0	-2	0	0	0	4	2	0
χ_{trans}	3	1	-1	-1	-1	-3	-1	1	1	1
χ_{rot}	3	1	-1	-1	-1	3	1	-1	-1	-1

Let us apply the above procedure to the X_4 molecule of Fig. 10.1. Note that many operations give $N_R = 0$, which greatly simplifies the calculation of $\chi(R)$. The result is given in Table 10.2. Decomposing the character obtained into the irreducible characters of D_{4h} given in Appendix B, we obtain

$$\begin{aligned} &A_{1g} + B_{1g} + B_{2g} + B_{2u} + E_u \\ &\quad + \text{translation } (A_{2u} + E_u) + \text{rotation } (A_{2g} + E_g) \; . \end{aligned} \tag{10.16}$$

Thus we have established that suitable linear combinations of $u_{k\alpha}$ will yield normal coordinates that form bases for these irreducible representations.

Explicit expressions for the linear combinations may be obtained by referring to the character table in simple cases. For example, the normal coordinate for the totally symmetric mode A_{1g} becomes

$$Q(A_{1g}) = u_{1x} + u_{2y} - u_{3x} - u_{4y} \; ,$$

apart from a constant factor. Comparing this expression with (10.8), we find the eigenvector $e_{k\alpha}$ for the A_{1g} mode, which gives the displacement pattern shown in Fig. 10.1.

A less intuitive but surer method for obtaining the normal coordinates is to use the projection operators (Sect. 6.6). For example, we obtain

$$Q(B_{1g}) = u_{1x} - u_{2y} - u_{3x} + u_{4y}$$

for the B_{1g} mode, operating the projection operator (6.35) on u_{1x} with use of the character of the B_{1g} representation.

Exercise 10.1. Use projection operators to find the normal modes shown in Fig. 10.1. *Hint*: For the E_u representation, construct the representation matrices using the basis $\{x, y\}$. Notice the orthogonality (10.6a) between the molecular vibrations and the translational motion.

As another example, let us consider a H_2O molecule, which has C_{2v} symmetry, the twofold axis being in the z direction in Fig. 10.3. The character for the displacements of the three atoms is calculated in Table 10.3, and its irreducible decomposition gives

$$2A_1 + B_1 + \text{translation} + \text{rotation} \; .$$

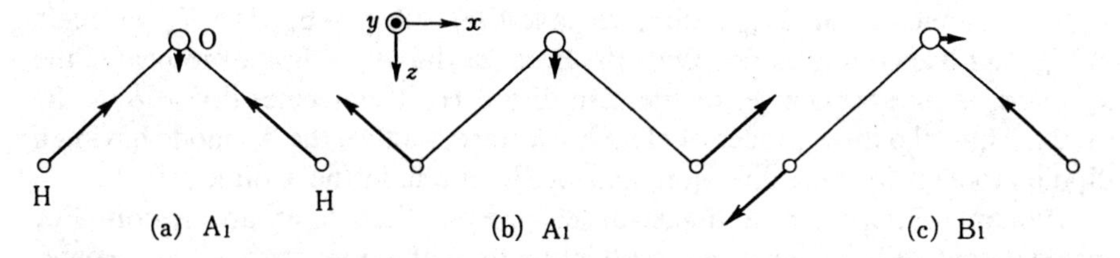

Fig. 10.3. Normal modes of the H_2O molecule. The arrows indicate the displacements

Table 10.3. Character of the representation based on the displacement vectors of the H_2O molecule. The molecule lies in the mirror plane σ_y

C_{2v}	E	C_2	σ_y	σ_x
N_R	3	1	3	1
$\chi^{(1)}(R)$	3	−1	1	1
$\chi(R)$	9	−1	3	1
χ_{trans}	3	−1	1	1
χ_{rot}	3	−1	−1	−1

We have two A_1 modes and one B_1 mode, both one dimensional. Figure 10.3 shows the displacements for these vibration modes. Since we have two A_1 modes, the actual displacement patterns of the normal modes are linear combinations of these two configurations; symmetry arguments cannot uniquely determine the normal modes in such a case. The frequencies of these three modes are found to be 3652, 1595, and 3756 cm^{-1}. The low frequency of the second mode originates in the weaker restoring force of bond bending as compared with bond stretching.

10.3 Selection Rules for Infrared Absorption and Raman Scattering

To determine frequencies of molecular vibrations, light absorption and light scattering experiments are often carried out. It is important to note, however, that the vibration modes that can interact with light are limited. In the case of infrared absorption, only those modes which give rise to an electric dipole moment $\boldsymbol{\mu}$ by molecular deformation can interact with light. Such modes are said to be infrared-active, and the other modes, infrared-inactive.

The moment $\boldsymbol{\mu}$ transforms as a vector under rotations, and so it belongs to the vector representation $\Gamma(\boldsymbol{r})$ based on the three components x, y, and z of a vector $\boldsymbol{r}$. Normal modes belonging to the vector representation $\Gamma(\boldsymbol{r})$ are infrared-active, while the other modes, having no vector components in their basis functions, do not cause electric dipole moments, and hence are infrared-inactive.

For instance, in the D_{4h} group, we have $\Gamma(\boldsymbol{r}) = A_{2u} + E_u$. The X_4 molecule of Fig. 10.1 has no A_{2u} modes. Only the E_u mode, having a dipole moment in the xy-plane, is infrared-active. In the case of the H_2O molecule, $\Gamma(\boldsymbol{r}) = A_1 + B_1 + B_2$. Thus, the three modes of H_2O are infrared-active, the A_1 mode having a dipole moment in the z direction, and the B_1 mode in the x direction.

Raman scattering is a higher-order process than light absorption. For incident light with frequency ω_0, scattered light with frequencies $\omega_0 \pm \omega_s$ comes out, where ω_s stands for the frequency of molecular vibrations. The mechanism of the frequency shift may be understood as follows: The electric field vector $\boldsymbol{E}_0 \cos \omega_0 t$ of the incident light causes a dipole moment $\boldsymbol{\mu} = \alpha \boldsymbol{E}_0 \cos \omega_0 t$ within the molecule, where α denotes the polarizability tensor of the molecule. The dipole moment, oscillating in time, emits light. Now, the vibrating molecule yields a polarizability oscillating with the molecular vibration frequency ω_s. The modulated moment $\boldsymbol{\mu}$ will then have frequencies $\omega_0 \pm \omega_s$ and the light emitted will have those shifted frequencies. This process is not allowed for all the vibration modes. Only those modes which can generate components of the polarizability tensor by molecular deformation can be Raman-active. The polarizability, being a symmetric tensor of rank two, forms a basis for a representation as the following six functions do:

$$x^2,\ y^2,\ z^2,\ xy,\ yz,\ zx\ .$$

Normal modes belonging to such a representation are Raman-active. The other modes, yielding no polarizabilities, are Raman-inactive.

For example, in D_{4h}, the above representation is reduced as follows:

$$A_{1g}:\ z^2, x^2 + y^2 \qquad B_{2g}:\ xy$$

$$B_{1g}:\ x^2 - y^2 \qquad E_g:\ \{zx, zy\}$$

Then we find that the X_4 molecule has Raman-active A_{1g}, B_{1g}, and B_{2g} modes. The B_{2g} mode, for instance, generates the polarizability α_{xy}. Detection of x-polarized scattered light under y-polarized incident light will discriminate the B_{2g} mode. Such an experiment, although meaningless in gases and liquids because the molecular orientations are not fixed, is possible in the solid state.

Exercise 10.2. Examine the selection rule for Raman scattering by a H_2O molecule.

10.4 Interaction of Electrons with Atomic Displacements

At the beginning of this chapter, we assumed the expression (10.1) for the energy of a molecule for small deformations from the equilibrium configuration. As a matter of fact, it is based on the assumption that the electronic state is nondegenerate, which is well satisfied in molecular ground states. We wish to study below how the interaction of electrons with atomic (nuclear) displacements is to be treated in such circumstances.

Molecules are composed of nuclei and electrons. On account of the large ratio ($\sim 10^4$) of their masses, nuclear motion is much slower than electron motion. The nuclei can then be regarded as at rest when electronic wavefunctions are considered: the energy eigenvalue $E_\nu(Q)$ of the νth electronic level is obtained for fixed nuclear displacements Q. This approximation is called the *adiabatic approximation* or Born–Oppenheimer approximation. The total effective potential experienced by the nuclei are obtained by adding the bare internuclear potential $U(Q)$ to $E_\nu(Q)$,

$$W_\nu(Q) = E_\nu(Q) + U(Q) \ , \tag{10.17}$$

which is called the *adiabatic potential.* Expansion of (10.17) in quadratic form in terms of Q about the equilibrium configuration for the molecular ground state ($\nu = 0$) gives the potential energy of (10.1) assumed in Sect. 10.1.

The adiabatic approximation breaks down when the adiabatic potential $W_\nu(Q)$ for the ground state gets close to another adiabatic potential $W_{\nu'}(Q)$ associated with a different ν. When the energy difference between the two potentials becomes of the order of the energy quantum $\hbar\omega$ of vibration, the nuclear motion can give rise to an electronic transition $\nu \to \nu'$, which means breakdown of the adiabatic approximation. Such proximity of the potentials occurs for degenerate electronic states ν. In the absence of degeneracy, the ground state is usually well separated from the excited states and the adiabatic approximation is valid.

The arguments developed above are now formulated below with symmetry considerations in mind. We neglect electron spin for the moment, and consider real wavefunctions. Nuclear displacements are taken to be small, and various quantities will be expanded to second order in the displacements when necessary.

Apart from the kinetic energy of the nuclei, the Hamiltonian of the molecule may be expanded

$$H = H_0 + H_1 + H_2 \tag{10.18}$$

in powers of the displacements. The first term H_0 represents the Hamiltonian of the electron system in the equilibrium molecular configuration. Its eigenvalues E_ν and eigenstates $|\nu\rangle$ satisfy

$$H|\nu\rangle = E_\nu|\nu\rangle \ .$$

The first-order term H_1 may be written as

$$H_1 = \sum_s V_s Q_s \tag{10.19}$$

using normal coordinates Q_s. Here V_s is a real Hermitian operator containing only the electron coordinates. The above H_1 arises from the interaction of electrons with nuclei. Consider now the effect of symmetry operations R on the Hamiltonian H_1. The operations R belonging to the symmetry group of the molecule simultaneously transform the nuclear and electron coordinates. The entire configuration (including the electron coordinates) remains geometrically equivalent after such operations. Therefore, H_1 should be left invariant by such operations. To satisfy this requirement, the electron operator V_s should transform in exactly the same way as Q_s, i.e., V_s should belong to the same irreducible representation and the same partner as Q_s.

The second-order term

$$H_2 = \sum_{ss'} V_{ss'} Q_s Q_{s'} \tag{10.20}$$

must be also invariant. It derives from two sources: the electron–nucleus interaction and the internuclear potential $U(Q)$. The latter is independent of the electron coordinates.

From the Hamiltonian (10.18), the energy $W_\nu(Q)$ for a nondegenerate state ν can be calculated to second order in Q_s using perturbation theory

$$W_\nu(Q) = E_\nu + \sum_s \langle\nu|V_s|\nu\rangle Q_s$$
$$+ \sum_{ss'} Q_s Q_{s'} \left[\langle\nu|V_{ss'}|\nu\rangle + \sum_{\varrho(\neq\nu)} \frac{\langle\nu|V_s|\varrho\rangle\langle\varrho|V_{s'}|\nu\rangle}{E_\nu - E_\varrho} \right]. \tag{10.21}$$

This is just the adiabatic potential (10.17) expanded in Q_s. The terms bilinear in Q_s consist of two contributions: first-order perturbation of H_2 and second-order perturbation of H_1. The latter represents the internuclear potential mediated by the electrons, and the screening effects are embodied in it.

Nondegeneracy of the ν state simplifies calculation of the matrix elements appearing in (10.21). The matrix element of the linear term

$$\langle\nu|V_s|\nu\rangle = \int \Psi_\nu V_s \Psi_\nu d\tau \tag{10.22}$$

vanishes unless V_s belongs to the identity representation, for Ψ_ν^2 becomes totally symmetric because Ψ_ν belongs to a one-dimensional representation. The remaining linear terms involving the totally symmetric modes can also be made to vanish by choosing the origin of those coordinates Q_s at the equilibrium configuration. As to the bilinear terms, they remain nonvanishing only when Q_s and $Q_{s'}$ have the same symmetry. Diagonalizing the quadratic form thus obtained, and adding the kinetic energy, we arrive at (10.10). The above formulation explicitly shows how the expressions (10.1) and (10.10) are derived on the basis of the adiabatic approximation.

Let us next consider electronic states $\{\nu\}$ degenerate at $Q = 0$. The effective Hamiltonian H_{eff} for this set $\{\nu\}$ of degenerate electronic states has the following matrix elements:

$$\langle v|H_{\text{eff}}|v'\rangle = E_v\delta_{vv'} + \sum_s \langle v|V_s|v'\rangle Q_s$$
$$+ \sum_{ss'} Q_s Q_{s'}\left[\langle v|V_{ss'}|v'\rangle + \sum_{\varrho \neq \{v\}} \frac{\langle v|V_s|\varrho\rangle\langle\varrho|V_{s'}|v'\rangle}{E_v - E_\varrho}\right]$$

to second order in Q. H_{eff} has off-diagonal elements in both linear and quadratic terms, which cause $v \to v'$ electronic transitions. It is for this reason that the adiabatic approximation breaks down in the presence of degeneracy.

The off-diagonal elements influence the stable molecular configurations as well. We will focus below on the linear terms of the effective Hamiltonian. In the absence of degeneracy, we have already seen that only totally symmetric Q_s can give nonvanishing linear terms. We have no such restrictions in the presence of degeneracy. The selection rule for "diagonal" matrix elements

$$\langle v|V_s|v'\rangle = \int \Psi_v V_s \Psi_{v'} d\tau$$

becomes as follows (Sect. 6.5): When V_s belongs to the irreducible representation $D^{(s)}$, and Ψ_v to $D^{(\text{el})}$, then the above matrix elements are nonvanishing if the symmetric product representation $[D^{(\text{el})} \times D^{(\text{el})}]$ contains $D^{(s)}$.

As an example, let us examine the selection rule for the X_4 molecule of Sect. 10.2, and take a twofold degenerate electronic level with E_u symmetry, whose wavefunctions transform like x and y. In this case, we have

$$[E_u \times E_u] = A_{1g} + B_{1g} + B_{2g} \ .$$

The linear terms are nonvanishing for normal coordinates Q_1, Q_2, and Q_3 belonging respectively to A_{1g}, B_{1g}, and B_{2g}. The effective Hamiltonian then has the following matrix elements:

$$\begin{array}{c} |x\rangle \\ |y\rangle \end{array}\begin{bmatrix} a_1Q_1 + a_2Q_2 & a_3Q_3 \\ a_3Q_3 & a_1Q_1 - a_2Q_2 \end{bmatrix}, \tag{10.23}$$

where a_1, a_2, and a_3 are the matrix elements of V_s. The diagonal term a_1Q_1 only shifts the equilibrium position of Q_1. Such a shift in Q_1 will not change the symmetry of the molecule, nor lift the degeneracy of the electronic level. The other terms, a_2Q_2 and a_3Q_3, have a profound effect on the stable molecular configuration. The displacements Q_2 and Q_3 can lower the energy by lifting the degeneracy. If we assume $a_2 > 0$ and $a_3 = 0$ for simplicity, a displacement of positive Q_2 will lower the energy of the $|y\rangle$ state. Therefore, the molecule tends to stabilize itself by spontaneously generating the B_{1g} deformation and lifting the degeneracy. In short, orbital degeneracy cannot survive asymmetric deformations. This phenomenon is called the *Jahn–Teller effect.*

For the Jahn–Teller effect to take place, the representation $[D^{(\text{el})} \times D^{(\text{el})}]$ must contain an asymmetric (other than the identity) representation formed by the normal modes. *Jahn* and *Teller* obtained normal modes for all symmetric molecules (done by considering the minimum number of points (nuclei) required to reproduce the symmetry for each point group), and found that, in any molecules except linear ones, symmetric products $[D^{(el)} \times D^{(el)}]$ of more-than-one-dimensional representations $D^{(el)}$ contain at least one such non-identity representations corresponding to the asymmetric modes. The result is stated as *the Jahn–Teller theorem*: "All nonlinear nuclear configurations are unstable for an orbitally degenerate electronic state" [10.2]. The theorem implies that any molecule, except linear ones, having a degenerate ground level will lower its symmetry by lifting the degeneracy. The ground electronic states of stable molecules will then have no orbital degeneracy.

In the above discussion, we have neglected electron spin and considered only orbital degeneracy. When we take account of electron spin, systems consisting of an odd number of electrons have at least twofold degeneracy by virtue of the time-reversal symmetry. This degeneracy, called Kramers degeneracy, can only be lifted by external magnetic fields: the degeneracy due to spin cannot cause the Jahn–Teller effect.

The idea of the Jahn–Teller effect was originally suggested by *Landau*. *Jahn* and *Teller* proved the theorem by examining the point groups one by one. For a more elegant, constructive proof involving induced representations, see [10.3–5].

*10.4.1 Kramers Degeneracy

For electron wavefunctions Ψ_ν containing spin, the criterion for the matrix elements

$$(\Psi_\nu, V_s \Psi_{\nu'}) \tag{10.24}$$

to vanish or not to vanish depends on the number of electrons N. In even-number systems, Ψ_ν belongs to a single-valued representation $D^{(el)}$. The matrix element (10.24) is nonvanishing if $[D^{(el)} \times D^{(el)}]$ contains $D^{(s)}$. In odd-number systems, Ψ_ν belongs to a double-valued representation $D^{(el)}$. Then (10.24) is nonvanishing if $\{D^{(el)} \times D^{(el)}\}$ contains $D^{(s)}$.

The above-mentioned selection rule for diagonal matrix elements (10.24) is derived in the following way: The set $\{\Psi_\nu\}$ of electronic states should contain the time-reversed states $\theta\Psi_\nu$, because the latter are degenerate to the same energy level. So, if the matrix elements (10.24) vanish by symmetry for all combinations of ν and ν', then

$$V_{\nu\nu'} \equiv (\theta\Psi_\nu, V_s \Psi_{\nu'}) \tag{10.25}$$

should also vanish (see Sect. 13.1 for the time-reversal operator θ and its properties). This matrix element becomes

$$\begin{aligned} V_{\nu\nu'} &= (\theta V_s \Psi_{\nu'}, \theta^2 \Psi_\nu) \\ &= (-1)^N (\theta\Psi_{\nu'}, V_s^\dagger \Psi_\nu) = (-1)^N V_{\nu'\nu} \ , \end{aligned} \tag{10.26}$$

Table 10.4. Modes causing the Jahn–Teller effect

$D^{(el)}$	Modes
E_g, E_u	E_g
T_{1g}, T_{1u}, T_{2g}, T_{2u}	E_g, T_{2g}
$E_{1/2g}$, $E_{1/2u}$, $E_{5/2g}$, $E_{5/2u}$	none
$G_{3/2g}$, $G_{3/2u}$	E_g, T_{2g}

where we have assumed that the Hermitian operator V_s commutes with the time-reversal operator θ. The matrix element is then symmetric or antisymmetric according as $(-1)^N = +1$ or -1. Now the arguments given at the end of Sect. 6.5 apply, and we obtain the selection rule mentioned above.

Exercise 10.3. Show that in molecules having inversion symmetry, only even ("g" representation) modes give rise to the Jahn–Teller effect.

Exercise 10.4. For a two-dimensional double-valued irreducible representation D, show that the antisymmetric square $\{D \times D\}$ is the identity representation.

Exercise 10.5. The MX_6 molecule of Fig. 9.6 has the symmetry group O_h. The normal modes of this molecule are A_{1g}, E_g, T_{2g}, $2T_{1u}$, and T_{2u}. Find the normal coordinates for these modes.

Exercise 10.6. Show that when a molecule with O_h symmetry is in an electronic state $D^{(el)}$, the Jahn–Teller effect is caused by the modes given in Table 10.4.

11. Space Groups

When we regard crystals as macroscopic continua, we have only to consider their directional (rotational) symmetry properties; point groups suffice for that purpose. In microscopic investigation of crystals, however, periodicity (translational symmetry) of the crystal structure plays an important part. Space groups describe the full microscopic symmetry of crystals. In this chapter, we discuss various properties and the notation of space groups and study their irreducible representations. Interested readers are invited then to proceed to Chap. 5, where space group representations are discussed using induced representations and ray representations.

11.1 Translational Symmetry of Crystals

The periodic structure of a crystal is specified by a set of three fundamental period vectors $\boldsymbol{t}_1$, $\boldsymbol{t}_2$ and $\boldsymbol{t}_3$. Figure 11.1 illustrates these vectors for the simple cubic (sc) and face-centered cubic (fcc) lattices. Using the unit vectors $\hat{x}$, $\hat{y}$ and $\hat{z}$ in the x, y and z directions, the period vectors are written as

$$\boldsymbol{t}_1 = a\hat{x} \;, \quad \boldsymbol{t}_2 = a\hat{y} \;, \quad \boldsymbol{t}_3 = a\hat{z} \tag{11.1}$$

for the sc lattice, and

$$\boldsymbol{t}_1 = \frac{a}{2}(\hat{y} + \hat{z}) \;, \quad \boldsymbol{t}_2 = \frac{a}{2}(\hat{z} + \hat{x}) \;, \quad \boldsymbol{t}_3 = \frac{a}{2}(\hat{x} + \hat{y}) \tag{11.2}$$

for the fcc lattice, where a stands for the edge length of the cubes. Linear combinations of these three vectors

$$\boldsymbol{t}_n = n_1\boldsymbol{t}_1 + n_2\boldsymbol{t}_2 + n_3\boldsymbol{t}_3$$

with integers n_1, n_2 and n_3 are called the *primitive translation vectors*: they describe the complete translational symmetry of the crystal.

The parallelepiped formed by the three vectors $\boldsymbol{t}_1$, $\boldsymbol{t}_2$ and $\boldsymbol{t}_3$ is called the *unit cell*, which is the minimum unit of the periodic structure. In the sc lattice, the cube shown in Fig. 11.1a is the unit cell, while in the fcc lattice, the unit cell is elongated in the [111] direction and does not have a symmetric shape. The large cube, which contains four unit cells, is called the conventional unit cell.

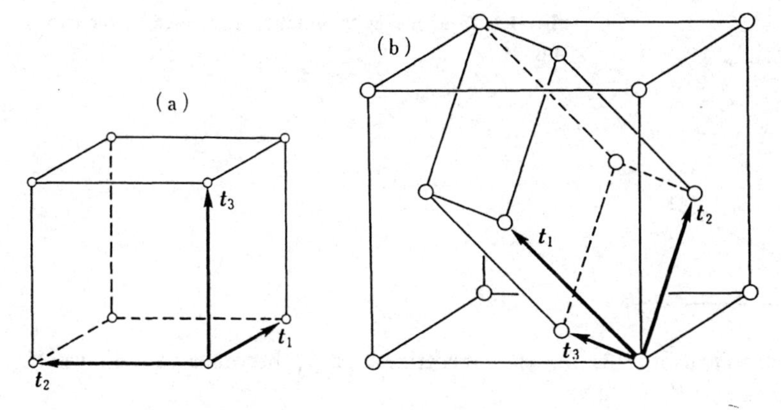

Fig. 11.1. (a) Simple cubic lattice. **(b)** Face-centered cubic lattice

11.2 Symmetry Operations in Space Groups

Let $\{\varepsilon|\,\boldsymbol{b}\}$ denote a translational operation that takes a point at $\boldsymbol{r}$ to another point $\boldsymbol{r}' = \boldsymbol{r} + \boldsymbol{b}$. Such an effect is written as

$$\{\varepsilon|\boldsymbol{b}\}\boldsymbol{r} = \boldsymbol{r} + \boldsymbol{b} \ .$$

Crystals are invariant with respect to the primitive translations $\{\varepsilon|\boldsymbol{t}_n\}$.

In fact, real crystals are finite, so that surfaces of the crystals cannot remain invariant. However, so long as the bulk properties of crystals are concerned, surface effects do not matter. To avoid mathematical difficulties due to the presence of surfaces, the *cyclic boundary condition*

$$\{\varepsilon|N\boldsymbol{t}_1\} = \{\varepsilon|N\boldsymbol{t}_2\} = \{\varepsilon|N\boldsymbol{t}_3\} = \{\varepsilon|0\} \tag{11.3}$$

is adopted, where N is assumed to be a huge integer, Na being of the order of a macroscopic dimension. We then have N^3 primitive translation vectors. The set of such N^3 translations $\{\varepsilon|\boldsymbol{t}_n\}$ will form a group T, the *translation group*. Its order is N^3.

As may be evident from the examples of Fig. 11.1, crystals can have rotational symmetry in addition to translational symmetry. To include such rotational symmetry operations, we introduce the symbol $\{\alpha|\boldsymbol{b}\}$ after Seitz for the operation which takes a point at $\boldsymbol{r}$ to $\boldsymbol{r}' = \alpha\boldsymbol{r} + \boldsymbol{b}$ (Fig. 11.2). Then,

$$\{\alpha|\boldsymbol{b}\}\boldsymbol{r} = \alpha\boldsymbol{r} + \boldsymbol{b} \ . \tag{11.4}$$

Here the 3×3 orthogonal matrix α rotates the vector $\boldsymbol{r}$. In the case of simple translations, α is replaced by the unit matrix ε.

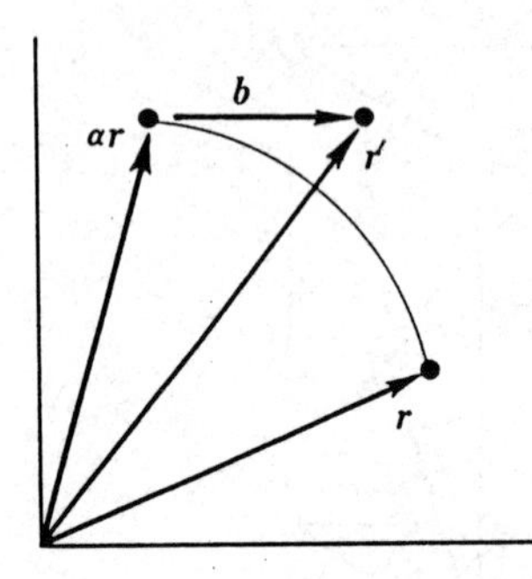

Fig. 11.2. $\{\alpha|\boldsymbol{b}\}$ means the rotation α followed by the translation $\boldsymbol{b}$

The multiplication rule for the operations $\{\alpha|\boldsymbol{b}\}$ becomes as follows: If $\{\alpha_1|\boldsymbol{b}_1\}$ takes $\boldsymbol{r}$ to $\boldsymbol{r}'$, we have

$$\boldsymbol{r}' = \{\alpha_1|\boldsymbol{b}_1\}\boldsymbol{r} = \alpha_1\boldsymbol{r} + \boldsymbol{b}_1 \ .$$

Operating $\{\alpha_2|\boldsymbol{b}_2\}$ on $\boldsymbol{r}'$ gives

$$\boldsymbol{r}'' = \{\alpha_2|\boldsymbol{b}_2\}\boldsymbol{r}' = \alpha_2\boldsymbol{r}' + \boldsymbol{b}_2 \ .$$

Eliminating $\boldsymbol{r}'$ from these two equations, we obtain

$$\{\alpha_2|\boldsymbol{b}_2\}\{\alpha_1|\boldsymbol{b}_1\}\boldsymbol{r} = \alpha_2\alpha_1\boldsymbol{r} + \alpha_2\boldsymbol{b}_1 + \boldsymbol{b}_2 \ ,$$

which leads to the multiplication rule

$$\{\alpha_2|\boldsymbol{b}_2\}\{\alpha_1|\boldsymbol{b}_1\} = \{\alpha_2\alpha_1|\alpha_2\boldsymbol{b}_1 + \boldsymbol{b}_2\} \ . \tag{11.5}$$

Application of this rule gives the expression for the inverse operation

$$\{\alpha|\boldsymbol{b}\}^{-1} = \{\alpha^{-1}| -\alpha^{-1}\boldsymbol{b}\} \ . \tag{11.6a}$$

The operation $\{\alpha|\boldsymbol{b}\}$ has been understood to refer to a given origin O. When the same operation is carried out about a new origin separated by $\boldsymbol{\eta}$ from O, its effect will be

$$\{\varepsilon|\boldsymbol{\eta}\}\{\alpha|\boldsymbol{b}\}\{\varepsilon| -\boldsymbol{\eta}\} = \{\alpha|\boldsymbol{b} + \boldsymbol{\eta} - \alpha\boldsymbol{\eta}\} \ . \tag{11.6b}$$

We have so far considered effects of the symmetry operations on the coordinates $\boldsymbol{r}$. In applications to, say, quantum mechanics, however, it becomes necessary to define the transformation of functions $\psi(\boldsymbol{r})$. This definition is given in parallel with the rotational operations (Sect. 4.3): For a given function $\psi(\boldsymbol{r})$, we define a new function $\psi'(\boldsymbol{r})$ by

$$\psi'(\boldsymbol{r}) = \psi(\{\alpha|\boldsymbol{b}\}^{-1}\boldsymbol{r}) = \psi(\alpha^{-1}\boldsymbol{r} - \alpha^{-1}\boldsymbol{b}) \ . \tag{11.7}$$

The function ψ' is now written as

$$\psi' = \{\alpha|\boldsymbol{b}\}\psi \tag{11.8}$$

in the sense that ψ' is obtained by operating $\{\alpha|\boldsymbol{b}\}$ on ψ. Note that the symbol $\{\alpha|\boldsymbol{b}\}$ appearing in (11.8) is an operator that operates on functions, so that it is different from the $\{\alpha|\boldsymbol{b}\}$ in (11.4, 7). (In principle, we should distinguish between them more clearly by using different symbols for them as was done in Sect. 4.3.) Equations (11.7, 8) give

$$[\{\alpha|\boldsymbol{b}\}\psi](\boldsymbol{r}) = \psi(\alpha^{-1}\boldsymbol{r} - \alpha^{-1}\boldsymbol{b}) \ . \tag{11.9}$$

The operators $\{\alpha|\boldsymbol{b}\}$ defined by (11.9) satisfy the multiplication rule (11.5).

11.3 Structure of Space Groups

Symmetry operations that leave a crystal invariant form a group, which is called the *space group*. Before discussing general properties of space groups, let us examine an example. Figure 11.3 shows the crystal structure of TiO_2, called the rutile structure. The three fundamental period vectors of this structure are obviously

$$\boldsymbol{t}_1 = a\hat{\boldsymbol{x}} \ , \qquad \boldsymbol{t}_2 = a\hat{\boldsymbol{y}}, \qquad \boldsymbol{t}_3 = c\hat{\boldsymbol{z}}$$

and these vectors form a simple tetragonal lattice. The unit cell contains two Ti and four O atoms. Apart from the translations specified by the above period vectors, the following operations bring this structure into coincidence with itself:

$\{\varepsilon|0\}$ Identity operation.
$\{C_2|0\}$ Rotation through π about the z-axis.

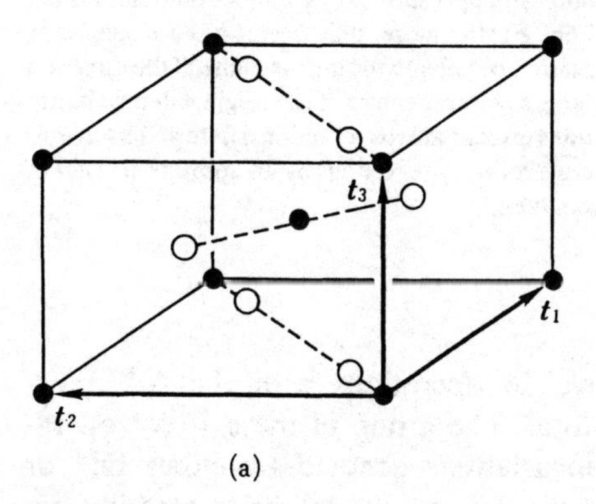

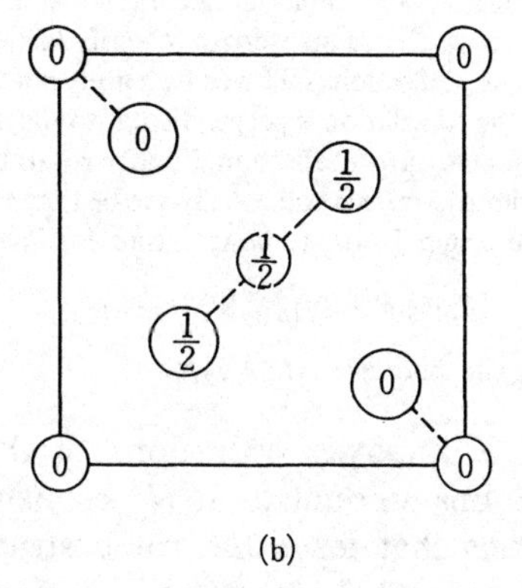

Fig. 11.3a, b. The rutile structure. (**a**) The filled and open circles represent Ti and O atoms respectively. (**b**) Projection onto the horizontal plane. The numbers indicate the heights in units of c above the base plane

$\{C_{2\bar{x}}|0\}$,
$\{C_{2\bar{y}}|0\}$ Rotations through π about the [110] and [1$\bar{1}$0] axes.
$\{I|0\}$ Inversion.
$\{\sigma_h|0\}$ Mirror reflection in the horizontal plane.
$\{\sigma_{d\bar{x}}|0\}$,
$\{\sigma_{d\bar{y}}|0\}$ Mirror reflections in the (110) and (1$\bar{1}$0) planes.

Here, we take the origin of the rotational operations at a Ti atom.

In addition to the above eight operations, the following eight operations that involve a nonprimitive translation

$$\boldsymbol{\tau} = (\boldsymbol{t}_1 + \boldsymbol{t}_2 + \boldsymbol{t}_3)/2 \qquad (11.10)$$

leave this crystal structure invariant:

$\{C_4|\boldsymbol{\tau}\}, \{C_4^{-1}|\boldsymbol{\tau}\}$ Rotations through $\pi/2$ about the z-axis followed by the $\boldsymbol{\tau}$ translation.
$\{S_4|\boldsymbol{\tau}\}, \{S_4^{-1}|\boldsymbol{\tau}\}$ $\pi/2$ rotatory reflections about the z-axis followed by the $\boldsymbol{\tau}$ translation.
$\{C_{2x}|\boldsymbol{\tau}\}, \{C_{2y}|\boldsymbol{\tau}\}$ Rotations through π about the x- and y-axes followed by the $\boldsymbol{\tau}$ translation
$\{\sigma_{vx}|\boldsymbol{\tau}\}, \{\sigma_{vy}|\boldsymbol{\tau}\}$ Mirror reflections in the vertical yz- and xz-planes followed by the $\boldsymbol{\tau}$ translation.

The last eight operations, which were absent in the point groups, are characteristic of space groups. Rotations followed by the $\boldsymbol{\tau}$ translation like $\{C_4|\boldsymbol{\tau}\}$ and $\{C_{2x}|\boldsymbol{\tau}\}$ operations are called *screws*, while mirror reflections followed by the $\boldsymbol{\tau}$ translation like $\{\sigma_{vx}|\boldsymbol{\tau}\}$ are called *glides.*

To be more specific, screw operations are defined as rotations followed by a nonprimitive translation in the direction of the rotation axis. If the direction of the translation is perpendicular to the rotation axis, a suitable choice of the origin will turn the composite operation into a simple rotation. It is left to the reader as an exercise to verify this using (11.6b). Furthermore, glide operations are defined as mirror reflections followed by a nonprimitive translation parallel to the mirror plane. If the direction of the translation is perpendicular to the mirror plane, a suitable choice of the origin will turn it into a simple mirror reflection. Finally, rotatory reflections (except mirror reflections) followed by a nonprimitive translation can always be turned into simple rotatory reflections by an appropriate shift of the origin. In the rutile structure, for instance, we have

$$\{S_4|\boldsymbol{\tau}\} = \{\varepsilon|\boldsymbol{\eta}\}\{S_4|0\}\{\varepsilon|-\boldsymbol{\eta}\}$$

by choosing $\boldsymbol{\eta} = \boldsymbol{t}_2/2 + \boldsymbol{t}_3/4$.

Composite operations of the above 16 operations with the N^3 translations amount to $16N^3$ elements in total. The group of these $16N^3$ operations that leave the rutile structure invariant is denoted by either D_{4h}^{14} or P(4_2/m)(2_1/n)(2/m). The nomenclature of space groups will be treated in Sect. 11.5. The above example for the rutile structure will aid in understanding general properties of space groups to be discussed below.

Rotational parts α of the symmetry operations $\{\alpha|\boldsymbol{b}\}$ of a space group G will form a group G_0. The group G_0 is evidently a crystallographic point group. In the rutile structure, the group G_0 is D_{4h}, which consists of the rotations of the above 16 operations.

The space group G contains the translation group T as its subgroup, and hence can be decomposed into cosets,

$$G = TR_1 + TR_2 + \cdots + TR_g \ , \tag{11.11}$$

where R_1 is the unit element $\{\varepsilon|0\}$. The coset representatives $R_1, R_2, \ldots, R_g$ represent operations with the same rotational parts. In the case of the rutile structure, we have $g = 16$, and the 16 elements mentioned above are just these coset representatives. For instance, the coset TR_1 consists of N^3 elements of the form $\{\varepsilon|\boldsymbol{t}_n\}$, and the coset TR_2 of N^3 elements $\{C_2|\boldsymbol{t}_n\}$.

Now, it is important to note that the translation group T forms an invariant subgroup of the space group G. According to the definition (2.31), the translation group T is an invariant subgroup if

$$\{\alpha|\boldsymbol{b}\}\{\varepsilon|\boldsymbol{t}_n\}\{\alpha|\boldsymbol{b}\}^{-1} \tag{11.12}$$

belongs to T for an arbitrary operation $\{\alpha|\boldsymbol{b}\}$ of the group G. The above product can be calculated using (11.5, 6) with the result $\{\varepsilon|\alpha\boldsymbol{t}_n\}$. Since it should leave the crystal structure unchanged, and since it is a simple translation, it must belong to the translation group T. We have thus found that the subgroup T is conjugate to itself, or T is an invariant subgroup of G. This is the fundamental property of space groups. In consequence, the above-decomposed g cosets will form a factor G/T isomorphic to the point group G_0.

Space group operations can contain essential non-primitive translations which can never be made to vanish by shifting the origin. The space groups which have such essential screws and/or glide operations are called *non-symmorphic* space groups. On the other hand, the space groups in which an appropriate choice of the origin can bring all the translations to primitive lattice translations are called *symmorphic* space groups, in the sense that such space groups have a structure similar to their point groups G_0. In symmorphic space groups, one can, in fact, choose simple rotations $\{\alpha|0\}$ as the coset representatives in (11.11), and then the set of R_i will form G_0. The 230 space groups divide into 73 symmorphic ones and 157 nonsymmorphic ones.

11.4 Bravais Lattices

Successive applications of primitive translations starting at the origin will generate a set of lattice points called the *lattice* or the *Bravais lattice.* The Bravais lattice can have some rotational symmetry in general, but compatibility with the translational symmetry imposes a restriction on it. As we have seen in

(11.12), for arbitrary primitive translation vectors t_n, rotated vectors αt_n must also be primitive translation vectors. This condition limits α to $2\pi/n$ rotations with $n = 1, 2, 3, 4$ and 6, as shown below.

Consider some primitive translation vector T_1 and rotate it through $2\pi/n$. Because of the discrete structure of the lattice, n successive rotations will bring it back to T_1 (see Fig. 11.4 for the $n = 4$ case). The n vectors $T_1, T_2, \ldots, T_n$ are primitive translation vectors equal in length. As is obvious from the figure, the vector $T_1 + T_2 + \cdots + T_n$ lies in the direction of the rotation axis, which means that *the rotation axis must be directed toward one of the primitive translation vectors.* Furthermore, the difference vectors $T_1 - T_2, T_2 - T_3, \ldots, T_n - T_1$ are primitive translation vectors perpendicular to the rotation axis, so that *the rotation axis is perpendicular to some of the primitive translation vectors.* Take then the shortest primitive vector T perpendicular to the C_n axis (Fig. 11.5). Since $C_n T$ and $C_n^{-1} T$ must be primitive vectors,

$$C_n T + C_n^{-1} T = 2T\cos(2\pi/n)$$

must also be a primitive vector, which requires

$$2\cos(2\pi/n) = \text{integer} \ .$$

This condition is satisfied only when $n = 1, 2, 3, 4$, or 6.

From the above-mentioned compatibility of rotational and translational symmetries, we have 14 different kinds of Bravais lattices. They are divided into

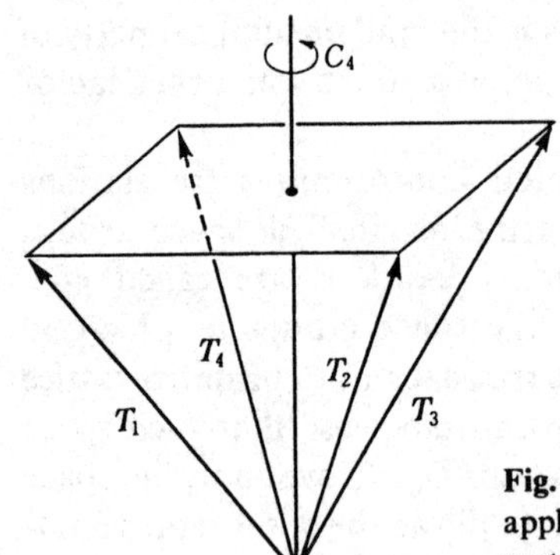

Fig. 11.4. Primitive translation vectors generated by successive applications of an n-fold rotation to the original translation vector T_1

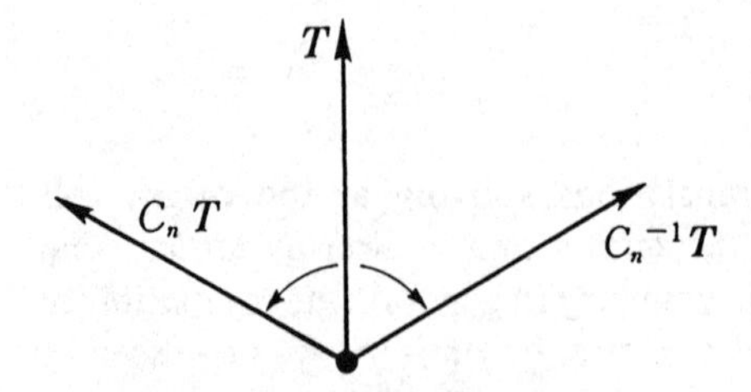

Fig. 11.5. The primitive translation vector T that is the shortest and perpendicular to the C_n axis is rotated clockwise and anticlockwise

seven crystal systems, and their correspondence with the crystallographic point groups can be given through the crystal systems as shown in Table 11.1. The Bravais lattices are characterized by the following symbols:

P:	simple lattice (except trigonal lattice)
R:	simple trigonal lattice
C, B, A:	base-centered lattice
F:	face-centered lattice
I:	body-centered lattice

Figure 11.6 shows the 14 Bravais lattices.

The actual crystal structures are obtained by assigning some structural unit (called a *basis*) to every lattice point. When the basis consists of a single atom, we have simple crystal structures, as exemplified by aluminum (face-centered cubic structure) and iron (body-centered cubic structure). In this case, the crystal structure has the same symmetry as the Bravais lattice. In general, the basis has more atoms. For instance, in TiO_2, the basis consists of two Ti atoms, at (000) and $(aac)/2$, and four O atoms, at $(dd0)$, $(\bar{d}\bar{d}0)$, $(d\bar{d}0) + \tau$, $(\bar{d}d0) + \tau$. The basis in general lowers the symmetry of the crystal structure compared to that of the Bravais lattice.

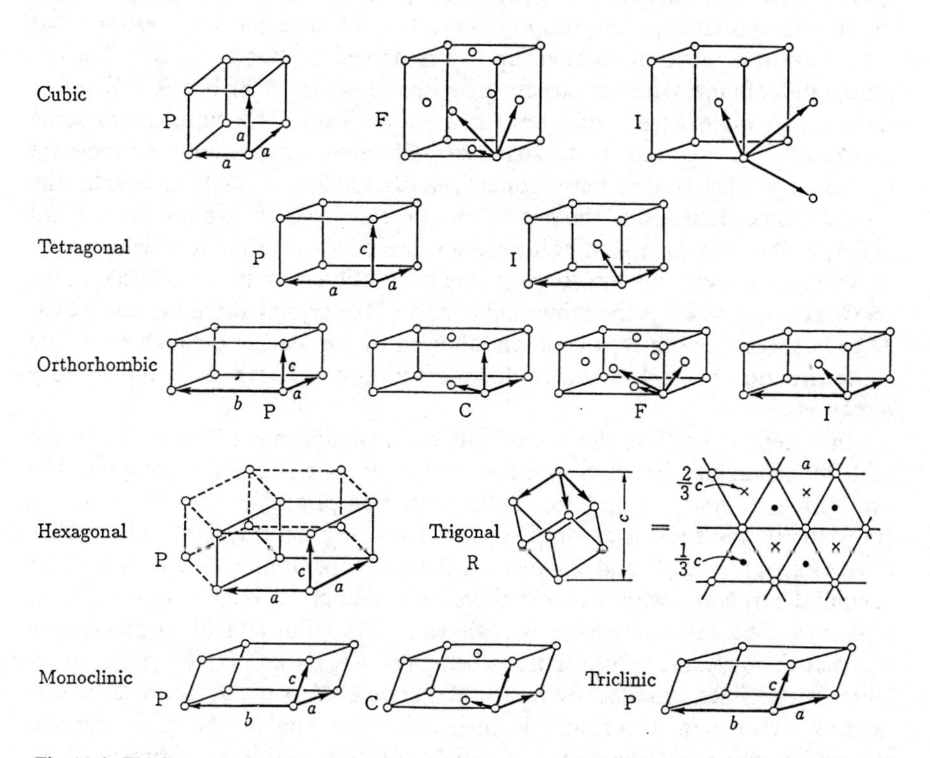

Fig. 11.6. The fourteen Bravais lattices. The arrows show the fundamental period vectors commonly used

Table 11.1. The fourteen Bravais lattices and the crystal systems. "Allowed point groups" means the point groups compatible with the Bravais lattice. Space groups corresponding to D_{3d}, D_3, C_{3v}, C_{3i} and C_3 can have either the hexagonal or trigonal Bravais lattice

Crystal system	Type of lattice	Allowed point groups
Cubic	P, F, I	O_h, O, T_d, T_h, T
Tetragonal	P, I (= F)	D_{4h}, D_4, D_{2d}, C_{4v}, C_{4h}, S_4, C_4
Orthorhombic	P, C, F, I	D_{2h}, D_2, C_{2v}
Hexagonal	P	D_{6h}, D_6, D_{3h}, C_{6v}, C_{6h}, C_{3h}, C_6
Trigonal	R	D_{3d}, D_3, C_{3v}, C_{3i}, C_3
Monoclinic	P, C	C_{2h}, C_2, C_{1h}
Triclinic	P	C_i, C_1

11.5 Nomenclature of Space Groups

Just as in point groups, we have Schönflies and international symbols for denoting space groups. The Schönflies symbols are more compact and simple: space groups that have the same point group are distinguished by superscripts to the corresponding point-group symbols. For instance, ten space groups that belong to the octahedral point group O_h are named O_h^1, O_h^2, . . . , O_h^{10}. The ten groups differ in their Bravais lattices (primitive (1–4); fcc (5–8); bcc (9, 10)) and in existence or nonexistence of screws and glides. Table 11.2 summarizes some typical space groups and crystal structures, where the international symbols are written in parallel. In the international symbols, the first character stands for the type of Bravais lattice, and the rest follows the rule for point groups. Screws and glides, which were absent in point groups, are denoted in the following way.

In a screw operation, the nonprimitive translation must be directed along the screw axis. Suppose c is the minimum period of the crystal along the screw axis. Then n_p stands for the $2\pi/n$ rotation followed by translation through pc/n. The rutile structure has a 4_2 screw in the z direction and 2_1 screws in the x and y directions.

In a glide operation, the nonprimitive translation must be parallel to the glide plane, and its direction specifies the type (a, b, c, n, or d) of the glide. The symbols a, b, and c stand for glides with non-primitive $a/2$, $b/2$, and $c/2$ translations, while n stands for glides with face-diagonal translations $(a + b)/2$, $(b + c)/2$, or $(c + a)/2$; and d stands for the glide operation characteristic of the diamond structure with non-primitive translations $(a + b)/4$, $(b + c)/4$, or $(c + a)/4$. The rutile structure has glides in the (100) and (010) planes, the associated nonprimitive translations being $(\boldsymbol{t}_2 + \boldsymbol{t}_3)/2$ and $(\boldsymbol{t}_1 + \boldsymbol{t}_3)/2$ respectively. Such glides, denoted by n, are perpendicular to the 2_1 screw axes and hence we write 2_1/n. As a result, the international symbol for the space group of the rutile structure becomes $P(4_2/m)(2_1/n)(2/m)$, which is abbreviated as $P4_2/mnm$.

Table 11.2. Typical space groups and crystal structures

Schönflies symbol	International symbol (abbreviated)	References for the irreducible representations	Crystal structure	Typical materials
O_h^1	$P\frac{4}{m}\bar{3}\frac{2}{m}$ (Pm3m)	[11.1, 3]	Simple cubic structure	α-Po
			CsCl structure	CsCl, TlCl
			Perovskite structure	$SrTiO_3$
O_h^5	$F\frac{4}{m}\bar{3}\frac{2}{m}$ (Fm3m)	[11.1, 3]	fcc structure	Al, Ar, Ni
			Fluorite structure	CaF_2
			NaCl structure	NaCl, PbTe
O_h^7	$F\frac{4_1}{d}\bar{3}\frac{2}{m}$ (Fd3m)	[11.2, 3]	Diamond structure	Diamond, Si, Ge
O_h^9	$I\frac{4}{m}\bar{3}\frac{2}{m}$ (Im3m)	[11.1, 3]	bcc structure	Na, Fe
T_d^2	$F\bar{4}3m$	[11.4]	Zinc-blende structure	GaAs, CuCl
C_{6v}^4	$P6_3mc$	[11.5–7]	Wurtzite structure	CdS
D_{6h}^4	$P\frac{6_3}{m}\frac{2}{m}\frac{2}{c}$ ($P6_3/mmc$)	[11.2, 3]	hcp structure	Mg

11.6 The Reciprocal Lattice and the Brillouin Zone

Consider a periodic function $U(\boldsymbol{r})$ with the same periods as the crystal lattice. It satisfies

$$U(\boldsymbol{r}-\boldsymbol{t}_n)=U(\boldsymbol{r}) \tag{11.13}$$

for any primitive translation vector $\boldsymbol{t}_n$. When this function is expanded in a triple Fourier series as

$$U(\boldsymbol{r})=\sum_{\boldsymbol{K}} U_{\boldsymbol{K}}\exp(\mathrm{i}\boldsymbol{K}\cdot\boldsymbol{r}) \ , \tag{11.14}$$

the periodicity (11.13) requires that the vectors $\boldsymbol{K}$ satisfy

$$\exp(\mathrm{i}\boldsymbol{K}\cdot\boldsymbol{t}_n)=1 \ . \tag{11.15}$$

To obtain a general expression for such $\boldsymbol{K}$ vectors, we seek three independent vectors $\boldsymbol{K}_1$, $\boldsymbol{K}_2$ and $\boldsymbol{K}_3$ that satisfy

$$\boldsymbol{K}_i\cdot\boldsymbol{t}_j=2\pi\delta_{ij} \ , \quad i,j=1,2,3 \ . \tag{11.16}$$

Equation (11.16) has the solution

$$K_1 = \frac{2\pi}{\Omega}(t_2 \times t_3) ,$$

$$K_2 = \frac{2\pi}{\Omega}(t_3 \times t_1) , \tag{11.17}$$

$$K_3 = \frac{2\pi}{\Omega}(t_1 \times t_2) ,$$

where $\Omega = t_1 \cdot (t_2 \times t_3)$ denotes the volume of the unit cell. The general expression for K satisfying (11.15) can then be written as

$$K_m = m_1 K_1 + m_2 K_2 + m_3 K_3 \tag{11.18}$$

with integers m_1, m_2 and m_3. The lattice constructed by these vectors is called the *reciprocal lattice*, and the vectors K_m are called the reciprocal lattice vectors.

For example, in the case of the simple cubic lattice, substitution of (11.1) into (11.17) gives

$$K_1 = (2\pi/a)\hat{x} , \quad K_2 = (2\pi/a)\hat{y} , \quad K_3 = (2\pi/a)\hat{z} ,$$

which will form the simple cubic reciprocal lattice. The face-centered cubic lattice generated by the vectors (11.2) has

$$K_1 = (2\pi/a)(-\hat{x} + \hat{y} + \hat{z}) ,$$

$$K_2 = (2\pi/a)(\hat{x} - \hat{y} + \hat{z}) ,$$

$$K_3 = (2\pi/a)(\hat{x} + \hat{y} - \hat{z}) ,$$

which will form the body-centered cubic reciprocal lattice. In general, Bravais and reciprocal lattices are related to each other in the following way: Reciprocal lattices of primitive and base-centered Bravais lattices are primitive and base-centered respectively, while the face-centered and body-centered lattices are interchanged.

The unit cell of the reciprocal lattice has the volume

$$\Omega' = K_1 \cdot (K_2 \times K_3) .$$

Substituting (11.17), we obtain

$$\Omega' = \left(\frac{2\pi}{\Omega}\right)^3 (t_2 \times t_3) \cdot [(t_3 \times t_1) \times (t_1 \times t_2)]$$

$$= \left(\frac{2\pi}{\Omega}\right)^3 [(t_2 \times t_3) \cdot t_1][(t_3 \times t_1) \cdot t_2]$$

$$= (2\pi)^3/\Omega , \tag{11.19}$$

which means that the volume of the reciprocal unit cell Ω' is $(2\pi)^3$ times the inverse of the unit cell volume Ω. As a candidate for the reciprocal unit cell having the volume Ω', one can choose the parallelepiped formed by the three vectors $\boldsymbol{K}_1$, $\boldsymbol{K}_2$ and $\boldsymbol{K}_3$ just as in the Bravais lattice. However, such a choice is unsuitable, because the resulting reciprocal unit cell does not embody the full symmetry of the reciprocal lattice. To obtain a symmetrical unit cell, we draw

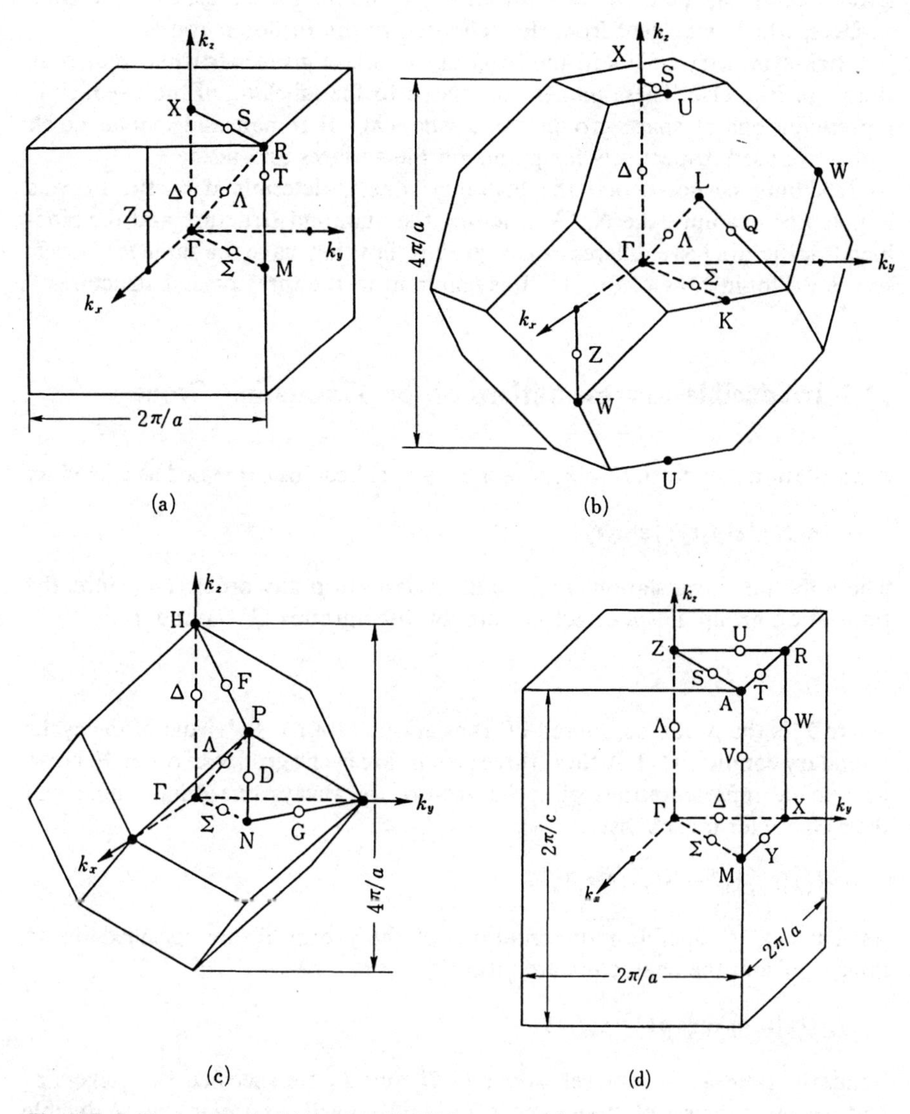

Fig. 11.7. Brillouin zones for (**a**) simple cubic, (**b**) face-centered cubic, (**c**) body-centered cubic, and (**d**) simple tetragonal lattices

planes that bisect the reciprocal lattice vectors perpendicularly. The innermost region surrounded by those planes, having the required symmetry, is called the first Brillouin zone, or simply the *Brillouin zone.* Figure 11.7 shows the Brillouin zones for the simple cubic, face-centered cubic, body-centered cubic and simple tetragonal Bravais lattices. The somewhat complicated shape (truncated octahedron) for the fcc lattice is made of the planes bisecting the $(2\pi/a)$ (111) and $(2\pi/a)$(200) reciprocal lattice vectors. The volumes of these polyhedra can be shown to be equal to Ω'. In addition, no two points differing only by a reciprocal lattice vector can be contained within the Brillouin zone except on the zone surfaces, which is evident from the definition of the Brillouin zone.

High-symmetry points in the Brillouin zone are given particular names as shown in Fig. 11.7. These names are related to the labelling of the irreducible representations of space groups. As a rule, capital roman and capital greek letters are used respectively for points on the surfaces and inside.

It should be noted that the Brillouin zone is determined by the Bravais lattice. For example, the NaCl structure, the diamond structure and the zinc-blende structure have different space groups, but they have the same fcc lattice, so the Brillouin zone of Fig. 11.7b is common to the three crystal structures.

11.7 Irreducible Representations of the Translation Group

A translational operation $\{\varepsilon|n_1\boldsymbol{t}_1 + n_2\boldsymbol{t}_2 + n_3\boldsymbol{t}_3\}$ can be expressed as a product

$$\{\varepsilon|n_1\boldsymbol{t}_1\}\{\varepsilon|n_2\boldsymbol{t}_2\}\{\varepsilon|n_3\boldsymbol{t}_3\} ,$$

where the three translations may be interchanged in any order. Therefore, the translation group T is a direct product of three groups T_1, T_2 and T_3:

$$T = T_1 \times T_2 \times T_3 , \tag{11.20}$$

where T_i is the group composed of translations $\{\varepsilon|n_i\boldsymbol{t}_i\}$. By virtue of the cyclic boundary condition (11.3), these three groups are cyclic groups of order N. Now, irreducible representations of cyclic groups are always one dimensional and their character must satisfy

$$[\chi(\{\varepsilon|\boldsymbol{t}_1\})]^N = \chi(\{\varepsilon|N\boldsymbol{t}_1\}) = 1 .$$

As a result, irreducible representations of the group T_1 are specified by an integer p_1 and the characters are given by

$$\chi_{p_1}(\{\varepsilon|\boldsymbol{t}_1\}) = \exp(2\pi i p_1/N) .$$

Similarly, irreducible representations of T_2 and T_3 are specified by integers p_2 and p_3. Then the set of three integers (p_1, p_2, p_3) will determine the irreducible representation of the translation group T. Rather than dealing with the integers

(p_1, p_2, p_3), we define a vector

$$\boldsymbol{k} = (p_1\boldsymbol{K}_1 + p_2\boldsymbol{K}_2 + p_3\boldsymbol{K}_3)/N \tag{11.21}$$

in reciprocal space, which may be used as a label to designate the irreducible representation of the translation group with the character

$$\chi^{\boldsymbol{k}}(\{\varepsilon|\boldsymbol{t}_n\}) = \exp(\mathrm{i}\boldsymbol{k}\cdot\boldsymbol{t}_n) \ . \tag{11.22}$$

The vector $\boldsymbol{k}$ may be regarded as a virtually continuous vector rather than a discrete one since N is a huge integer.

Replacement of $\boldsymbol{k}$ in (11.22) by $\boldsymbol{k} + \boldsymbol{K}$, which differs from $\boldsymbol{k}$ only by a reciprocal lattice vector $\boldsymbol{K}$, does not change the value of the character, which means that $\boldsymbol{k} + \boldsymbol{K}$ points to the same irreducible representation as $\boldsymbol{k}$. To avoid multiple counting, $\boldsymbol{k}$ is usually confined to within the Brillouin zone. The number of irreducible representations (or the number of independent $\boldsymbol{k}$ vectors in the Brillouin zone) is N^3, being equal to the order of the translation group T.

According to (11.9), the effect of translation $\{\varepsilon|\boldsymbol{t}_n\}$ on a function $\psi(\boldsymbol{r})$ is described by

$$\{\varepsilon|\boldsymbol{t}_n\}\psi(\boldsymbol{r}) = \psi(\boldsymbol{r} - \boldsymbol{t}_n) \ . \tag{11.23}$$

The operator $\{\varepsilon|\boldsymbol{t}_n\}$ can be written explicitly as

$$\{\varepsilon|\boldsymbol{t}_n\} = \exp(-\mathrm{i}\boldsymbol{p}\cdot\boldsymbol{t}_n/\hbar) \tag{11.24}$$

using the momentum operator $\boldsymbol{p} = -\mathrm{i}\hbar\boldsymbol{\nabla}$.

Now the basis function $\psi_{\boldsymbol{k}}(\boldsymbol{r})$ of the irreducible representation $\boldsymbol{k}$ should satisfy

$$\{\varepsilon|\boldsymbol{t}_n\}\psi_{\boldsymbol{k}} = \chi^{\boldsymbol{k}}(\{\varepsilon|\boldsymbol{t}_n\})\psi_{\boldsymbol{k}} = \exp(\mathrm{i}\boldsymbol{k}\cdot\boldsymbol{t}_n)\psi_{\boldsymbol{k}} \ . \tag{11.25}$$

From (11.23, 25), we obtain

$$\psi_{\boldsymbol{k}}(\boldsymbol{r} - \boldsymbol{t}_n) = \exp(\mathrm{i}\boldsymbol{k}\cdot\boldsymbol{t}_n)\psi_{\boldsymbol{k}}(\boldsymbol{r}) \ , \tag{11.26}$$

which is known as the *Bloch theorem*. The functions $\psi_{\boldsymbol{k}}(\boldsymbol{r})$ satisfying (11.26) are called *Bloch functions*. If we put

$$\psi_{\boldsymbol{k}}(\boldsymbol{r}) = \exp(-\mathrm{i}\boldsymbol{k}\cdot\boldsymbol{r})u_{\boldsymbol{k}}(\boldsymbol{r}) \tag{11.27}$$

and substitute it in (11.26), we find

$$u_{\boldsymbol{k}}(\boldsymbol{r} - \boldsymbol{t}_n) = u_{\boldsymbol{k}}(\boldsymbol{r}) \ ,$$

i.e., $u_{\boldsymbol{k}}(\boldsymbol{r})$ is a periodic function.

Bloch functions may also be constructed in the form

$$\psi_{\boldsymbol{k}}(\boldsymbol{r}) = \sum_n \exp(-\mathrm{i}\boldsymbol{k}\cdot\boldsymbol{t}_n)a(\boldsymbol{r} - \boldsymbol{t}_n) \ . \tag{11.28}$$

The above forms for $\psi_k(r)$ tell us that the vector k stands for the wavevector of Bloch waves propagating in the crystal.

11.8 The Group of the Wavevector k and Its Irreducible Representations

As we have seen in (11.11), a space group G can be decomposed into cosets. The coset decomposition is characterized by the g coset representatives

$$R_i = \{\alpha_i | b_i\} , \quad i = 1, 2, \ldots, g , \tag{11.29}$$

where R_1 is the unit element $\{\varepsilon|0\}$.

Let us consider a basis function ψ_k for the irreducible representation k of the translation group T and examine what kind of functions are generated by operating R_i on ψ_k. If we define

$$\phi_k = \{\alpha | b\} \psi_k , \tag{11.30}$$

then the multiplication rule (11.5) yields

$$\begin{aligned} \{\varepsilon|t_n\}\phi_k &= \{\varepsilon|t_n\}\{\alpha|b\}\psi_k \\ &= \{\alpha|b\}\{\varepsilon|\alpha^{-1}t_n\}\psi_k \\ &= \exp(\mathrm{i}k\cdot\alpha^{-1}t_n)\phi_k \\ &= \exp(\mathrm{i}\alpha k\cdot t_n)\phi_k , \end{aligned} \tag{11.31}$$

where the last equality depends on the orthogonality of the 3×3 matrix α. This result shows that the function ϕ_k defined by (11.30) belongs to the irreducible representation αk of the translation group.

Thus, if we start from some k vector and apply the g operations R_i, we shall have g vectors $\alpha_i k$. The set of different vectors $\alpha_i k$ obtained by application of the rotations α_i on k is called the *star* of k and is designated as star$\{k\}$ or simply as $\{k\}$. In constructing star$\{k\}$, those vectors which differ only by a reciprocal lattice vector must be regarded as the same vector because they give the same irreducible representation of the translation group. The number of wavevectors in a star is called the order of the star.

Let us observe some examples of the stars in the rutile structure. The Brillouin zone for this structure is depicted in Fig. 11.7d, where high-symmetry points are given particular names. If we take a general point $k = (k_x, k_y, k_z)$ in the Brillouin zone and apply the 16 operations R_i, we obtain 16 vectors of the form $(\pm k_x, \pm k_y, \pm k_z)$ and $(\pm k_y, \pm k_x, \pm k_z)$. Star$\{k\}$ then consists of these 16 vectors, and the order of the star is equal to the order of the point group. On the other hand, at the origin $k_\Gamma \equiv (000)$, the star consists of this single vector. The star of $k_Z \equiv (0, 0, \pi/c)$ also consists of this single vector: although the mirror

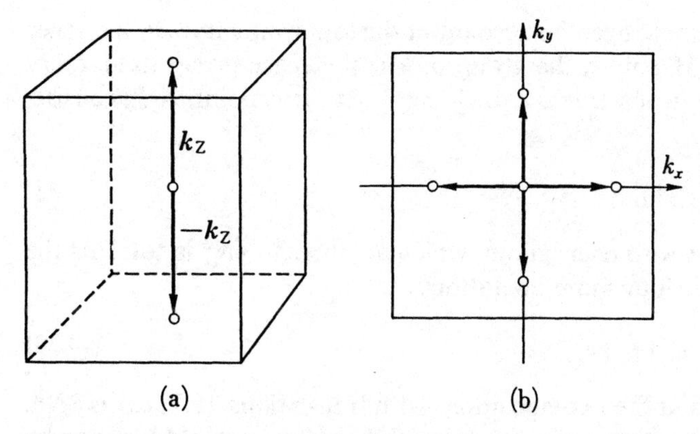

Fig. 11.8. (a) The star of Z consists of a single wavevector $\boldsymbol{k}_Z$. (b) Four $\boldsymbol{k}$ vectors constitute the star of Δ

reflection σ_h sends $\boldsymbol{k}_Z$ to $-\boldsymbol{k}_Z$, as shown in Fig. 11.8a, the latter vector cannot be counted as independent of the former since they are connected by a reciprocal lattice vector $(0, 0, 2\pi/c)$. Similarly, the order of the star is 1 at A and M points. The star of $\boldsymbol{k}_\Delta \equiv (0, k, 0)$ is composed of four vectors $(0, \pm k, 0)$ and $(\pm k, 0, 0)$, as shown in Fig. 11.8b. When the Δ point is moved onto the zone boundary point X, the star of X consists of two vectors $(0, \pi/a, 0)$ and $(\pi/a, 0, 0)$.

As may be seen in the above examples, at high-symmetry points in the Brillouin zone, some operations leave $\boldsymbol{k}$ unchanged. To examine such cases in more detail, let us consider those symmetry operations $\{\beta|\boldsymbol{b}\}$ whose rotational part β leaves the $\boldsymbol{k}$ vector unchanged,

$$\beta\boldsymbol{k} \doteq \boldsymbol{k} \ , \tag{11.32}$$

where the dot above the equality sign means "modulo reciprocal lattice vectors," i.e., $\beta\boldsymbol{k}$ and $\boldsymbol{k}$ may differ by a reciprocal lattice vector. For such a rotation β, we have

$$\exp(\mathrm{i}\beta\boldsymbol{k}\cdot\boldsymbol{t}_n) = \exp(\mathrm{i}\boldsymbol{k}\cdot\boldsymbol{t}_n) \ ,$$

and hence from (11.31)

$$\{\varepsilon|\boldsymbol{t}_n\}\{\beta|\boldsymbol{b}\}\psi_{\boldsymbol{k}} = \exp(\mathrm{i}\boldsymbol{k}\cdot\boldsymbol{t}_n)\{\beta|\boldsymbol{b}\}\psi_{\boldsymbol{k}} \ ,$$

which means that the function $\{\beta|\boldsymbol{b}\}\psi_{\boldsymbol{k}}$ belongs to the same irreducible representation $\boldsymbol{k}$ of the translation group as $\psi_{\boldsymbol{k}}$ does. The set of such operations $\{\beta|\boldsymbol{b}\}$ forms a group, which is called *the group of the wavevector* $\boldsymbol{k}$ and is designated by $\mathscr{G}(\boldsymbol{k})$. The group of $\boldsymbol{k}$ is in general an assembly of several cosets of (11.11). It contains the translation group as an invariant subgroup.

Let us observe again some examples in the rutile structure. For a general point $\boldsymbol{k}$ in the Brillouin zone, the group of $\boldsymbol{k}$ is the translation group: no rotations can leave $\boldsymbol{k}$ unchanged on account of the low symmetry. On the other hand, at Γ, Z, A and M points, the group of $\boldsymbol{k}$ is the space group itself: every operation of the space group leaves $\boldsymbol{k}$ unchanged. At the Δ point, $\mathscr{G}(\boldsymbol{k})$ consists of

$$\{\varepsilon|0\},\ \{\sigma_{\mathrm{h}}|0\},\ \{C_{2y}|\boldsymbol{\tau}\},\ \{\sigma_{\mathrm{v}x}|\boldsymbol{\tau}\} \tag{11.33}$$

and their compositions with translations, which amounts to $4N^3$ in total. At the X point, $\mathscr{G}(\boldsymbol{k})$ contains four more operations

$$\{C_2|0\},\ \{I|0\},\ \{C_{2x}|\boldsymbol{\tau}\},\ \{\sigma_{\mathrm{v}y}|\boldsymbol{\tau}\} \tag{11.34}$$

in addition to (11.33) and their compositions with translations. Its order is $8N^3$.

As may be seen from the above examples, the order s of star$\{\boldsymbol{k}\}$, the order kN^3 of the group of $\boldsymbol{k}$, and the order g of the point group are related by

$$sk = g\ . \tag{11.35}$$

Let us now investigate irreducible representations[1] of the group of $\boldsymbol{k}$. Since the basis functions are required to satisfy (11.25), the representation matrix $\hat{D}^{\boldsymbol{k}}$ for translations should be

$$\hat{D}^{\boldsymbol{k}}(\{\varepsilon|\boldsymbol{t}_n\}) = \exp(\mathrm{i}\boldsymbol{k}\cdot\boldsymbol{t}_n)\hat{1}_d\ , \tag{11.36}$$

where the symbol $\hat{1}_d$ stands for the unit matrix with the dimension d of the representation. Irreducible representations of the group of $\boldsymbol{k}$ are readily obtained from those of point groups in the following two cases:

(a) $\boldsymbol{k}$ is located inside the Brillouin zone;
(b) $\boldsymbol{k}$ is located on the Brillouin zone boundary, and the group of $\boldsymbol{k}$ is symmorphic (i.e., does not contain essential screws and glides).

In contrast to this, special considerations are required when

(c) $\boldsymbol{k}$ is located on the Brillouin zone boundary, and the group of $\boldsymbol{k}$ is nonsymmorphic.

Let $\mathscr{G}_0(\boldsymbol{k})$ denote the point group composed of the rotations β of the elements $\{\beta|\boldsymbol{b}\}$ in $\mathscr{G}(\boldsymbol{k})$, and suppose Γ to be an irreducible representation of the point group $\mathscr{G}_0(\boldsymbol{k})$. Then it turns out that

$$\hat{D}^{\boldsymbol{k}}(\{\beta|\boldsymbol{b}\}) = \exp(\mathrm{i}\boldsymbol{k}\cdot\boldsymbol{b})\hat{\Gamma}(\beta) \tag{11.37}$$

form an irreducible representation of $\mathscr{G}(\boldsymbol{k})$ in the cases (a) and (b). Irreducibility of $\boldsymbol{D}^{\boldsymbol{k}}$ is apparent since Γ is irreducible. To confirm that (11.37) really forms a

[1] Irreducible representations of the group of $\boldsymbol{k}$ are called *small representations.*

representation, we calculate matrix products corresponding to the multiplication rule (11.5). The product of the two matrices is

$$\hat{D}^{k}(\{\beta_2|b_2\})\hat{D}^{k}(\{\beta_1|b_1\}) = \exp[\mathrm{i}k\cdot(b_1 + b_2)]\hat{\Gamma}(\beta_2\beta_1) \; . \tag{11.38}$$

On the other hand, the matrix for the product operation is

$$\hat{D}^{k}(\{\beta_2\beta_1|\beta_2 b_1 + b_2\}) = \exp[\mathrm{i}k\cdot(\beta_2 b_1 + b_2)]\hat{\Gamma}(\beta_2\beta_1) \; . \tag{11.39}$$

These two matrices are identical if and only if

$$\exp[\mathrm{i}(\beta_2^{-1}k - k)\cdot b_1] = 1 \; . \tag{11.40}$$

In the case (a), we always have $\beta_2^{-1}k = k$, and hence (11.40) holds. For k on the zone boundary, $\beta_2^{-1}k - k$ does not necessarily vanish; it is equal to a reciprocal lattice vector. Even so, if b_1 is a primitive translation vector [case (b)], (11.40) again holds. Consequently, (11.37) gives irreducible representations of $\mathcal{G}(k)$ in the cases (a) and (b) mentioned above. Since the irreducible representations are well known for point groups, we can readily construct irreducible representations of $\mathcal{G}(k)$ in such cases.

In the rutile structure, small representations for k inside the Brillouin zone are obtained by this means:

Γ: This is the point $k = 0$. The group $\mathcal{G}_0(k)$ is D_{4h}. The irreducible representations of D_{4h} may be taken over to the small representations. Translations are represented by the unit matrix.

Λ: For general points on the k_z axis, $\mathcal{G}_0(k)$ is C_{4v}. Small representation matrices are obtained by multiplying the irreducible representation matrices of C_{4v} with the phase factor $\exp(\mathrm{i}k\cdot b)$.

Δ: For general points on the k_y (or k_x) axis, $\mathcal{G}_0(k)$ is C_{2v}.

The remaining case (c) has to be treated separately. In this case, we have two methods available to obtain small representations: one is to use the theory of ray representations (Sects. 5.3, 5.4), and the other is to use Herring's method [11.2]. We will demonstrate here the latter method for the particular example of the X point of the rutile structure.

The group of the wavevector k at the X point is composed of the $8N^3$ elements given in (11.33, 34). To obtain irreducible representations of this nonsymmorphic group, we begin by considering those translations which satisfy

$$\exp(\mathrm{i}k\cdot t_n) = 1 \; . \tag{11.41}$$

Since $k = (0, \pi/a, 0)$, such translations are written as

$$t_n = n_1 t_1 + 2n_2 t_2 + n_3 t_3 \tag{11.42}$$

using integers n_1, n_2 and n_3. Let $T(k)$ denote the group of those translations of order $N^3/2$, and decompose the group $\mathcal{G}(k)$ with respect to $T(k)$:

$$\mathcal{G}(k) = T(k)P_1 + T(k)P_2 + \cdots + T(k)P_{16} \; . \tag{11.43}$$

The 16 representative elements P_1–P_{16} are the eight elements given in (11.33, 34) plus their compositions with $\{\varepsilon|\boldsymbol{t}_2\}$. The above cosets form the factor group $\mathscr{G}(\boldsymbol{k})/T(\boldsymbol{k})$, since $T(\boldsymbol{k})$ is an invariant subgroup. Irreducible representations of this factor group will give the required small representations since the translations $\{\varepsilon|\boldsymbol{t}_n\}$ belonging to $T(\boldsymbol{k})$ are represented by the unit matrix because of (11.41). The first step to this end is to classify the cosets into classes. For example, from

$$\{C_{2x}|\boldsymbol{\tau}\}^{-1}\{C_2|0\}\{C_{2x}|\boldsymbol{\tau}\} = \{C_2^{-1}|C_{2x}(C_2\boldsymbol{\tau} - \boldsymbol{\tau})\} = \{C_2|\boldsymbol{t}_2 - \boldsymbol{t}_1\} ,$$

we find that the coset represented by $\{C_2|0\}$ is conjugate to the coset represented by $\{C_2|\boldsymbol{t}_2\}$. After similar calculations for the other cosets, we find that the factor group has the ten classes shown in the left column of Table 11.3. It should then have ten irreducible representations, of which eight are 1-dimensional and two are 2-dimensional ($1^2 + 1^2 + 1^2 + 1^2 + 1^2 + 1^2 + 1^2 + 1^2 + 2^2 + 2^2 = 16$). As a matter of fact, this factor group is isomorphic to the point group D_{4h}. So, in Table 11.3, we have taken over the irreducible characters of D_{4h}. Now, irreducible representations of the group of $\boldsymbol{k}$ have to satisfy (11.36), i.e., their characters must satisfy

$$\chi^{\boldsymbol{k}}(\{\varepsilon|\boldsymbol{t}_2\}) = \exp(\mathrm{i}\boldsymbol{k}\cdot\boldsymbol{t}_2)d = -d \tag{11.44}$$

for the $\{\varepsilon|\boldsymbol{t}_2\}$ translation. Among the representations found in Table 11.3, only those ones that satisfy (11.44) are permissible as the small representations. As a result, we have only E_g and E_u representations, which will be designated by X_1 and X_2 as small representations of $\mathscr{G}(\boldsymbol{k})$. Table 11.4 shows the matrices and characters for these representations. Their basis functions may be chosen, for

Table 11.3. Classification of the cosets of the factor group $\mathscr{G}(\boldsymbol{k})/T(\boldsymbol{k})$ and its irreducible representations. The five "g" representations given below should be supplemented with five more "u" representations

Representative elements	Elements of D_{4h}	A_{1g}	A_{2g}	B_{1g}	B_{2g}	E_g
$\{\varepsilon|0\}$	E	1	1	1	1	2
$\{\varepsilon|\boldsymbol{t}_2\}$	C_4^2	1	1	1	1	−2
$\{C_{2y}|\boldsymbol{\tau}, \boldsymbol{\tau}+\boldsymbol{t}_2\}$	$2C_4$	1	1	−1	−1	0
$\{C_2|0, \boldsymbol{t}_2\}$	$2C_2'$	1	−1	1	−1	0
$\{C_{2x}|\boldsymbol{\tau}, \boldsymbol{\tau}+\boldsymbol{t}_2\}$	$2C_2''$	1	−1	−1	1	0
$\{\sigma_h|0\}$	I	1	1	1	1	2
$\{\sigma_h|\boldsymbol{t}_2\}$	σ_h	1	1	1	1	−2
$\{\sigma_{vx}|\boldsymbol{\tau}, \boldsymbol{\tau}+\boldsymbol{t}_2\}$	$2IC_4$	1	1	−1	−1	0
$\{I|0, \boldsymbol{t}_2\}$	$2\sigma_v$	1	−1	1	−1	0
$\{\sigma_{vy}|\boldsymbol{\tau}, \boldsymbol{\tau}+\boldsymbol{t}_2\}$	$2\sigma_d$	1	−1	−1	1	0

Table 11.4. Irreducible representation matrices and characters of the group of $\boldsymbol{k}$ at the X point of the rutile structure

	$\{\varepsilon\|0\}$	$\{C_{2y}\|\boldsymbol{\tau}\}$	$\{C_2\|0\}$	$\{C_{2x}\|\boldsymbol{\tau}\}$	$\{\sigma_h\|0\}$	$\{\sigma_{vx}\|\boldsymbol{\tau}\}$	$\{I\|0\}$	$\{\sigma_{vy}\|\boldsymbol{\tau}\}$
X_1	$\begin{bmatrix}1 & \\ & 1\end{bmatrix}$	$\begin{bmatrix} & -1\\ 1 & \end{bmatrix}$	$\begin{bmatrix}1 & \\ & -1\end{bmatrix}$	$\begin{bmatrix} & 1\\ 1 & \end{bmatrix}$	$\begin{bmatrix}1 & \\ & 1\end{bmatrix}$	$\begin{bmatrix} & -1\\ 1 & \end{bmatrix}$	$\begin{bmatrix}1 & \\ & -1\end{bmatrix}$	$\begin{bmatrix} & 1\\ 1 & \end{bmatrix}$
	2	0	0	0	2	0	0	0
X_2	$\begin{bmatrix}1 & \\ & 1\end{bmatrix}$	$\begin{bmatrix} & -1\\ 1 & \end{bmatrix}$	$\begin{bmatrix}1 & \\ & -1\end{bmatrix}$	$\begin{bmatrix} & 1\\ 1 & \end{bmatrix}$	$\begin{bmatrix}-1 & \\ & -1\end{bmatrix}$	$\begin{bmatrix} & 1\\ -1 & \end{bmatrix}$	$\begin{bmatrix}-1 & \\ & 1\end{bmatrix}$	$\begin{bmatrix} & -1\\ -1 & \end{bmatrix}$
	2	0	0	0	-2	0	0	0

example, as

$$X_1: \ \{\cos(\pi y/a), \sin(\pi y/a)\} \ ,$$

$$X_2: \ \{\cos(\pi y/a)\sin(2\pi z/c), \sin(\pi y/a)\sin(2\pi z/c)\} \ . \tag{11.45}$$

For the familiar crystal structures shown in Table 11.2, the notation for the irreducible representations is established; the table also shows the references for the original definitions of the irreducible representations. Some of these original papers are reprinted in a useful book by *Knox* and *Gold* [11.8]. The compilation of *Koster* [11.9], though often cited, contains some important discrepancies from the original papers as warned by *Warren* [11.10]. A complete tabulation of the irreducible representations of the 230 space groups is given by *Bradley* and *Cracknell* [11.11].

11.9 Irreducible Representations of a Space Group

We have already considered the group of $\boldsymbol{k}$ and its irreducible representations (small representations) in the preceding section. We now wish to investigate the irreducible representations of the entire space group. As will be seen in due course, knowledge of the small representations suffices to derive irreducible representations of the space group.

Basis functions $\psi_{\boldsymbol{k}\nu}$ for a d-dimensional small representation $D^{\boldsymbol{k}}$ of the group $\mathscr{G}(\boldsymbol{k})$ are transformed as

$$\{\beta|\boldsymbol{b}\}\psi_{\boldsymbol{k}\nu} = \sum_{\mu=1}^{d} \psi_{\boldsymbol{k}\mu} D^{\boldsymbol{k}}_{\mu\nu}(\{\beta|\boldsymbol{b}\}) \tag{11.46}$$

by symmetry elements $\{\beta|\boldsymbol{b}\}$ of $\mathscr{G}(\boldsymbol{k})$. Application of symmetry operations which do not belong to $\mathscr{G}(\boldsymbol{k})$ to the above basis functions will give new functions. In order to classify such functions, let us decompose the space group in terms of the group of $\boldsymbol{k}$:

$$G = \mathscr{G}(\boldsymbol{k}) + \{\alpha_2|\boldsymbol{b}_2\}\mathscr{G}(\boldsymbol{k}) + \cdots + \{\alpha_s|\boldsymbol{b}_s\}\mathscr{G}(\boldsymbol{k}) . \tag{11.47}$$

Here, $\{\alpha_j|\boldsymbol{b}_j\}$ represents the jth coset. Operating these representative elements on the above basis functions, we obtain

$$\psi_{\boldsymbol{k}_j,\nu} \equiv \{\alpha_j|\boldsymbol{b}_j\}\psi_{\boldsymbol{k}\nu} , \quad j = 1, 2, \ldots, s; \nu = 1, 2, \ldots, d \tag{11.48}$$

where $\{\alpha_1|\boldsymbol{b}_1\}$ is the unit element $\{\varepsilon|0\}$. The vectors

$$\boldsymbol{k}_j \equiv \alpha_j\boldsymbol{k} , \quad j = 1, 2, \ldots, s \tag{11.49}$$

constitute star$\{\boldsymbol{k}\}$, and s is the order of the star.

Examples of the coset decomposition (11.47) can again be found in the rutile structure.

Γ: At $\boldsymbol{k} = 0$, we have $G = \mathscr{G}(\boldsymbol{k})$. The small representations can be taken over as the irreducible representations of the entire space group G. The situation is the same at Z, A and M.

Δ: Coset representatives are $\{\varepsilon|0\}$, $\{C_4|\boldsymbol{\tau}\}$, $\{C_2|0\}$, and $\{C_4^{-1}|\boldsymbol{\tau}\}$, which generate the star shown in Fig. 11.8b.

X: Coset representatives are $\{\varepsilon|0\}$ and $\{\sigma_{d\bar{y}}|0\}$.

As we have shown in (11.30, 31), the functions (11.48) belong to the irreducible representation $\boldsymbol{k}_j$ of the translation group. Furthermore, they form a basis for a small representation of the group $\mathscr{G}(\boldsymbol{k}_j)$. To see this, suppose $\{\beta|\boldsymbol{b}\}$ is an element of $\mathscr{G}(\boldsymbol{k})$ and consider the product

$$\{\alpha_j|\boldsymbol{b}_j\}\{\beta|\boldsymbol{b}\}\{\alpha_j|\boldsymbol{b}_j\}^{-1} . \tag{11.50}$$

This product is an element of $\mathscr{G}(\boldsymbol{k}_j)$, the group of $\boldsymbol{k}_j$, because, from (11.32 and 49),

$$\alpha_j\beta\alpha_j^{-1}\boldsymbol{k}_j \doteq \boldsymbol{k}_j .$$

Every element in $\mathscr{G}(\boldsymbol{k}_j)$ can be expressed in the form (11.50). Operating (11.50) on (11.48), we find

$$\{\alpha_j|\boldsymbol{b}_j\}\{\beta|\boldsymbol{b}\}\{\alpha_j|\boldsymbol{b}_j\}^{-1}\psi_{\boldsymbol{k}_j,\nu} = \sum_{\mu=1}^{d} \psi_{\boldsymbol{k}_j,\mu}D^{\boldsymbol{k}}_{\mu\nu}(\{\beta|\boldsymbol{b}\}) , \tag{11.51}$$

which shows that the $\psi_{\boldsymbol{k}_j,\nu}$'s form a basis for a representation of the group $\mathscr{G}(\boldsymbol{k}_j)$. Therefore, the representation matrices of $\mathscr{G}(\boldsymbol{k}_j)$ are given by

$$\hat{D}^{\boldsymbol{k}_j}(\{\beta|\boldsymbol{b}\}) = \hat{D}^{\boldsymbol{k}}(\{\alpha_j|\boldsymbol{b}_j\}^{-1}\{\beta|\boldsymbol{b}\}\{\alpha_j|\boldsymbol{b}_j\}) , \quad \text{for } \{\beta|\boldsymbol{b}\} \in \mathscr{G}(\boldsymbol{k}_j) . \tag{11.52}$$

The representation matrices of the entire space group are constructed from the basis functions (11.48). The number of the basis functions

$$n = sd \tag{11.53}$$

stands for the dimension of the representation. This n-dimensional representation is irreducible, for, because of the irreducibility of D^{k}, it is impossible to reduce the representation by taking appropriate linear combinations of basis functions having the same $\boldsymbol{k}$ vector. Furthermore, linear combinations of basis functions with different $\boldsymbol{k}$ cannot reduce the representation, since the basis functions are given different phase factors $\exp(\mathrm{i}\boldsymbol{k}\cdot\boldsymbol{t}_n)$ under translations. Therefore, the representation constructed by the n basis functions is irreducible.

The representation matrix takes the following form for translations:

$$\hat{D}(\{\varepsilon|\boldsymbol{t}_n\}) = \begin{bmatrix} \hat{D}^{k_1}(\{\varepsilon|\boldsymbol{t}_n\}) & 0 & \cdots & 0 \\ 0 & \hat{D}^{k_2}(\{\varepsilon|\boldsymbol{t}_n\}) & \cdots & 0 \\ \vdots & \vdots & & \vdots \\ 0 & 0 & \cdots & \hat{D}^{k_s}(\{\varepsilon|\boldsymbol{t}_n\}) \end{bmatrix} .$$

It consists of s diagonal block matrices of dimension d,

$$\hat{D}^{k_j}(\{\varepsilon|\boldsymbol{t}_n\}) = \exp(\mathrm{i}\boldsymbol{k}_j\cdot\boldsymbol{t}_n)\hat{1}_d .$$

For a general element $\{\alpha|\boldsymbol{b}\}$, the representation matrix is constructed from $d \times d$ block matrices, as exemplified below:

$$\hat{D}(\{\alpha|\boldsymbol{b}\}) = \quad [\text{block diagram}] \quad . \tag{11.54}$$

Only the shaded portions are nonvanishing. When $\{\alpha|\boldsymbol{b}\}$ belongs to $\mathcal{G}(\boldsymbol{k}_j)$, the jth diagonal block is nonvanishing, and that block matrix is given by (11.52). When $\{\alpha|\boldsymbol{b}\}$ does not belong to $\mathcal{G}(\boldsymbol{k}_j)$ and satisfy

$$\boldsymbol{k}_l \doteq \alpha\boldsymbol{k}_j ,$$

the l–j block matrix is given by

$$\hat{D}^{k}(\{\alpha_l|\boldsymbol{b}_l\}^{-1}\{\alpha|\boldsymbol{b}\}\{\alpha_j|\boldsymbol{b}_j\}) . \tag{11.55}$$

Since the rotation α interchanges s vectors in the star, the large matrix (11.54) has only one nonvanishing block matrix in each row and each column.

To conclude, an irreducible representation of the space group is characterized by star$\{\boldsymbol{k}\}$ and a small representation of $\mathscr{G}(\boldsymbol{k})$.

Exercise 11.1. Using (11.52, 55) and Table 11.4, derive the following representation matrices (for the generating elements) for the irreducible representation X_1 of the space group D_{4h}^{14}.

$$\hat{D}(\{\sigma_{d\bar{y}}|0\}) = \begin{bmatrix} & & 1 & \\ & & & 1 \\ 1 & & & \\ & 1 & & \end{bmatrix}$$

$$\hat{D}(\{C_{2y}|\boldsymbol{\tau}\}) = \begin{bmatrix} & -1 & & \\ 1 & & & \\ & & & 1 \\ & & 1 & \end{bmatrix}$$

$$\hat{D}(\{C_4|\boldsymbol{\tau}\}) = \begin{bmatrix} & & & -1 \\ & & 1 & \\ & 1 & & \\ 1 & & & \end{bmatrix}$$

11.10 Double Space Groups

When we introduce electron spin functions, spatial rotation through 4π becomes the unit element $\{\varepsilon|0\}$; the 2π rotation $\{\bar{\varepsilon}|0\}$ has to be distinguished from $\{\varepsilon|0\}$ because it changes the sign of the spin functions. We then have to include the barred operations

$$\{\bar{\alpha}|\boldsymbol{b}\} = \{\bar{\varepsilon}|0\}\{\alpha|\boldsymbol{b}\}$$

as group elements, as we did in point groups (Sect. 8.5). The resulting enlarged group is called the *double space group.*

The structure of double space groups is similar to that of the usual space groups (single space groups). They contain the translation group T as an invariant subgroup, and the coset decomposition (11.11) can be performed. The number of cosets is doubled by virtue of the inclusion of the barred operations.

Let us next proceed to the irreducible representations of double space groups. The definition of the group of $\boldsymbol{k}$ parallels that of the single space groups: $\mathscr{G}(\boldsymbol{k})$ is the set of operations $\{\beta|\boldsymbol{b}\}$ that satisfy $\beta\boldsymbol{k} \doteq \boldsymbol{k}$. Since $\mathscr{G}(\boldsymbol{k})$ contains barred elements $\{\bar{\beta}|\boldsymbol{b}\}$ as well, the order of $\mathscr{G}(\boldsymbol{k})$ is doubled. In the rutile structure, for instance, the group of X contains barred elements

$$\begin{aligned} &\{\bar{\varepsilon}|0\}\,, \quad \{\bar{\sigma}_h|0\}\,, \quad \{\bar{C}_{2y}|\boldsymbol{\tau}\}\,, \quad \{\bar{\sigma}_{vx}|\boldsymbol{\tau}\}\,, \\ &\{\bar{C}_2|0\}\,, \quad \{\bar{I}|0\}\,, \quad \{\bar{C}_{2x}|\boldsymbol{\tau}\}\,, \quad \{\bar{\sigma}_{vy}|\boldsymbol{\tau}\}\,, \end{aligned} \qquad (11.56)$$

in addition to the unbarred ones given in (11.33, 34), as well as their combinations with t_n translations.

Just as in the single space groups, small representations are readily constructed by (11.37) in the cases (a) and (b) of Sect. 11.8. The point group $\mathscr{G}_0(k)$ is here a double point group and Γ in (11.37) is a double-valued representation. The proof given below (11.37) holds for double space groups as well.

Special treatment is needed, however, in the case (c). We use below Herring's method to obtain double-valued small representations at the X point of the rutile structure. The group $T(k)$ is defined in the same way as in the single group. The factor group $\mathscr{G}(k)/T(k)$ has now 32 cosets, which are classified into the 14 classes shown in the first column of Table 11.5. [Note that the double factor group $\mathscr{G}(k)/T(k)$ has no isomorphism to the double point group D'_{4h} because of the different class structure.] For example,

$$\{C_{2x}|\tau\}^{-1}\{C_2|0\}\{C_{2x}|\tau\} = \{\bar{C}_2|t_2 - t_1\}$$

means that the coset represented by $\{C_2|0\}$ is conjugate to the coset represented by $\{\bar{C}_2|t_2\}$. The factor group has 14 irreducible representations – eight 1-dimensional and six 2-dimensional ones ($8 \times 1^2 + 6 \times 2^2 = 32$). Of these 14 representations, 10 representations are single-valued and have already appeared in Table 11.3. The other, double-valued ones are shown in Table 11.5. The first and second orthogonalities of irreducible characters hold among the 14 representations. Now the character must satisfy (11.44), which X_5 and X_6 violate. Finally, the irreducible representations of the double group of k at X are summarized in Table 11.6. Here, X_1 and X_2 are single valued, while X_3 and X_4 are double valued. The basis functions for the double-valued representations

Table 11.5. Classification of the cosets of the double factor group $\mathscr{G}(k)/T(k)$ at the X point of the rutile structure and the irreducible characters

Representative elements	X_3	X_4	(X_5)	(X_6)
$\{\varepsilon\|0\}$	2	2	2	2
$\{\bar{\varepsilon}\|0\}$	−2	−2	−2	−2
$\{\varepsilon\|t_2\}$	−2	−2	2	2
$\{\bar{\varepsilon}\|t_2\}$	2	2	−2	−2
$\{C_{2y}, \bar{C}_{2y}\|\tau, \tau + t_2\}$	0	0	0	0
$\{C_2\|0\}, \{\bar{C}_2\|t_2\}$	2i	−2i	0	0
$\{\bar{C}_2\|0\}, \{C_2\|t_2\}$	−2i	2i	0	0
$\{C_{2x}, \bar{C}_{2x}\|\tau, \tau + t_2\}$	0	0	0	0
$\{\sigma_h, \bar{\sigma}_h\|0\}$	0	0	0	0
$\{\sigma_h, \bar{\sigma}_h\|t_2\}$	0	0	0	0
$\{\sigma_{vx}, \bar{\sigma}_{vx}\|\tau, \tau + t_2\}$	0	0	0	0
$\{I\|0, t_2\}$	0	0	2	−2
$\{\bar{I}\|0, t_2\}$	0	0	−2	2
$\{\sigma_{vy}, \bar{\sigma}_{vy}\|\tau, \tau + t_2\}$	0	0	0	0

Table 11.6. Single-valued and double-valued irreducible representations of the group of $\boldsymbol{k}$ at the X point of the rutile structure. The double-valued representations X_3 and X_4 are degenerate by time-reversal symmetry

	X_1	X_2	X_3	X_4
$\{\varepsilon \mid 0\}$	2	2	2	2
$\{\bar{\varepsilon} \mid 0\}$	2	2	-2	-2
$\{C_{2y} \mid \tau\}$	0	0	0	0
$\{C_2 \mid 0\}$	0	0	2i	-2i
$\{C_{2x} \mid \tau\}$	0	0	0	0
$\{\sigma_h \mid 0\}$	2	-2	0	0
$\{\sigma_{vx} \mid \tau\}$	0	0	0	0
$\{I \mid 0\}$	0	0	0	0
$\{\sigma_{vy} \mid \tau\}$	0	0	0	0

may be chosen as

$$X_3: \{\alpha \sin(\pi y/a)\ ,\quad \beta \cos(\pi y/a)\}\ ,$$

$$X_4: \{\alpha \cos(\pi y/a)\ ,\quad \beta \sin(\pi y/a)\}\ ,$$

where α and β are spin functions quantized along the z-axis.

Once the small representations are obtained, irreducible representations of the entire double space group can be constructed in exactly the same way as in the single space group (Sect. 11.9).

Exercise 11.2. Find the representation matrices for X_3 and X_4 using the basis functions given above. For the transformation of spin functions, see Table 8.2 of Sect. 8.6

12. Electronic States in Crystals

The energy band theory assumes that electrons in a crystal move independently of the other electrons in a potential $V(\boldsymbol{r})$ that has the same symmetry as the crystal lattice. In this chapter, we investigate how the crystal symmetry influences the energy spectra and wavefunctions (Bloch functions) of electrons. Selection rules associated with various electronic processes are also discussed. Finally, Frenkel excitons in molecular crystals are considered, which present another interesting application of the space group representations.

12.1 Bloch Functions and $E(\boldsymbol{k})$ Spectra

Electrons moving in a crystal potential $V(\boldsymbol{r})$ obey the one-electron Hamiltonian

$$H = \boldsymbol{p}^2/2m + V(\boldsymbol{r}) \ . \tag{12.1}$$

Here, the potential $V(\boldsymbol{r})$ is assumed to have the same symmetry as the crystal; in other words, $V(\boldsymbol{r})$ commutes with the symmetry operations R of the space group G of the crystal. Since the R's commute with $\boldsymbol{p}^2 = -\hbar^2 \nabla^2$ as well, they commute with the Hamiltonian H:

$$RHR^{-1} = H \ , \quad R \in G \ . \tag{12.2}$$

The commutativity may be verified explicitly in the following way: If the operation $R = \{\alpha|\boldsymbol{b}\}$ consists of a rotation through θ about the axis $\boldsymbol{n}$ followed by a translation through $\boldsymbol{b}$, we can write it as

$$R = \exp(-\mathrm{i}\boldsymbol{p}\cdot\boldsymbol{b}/\hbar)\exp(-\mathrm{i}\theta\boldsymbol{l}\cdot\boldsymbol{n}) \tag{12.3}$$

in terms of the momentum $\boldsymbol{p}$ and angular momentum $\boldsymbol{l}$ operators. Using the above expression for R, we have

$$\begin{aligned} RV(\boldsymbol{r})R^{-1} &= \exp(-\mathrm{i}\boldsymbol{p}\cdot\boldsymbol{b}/\hbar)\, V(\alpha^{-1}\boldsymbol{r})\exp(\mathrm{i}\boldsymbol{p}\cdot\boldsymbol{b}/\hbar) \\ &= V(\alpha^{-1}(\boldsymbol{r}-\boldsymbol{b})) = V(\{\alpha|\boldsymbol{b}\}^{-1}\boldsymbol{r}) = V(\boldsymbol{r}). \end{aligned}$$

In addition, we have $R\boldsymbol{p}^2R^{-1} = \boldsymbol{p}^2$, because $\boldsymbol{l}$ commutes with $\boldsymbol{p}^2$.

By virtue of the invariance (12.2), the general theory (Sect. 6.2) tells us that eigenfunctions of the Hamiltonian H form bases for irreducible representations of the space group G. Several important results follow from this. As we have seen in the preceding chapter, an irreducible representation of the space group G is characterized by a wavevector $\boldsymbol{k}$ and a small representation (an irreducible representation of the group of $\boldsymbol{k}$). Let $\psi_{\boldsymbol{k}\nu}$ denote eigenfunctions and $E(\boldsymbol{k})$ the corresponding eigenvalues. Then we have

$$H\psi_{\boldsymbol{k}\nu} = E(\boldsymbol{k})\psi_{\boldsymbol{k}\nu} \; . \tag{12.4}$$

First of all, translational symmetry of the crystal requires that the functions $\psi_{\boldsymbol{k}\nu}$ be Bloch functions satisfying (11.26). Furthermore, inasmuch as $\boldsymbol{k}$ and $\boldsymbol{k} + \boldsymbol{K}_n$ are regarded as the same irreducible representation (Sect. 11.7), we have

$$E(\boldsymbol{k} + \boldsymbol{K}_n) = E(\boldsymbol{k}) \; , \tag{12.5}$$

i.e., the energy spectra $E(\boldsymbol{k})$ are periodic in the reciprocal lattice vectors if we cease confining $\boldsymbol{k}$ within the Brillouin zone.

Rotational symmetry of the crystal brings further important properties. The operations $\{\beta|\boldsymbol{b}\}$ that leave $\boldsymbol{k}$ unchanged form the group of the wavevector $\boldsymbol{k}$, and the d-fold degenerate functions $\psi_{\boldsymbol{k}\nu}$ $(\nu = 1, 2, \ldots, d)$ form a basis for a d-dimensional irreducible representation of the group of $\boldsymbol{k}$, transforming like (11.46) among themselves.

To consider the effects of symmetry operations which do not belong to the group of $\boldsymbol{k}$, we decompose the space group G into cosets as shown in (11.47). Then the set of functions $\psi_{\boldsymbol{k}_j,\nu}$ $(j = 1, 2, \ldots, s;\; \nu = 1, 2, \ldots, d)$ defined by (11.48) form a basis for an irreducible representation of the space group G with dimension $n = sd$. These functions are transformed into themselves by the symmetry operations of the space group. For instance, operation of $\{\alpha|\boldsymbol{b}\}$ on $\psi_{\boldsymbol{k}\nu}$ will give a linear combination of $\psi_{\alpha\boldsymbol{k},\mu}$ over μ. The set of s vectors $\{\boldsymbol{k}, \boldsymbol{k}_2, \ldots, \boldsymbol{k}_s\}$ forms the star of $\boldsymbol{k}$. Since the sd functions all belong to the same energy eigenvalue, we have

$$E(\boldsymbol{k}) = E(\boldsymbol{k}_2) = \cdots = E(\boldsymbol{k}_s) \; , \tag{12.6}$$

which means that the $E(\boldsymbol{k})$ spectra have the full point-group symmetry.

12.2 Examples of Energy Bands: Ge and TlBr

In this and the succeeding sections, we choose Ge and TlBr as the exemplary materials. Figure 12.1 shows energy bands of these two materials along the $\langle 111 \rangle$ and $\langle 100 \rangle$ directions in the Brillouin zone. The global features of these energy bands may be understood as follows: A Ge atom has the configuration $(4s)^2(4p)^2$ in the outermost shell. According to the chemical-bond picture, these atomic orbitals form sp^3 hybridized orbitals, whose bonding orbitals, occupied by the valence electrons, correspond to the valence bands, while the antibonding orbitals correspond to the conduction bands. On the other hand, TlBr is an ionic crystal consisting of Tl^+ ions with the $(6s)^2$ configuration and Br^- ions

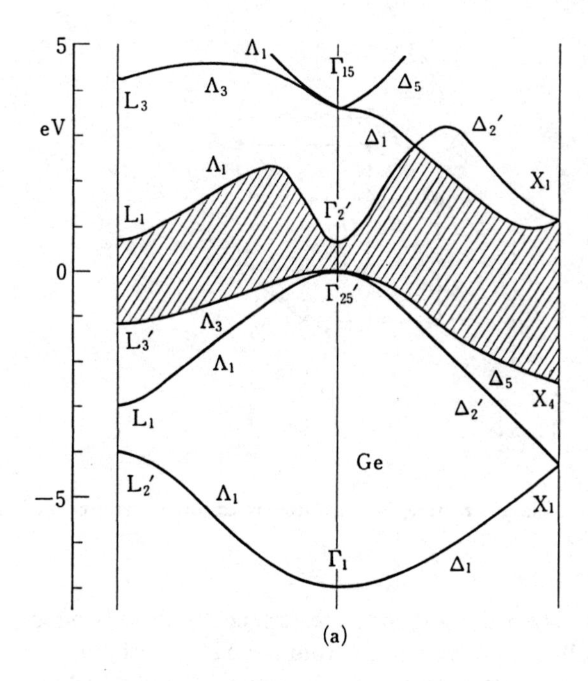

(a)

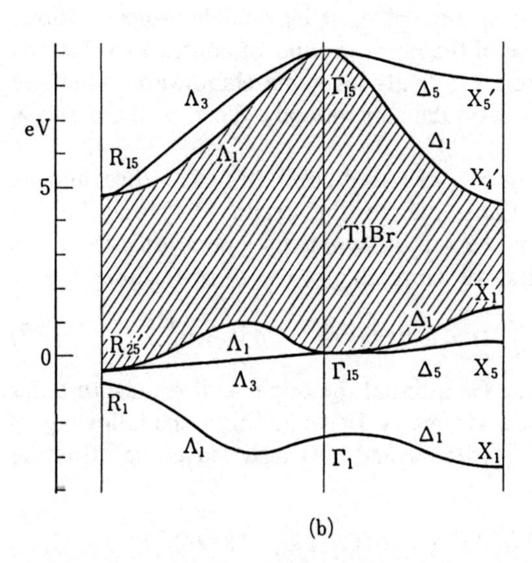

(b)

Fig. 12.1. Energy bands in (**a**) Ge and (**b**) TlBr. The spin-orbit interaction is neglected. The energy bands above the shaded forbidden bands are conduction bands and those below are valence bands. The small representations for TlBr are for rotations about the Tl nucleus

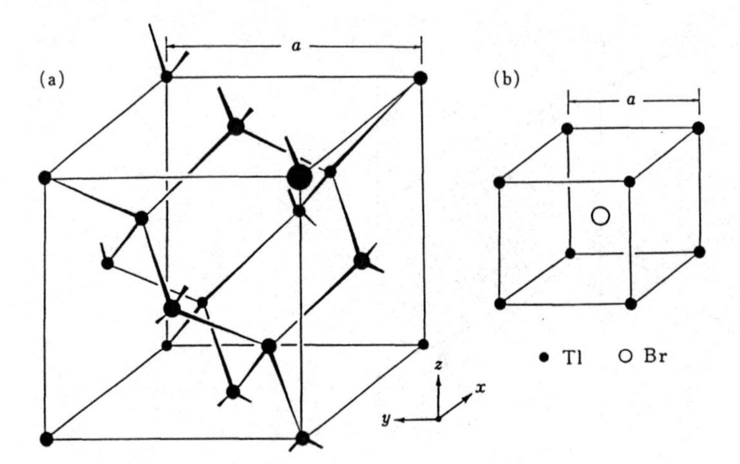

Fig. 12.2. (**a**) Crystal structure of Ge (diamond structure). Bravais lattice is fcc. (**b**) Crystal structure of TlBr (CsCl structure). Bravais lattice is sc

with the $(4p)^6$ configuration. The valence bands are formed by these atomic orbitals, and the conduction bands derive mainly from the Tl^+ $6p$ orbitals.

As to the crystal structure of these materials, Ge has the diamond structure (space group O_h^7) and TlBr has the CsCl structure (space group O_h^1), as shown in Fig. 12.2. Since the latter group is symmorphic, its irreducible representations are readily constructed from those of the point groups. In contrast to this, the former, the nonsymmorphic group O_h^7, calls for some elaboration when we consider small representations for $\boldsymbol{k}$ on the zone boundary, e.g., at the L and X points.

The diamond structure has the face-centered cubic Bravais lattice, and its unit cell consists of two Ge atoms, one at the origin (000) and one at $\boldsymbol{\tau} = (a/4, a/4, a/4)$. The following two sets of symmetry operations leave this crystal structure invariant: the first set, consisting of the 24 operations

$$\{\varepsilon|0\}\ , \quad 6\{IC_4|0\}\ , \quad 3\{C_4^2|0\}\ , \quad 6\{\sigma_d|0\}\ , \quad 8\{C_3|0\}\ , \tag{12.7}$$

which form the T_d group, leave the Ge atom at the origin as it stands. In other words, every Ge atom has the local symmetry T_d. In addition, the following 24 composite operations with $\{I|\boldsymbol{\tau}\}$ (the second set) also leave the structure unchanged:

$$\{I|\boldsymbol{\tau}\}\ , \quad 6\{C_4|\boldsymbol{\tau}\}\ , \quad 3\{\sigma_h|\boldsymbol{\tau}\}\ , \quad 6\{C_2'|\boldsymbol{\tau}\}\ , \quad 8\{IC_3|\boldsymbol{\tau}\}\ . \tag{12.8}$$

The space group O_h^7 consists of these 48 operations and their compositions with translations, which amount to $48N^3$ in total.

Of these group elements, $\{C_4|\boldsymbol{\tau}\}$ is an essential screw 4_1, and $\{\sigma_h|\boldsymbol{\tau}\}$ is an essential glide d: the nonprimitive translation $\boldsymbol{\tau}$ in the other elements can be suppressed by shifting the origin of the symmetry operations. For example, if we displace the origin to $\tau/2 = (a/8, a/8, a/8)$, then $\{I|\tau\}$ becomes the simple inversion $\{I|0\}$:

$$\{I|\tau\} = \{\varepsilon|\tau/2\}\{I|0\}\{\varepsilon|-\tau/2\} ,$$

which means that the Ge crystal has a center of inversion symmetry at the midpoints of two neighboring Ge atoms.

Of the small representations appearing in Fig. 12.1a, those for Γ, Δ, and Λ, which are located inside the Brillouin zone, are obtained from the irreducible representations of the point groups O_h, C_{4v}, and C_{3v}, respectively.

At the L point, where $\boldsymbol{k} = (2\pi/a)(1/2, 1/2, 1/2)$, the group of $\boldsymbol{k}$ has the representative elements

$$\{\varepsilon|0\} ,\quad 2\{C_3|0\} ,\quad 3\{\sigma_d|0\} ,$$
$$\{I|\tau\} ,\quad 2\{IC_3|\tau\} ,\quad 3\{C_2'|\tau\} .$$

Although this group looks nonsymmorphic at the first glance, it is essentially symmorphic, because displacement of the origin to $\tau/2$ suppresses the τ translation. For example, we have

$$\{C_2'|\tau\} = \{\varepsilon|\tau/2\}\{C_2'|0\}\{\varepsilon|-\tau/2\}$$

for the 180° rotations about the $[1\bar{1}0]$, $[\bar{1}01]$ and $[01\bar{1}]$ axes, which means that

Table 12.1. Characters and basis functions of the small representations at the X point of the diamond structure. X_5 is double valued. C'_{2xy} and $C'_{2\bar{x}y}$ stand for the π rotations about the [110] and $[1\bar{1}0]$ axes

	X_1	X_2	X_3	X_4	X_5
$\{\varepsilon\|0\}$	2	2	2	2	4
$\{\bar{\varepsilon}\|0\}$	2	2	2	2	−4
$\{C_{4x}^2, C_{4y}^2\|0\}$	0	0	0	0	0
$\{C_4^2\|0\}$	2	2	−2	−2	0
$2\{C_4\|\tau\}$	0	0	0	0	0
$\{C'_{2xy}\|\tau\}$	0	0	2	−2	0
$\{C'_{2\bar{x}y}\|\tau\}$	0	0	−2	2	0
$\{I\|\tau\}$	0	0	0	0	0
$\{\sigma_x, \sigma_y\|\tau\}$	0	0	0	0	0
$\{\sigma_h\|\tau\}$	0	0	0	0	0
$2\{S_4\|0\}$	0	0	0	0	0
$2\{\sigma_d\|0\}$	2	−2	0	0	0

X_1: $\{\cos Z, \sin Z\}$, $\{\cos X \cos Y, \sin X \sin Y\}$,
X_2: $\{(\cos 2X - \cos 2Y)\cos Z, (\cos 2X - \cos 2Y)\sin Z\}$,
X_3: $\{\cos X \sin Y \cos 2Z, \sin X \cos Y \cos 2Z\}$,
X_4: $\{\cos X \sin Y, \sin X \cos Y\}$,
X_5: $\{\alpha \cos Z, \alpha \sin Z, \beta \cos Z, \beta \sin Z\}$,
where $X = 2\pi x/a$, $Y = 2\pi y/a$, and $Z = 2\pi z/a$.

$\{C_2'|\tau\}$ are the 180° rotations about the midpoints. Thus the small representations at L are derived from the irreducible representations of the point group D_{3d}. The displacement of the origin influences the basis functions, however (Sect. 12.4).

At the X point, where $\boldsymbol{k} = (2\pi/a)(001)$, the group of $\boldsymbol{k}$ is a nonsymmorphic group containing essential screws and glides. In this case, small representations are obtained by means of Herring's method, as explained in Sect. 11.8 for the rutile structure, or else by using the theory of ray representations (Sects. 5.3, 5.4). The result is shown in Table 12.1 together with the basis functions. In this table, C_{2xy}' and $C_{2\bar{x}y}'$ stand for 180° rotations about the [110] and [$1\bar{1}0$] axes. Notice that the characters for X_3 and X_4 representations differ only in sign for these two operations.

12.3 Compatibility or Connectivity Relations

When we slightly move the wavevector from some high-symmetry point $\boldsymbol{k}_0$ in the Brillouin zone to a lower-symmetry point $\boldsymbol{k}$, then $\mathscr{G}(\boldsymbol{k})$ becomes a subgroup of $\mathscr{G}(\boldsymbol{k}_0)$. The symmetry lowering is accompanied by a lifting of the degeneracy of the $E(\boldsymbol{k})$ spectra as well as the symmetry lowering of the corresponding Bloch functions.

Let us study this problem for a specific example. In Ge, the top of the valence band Γ_{25}' occurs at the Γ point. If we move the $\boldsymbol{k}$ vector in the $\langle 001 \rangle$ direction, the group of the wavevector changes from $\mathscr{G}_\Gamma$ to $\mathscr{G}_\Delta$, the latter being a subgroup of the former. Upon an infinitesimal change in the $\boldsymbol{k}$ vector, the Bloch functions belonging to the Γ_{25}' representation form a basis for a representation of $\mathscr{G}_\Delta$, which is not necessarily irreducible in general. Therefore we reduce Γ_{25}' into irreducible representations of $\mathscr{G}_\Delta$. Comparison of the character tables tells us

$$\Gamma_{25}' \rightarrow \Delta_2' + \Delta_5 \ , \tag{12.9}$$

indicating that the threefold degenerate energy level Γ_{25}' is split into the nondegenerate level Δ_2' and the twofold degenerate level Δ_5. Relations like (12.9) are called *compatibility* or *connectivity relations*.

Generally speaking, connectivity relations are obtained by restricting the small representations of $\mathscr{G}(\boldsymbol{k}_0)$ to a lower-symmetry group $\mathscr{G}(\boldsymbol{k})$ and reducing them into small representations of $\mathscr{G}(\boldsymbol{k})$. More examples may be found in the $E(\boldsymbol{k})$ spectra of Fig. 12.1.

12.4 Bloch Functions Expressed in Terms of Plane Waves

One-electron wavefunctions $\psi_{\boldsymbol{k}}(\boldsymbol{r})$ with wavevector $\boldsymbol{k}$ are expressible as

$$\psi_{\boldsymbol{k}}(\boldsymbol{r}) = \exp(-\mathrm{i}\boldsymbol{k}\cdot\boldsymbol{r})u_{\boldsymbol{k}}(\boldsymbol{r}) \ , \tag{12.10}$$

where $u_k(r)$ are periodic functions. Fourier expansion of such periodic functions gives only the Fourier components with reciprocal lattice vectors $\boldsymbol{K}_n$:

$$\psi_k(r) = \sum_n c_n \exp(-\mathrm{i}(\boldsymbol{k} + \boldsymbol{K}_n)\cdot \boldsymbol{r}) \ . \tag{12.11}$$

In this section, we wish to see how the symmetry-adapted functions $\psi_k(r)$ of small representations may be constructed as (12.11) from plane waves.

To see this, we consider the extreme limit of a vanishing crystal potential $V(\boldsymbol{r})$, which is called the empty lattice. The electrons, being free, have the energy spectra

$$E(\boldsymbol{k}) = \hbar^2 k^2/2m \ .$$

Since we are considering the *limit* of a vanishing potential, the translational symmetry (12.5) must still hold. Then the $E(\boldsymbol{k})$ spectra are obtained by bringing together the free-electron spectra $\hbar^2(\boldsymbol{k} + \boldsymbol{K}_n)^2/2m$ about all the reciprocal lattice points. Figure 12.3 shows the $E(\boldsymbol{k})$ curves for the empty face-centered cubic lattice.

Introduction of a weak crystal potential $V(\boldsymbol{r})$ will lift the degeneracies and give rise to the Bloch functions, which can be written as linear combinations of plane waves.

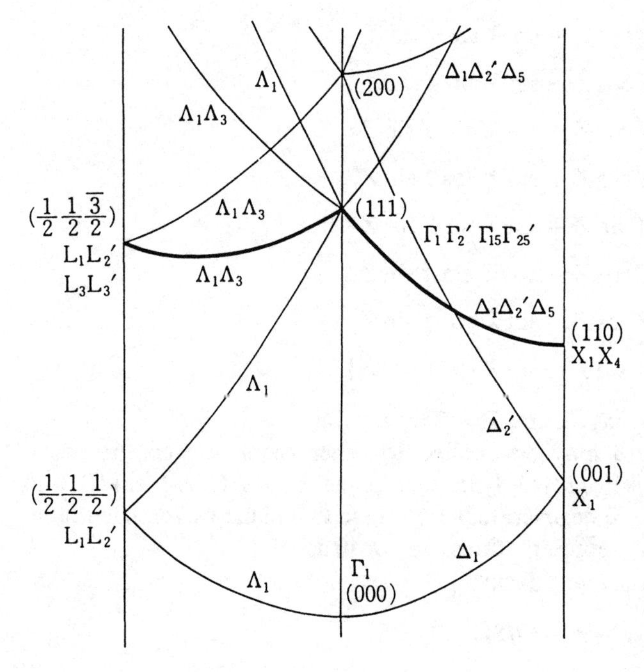

Fig. 12.3. $E(\boldsymbol{k})$ spectra for the empty fcc lattice. Introduction of a nonvanishing crystal potential lifts the degeneracy of the thick line. The resulting energy gap forms the forbidden band. The numbers in parentheses stand for wavevectors in units of $2\pi/a$

Let us begin with the Γ point. The plane wave with $\boldsymbol{k} = (000)$ belongs to the identity representation Γ_1. (The names and characters of the small representations are given in Appendix B for the O_h point group.) We next take the eight plane waves with wavevectors $(2\pi/a)(\pm 1, \pm 1, \pm 1)$, which will form a basis for a reducible representation. The plane waves are transformed by symmetry operations according to

$$\{\alpha|\boldsymbol{b}\}\exp(\mathrm{i}\boldsymbol{k}\cdot\boldsymbol{r}) = \exp(\mathrm{i}\alpha\boldsymbol{k}\cdot(\boldsymbol{r}-\boldsymbol{b})) \ , \tag{12.12}$$

as seen from (11.9). Using (12.12), we find the character of the eight-dimensional representation based on the eight plane waves shown in Table 12.2. Reduction of this representation gives $\Gamma_1 + \Gamma_2' + \Gamma_{15} + \Gamma_{25}'$, which means that the crystal potential lifts the eightfold degeneracy, giving rise to these four levels. Actually, Γ_{25}' is the top of the valence band, and Γ_2' and Γ_{15} are seen in the conduction bands in Fig. 12.1a.

Explicit expressions for the symmetry-adapted plane waves can be obtained by using the projection operators

$$\sum_R D^{\boldsymbol{k}}_{\nu\nu}(R)^* R \ , \tag{12.13}$$

where $\hat{D}^{\boldsymbol{k}}(R)$ stands for small representation matrices and the summation runs over the representative elements R of the group of $\boldsymbol{k}$. Operating (12.13) on an arbitrary function results in a symmetry-adapted function $\psi_{\boldsymbol{k}\nu}$. In the case of the eight plane waves mentioned above, we obtain the result given below. [Explicit expressions for the projection operators (12.13) at the Γ point may be found in the answer to Exercise 10.5.]

$$\begin{aligned}
&\Gamma_1 && \cos X\cos Y\cos Z + \sin X\sin Y\sin Z \ ,\\
&\Gamma_2' && \cos X\cos Y\cos Z - \sin X\sin Y\sin Z \ ,\\
&\Gamma_{25}' && \{\sin X\cos Y\cos Z + \cos X\sin Y\sin Z \ ,\\
& && \cos X\sin Y\cos Z + \sin X\cos Y\sin Z \ ,\\
& && \cos X\cos Y\sin Z + \sin X\sin Y\cos Z\}\\
&\Gamma_{15} && \{\sin X\cos Y\cos Z - \cos X\sin Y\sin Z \ ,\\
& && \cos X\sin Y\cos Z - \sin X\cos Y\sin Z \ ,\\
& && \cos X\cos Y\sin Z - \sin X\sin Y\cos Z\}
\end{aligned}$$

where $X = 2\pi x/a$, $Y = 2\pi y/a$, and $Z = 2\pi z/a$.

The same procedure may be applied to other cases as well. At the L point, where $\boldsymbol{k} = (2\pi/a)(1/2, 1/2, 1/2)$, two plane waves $(1/2, 1/2, 1/2)$ and $(-1/2, -1/2, -1/2)$ are degenerate. From these two plane waves, basis functions for the L_1 and L_2' representations are constructed:

$$\begin{aligned}
&L_1 && \cos[(\pi/a)(x+y+z-3a/8)] \ ,\\
&L_2' && \sin[(\pi/a)(x+y+z-3a/8)] \ .
\end{aligned}$$

Table 12.2. Character of the representation based on the eight plane waves at the Γ point

R	E	$6C_4$	$3C_4^2$	$6C_2'$	$8C_3$	I	$6S_4$	$3\sigma_h$	$6\sigma_d$	$8S_6$
$\chi(R)$	8	0	0	0	2	0	0	0	4	0

The shift of the origin mentioned in Sect. 12.2 gives rise to the term $3a/8$ in the arguments of these functions.

The symmetrized plane waves at the X point are shown in Table 12.1.

12.5 Choice of the Origin

In some crystals, it happens that two points in the unit cell are equally entitled to be chosen as the origin of symmetry operations. In principle, it does not matter which one is chosen as the origin. However, when we examine symmetry properties of given wavefunctions, it becomes necessary to specify which point has been chosen. The situation will be readily understood from the following example.

In TlBr (CsCl structure), the symmetry about a Tl atom is exactly the same as that about a Br atom. Therefore, both atoms have an equal right to be chosen as the origin. Let us choose the origin at the Tl nucleus and consider the Bloch function

$$\psi(\boldsymbol{r}) = \sin(\pi z/a) ,$$

which has $\boldsymbol{k} = (\pi/a)(001)$ and belongs to the small representation X_4'. Now look at the same function from the Br nucleus, which is separated from the Tl nucleus by $\boldsymbol{\eta} = (a/2, a/2, a/2)$. Then we find $\psi(\boldsymbol{r}) = \cos((\pi/a)(z - a/2))$, which belongs to the X_1 representation. This situation is illustrated in Fig. 12.4. As this example shows, one has to specify the origin of symmetry operations when speaking of labels of irreducible representations.

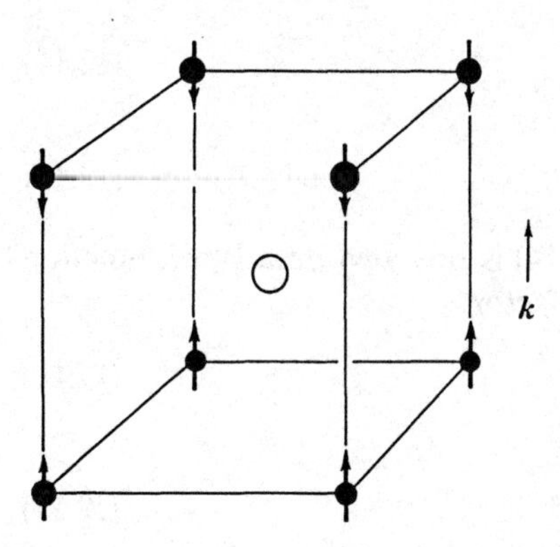

Fig. 12.4. Bloch functions having X_4' symmetry about the Tl nuclei behave like X_1 about the Br nuclei

Table 12.3a,b. Correspondence of the symmetry properties of Bloch functions about the Cs (Na) and Cl nuclei

(a) CsCl structure

X_1	X_2	X_3	X_4	X_5	X_6^+	X_7^+	T_1	T_2	T_5	T_6	M_1	M_2	M_5
X_4'	X_3'	X_2'	X_1'	X_5'	X_6^-	X_7^-	T_2'	T_1'	T_5	T_7	M_3	M_4	M_5

M_1'	M_2'	M_5'	M_6^+	M_6^-	S_1	S_2	S_5	R_1	R_2	R_{12}	R_{15}'	R_{25}'
M_3'	M_4'	M_5'	M_7^+	M_7^-	S_3	S_4	S_5	R_2'	R_1'	R_{12}'	R_{25}	R_{15}

R_6^+	R_7^+	R_8^+	Z_1	Z_2	Z_5
R_7^-	R_6^-	R_8^-	Z_3	Z_4	Z_5

(b) NaCl structure

L_1	L_2	L_3	L_6^+	L_4^+	L_5^+	Q_+	W_1	W_2	W_3	W_6
L_2'	L_1'	L_3'	L_6^-	L_5^-	L_4^-	Q_-	W_2'	W_1'	W_3	W_7

Of the crystal structures given in Table 11.2, the above problem arises in the NaCl, zinc-blende, wurtzite and perovskite structures as well as in the CsCl structure. These structures have two atoms that can equally well be chosen as the origin. Table 12.3 summarizes the correspondence of irreducible representation labels about the two nuclei for CsCl (including perovskite) and NaCl structures.

12.5.1 Effect of the Choice of Origin on Bloch Wavefunctions

Let us study the above problem in greater detail. When we understand $\{\alpha|\boldsymbol{t}_n\}$ as a symmetry operation about a Tl nucleus belonging to the group of wavevector $\boldsymbol{k}$, the same operation about a Br nucleus becomes

$$\begin{aligned}\{\alpha|\boldsymbol{t}_n\}_{\mathrm{Br}} &= \{\varepsilon|\boldsymbol{\eta}\}\{\alpha|\boldsymbol{t}_n\}\{\varepsilon|-\boldsymbol{\eta}\} \\ &= \{\varepsilon|-\boldsymbol{t}(\alpha)\}\{\alpha|\boldsymbol{t}_n\}\ ,\end{aligned} \tag{12.14}$$

where

$$\boldsymbol{t}(\alpha) = \alpha\boldsymbol{\eta} - \boldsymbol{\eta} \tag{12.15}$$

is a primitive translation vector. When (12.14) is operated on a Bloch function with wavevector $\boldsymbol{k}$, it gives rise to a phase factor

$$\exp(-\mathrm{i}\boldsymbol{k}\cdot\boldsymbol{t}(\alpha)) = \exp(-\mathrm{i}\boldsymbol{K}(\alpha)\cdot\boldsymbol{\eta})\ , \tag{12.16}$$

where

$$\boldsymbol{K}(\alpha) = \alpha^{-1}\boldsymbol{k} - \boldsymbol{k}\ . \tag{12.17}$$

Since $\{\alpha|\boldsymbol{t}_n\}$ belongs to the group of $\boldsymbol{k}$, we have $\boldsymbol{K}(\alpha) \doteq 0$. In particular, when $\boldsymbol{k}$ lies inside the Brillouin zone, $\boldsymbol{K}(\alpha)$ vanishes and the phase factor (12.16) is equal to unity. In this case, the two choices of the origin make no difference. For $\boldsymbol{k}$ on the zone boundary, however, $\boldsymbol{K}(\alpha)$ is, in general, a nonvanishing reciprocal lattice vector. Then the phase factor is not necessarily equal to unity, and the symmetry properties of Bloch functions become different about the two nuclei. The correspondence between the symmetries about the two nuclei may be found by multiplying the irreducible characters with the phase factors (12.16), which take on -1 for $\alpha = C_2', C_2'', I, IC_4$, and σ_h. Comparison of the characters tells us that X_4' corresponds to X_1.

Exercise 12.1. Draw a figure like Fig. 12.4 for the M_4' symmetry about the Tl nucleus. For what operations does the phase factor (12.16) take on the value -1 at the M point?

12.6 Bloch Functions Expressed in Terms of Atomic Orbitals

Bloch functions can also be constructed as

$$\psi_{\boldsymbol{k}}(\boldsymbol{r}) = \sum_n \exp(-\mathrm{i}\boldsymbol{k}\cdot\boldsymbol{t}_n)\phi(\boldsymbol{r}-\boldsymbol{t}_n)$$

from atomic orbitals $\phi(\boldsymbol{r}-\boldsymbol{t}_n)$ localized at the lattice site $\boldsymbol{t}_n$. Since the unit cell of the crystal can contain more than one atom, the general expression for the Bloch functions constructed from the atomic orbital ϕ_λ located at $\boldsymbol{\eta}$ in the unit cell becomes

$$\psi_{\boldsymbol{k}}(\boldsymbol{\eta}, \lambda; \boldsymbol{r}) = \sum_n \exp(-\mathrm{i}\boldsymbol{k}\cdot\boldsymbol{t}_n)\phi_\lambda(\boldsymbol{r}-\boldsymbol{t}_n-\boldsymbol{\eta}) \ . \tag{12.18}$$

Actual eigenfunctions of the Hamiltonian should be linear combinations of these functions over various $\boldsymbol{\eta}$ in the unit cell, and when the ϕ_λ orbitals have degeneracy like p and d orbitals, they should be further linear combinations over λ of (12.18).

To see what kinds of wavefunctions are constructed from the atomic orbitals, one has to decompose the reducible representation based on those functions (12.18). After having found what small representations result from the atomic orbitals, one applies the projection operators (12.13) and obtains symmetry-adapted wavefunctions. In doing such calculations, one can use

$$\begin{aligned}\{\beta|\boldsymbol{b}\}\psi_{\boldsymbol{k}}(\boldsymbol{\eta}, \lambda; \boldsymbol{r}) &= \sum_n \exp(-\mathrm{i}\boldsymbol{k}\cdot\boldsymbol{t}_n)\phi_\lambda(\beta^{-1}\boldsymbol{r}-\beta^{-1}\boldsymbol{b}-\boldsymbol{t}_n-\boldsymbol{\eta}) \\ &= \sum_n \exp(-\mathrm{i}\boldsymbol{k}\cdot\boldsymbol{t}_n)[\beta\phi_\lambda](\boldsymbol{r}-\boldsymbol{t}_n-\beta\boldsymbol{\eta}-\boldsymbol{b}) \ ,\end{aligned}$$

for operations $\{\beta|\boldsymbol{b}\}$ in the group of $\boldsymbol{k}$. In particular, if $\beta\boldsymbol{\eta}-\boldsymbol{\eta}$ is a primitive translation vector, the right-hand side becomes

$$\exp(\mathrm{i}\boldsymbol{K}(\beta)\cdot\boldsymbol{\eta})\sum_n \exp(-\mathrm{i}\boldsymbol{k}\cdot\boldsymbol{t}_n)[\beta\phi_\lambda](\boldsymbol{r}-\boldsymbol{t}_n-\boldsymbol{\eta}-\boldsymbol{b})\ , \qquad (12.19)$$

where the reciprocal lattice vector $\boldsymbol{K}(\beta)$ is defined by (12.17). The above-mentioned procedure parallels the case of a related problem – Frenkel excitons in molecular crystals, expounded in Sect. 12.11.

Let us now observe some examples in germanium. The unit cell of the germanium crystal has two atoms, at $\boldsymbol{\eta}=0$ and $\boldsymbol{\eta}=\boldsymbol{\tau}$. Therefore, from the 4$s$ atomic orbitals of Ge, we can construct $\psi_{\boldsymbol{k}}(0, 4s)$ and $\psi_{\boldsymbol{k}}(\boldsymbol{\tau}, 4s)$. Reduction of the representation based on these two functions will tell us what small representations result from the 4s atomic orbitals. The operations containing the nonprimitive translation $\boldsymbol{\tau}$ interchange the functions with different $\boldsymbol{\eta}$, and hence the corresponding values of the character vanish. Therefore, it suffices to find the character only for the elements given in (12.7). At $\boldsymbol{k}=0$, the values of the character are commonly 2. Irreducible decomposition of this character gives $\Gamma_1+\Gamma_2'$. The basis function for the Γ_1 representation is the bonding orbital $\psi_0(0, 4s)+\psi_0(\boldsymbol{\tau}, 4s)$, and that for Γ_2' is the antibonding orbital $\psi_0(0, 4s)-\psi_0(\boldsymbol{\tau}, 4s)$, both of which can be found in the energy bands of Fig. 12.1a.

From the Ge 4p orbitals, we have six functions in total, because λ takes on p_x, p_y, and p_z. At $\boldsymbol{k}=0$, we obtain Γ_{15} and Γ_{25}' representations with the basis functions

$$\psi_0(0, 4p_z)+\psi_0(\boldsymbol{\tau}, 4p_z) \quad \text{and} \quad \psi_0(0, 4p_z)-\psi_0(\boldsymbol{\tau}, 4p_z)$$

together with the ones in which p_z is replaced by p_x and p_y. Figure 12.5 shows these antibonding (Γ_{15}) and bonding (Γ_{25}') wavefunctions.

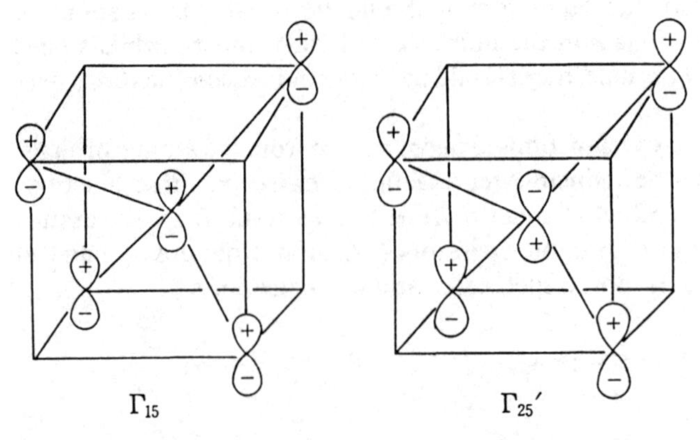

Fig. 12.5. Wavefunctions for the Γ_{15} and Γ_{25}' states in germanium. The signs in the figure show the signs of the p orbitals. Note that neighboring p orbitals have larger overlap in the bonding state Γ_{25}'

Table 12.4a, b, c. Symmetries of Bloch functions constructed from atomic orbitals. $d\gamma$ and $d\varepsilon$ stand for the twofold and threefold degenerate levels of a d electron placed in a cubic field. The rows for the p orbitals may also be interpreted as the normal lattice vibration modes

(a) Diamond structure

s	$\Gamma_1\Gamma_2'$	X_1	L_1L_2'
p	$\Gamma_{15}\Gamma_{25}'$	$X_1X_3X_4$	$L_1L_2'L_3L_3'$

(b) CsCl structure (origin at Cs)

Cs	s	Γ_1	R_1	X_1	M_1
	p	Γ_{15}	R_{15}	$X_4'X_5'$	$M_4'M_5'$
	$d\gamma$	Γ_{12}	R_{12}	X_1X_2	M_1M_2
	$d\varepsilon$	Γ_{25}'	R_{25}'	X_3X_5	M_3M_5
Cl	s	Γ_1	R_2'	X_4'	M_3
	p	Γ_{15}	R_{25}'	X_1X_5	$M_2'M_5'$
	$d\gamma$	Γ_{12}	R_{12}'	$X_4'X_3'$	M_3M_4
	$d\varepsilon$	Γ_{25}'	R_{15}	$X_2'X_5'$	M_1M_5

(c) NaCl structure (origin at Na)

Na	s	Γ_1	X_1	W_1	L_1
	p	Γ_{15}	$X_4'X_5'$	$W_2'W_3$	$L_2'L_3'$
	$d\gamma$	Γ_{12}	X_1X_2	W_1W_2'	L_3
	$d\varepsilon$	Γ_{25}'	X_3X_5	$W_1'W_3$	L_1L_3
Cl	s	Γ_1	X_1	W_2'	L_2'
	p	Γ_{15}	$X_4'X_5'$	W_1W_3	L_1L_3
	$d\gamma$	Γ_{12}	X_1X_2	W_1W_2'	L_3'
	$d\varepsilon$	Γ_{25}'	X_3X_5	W_2W_3	$L_2'L_3'$

In Table 12.4, we summarize the symmetries of Bloch functions constructed from atomic orbitals at high-symmetry points in the Brillouin zone for the diamond, CsCl and NaCl structures.

Exercise 12.2. Compare Table 12.4b with the energy bands of TlBr shown in Fig. 12.1b, and confirm that the valence bands are really composed of the Br 4p orbitals as well as of the Tl 6s orbitals, and that the conduction bands are composed of the Tl 6p orbitals.

12.7 Lattice Vibrations

Normal vibrations in crystals may be approached in essentially the same way as in molecules (Chap. 10). However, the large number of atoms in a crystal inevitably make the group-theoretic treatment of lattice vibrations rather unwieldy. We refer readers to [12.1–3] for a complete treatment and here choose to take a short cut.

As a matter of fact, we can use the results of the preceding section to see what irreducible representations are allowed as normal modes and how the atoms are displaced in those representations, for each atom in the crystal has three components of displacement vectors which can correspond to the p_x, p_y, and p_z atomic orbitals. In consequence, the small representations formed by the p orbitals give the symmetries of the normal modes. For the crystal structures shown in Table 12.4, the small representations appearing in the rows for the p orbitals determine the normal modes of the lattice vibrations. Displacement patterns of the modes can be read from the symmetry-adapted Bloch functions.

For example, germanium has the Γ_{15} acoustic mode in which the two Ge atoms at (000) and $\boldsymbol{\tau}$ move in phase, and the Γ'_{25} optic mode in which they move out of phase.

At the X point, we have X_1, X_3, and X_4 modes. Atomic displacements are more complicated in these modes, but we can set them up from the symmetry-adapted basis functions shown in Table 12.1, relying on the above-mentioned equivalence of the atomic displacements to p functions. Take, for instance, the basis function $\sin X \cos Y \cos 2Z$ for the X_3 representation, which behaves like x about the Ge atom at (000), and like y about the Ge atom at $\boldsymbol{\tau}$. This may be translated as follows: "The atom at the origin moves in the $+x$ direction, and the atom at $\boldsymbol{\tau}$ in the $+y$ direction by the same amount," as depicted in Fig. 12.6. This mode, having a rather low frequency, is a bond-bending mode. Displacement patterns for the X_1 and X_4 modes are obtained by the same means. If the symmetry-adapted function has no p components about an atom, then "such an atom does not move," as happens in the X_1 representation.

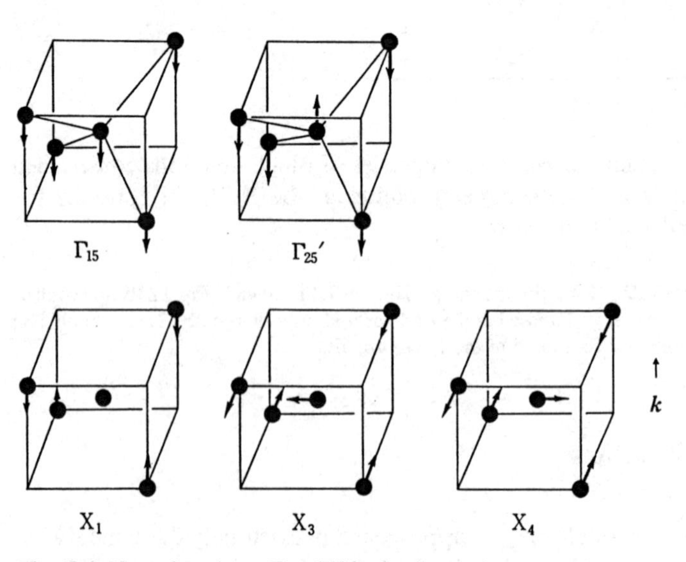

Fig. 12.6. Normal modes at Γ and X in the diamond structure

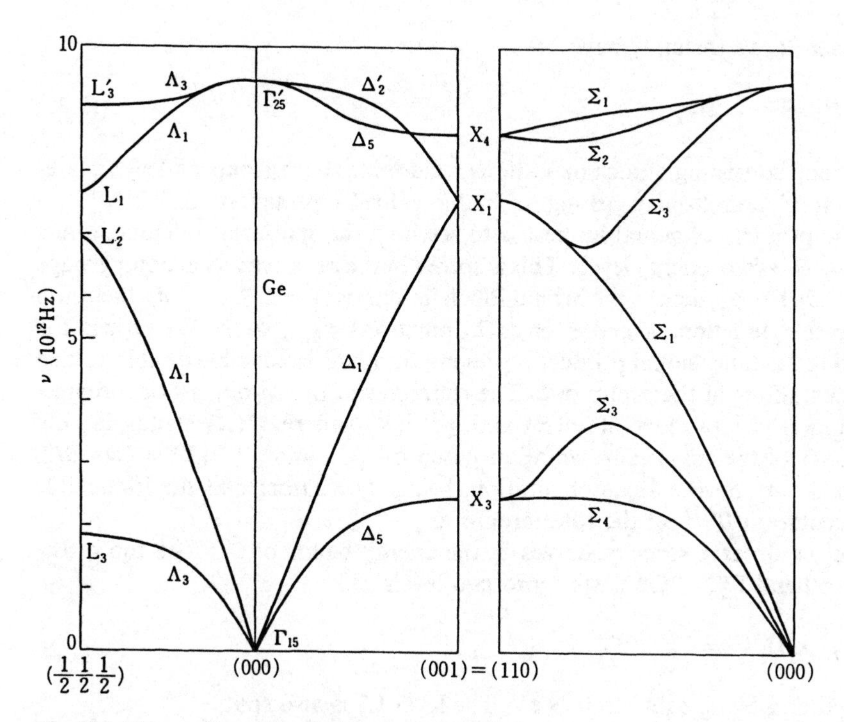

Fig. 12.7. Phonon dispersion relations in Ge [12.4]

Figure 12.7 shows the dispersion relations of lattice vibrations in Ge determined by neutron inelastic scattering experiments. Note that the connectivity relations hold just as in the case of the electron energy bands.

12.8 The Spin-Orbit Interaction and Double Space Groups

In crystals composed of heavy atoms, effects of the spin-orbit interaction

$$H_{SO} = \frac{\hbar}{2m^2c^2}\, \boldsymbol{s}\cdot(\operatorname{grad} V(\boldsymbol{r}) \times \boldsymbol{p}) \tag{12.20}$$

become considerable for valence electrons. Because of the translational invariance of the crystal potential $V(\boldsymbol{r})$, the Hamiltonian H_{SO} is invariant with respect to the translations $\boldsymbol{t}_n$; the spin angular-momentum $\boldsymbol{s}$ undergoes no changes under the translations. For rotational operations, spin must be rotated as well to keep H_{SO} invariant. Thus the expression (12.3) for space group operations has to be generalized to

$$R = \exp(-\mathrm{i}\boldsymbol{p}\cdot\boldsymbol{b}/\hbar)\exp[-\mathrm{i}\theta(\boldsymbol{l}+\boldsymbol{s})\cdot\boldsymbol{n}] \ . \tag{12.21}$$

For such R, we certainly have

$$RH_{\mathrm{SO}}R^{-1} = H_{\mathrm{SO}} \; . \tag{12.22}$$

The group consisting of such operations is a double space group, and eigenstates of the total Hamiltonian belong to double-valued representations.

The problem of actual interest is to see how the spin-orbit interaction lifts the degeneracy of energy levels. This is solved in the same way as in point groups (Sect. 8.5). Let $\psi_{k\nu}$ denote the orbital Bloch functions ($\nu = 1, 2, \ldots, d$). Inclusion of the spin functions α and β gives $2d$ functions $\psi_{k\nu}\alpha$, $\psi_{k\nu}\beta$. We reduce the resulting $2d$-dimensional product representation into irreducible double-valued representations of the group of $\boldsymbol{k}$. The character of the $2d$-dimensional representation is the product of $\chi^{k}(R)$ and $\chi^{(1/2)}(R)$, where $\chi^{k}(R)$ stands for the character of the small representation based on $\psi_{k\nu}$, and $\chi^{(1/2)}(R) = 2\cos(\theta/2)$ depends only on the angle of rotation, being the character of the irreducible representation $D^{(1/2)}$ of the rotation group.

Let us observe some examples in the energy bands of Ge. The top of the valence band Γ'_{25} of Ge is split into two levels,

$$\Gamma'_{25} \times D^{(1/2)} = \Gamma_8^+ + \Gamma_7^+ \; ,$$

the splitting being equal to 0.29 eV. The level L'_3 is also split,

$$\mathrm{L}'_3 \times D^{(1/2)} = \mathrm{L}_4^- + \mathrm{L}_5^- + \mathrm{L}_6^- \; .$$

Here, L_4^- and L_5^- are degenerate owing to the time-reversal symmetry of the Hamiltonian. Energy levels at the X point have no spin-orbit splitting since they all belong to the four-dimensional representation X_5.

12.9 Scattering of an Electron by Lattice Vibrations

Just as in molecules (Sect. 10.4), the interaction Hamiltonian of an electron with lattice vibrations can be written in the form

$$H_{\mathrm{ep}} = \sum_s V_s(\boldsymbol{r}) Q_s \tag{12.23}$$

to first order in the displacements. Here Q_s stands for the normal coordinates of the lattice vibrations. The suffix s specifies the $\boldsymbol{k}$ vector and the small representation (as well as the partner when the dimension of the small representation is larger than one). $V_s(\boldsymbol{r})$ contains the electron coordinates and belongs to the same representation as Q_s. The Hamiltonian H_{ep} is invariant when symmetry operations are performed simultaneously on both electron coordinates and atomic displacements.

This Hamiltonian gives rise to electron–phonon scattering, in which an electron in the initial state ψ_i is scattered to the final state ψ_f. The matrix element

$$\int \psi_f^* V_s \psi_i dr \tag{12.24}$$

remains nonvanishing only when the product $\psi_f^* \psi_i$ contains irreducible components that transform like V_s. The selection rule for such a process may be approached by means of the general theory expounded in Sect. 12.12. Here, we choose to study the problem in a more elementary fashion using explicit expressions for the basis functions.

Let us consider the electron–phonon scattering process in Ge: a conduction electron at the bottom L_1 of the conduction band is scattered to equivalent L_1 states. What kinds of phonons will allow such intervalley scattering? The four L points shown in Fig. 12.8 are generated by successively operating IC_4 on $\boldsymbol{k}_1 = (2\pi/a)(1/2, 1/2, 1/2)$. Since the basis function of the small representation L_1 for $\mathscr{G}(\boldsymbol{k}_1)$ is given by (Sect. 12.4)

$$\psi_1 = \cos[(\pi/a)(x + y + z - 3a/8)] ,$$

successive applications of $\{IC_4|0\}$ to it give

$$\psi_2 = \{IC_4|0\}\psi_1 = \cos[(\pi/a)(x - y - z - 3a/8)] ,$$
$$\psi_3 = \{C_4^2|0\}\psi_1 = \cos[(\pi/a)(-x - y + z - 3a/8)] ,$$
$$\psi_4 = \{IC_4^3|0\}\psi_1 = \cos[(\pi/a)(-x + y - z - 3a/8)] .$$

These four functions form the basis for the irreducible representation L_1 of the space group O_h^7. Phonon scattering takes place among these four states. Products of these real functions turn out to have the following wavevectors:

$$\psi_1\psi_2 \text{ and } \psi_3\psi_4 \text{ have } \boldsymbol{k} = (2\pi/a)(100) ,$$

$$\psi_1\psi_4 \text{ and } \psi_2\psi_3 \text{ have } \boldsymbol{k} = (2\pi/a)(010) ,$$

$$\psi_1\psi_3 \text{ and } \psi_2\psi_4 \text{ have } \boldsymbol{k} = (2\pi/a)(001) .$$

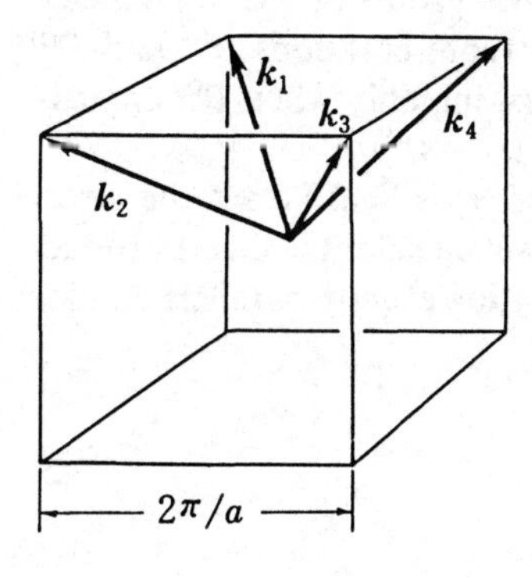

Fig. 12.8. Four wavevectors constituting the star of L. $-\boldsymbol{k}_i$ is regarded as identical to $\boldsymbol{k}_i$, since they differ only by a reciprocal lattice vector

Furthermore, $\{\psi_1\psi_3 + \psi_2\psi_4, \quad \psi_1\psi_3 - \psi_2\psi_4\}$ is seen to transform according to the small representation X_1. Consequently, the scattering between the L_1 valleys can take place through X_1 phonons.

Exercise 12.3. In fact, we have found above that $[L_1 \times L_1] = X_1$. Consider antisymmetric basis functions like $\psi_1(\boldsymbol{r})\psi_3(\boldsymbol{r}') - \psi_3(\boldsymbol{r})\psi_1(\boldsymbol{r}')$ and show that $\{L_1 \times L_1\} = X_3$. Derive the same results using the general theory of Sect. 12.12.

12.10 Interband Optical Transitions

Electrons undergo optical transitions through the interaction Hamiltonian

$$H_{\mathrm{opt}} = -(e/mc)\boldsymbol{A}\cdot\boldsymbol{p} \ , \tag{12.25}$$

where $\boldsymbol{p}$ is the electron momentum operator, and the vector potential

$$\boldsymbol{A} = \boldsymbol{A}_0 \cos(\boldsymbol{q}\cdot\boldsymbol{r} - \omega t) \tag{12.26}$$

of the radiation field has wavevector $\boldsymbol{q}$ and angular frequency ω. Normal light has very small wavenumbers as compared with reciprocal lattice vectors: for instance, for $\hbar\omega = 1$ eV, we have $q = \omega/c \sim 5 \times 10^4\ \mathrm{cm}^{-1} \ll 10^8\ \mathrm{cm}^{-1}$. Therefore, the wavevector $\boldsymbol{q}$ in (12.26) can be put to zero. The vector potential $\boldsymbol{A}$ then no longer depends on the electron coordinates, and the optical transition from the initial state ψ_{i} to the final state ψ_{f} is governed by the matrix element

$$\int \psi_{\mathrm{f}}^* \boldsymbol{p} \psi_{\mathrm{i}} d\boldsymbol{r} \ . \tag{12.27}$$

Such a transition is called a *direct optical transition.* Since we have put $\boldsymbol{q} = 0$, the wavenumber of the electron is conserved:

$$\boldsymbol{k}_{\mathrm{f}} = \boldsymbol{k}_{\mathrm{i}} \ .$$

The momentum operator $\boldsymbol{p}$, being a vector having vanishing wavenumber, forms the basis for the vector representation $\Gamma(\boldsymbol{r})$ of the group of the wavevector $\boldsymbol{k} = 0$. If the wavefunctions ψ_{f} and ψ_{i} belong to representations $D^{(\mathrm{f})}$ and $D^{(\mathrm{i})}$ respectively, the matrix element (12.27) is nonvanishing only when $D^{(\mathrm{f})}$ appears in the re duction of the product representation $\Gamma(\boldsymbol{r}) \times D^{(\mathrm{i})}$.

In Ge, for instance, the vector representation $\Gamma(\boldsymbol{r})$ is Γ_{15}. To see the direct transitions from the top of the valence band Γ'_{25}, we consider the direct product $\Gamma_{15} \times \Gamma'_{25}$. Reduction of such a product representation exactly parallels that for point groups, and we find

$$\Gamma_{15} \times \Gamma'_{25} = \Gamma'_2 + \Gamma'_{12} + \Gamma_{15} + \Gamma_{25} \ .$$

Only these states are allowed as the final states.

Table 12.5. Calculation of the character of the product representation $\Gamma_{15} \times L_3'$

	Γ_{15}	L_3'	$\Gamma_{15} \times L_3'$
$\{\varepsilon\|0\}$	3	2	6
$2\{C_3\|0\}$	0	-1	0
$3\{C_2'\|\tau\}$	-1	0	0
$\{I\|\tau\}$	-3	-2	6
$2\{IC_3\|\tau\}$	0	1	0
$3\{\sigma_d\|0\}$	1	0	0

Let us next consider the direct transition from the L_3' state. In thise case, we have to reduce $\Gamma_{15} \times L_3'$. Since the transition preserves the wavevector, this product representation belongs to the same point L. Now the group of L is a subgroup of the group of Γ. Therefore, in calculating the character of $\Gamma_{15} \times L_3'$, one has only to restrict Γ_{15} to the group of L and consider only the elements of the group of L. Irreducible decomposition of the character obtained in Table 12.5 gives

$$\Gamma_{15} \times L_3' = L_1 + L_2 + 2L_3 \ .$$

Only these states are allowed as the final states. As this example shows, products of $\boldsymbol{k} = 0$ representations with other representations are easily calculated.

Exercise 12.4. Find the selection rule for direct optical transitions from the X_4 state in the diamond structure.

Participation of lattice vibrations can give rise to *indirect optical transitions*, which are higher-order processes involving the electron–phonon interaction (12.23). The indirect transitions occur through

$$\sum_l \langle \mathrm{f}|H_{\mathrm{ep}}|l\rangle (E_\mathrm{i} - E_l)^{-1} \langle l|H_{\mathrm{opt}}|\mathrm{i}\rangle + (H_{\mathrm{ep}} \leftrightarrow H_{\mathrm{opt}}) \ . \tag{12.28}$$

In examining the selection rule for such a process, the sum over the intermediate states l,

$$\sum_l |l\rangle (E_\mathrm{i} - E_l)^{-1} \langle l|$$

may be regarded as belonging to the identity representation. The matrix element (12.28) is nonvanishing if the representation $D^{(f)}$ appears in the reduction of $D^{(s)} \times \Gamma(\boldsymbol{r}) \times D^{(i)}$, where $D^{(s)}$ stands for the representation of the normal coordinate Q_s in (12.23).

Germanium has the bottom of the conduction band at L_1. Therefore, the lowest-energy electronic excitation is the indirect $\Gamma_{25}' \rightarrow L_1$ transition, which becomes possible only with the help of phonons having the L wavevector. Since $D^{(f)} = L_1$ and $\Gamma(\boldsymbol{r}) = \Gamma_{15}$, only those normal modes that appear in the reduction

$$L_1 \times \Gamma_{15} \times \Gamma'_{25} = L'_1 + 2L'_2 + 3L'_3$$

can participate in the indirect process. Actually, germanium has L'_2 and L'_3 normal modes.

Exercise 12.5. TlBr also has the indirect band gap $X_1 \rightarrow R_{15}$. The indirect transitions need phonons at the M point. Show that no normal modes at the M point can participate in those indirect transitions. As a result, the band gap of TlBr is indirect-forbidden (forbidden even with the help of phonons).

12.11 Frenkel Excitons in Molecular Crystals

In molecular crystals, the electronic states of the molecules are not greatly changed from those in free molecules. However, the lowered symmetry of the environment gives rise to a shift or splitting of the energy levels. Furthermore, weak intermolecular interactions allow propagation of molecular excitations. Such a propagating wave, when regarded as a particle, is called a *Frenkel exciton.*

Suppose that the αth molecule in the mth unit cell is in the excited state $\psi_{m\alpha\lambda}$, while the other molecules are in the ground state $\psi_{n\beta}$. The wavefunction for such an excited state is written as

$$\Psi_{m\alpha\lambda} = \psi_{m\alpha\lambda} \prod_{n\beta}{}' \psi_{n\beta} \;, \tag{12.29}$$

where the prime excludes $\psi_{m\alpha}$ from the product. Equivalent molecules in the crystal have the same local symmetry characterized by some point group G_s, which will be called the *site group.* The ground-state wavefunction of the molecule usually belongs to the identity representation of the site group G_s. The excited-state wavefunctions $\psi_{m\alpha\lambda}$ also belong to some irreducible representation of G_s. The index λ is meant to specify the representation ν and the partner μ of the basis.

The linear combination of the above excited states

$$\Psi_{k\alpha\lambda} = N^{-3/2} \sum_m \exp(-i\boldsymbol{k}\cdot\boldsymbol{t}_m)\Psi_{m\alpha\lambda} \;, \tag{12.30}$$

which satisfies (11.28), represents the excitation wave propagating with the wavevector $\boldsymbol{k}$. Here, N^3 stands for the number of unit cells in the crystal. When the unit cell contains more than one molecule, the actual wavefunctions are obtained by taking appropriate linear combinations

$$\Psi_{k\sigma} = \sum_{\alpha\mu} C_{\alpha\mu}(k\sigma)\Psi_{k\alpha\nu\mu} , \tag{12.31}$$

where we have explicitly replaced the suffix λ by $\nu\mu$.

The coefficients $C_{\alpha\mu}(k\sigma)$ are determined by diagonalizing the Hamiltonian. However, for k vectors at high-symmetry points in the Brillouin zone, we can determine the coefficients from symmetry arguments without invoking direct diagonalization. We shall consider below only equivalent molecules (which are connected by symmetry operations), since the coupling of inequivalent molecules can only be determined by diagonalization of the Hamiltonian.

The character χ^k of the representation of $\mathscr{G}(k)$ based on $\Psi_{k\alpha\nu\mu}$ is given by

$$\chi^k(R) = \sum_{\alpha\mu} (\Psi_{k\alpha\nu\mu}, R\Psi_{k\alpha\nu\mu}) , \tag{12.32}$$

for representative elements R of $\mathscr{G}(k)$. The right-hand side will be nonvanishing only for those operations R that do not change α, i.e., that bring the sublattice α into coincidence with itself. Since screws and glides lack this condition, we can set $R = \{\beta|0\}$. This rotation moves the αth molecule in the $m = 0$ unit cell from η_α to $\beta\eta_\alpha$. The difference

$$t_\alpha(\beta) = \beta\eta_\alpha - \eta_\alpha$$

is a primitive translation vector. Using (12.30) in (12.32), we obtain

$$\chi^k(\{\beta|0\}) = \sum_{\alpha\mu} \exp(\mathrm{i}k \cdot t_\alpha(\beta))(\{\varepsilon|t_\alpha(\beta)\}\Psi_{0\alpha\nu\mu}, \{\beta|0\}\Psi_{0\alpha\nu\mu}) . \tag{12.33}$$

The matrix element on the right-hand side is equal to

$$(\Psi_{0\alpha\nu\mu}, \{\varepsilon| -t_\alpha(\beta)\}\{\beta|0\}\Psi_{0\alpha\nu\mu}) . \tag{12.34}$$

Now

$$\{\varepsilon| -t_\alpha(\beta)\}\{\beta|0\} = \{\varepsilon|\eta_\alpha\}\{\beta|0\}\{\varepsilon|\eta_\alpha\}^{-1}$$

stands for the rotation β about the α site (cf. (11.6b)), and hence it is an element of the site group G_s. Using the reciprocal lattice vector defined by (12.17), we obtain

$$\chi^k(\{\beta|0\}) = \sum_{\alpha} \exp(\mathrm{i}K(\beta) \cdot \eta_\alpha)\chi^{(\alpha\nu)}(\beta) , \tag{12.35}$$

where

$$\chi^{(\alpha\nu)}(\beta) = \sum_{\mu} (\Psi_{0\alpha\nu\mu}, \beta\Psi_{0\alpha\nu\mu}) = \sum_{\mu} (\psi_{0\alpha\nu\mu}, \beta\psi_{0\alpha\nu\mu})$$

is the character of the irreducible representation ν of the site group G_s of the αth molecule.

For wavevectors $\boldsymbol{k}$ inside the Brillouin zone, $\boldsymbol{K}(\beta)$ vanishes, and we have

$$\chi^{\boldsymbol{k}}(\{\beta|0\}) = \sum_{\alpha} \chi^{(\alpha\nu)}(\beta) \quad \text{for } \boldsymbol{k} \text{ inside the BZ} . \tag{12.36}$$

Moreover, if the crystal is composed of p equivalent sublattices with the same site group G_s, the character is common to all α, and in consequence,

$$\chi^{\boldsymbol{k}}(\{\beta|0\}) = p\chi^{(\nu)}(\beta) \quad \text{for } \boldsymbol{k} \text{ inside the BZ.} \tag{12.37}$$

The character vanishes for the elements R that do not belong to G_s.

We confine ourselves below to the case of $\boldsymbol{k} = 0$ excitons, which can be excited by light absorption. Then the group of $\boldsymbol{k}$ is the space group G itself, and its small representations Γ are derived from the irreducible representations of the corresponding point group G_0. The number of times a representation Γ appears in the reduction of the above-obtained representation is given by

$$w_{\nu\Gamma} = \frac{1}{g} \sum_{\{\beta|0\}} \chi^{\Gamma}(\{\beta|0\})^* \chi^{\boldsymbol{k}=0}(\{\beta|0\}) ,$$

where χ^{Γ} stands for the character of the representation Γ, and g for the order of G_0. Substitution of (12.37) brings the right-hand side into

$$w_{\nu\Gamma} = \frac{p}{g} \sum_{\beta} \chi^{\Gamma}(\{\beta|0\})^* \chi^{(\nu)}(\beta) . \tag{12.38}$$

The site group G_s is a subgroup of G_0 and its order is equal to g/p. The right-hand side of (12.38) is the number $w_{\Gamma\nu}$ of times the representation ν appears in the reduction of $\Gamma \downarrow G_s$, and so we have[1]

$$w_{\nu\Gamma} = w_{\Gamma\nu} . \tag{12.39}$$

If $w_{\nu\Gamma} = 1$ for a given ν, i.e., if the representation Γ appears at most once in the above reduction, we can construct symmetry-adapted exciton wavefunctions

$$\Psi_{\boldsymbol{k}=0,\nu\Gamma\gamma} = \sum_{\alpha\mu} C_{\alpha\mu}(\Gamma\gamma) \Psi_{\boldsymbol{k}=0,\alpha\nu\mu} \tag{12.40}$$

without invoking diagonalization of the Hamiltonian. In particular, when Γ is a

---

[1] In the language of induced representations (Sect. 5.1), the functions (12.30) form a basis for the representation $\nu \uparrow G$ induced from the irreducible representation ν of the site group G_s. Equation (12.39) is nothing but the Frobenius reciprocity (5.10).

one-dimensional representation, application of the projection operator gives

$$\Psi_{k=0,\nu\Gamma} = \frac{1}{\sqrt{p}} \sum_{\alpha=0}^{p-1} \chi^{\Gamma}(Q_\alpha)^* \Psi_{k=0,\alpha\nu} , \qquad (12.41)$$

with the operations Q_α generating p equivalent molecules in the unit cell. The energy of the exciton becomes in such a case (see Exercise 6.10)

$$\begin{aligned} E_{\nu\Gamma}(k=0) &= (\Psi_{k=0,\nu\Gamma}, H\Psi_{k=0,\nu\Gamma}) \\ &= E_\nu + \sum_{\alpha=0}^{p-1} \chi^{\Gamma}(Q_\alpha)(\Psi_{k=0,\alpha\nu}, H\Psi_{k=0,0\nu}) , \end{aligned} \qquad (12.42)$$

$$\begin{aligned} (\Psi_{k=0,\alpha\nu}, H\Psi_{k=0,\alpha'\nu}) &= \sum_n (\Psi_{0\alpha\nu}, H\Psi_{n\alpha'\nu}) \\ &= \sum_n (\psi_{0\alpha\nu}\psi_{n\alpha'} | W_{0\alpha,n\alpha'} | \psi_{0\alpha}\psi_{n\alpha'\nu}) . \end{aligned}$$

The exciton propagates in the crystal owing to the intermolecular interaction $W_{m\alpha,n\alpha'}$, which simultaneously excites the $m\alpha$ molecule and de-excites the $n\alpha'$ molecule. This interaction gives rise to a splitting of the originally p-fold degenerate level E_ν of the p equivalent molecules, which is called the *Davydov splitting.*

The exciton energies are determined by optical absorption measurements. If the electric dipole moment P_ξ belongs to the small representation Γ_0, then the light polarized in the ξ direction can excite only the exciton with the Γ_0 symmetry. Therefore optical measurements can determine the symmetries as well as energies of the exciton states.

Let us apply the above theory to the benzene crystal. Benzene crystallizes in the orthorhombic D_{2h}^{15} or $P\frac{2_1}{b}\frac{2_1}{c}\frac{2_1}{a}$ space group. Its unit cell ($a = 7.29$, $b = 9.47$, $c = 6.74$ Å at -195°C) contains four equivalent benzene molecules. The above space group has the elements

$$\begin{aligned} &\{\varepsilon|0\}, \quad \{C_{2x}|\tau_3\}, \quad \{C_{2y}|\tau_1\}, \quad \{C_{2z}|\tau_2\} , \\ &\{I|0\}, \quad \{\sigma_{vx}|\tau_3\}, \quad \{\sigma_{vy}|\tau_1\}, \quad \{\sigma_h|\tau_2\} , \end{aligned}$$

where

$$\begin{aligned} \tau_3 &= (t_1 + t_2)/2 , \\ \tau_1 &= (t_2 + t_3)/2 , \\ \tau_2 &= (t_3 + t_1)/2 , \end{aligned}$$

and t_1, t_2, and t_3 stand for the fundamental periods along the a, b, and c

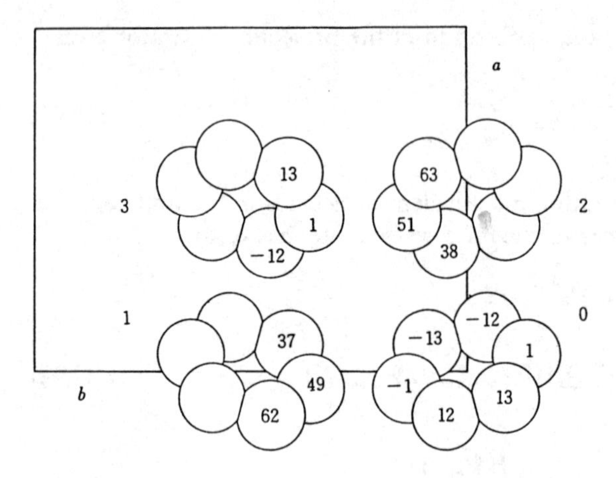

Fig. 12.9. Crystal structure of benzene. The numbers on the atoms indicate the heights in units of c

directions. Application of the above operations in the upper (and lower) row to the corner benzene molecule generates equivalent molecules ($0 \to 0, 0 \to 3, 0 \to 1, 0 \to 2$ in Fig. 12.9). Note that the site group G_s here is $C_i = \{E, I\}$, and the point group G_0 is D_{2h}.

The high symmetry D_{6h} of the free benzene molecule is lowered to C_i. Correspondingly, the "g" representations of D_{6h} become A_g under C_i, while the "u" ones become A_u. The exciton wavefunctions (12.30) are constructed from $\nu = A_g$ or $\nu = A_u$ excited states. The character (12.37) becomes

$$\chi^{k=0}(\{\varepsilon|0\}) = 4 \; , \quad \chi^{k=0}(\{I|0\}) = \pm 4 \; .$$

It vanishes for the other representative elements. Irreducible decomposition of this representation in the point group D_{2h} gives

$$A_g + B_{1g} + B_{2g} + B_{3g} \; , \quad A_u + B_{1u} + B_{2u} + B_{3u} \; ,$$

for the A_g and A_u excitations, respectively.

Since the dipole moments P_x, P_y, and P_z belong to the B_{3u}, B_{2u}, and B_{1u} representations, light polarized along the x, y, and z directions can generate excitons of B_{3u}, B_{2u}, and B_{1u} symmetry, respectively. The wavefunctions for the four excitons are

$$\begin{aligned}
\Psi(A) &= (\Psi_0 + \Psi_1 + \Psi_2 + \Psi_3)/2 \; , \\
\Psi(B_1) &= (\Psi_0 - \Psi_1 + \Psi_2 - \Psi_3)/2 \; , \\
\Psi(B_2) &= (\Psi_0 + \Psi_1 - \Psi_2 - \Psi_3)/2 \; , \\
\Psi(B_3) &= (\Psi_0 - \Psi_1 - \Psi_2 + \Psi_3)/2 \; .
\end{aligned}$$

Ψ_α on the right-hand sides stands for $\Psi_{k=0,\alpha\nu}$ ($\nu = A_g, A_u$). The corresponding energy eigenvalues are

$$
\begin{aligned}
E(\mathrm{A}) &= E_v + \Delta_1 + \Delta_2 + \Delta_3 \ ,\\
E(\mathrm{B}_1) &= E_v - \Delta_1 + \Delta_2 - \Delta_3 \ ,\\
E(\mathrm{B}_2) &= E_v + \Delta_1 - \Delta_2 - \Delta_3 \ ,\\
E(\mathrm{B}_3) &= E_v - \Delta_1 - \Delta_2 + \Delta_3 \ ,
\end{aligned}
$$

in terms of

$$\Delta_\alpha = (\Psi_{k=0,\alpha v}, H\Psi_{k=0,0v}) \ .$$

Free benzene molecules have the lowest excited level $^1\mathrm{B}_{2u}$ at 4.71 eV, which cannot be detected by optical means because the transition $\mathrm{A}_{1g} \rightarrow \mathrm{B}_{2u}$ is dipole forbidden in D_{6h}. Symmetry lowering in the crystal relaxes the selection rule, and the split energy levels are actually observed at 37803 cm^{-1} (B_{3u}), 37846 cm^{-1} (B_{2u}), and 37839 cm^{-1} (B_{1u}).

*12.12 Selection Rules in Space Groups

Selection rules in general have been discussed in Sect. 6.5. Matrix elements (6.26) remain non-vanishing if the product representation $D^{(\alpha)} \times D^{(\beta)}$ contains $D^{(\gamma)}$, i.e., if the number of times the representation $D^{(\gamma)}$ appears in the reduction of $D^{(\alpha)} \times D^{(\beta)}$

$$q_\gamma^{(\alpha\times\beta)} = \frac{1}{gN^3}\sum_R \chi^{(\gamma)}(R)^*\chi^{(\alpha)}(R)\chi^{(\beta)}(R) \tag{12.43}$$

does not vanish. Here, gN^3 stands for the order of the space group G, and N^3 for the order of the translation group T. The character $\chi^{(\alpha)}(R)$ is the trace

$$\chi^{(\alpha)}(R) = \sum_m D_{mm}^{(\alpha)}(R) \tag{12.44}$$

of the irreducible representation matrix $\hat{D}^{(\alpha)}(R)$ of the space group. In principle, it suffices to calculate the above $q_\gamma^{(\alpha\times\beta)}$. In fact, the large number of the space group elements makes that calculation rather complicated. In this section, we wish to derive an equivalent expression for $q_\gamma^{(\alpha\times\beta)}$ that will allow actual calculations solely in terms of the characters of the small representations.

Irreducible representations of the space group G are characterized by star$\{k\} = \{k_1, k_2, \ldots, k_s\}$ and the small representation Γ of the group $\mathcal{G}(k_1)$. The representation matrices have a structure like (11.54). The partners in the representation are identified by the wavevector k_p of the star and the partner σ in the small representation Γ. Therefore, the irreducible representation α and its partners m are identified by

$$\alpha = \{\boldsymbol{k}^{(\alpha)}\}\Gamma_\alpha \ ; \quad m = \boldsymbol{k}_p^{(\alpha)}, \sigma_p \ . \tag{12.45}$$

The wavevector $\boldsymbol{k}_p^{(\alpha)}$ is related to the first member $\boldsymbol{k}_1^{(\alpha)}$ of the star by[2]

$$\boldsymbol{k}_p^{(\alpha)} = \alpha_p \boldsymbol{k}_1^{(\alpha)} \tag{12.46}$$

through the representative element $\{\alpha_p|\boldsymbol{b}_p\}$ of the coset decomposition (11.47) of G with respect to $\mathscr{G}(\boldsymbol{k}^{(\alpha)})$. Since the diagonal elements $D_{mm}^{(\alpha)}(R)$ are nonvanishing only when $R \in \mathscr{G}(\boldsymbol{k}_p^{(\alpha)})$, we have

$$D_{mm}^{(\alpha)}(R) = D_{\sigma_p\sigma_p}^{\boldsymbol{k}_p^{(\alpha)}(\Gamma_\alpha)}(Q)\exp(\mathrm{i}\boldsymbol{k}_p^{(\alpha)} \cdot \boldsymbol{t}_n) \tag{12.47}$$

for an element

$$R = \{\varepsilon|\boldsymbol{t}_n\}Q \in \mathscr{G}(\boldsymbol{k}_p^{(\alpha)}) \ .$$

The elements Q are understood to be the representative elements of the group of $\boldsymbol{k}_p$, i.e., they are coset representatives of the factor group $\mathscr{G}(\boldsymbol{k}_p^{(\alpha)})/T$. The small representation $D^{\boldsymbol{k}_p^{(\alpha)}(\Gamma_\alpha)}$ has the character $\chi^{\boldsymbol{k}_p^{(\alpha)}(\Gamma_\alpha)}$, which is related to $\chi^{\boldsymbol{k}^{(\alpha)}(\Gamma_\alpha)}$ by

$$\chi^{\boldsymbol{k}_p^{(\alpha)}(\Gamma_\alpha)}(Q) = \chi^{\boldsymbol{k}^{(\alpha)}(\Gamma_\alpha)}(\{\alpha_p|\boldsymbol{b}_p\}^{-1}Q\{\alpha_p|\boldsymbol{b}_p\}) \ , \tag{12.48}$$

as seen from (11.52).

We now use (12.47) in (12.43) and sum over the primitive translations. We then find

$$q_\gamma^{(\alpha \times \beta)} = \frac{1}{g}\sum_{pqr} \delta(\boldsymbol{k}_r^{(\gamma)}, \boldsymbol{k}_p^{(\alpha)} + \boldsymbol{k}_q^{(\beta)}) \sum_{Q \in \mathscr{G}(\boldsymbol{k}_r^{(\gamma)})/T} \chi^{\boldsymbol{k}_r^{(\gamma)}(\Gamma_\gamma)}(Q)^* \, \chi^{\boldsymbol{k}_p^{(\alpha)}(\Gamma_\alpha)}(Q) \, \chi^{\boldsymbol{k}_q^{(\beta)}(\Gamma_\beta)}(Q) \ , \tag{12.49}$$

where Q must be elements common to $\mathscr{G}(\boldsymbol{k}_p^{(\alpha)})$ and $\mathscr{G}(\boldsymbol{k}_q^{(\beta)})$ for the product not to vanish. The symbol $\delta(\boldsymbol{k}, \boldsymbol{k}')$ is defined by

$$\delta(\boldsymbol{k}, \boldsymbol{k}') = \begin{cases} 1 & \text{if} \quad \boldsymbol{k} \doteq \boldsymbol{k}' \ , \\ 0 & \text{otherwise} \ . \end{cases}$$

When the summation over r is performed, (12.49) simplifies to

$$q_\gamma^{(\alpha \times \beta)} = \frac{s(\boldsymbol{k}^{(\gamma)})}{g}\sum_{pq} \delta(\boldsymbol{k}^{(\gamma)}, \boldsymbol{k}_p^{(\alpha)} + \boldsymbol{k}_q^{(\beta)})$$
$$\times \sum_{Q \in \mathscr{G}(\boldsymbol{k}^{(\gamma)})/T} \chi^{\boldsymbol{k}^{(\gamma)}(\Gamma_\gamma)}(Q)^* \, \chi^{\boldsymbol{k}_p^{(\alpha)}(\Gamma_\alpha)}(Q) \, \chi^{\boldsymbol{k}_q^{(\beta)}(\Gamma_\beta)}(Q) \ ,$$

---

[2] Do not confuse the rotation α_p with the suffix α that specifies irreducible representations. We shall write below simply $\boldsymbol{k}^{(\alpha)}$ for $\boldsymbol{k}_1^{(\alpha)}$.

using the order $s(\boldsymbol{k}^{(\gamma)})$ of star$\{\boldsymbol{k}^{(\gamma)}\}$. The factor $s(\boldsymbol{k}^{(\gamma)})/g$ on the right-hand side is equal to $[\boldsymbol{k}^{(\gamma)}]^{-1}$, if we let $[\boldsymbol{k}^{(\gamma)}] \times N^3$ denote the order of $\mathcal{G}(\boldsymbol{k}^{(\gamma)})$. The summation on the right-hand side is carried out in two steps:

$$q_\gamma^{(\alpha\times\beta)} = [\boldsymbol{k}^{(\gamma)}]^{-1} \sum_{Q\in\mathcal{G}(\boldsymbol{k}^{(\gamma)})/T} \chi^{\boldsymbol{k}^{(\gamma)}(\Gamma_\gamma)}(Q)^* \, \chi^{\boldsymbol{k}^{(\gamma)}(\alpha\times\beta)}(Q) \;, \tag{12.50}$$

and

$$\chi^{\boldsymbol{k}^{(\gamma)}(\alpha\times\beta)}(Q) = \sum_{p,q} \delta(\boldsymbol{k}^{(\gamma)}, \boldsymbol{k}_p^{(\alpha)} + \boldsymbol{k}_q^{(\beta)}) \chi^{\boldsymbol{k}_p^{(\alpha)}(\Gamma_\alpha)}(Q) \, \chi^{\boldsymbol{k}_q^{(\beta)}(\Gamma_\beta)}(Q) \;. \tag{12.51}$$

Roughly speaking, the latter character is generated by the basis functions of $\alpha \times \beta$ satisfying

$$\boldsymbol{k}^{(\gamma)} \doteq \boldsymbol{k}_p^{(\alpha)} + \boldsymbol{k}_q^{(\beta)} \;. \tag{12.52}$$

The summation in (12.51) runs over those pairs of $\boldsymbol{k}_p^{(\alpha)}$, $\boldsymbol{k}_q^{(\beta)}$ that satisfy $Q\in\mathcal{G}(\boldsymbol{k}_p^{(\alpha)})\cap\mathcal{G}(\boldsymbol{k}_q^{(\beta)})$ for the given representative element $Q\in\mathcal{G}(\boldsymbol{k}^{(\gamma)})/T$.

Evaluation of the sum in (12.51) further breaks into two steps. If (12.52) holds for some pair of $\boldsymbol{k}_p^{(\alpha)}$ and $\boldsymbol{k}_q^{(\beta)}$, then other pairs $S\boldsymbol{k}_p^{(\alpha)}$, $S\boldsymbol{k}_q^{(\beta)}$ will also satisfy (12.52), if $S\in\mathcal{G}(\boldsymbol{k}^{(\gamma)})$. The summation in (12.51) naturally contains contributions from such pairs. Let then $\{p, q\}$ denote the set of such distinct pairs $S\boldsymbol{k}_p^{(\alpha)}$, $S\boldsymbol{k}_q^{(\beta)}$ generated from the original pair satisfying (12.52). The summation (12.51) contains in general other pairs as well which cannot be reached by means of the elements of $\mathcal{G}(\boldsymbol{k}^{(\gamma)})$. Then (12.51) becomes a sum over such sets $\{p, q\}$:

$$\chi^{\boldsymbol{k}^{(\gamma)}(\alpha\times\beta)}(Q) = \sum_{\{p,q\}} \chi^{\boldsymbol{k}^{(\gamma)}(\boldsymbol{k}_p^{(\alpha)}(\Gamma_\alpha)\times\boldsymbol{k}_q^{(\beta)}(\Gamma_\beta))}(Q) \;, \tag{12.53}$$

$$\chi^{\boldsymbol{k}^{(\gamma)}(\boldsymbol{k}_p^{(\alpha)}(\Gamma_\alpha)\times\boldsymbol{k}_q^{(\beta)}(\Gamma_\beta))}(Q) = \sum_{S_\nu} \chi^{\boldsymbol{k}_p^{(\alpha)}(\Gamma_\alpha)}(S_\nu^{-1}QS_\nu)\chi^{\boldsymbol{k}_q^{(\beta)}(\Gamma_\beta)}(S_\nu^{-1}QS_\nu) \;. \tag{12.54}$$

The summation in (12.54) over $S_\nu\in\mathcal{G}(\boldsymbol{k}^{(\gamma)})/T$ is taken only for those S_ν that satisfy

$$S_\nu^{-1}QS_\nu\in\mathcal{G}(\boldsymbol{k}_p^{(\alpha)})\cap\mathcal{G}(\boldsymbol{k}_q^{(\beta)}) \tag{12.55}$$

for the given pair $\boldsymbol{k}_p^{(\alpha)}$, $\boldsymbol{k}_q^{(\beta)}$. If we write

$$\boldsymbol{k}_{p\nu}^{(\alpha)} = S_\nu\boldsymbol{k}_p^{(\alpha)} \;, \qquad \boldsymbol{k}_{q\nu}^{(\beta)} = S_\nu\boldsymbol{k}_q^{(\beta)} \;,$$

then (12.54) may be rewritten as

$$\chi^{\boldsymbol{k}^{(\gamma)}(\boldsymbol{k}_p^{(\alpha)}(\Gamma_\alpha)\times\boldsymbol{k}_q^{(\beta)}(\Gamma_\beta))}(Q) = \sum_{\nu=1}^{N} \chi^{\boldsymbol{k}_{p\nu}^{(\alpha)}(\Gamma_\alpha)}(Q)\chi^{\boldsymbol{k}_{q\nu}^{(\beta)}(\Gamma_\beta)}(Q) \;. \tag{12.56}$$

The right-hand side is to be summed over those $\boldsymbol{k}_{p\nu}^{(\alpha)}$ (and $\boldsymbol{k}_{q\nu}^{(\beta)}$) that are generated from $\boldsymbol{k}_p^{(\alpha)}$ ($\boldsymbol{k}_q^{(\beta)}$) by the application of elements $S\in\mathcal{G}(\boldsymbol{k}^{(\gamma)})/T$ *and* remain invariant under Q. The number of the terms N is equal to the number of such pairs, which may be designated as

$$\begin{aligned} N &= N(Q; \boldsymbol{k}^{(\gamma)}\ \mathrm{star}(\boldsymbol{k}_p^{(\alpha)})) \\ &= N(Q; \boldsymbol{k}^{(\gamma)}\ \mathrm{star}(\boldsymbol{k}_q^{(\beta)}))\ . \end{aligned} \tag{12.57}$$

Here, $\boldsymbol{k}^{(\gamma)}\ \mathrm{star}(\boldsymbol{k}_p^{(\alpha)})$ denotes the set of the wavevectors $\{S\boldsymbol{k}_p^{(\alpha)} | S \in \mathscr{G}(\boldsymbol{k}^{(\gamma)})\}$. The operation Q leaves N of these wavevectors invariant.

Lax and *Hopfield* [12.5] put (12.54) in the following form:

$$\chi^{\boldsymbol{k}^{(\gamma)}(\boldsymbol{k}_p^{(\alpha)}(\Gamma_\alpha) \times \boldsymbol{k}_q^{(\beta)}(\Gamma_\beta))}(Q) = N(Q; \boldsymbol{k}^{(\gamma)}\ \mathrm{star}(\boldsymbol{k}_p^{(\alpha)}))\, \langle \chi^{\boldsymbol{k}_p^{(\alpha)}(\Gamma_\alpha)}(Q)\, \chi^{\boldsymbol{k}_q^{(\beta)}(\Gamma_\beta)}(Q) \rangle_{\boldsymbol{k}^{(\gamma)}} \tag{12.58}$$

$$\langle \chi^{\boldsymbol{k}_p^{(\alpha)}(\Gamma_\alpha)}(Q)\, \chi^{\boldsymbol{k}_q^{(\beta)}(\Gamma_\beta)}(Q) \rangle_{\boldsymbol{k}^{(\gamma)}} \equiv \sum_{P \in \mathscr{C}(Q)} \chi^{\boldsymbol{k}_p^{(\alpha)}(\Gamma_\alpha)}(P)\, \chi^{\boldsymbol{k}_q^{(\beta)}(\Gamma_\beta)}(P) \Big/ \sum_{p \in \mathscr{C}(Q)} 1\ . \tag{12.59}$$

The summations in (12.59) are taken for those $P \in \mathscr{G}(\boldsymbol{k}^{(\gamma)})/T(\boldsymbol{k}^{(\gamma)})$ which satisfy $P \in \mathscr{G}(\boldsymbol{k}_p^{(\alpha)}) \cap \mathscr{G}(\boldsymbol{k}_q^{(\beta)})$; $\mathscr{C}(Q)$ stands for the class of the group $\mathscr{G}(\boldsymbol{k}^{(\gamma)})/T(\boldsymbol{k}^{(\gamma)})$ which Q belongs to. Formula (12.59) is convenient since the knowledge of the characters for small representations at $\boldsymbol{k}_p^{(\alpha)}$ and $\boldsymbol{k}_q^{(\beta)}$ is sufficient for the calculation of the required character (12.58). Here, $T(\boldsymbol{k})$ is the invariant subgroup of pure translations $\{\varepsilon | \boldsymbol{t}_n\}$ such that $\exp(\mathrm{i}\boldsymbol{k} \cdot \boldsymbol{t}_n) = 1$.

Example 1: In Sect. 12.10, we found that the optical transition $\mathrm{L}_3' \to \mathrm{L}_1$ is allowed in germanium by showing that the product representation $\Gamma_{15} \times \mathrm{L}_3'$ contains L_1. We can also show the same result by inspecting whether $\mathrm{L}_1 \times \mathrm{L}_3'$ contains Γ_{15}. (Note that we have $\mathrm{L}_1^* = \mathrm{L}_1$ since it is a real representation according to the Herring criterion.)

The wavevector at the L point is $\boldsymbol{k}_\mathrm{L} = (2\pi/a)(1/2, 1/2, 1/2)$. Starting from the initial pair of vectors

$$\boldsymbol{k}^{(\alpha)} = \boldsymbol{k}^{(\beta)} = \boldsymbol{k}_\mathrm{L}\ ,$$

which satisfy

$$\boldsymbol{k}^{(\alpha)} + \boldsymbol{k}^{(\beta)} \doteq \boldsymbol{k}^{(\gamma)} = 0\ ,$$

we can generate four pairs of vectors by successive applications of an element, say, $\{IC_4|0\}$ belonging to $\mathscr{G}(\boldsymbol{k}^{(\gamma)})$. The right-hand side of (12.53) then consists of a single term,

$$\chi^{\Gamma(\{\mathrm{L}\}\mathrm{L}_1 \times \{\mathrm{L}\}\mathrm{L}_3')}(Q) = \chi^{\Gamma(\mathrm{L}_1 \times \mathrm{L}_3')}(Q)\ , \tag{12.60}$$

and (12.54) gives

$$\chi^{\Gamma(\mathrm{L}_1 \times \mathrm{L}_3')}(Q) = \sum_{\nu=0}^{3} \chi^{(\mathrm{L}_1)}(S_\nu^{-1} Q S_\nu)\, \chi^{(\mathrm{L}_3')}(S_\nu^{-1} Q S_\nu')\ , \tag{12.61}$$

where

$$S_\nu = \{IC_4|0\}^\nu\ .$$

Table 12.6. Result of the character evaluation using (12.61) or (12.62). The second factors in the table may be interpreted either as the number of nonvanishing terms in (12.61) or as $N(Q; \{\mathrm{L}\})$ of (12.62). The characters of the small representations at L and Γ may be found in the tables of Appendix B

$Q \in \mathcal{G}_\Gamma/T$	$\chi^{\Gamma(\mathrm{L}_1 \times \mathrm{L}'_3)}(Q)$
$\{\varepsilon\|0\}$	2×4
$6\{C'_2\|\tau\}$	0×2
$8\{C_3\|0\}$	-1×1
$\{I\|\tau\}$	-2×4
$6\{\sigma_d\|0\}$	0×2
$8\{IC_3\|\tau\}$	1×1
Other elements	0

Or, equivalently, (12.58) gives

$$\chi^{\Gamma(\mathrm{L}_1 \times \mathrm{L}'_3)}(Q) = \langle \chi^{(\mathrm{L}_1)}(Q)\chi^{(\mathrm{L}'_3)}(Q) \rangle_\Gamma N(Q; \{\mathrm{L}\}) \ . \tag{12.62}$$

Either expression gives the result shown in Table 12.6. From this character, we find that

$$\mathrm{L}_1 \times \mathrm{L}'_3 = \Gamma'_{12} + \Gamma_{15} + \Gamma_{25} \ ,$$

which does actually contain Γ_{15}, and hence the transition $\mathrm{L}'_3 \to \mathrm{L}_1$ is allowed.

Example 2: Let us next consider the representations Γ and M of the space group T^1, or P23. The star of M consists of three points $\mathrm{M}(\pi/a, \pi/a, 0)$, $\mathrm{M}_+(0, \pi/a, \pi/a)$, and $\mathrm{M}_-(\pi/a, 0, \pi/a)$, as shown in Fig. 12.10. The small representations of this symmorphic space group are easily obtained from the irreducible representations of the point groups T and D_2. The characters are shown in Table 12.7.

Take for instance the product representation $\{\mathrm{M}\}\mathrm{B}_2 \times \{\mathrm{M}\}\mathrm{B}_3$. The characters of the small representations at M_+ and M_- are related to those at M

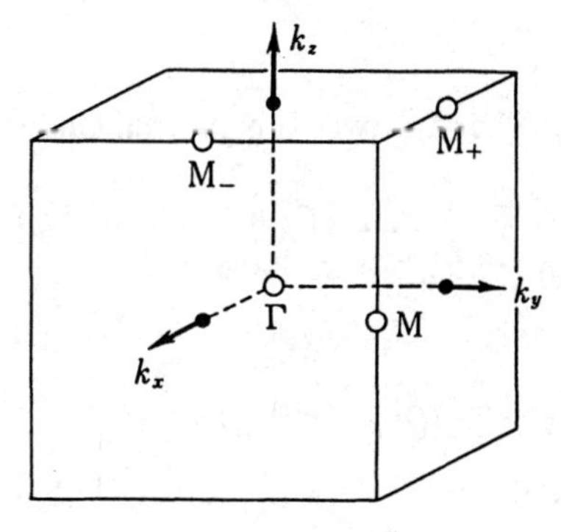

Fig. 12.10. The Brillouin zone for the space group P23. The three vectors M, M_+, and M_- constitute a star

Table 12.7. Characters of the small representations at Γ and M. $\omega = \exp(-2\pi i/3)$.

Γ	A	^{1}E	^{2}E	T
$\{\varepsilon\|0\}$	1	1	1	3
$3\{C_2\|0\}$	1	1	1	−1
$4\{C_3^+\|0\}$	1	ω	ω^2	0
$4\{C_3^-\|0\}$	1	ω^2	ω	0

M	A	B$_1$	B$_2$	B$_3$
$\{\varepsilon\|0\}$	1	1	1	1
$\{C_{2x}\|0\}$	1	−1	−1	1
$\{C_{2y}\|0\}$	1	−1	1	−1
$\{C_{2z}\|0\}$	1	1	−1	−1

Table 12.8. Characters of the small representations at the M$_+$ and M$_-$ points

Q	$\{C_3^+\|0\}^{-1}Q\{C_3^+\|0\}$	$\{C_3^-\|0\}^{-1}Q\{C_3^-\|0\}$	M$_+$(B$_2$)	M$_+$(B$_3$)	M$_-$(B$_2$)	M$_-$(B$_3$)
$\{\varepsilon\|0\}$	$\{\varepsilon\|0\}$	$\{\varepsilon\|0\}$	1	1	1	1
$\{C_{2x}\|0\}$	$\{C_{2z}\|0\}$	$\{C_{2y}\|0\}$	−1	−1	1	−1
$\{C_{2y}\|0\}$	$\{C_{2x}\|0\}$	$\{C_{2z}\|0\}$	−1	1	−1	−1
$\{C_{2z}\|0\}$	$\{C_{2y}\|0\}$	$\{C_{2x}\|0\}$	1	−1	−1	1

through (12.48) and their values are shown in Table 12.8. Since the combinations of the three vectors of star$\{$M$\}$ give

$$\boldsymbol{k}_{\mathrm{M}} + \boldsymbol{k}_{\mathrm{M}} \doteq \boldsymbol{k}_{\mathrm{M}_+} + \boldsymbol{k}_{\mathrm{M}_+} \doteq \boldsymbol{k}_{\mathrm{M}_-} + \boldsymbol{k}_{\mathrm{M}_-} \doteq 0$$

and

$$\boldsymbol{k}_{\mathrm{M}_+} + \boldsymbol{k}_{\mathrm{M}_-} \doteq \boldsymbol{k}_{\mathrm{M}_-} + \boldsymbol{k}_{\mathrm{M}_+} \doteq \boldsymbol{k}_{\mathrm{M}} \ ,$$

we have a three-dimensional representation at Γ and a two-dimensional small representation at M.

In the case of $\boldsymbol{k}^{(\gamma)} = 0$, application of $\{C_3^+|0\}$ and $\{C_3^-|0\}$ to $(\boldsymbol{k}_{\mathrm{M}}, \boldsymbol{k}_{\mathrm{M}})$ generates the other two pairs, and hence from (12.53, 54) we have

$$\begin{aligned}\chi^{\Gamma(\{\mathrm{M}\}\mathrm{B}_2 \times \{\mathrm{M}\}\mathrm{B}_3)}(Q) &= \chi^{\Gamma(\mathrm{M}(\mathrm{B}_2) \times \mathrm{M}(\mathrm{B}_3))}(Q) \\ &= \chi^{\mathrm{M}(\mathrm{B}_2)}(Q)\, \chi^{\mathrm{M}(\mathrm{B}_3)}(Q) + \chi^{\mathrm{M}_+(\mathrm{B}_2)}(Q)\, \chi^{\mathrm{M}_+(\mathrm{B}_3)}(Q) \\ &\quad + \chi^{\mathrm{M}_-(\mathrm{B}_2)}(Q)\, \chi^{\mathrm{M}_-(\mathrm{B}_3)}(Q) \ , \end{aligned} \tag{12.63}$$

Table 12.9. Result of the character evaluation using (12.63) or (12.64)

$Q \in \mathcal{G}_\Gamma/T$	(12.63)	$\langle \chi^{M(B_2)}(Q)\chi^{M(B_3)}(Q)\rangle_\Gamma$	$N(Q; \{M\})$
$\{\varepsilon\|0\}$	$1+1+1=3$	1	3
$\{C_{2x}\|0\}$	$-1+1-1=-1$	$-1/3$	3
$\{C_{2y}\|0\}$	$-1-1+1=-1$	$-1/3$	3
$\{C_{2z}\|0\}$	$1-1-1=-1$	$-1/3$	3
$4\{C_3^+\|0\}$	0		0
$4\{C_3^-\|0\}$	0		0

Table 12.10. Character evaluation using (12.65)

$Q \in \mathcal{G}_M/T$	
$\{\varepsilon\|0\}$	$1+1=2$
$\{C_{2x}\|0\}$	$1-1=0$
$\{C_{2y}\|0\}$	$1-1=0$
$\{C_{2z}\|0\}$	$1+1=2$

or else from (12.58) we have

$$= N(Q; \{M\})\langle \chi^{M(B_2)}(Q)\, \chi^{M(B_3)}(Q)\rangle_\Gamma \; . \tag{12.64}$$

From the result shown in Table 12.9, we obtain the T representation at Γ.

In the case of $k^{(\gamma)} = k_M$, (12.53) consists of two terms, for the two pairs of wavevectors (k_{M_+}, k_{M_+}) and (k_{M_-}, k_{M_-}) are not coupled by the operations of $\mathcal{G}$(M). Formula (12.53) now becomes

$$\begin{aligned}\chi^{M(\{M\}B_2 \times \{M\}B_3)}(Q) &= \chi^{M(M_+(B_2)\times M_-(B_3))}(Q) + \chi^{M(M_-(B_2)\times M_+(B_3))}(Q) \\ &= \chi^{M_+(B_2)}(Q)\,\chi^{M_-(B_3)}(Q) + \chi^{M_-(B_2)}(Q)\,\chi^{M_+(B_3)}(Q) \; , \end{aligned} \tag{12.65}$$

which may be calculated using Table 12.8. The result shown in Table 12.10 gives $A + B_1$. As a consequence, we find

$$\{M\}B_2 \times \{M\}B_3 = \{\Gamma\}T + \{M\}A + \{M\}B_1 \; . \tag{12.66}$$

12.12.1 Symmetric and Antisymmetric Product Representations

From the symmetric and antisymmetric combinations

$$\phi_1(k_1\mu_1)\phi_2(k_2\mu_2) \pm \phi_1(k_2\mu_2)\phi_2(k_1\mu_1) \tag{12.67}$$

of the products $\phi_1(k_1\mu_1)\phi_2(k_2\mu_2)$ of basis functions $\phi(k\mu)$ for the irreducible representation $\{k\}\Gamma$ of the space group, one can construct symmetric product

$[(\{\boldsymbol{k}\}\Gamma)^2]$ and antisymmetric product $\{(\{\boldsymbol{k}\}\Gamma)^2\}$ representations that have the wavevector $\boldsymbol{K} \doteq \boldsymbol{k}_1 + \boldsymbol{k}_2$. Calculation of the character for such representations can be done by replacing $\chi^{(\alpha)}\chi^{(\beta)}$ in (12.43) with

$$(\chi^{(\alpha)}(R)^2 \pm \chi^{(\alpha)}(R^2))/2 \ .$$

As a result of the calculation, we obtain

$$\left.\begin{matrix}\chi^{\boldsymbol{K}[(\{\boldsymbol{k}\}\Gamma)^2]}(Q) \\ \chi^{\boldsymbol{K}\{(\{\boldsymbol{k}\}\Gamma)^2\}}(Q)\end{matrix}\right\} = \frac{1}{2}\left(\chi^{\boldsymbol{K}(\{\boldsymbol{k}\}\Gamma \times \{\boldsymbol{k}\}\Gamma)}(Q) \pm \chi^{\boldsymbol{K}(\{\boldsymbol{k}\}\Gamma)}(Q^2)\right) . \tag{12.68}$$

The first term on the right-hand side is defined by (12.51), which may be calculated from (12.53) using either (12.54) or (12.58). The second term is

$$\chi^{\boldsymbol{K}(\{\boldsymbol{k}\}\Gamma)}(Q^2) \equiv \sum_{\boldsymbol{k} \in \{\boldsymbol{k}\}} \delta(\boldsymbol{k} + Q\boldsymbol{k}, \boldsymbol{K})\chi^{\boldsymbol{k}(\Gamma)}(Q^2) \ , \tag{12.69}$$

where $\chi^{\boldsymbol{k}(\Gamma)}$ stands for the character of the small representation Γ of $\mathcal{G}(\boldsymbol{k})$. The summation is automatically taken for those Q that satisfy $Q^2 \in \mathcal{G}(\boldsymbol{k})$ because of the δ factor. It goes without saying that one has to use (12.48) when calculating the character $\chi^{\boldsymbol{k}(\Gamma)}(Q^2)$.

For instance, consider the product representation $\{\mathrm{M}\}\mathrm{B}_3 \times \{\mathrm{M}\}\mathrm{B}_3$ in the space group T^1. A calculation similar to Example 2 gives

$$\{\mathrm{M}\}\mathrm{B}_3 \times \{\mathrm{M}\}\mathrm{B}_3 = \{\Gamma\}\mathrm{A} + \{\Gamma\}^1\mathrm{E} + \{\Gamma\}^2\mathrm{E} + 2\{\mathrm{M}\}\mathrm{B}_3 \ . \tag{12.70}$$

The representation $\{\mathrm{M}\}\mathrm{B}_3$ appears twice. If we calculate (12.69) as well, we find that one of the $\{\mathrm{M}\}\mathrm{B}_3$ is antisymmetric, the other being symmetric.

Exercise 12.6. In Example 2, one can choose the following basis functions for the B_2 and B_3 small representations:

$$\mathrm{M}(\mathrm{B}_2): \ \cos(\pi x/a)\sin(\pi y/a) \ ,$$

$$\mathrm{M}(\mathrm{B}_3): \ \sin(\pi x/a)\cos(\pi y/a) \ .$$

Basis functions at M_+ and M_- are obtained by operating $\{C_3^+|0\}$ and $\{C_3^-|0\}$ on them. Using these explicit expressions for the Bloch functions, construct the basis functions for the representations obtained on the right-hand side of (12.66).

13. Time Reversal and Nonunitary Groups

When the system under consideration is invariant with respect to time reversal, it is possible to regard it as having the symmetry of a nonunitary group which consists of time reversal θ in addition to the unitary symmetry operations considered so far. The purpose of this chapter is to discuss the representations (corepresentations) of such nonunitary groups and to examine the additional degeneracy brought in by including the operation θ. A well-known example of a nonunitary group is the magnetic space group. By considering this group, the symmetry of magnons (spin waves) and excitons in magnetic compounds and selection rules for their excitation can be treated in the same way as excitons in molecular crystals.

13.1 Time Reversal

Suppose the Hamiltonian H is independent of time t and the wavefunction $\Psi = \Psi(t)$ satisfies the Schrödinger equation

$$i\hbar \frac{\partial}{\partial t} \Psi = H\Psi \ . \tag{13.1}$$

Let us assume for the moment that H does not include spin operators so that Ψ is a function of only orbital coordinates. If we take the complex conjugate of (13.1), we have

$$-i\hbar \frac{\partial}{\partial t} \Psi^* = H^*\Psi^* \ .$$

In the absence of magnetic fields, the Hamiltonian H is a real operator containing no imaginary numbers, so that we have

$$-i\hbar \frac{\partial}{\partial t} \Psi^* = H\Psi^* \ .$$

Put $t' = -t$ here and we find

$$i\hbar \frac{\partial}{\partial t'} \Psi(-t')^* = H\Psi(-t')^* \ . \tag{13.2}$$

This implies that the function $\Psi(-t')^*$ satisfies the same equation (13.1) as $\Psi(t)$. The quantum-mechanical state described by $\Psi(-t)^*$ is obtained from $\Psi(t)$ by reversing the sense of motion or reversing the velocity vector. For example, the wavefunction

$$\Psi(t) = \mathrm{e}^{\mathrm{i}(\boldsymbol{k}\cdot\boldsymbol{r} - \omega t)}$$

represents the motion of a free particle with momentum $\hbar\boldsymbol{k}$, whereas the function

$$\Psi(-t)^* = \mathrm{e}^{\mathrm{i}(-\boldsymbol{k}\cdot\boldsymbol{r} - \omega t)}$$

describes that of the particle with the same magnitude of momentum but moving in the reverse direction. We thus call the state given by

$$\theta\Psi(t) = \Psi(-t)^* \tag{13.3}$$

the *time reversed* state of $\Psi(t)$. Perhaps what is meant by time reversal is much better expressed by calling it the velocity reversal. To put it another way, we can view the motion pictured by $\theta\Psi(t)$ by projecting in reversed motion the movie film that shot the action prescribed by $\Psi(t)$.

In a stationary state with definite energy E, we may put aside the phase factor $\exp(-\mathrm{i}Et/\hbar)$ in $\Psi(t)$, so that simple complex conjugation, which will be denoted by K hereafter, gives the time-reversed state:

$$\theta\Psi = K\Psi = \Psi^* \; . \tag{13.4}$$

Another simple example is to consider the time reversal of the state

$$\psi_{lm}(\boldsymbol{r}) = R(r)\, Y_{lm}(\vartheta, \varphi) \tag{13.5}$$

with definite value m of the angular momentum l_z about the z-axis. Since we have

$$K\psi_{lm}(\boldsymbol{r}) = R(r)^*\, Y_{lm}(\vartheta, \varphi)^* = (-1)^m R(r)^*\, Y_{l,-m}(\vartheta, \varphi) \; ,$$

the time-reversed state has $-m$ as the value of l_z. When ψ_{lm} represents a state rotating counterclockwise, $\theta\psi_{lm}$ corresponds to that of a clockwise rotation, which implies that for

$$l_z\psi_{lm} = m\psi_{lm} \tag{13.6}$$

we have

$$l_z\theta\psi_{lm} = -m\theta\psi_{lm} \; . \tag{13.7}$$

It then follows from (13.6 and 7) that

$$\theta l_z\psi_{lm} = m\theta\psi_{lm} = -l_z\theta\psi_{lm} \; ,$$

or rewriting the left-hand side

$$(\theta l_z\theta^{-1})\theta\psi_{lm} = -l_z\theta\psi_{lm} \; .$$

Let us call $\theta l_z \theta^{-1}$ the time reversal of l_z. We thus have

$$\theta \boldsymbol{l} \theta^{-1} = -\boldsymbol{l} \ . \tag{13.8}$$

A similar relation holds for the momentum operator $\boldsymbol{p}$:

$$\theta \boldsymbol{p} \theta^{-1} = -\boldsymbol{p} \ . \tag{13.9}$$

These relations correspond to the fact that both $\boldsymbol{l}$ and $\boldsymbol{p}$ are imaginary operators containing the factor i. Contrary to this, real operators like coordinate $\boldsymbol{r}$ and potential energy $V(\boldsymbol{r})$ remain invariant under time reversal.

It is quite natural to impose the relation (13.8) on the spin angular momentum $\boldsymbol{s}$ as well. Let us determine, accordingly, the operator θ acting on spin functions in such a way that

$$\theta \boldsymbol{s} \theta^{-1} = -\boldsymbol{s} \ , \tag{13.10}$$

or

$$\theta \boldsymbol{s} + \boldsymbol{s} \theta = 0 \tag{13.11}$$

is satisfied. It is easy to verify that the following choice fulfills the requirement:

$$\theta = -\mathrm{i}\hat{\sigma}_y K \ , \quad \hat{\sigma}_y = \begin{bmatrix} 0 & -\mathrm{i} \\ \mathrm{i} & 0 \end{bmatrix} . \tag{13.12}$$

Under the operation of θ, the product of orbital and spin functions (spin-orbital) is transformed as[1]

$$\theta(\psi\alpha) = \psi^*\beta \ , \quad \theta(\psi\beta) = -\psi^*\alpha \ . \tag{13.13}$$

Note that $-\mathrm{i}\hat{\sigma}_y$ is a real unitary operator.

This completes the definition of the time-reversal operator θ including spin. In contrast with unitary operators, this operator satisfies the relation

$$(\theta\Phi, \theta\Psi) = (K\Phi, K\Psi) = (\Phi, \Psi)^* = (\Psi, \Phi) \ . \tag{13.14}$$

Operators with this property are called antiunitary operators. The product $a = \theta u$ (or $u\theta$) of θ and a unitary operator u is antiunitary.

One is tempted to think that operating θ twice upon Ψ would lead to Ψ itself. The correct result, however, turns out to be

$$\theta^2 \Psi = (-1)^N \Psi \ , \tag{13.15}$$

depending upon the total number N of electrons in the system. Since $\theta^2\Psi$ and Ψ

---

[1] If we write the spin functions χ_m as $\chi_{1/2} = \alpha$ and $\chi_{-1/2} = \beta$, we have $\theta\chi_m = (-1)^{1/2-m}\chi_{-m}$, or $\theta\alpha = \beta$, $\theta\beta = -\alpha$.

differ only by a constant phase factor, they represent the same physical state. Equation (13.15) follows immediately from the definition of θ for a many-electron system:

$$\theta = \prod_{j=1}^{N} (-\mathrm{i}\hat{\sigma}_y^{(j)})K \ , \tag{13.16}$$

and $(-\mathrm{i}\hat{\sigma}_y)^2 = -\hat{1}$.

When N is even, (13.15) does not tell us whether $\theta\Psi$ and Ψ are linearly independent. When N is odd, they are orthogonal, because

$$(\Psi, \theta\Psi) = (\theta^2\Psi, \theta\Psi) = -(\Psi, \theta\Psi) \tag{13.17}$$

according to (13.14 and 15). If Ψ is an eigenfunction of the Hamiltonian H invariant to time reversal

$$\theta H \theta^{-1} = H \ , \tag{13.18}$$

the wavefunction $\theta\Psi$, which is independent of Ψ, is also an eigenfunction of H with the same energy. This means that every energy level of a system with an odd number of electrons is n-fold degenerate, where n is an even number ($\geqq 2$). This degeneracy is called *Kramers degeneracy*. Only external magnetic fields can lift this degeneracy, because the Hamiltonian then loses the time-reversal symmetry. In fact, for the Zeeman term $H_Z = \mu_B(\boldsymbol{L} + 2\boldsymbol{S})\cdot\boldsymbol{H}$, we have $\theta H_Z \theta^{-1} = -H_Z$.

The time-reversal operator θ commutes with the operators of rotation, inversion and translation. We have seen that rotation through an angle ω about an axis in the direction of a unit vector $\boldsymbol{\lambda}$ is given by

$$R = \exp(-\mathrm{i}\omega\boldsymbol{\lambda}\cdot\boldsymbol{j}) \ , \quad \boldsymbol{j} = \boldsymbol{l} + \boldsymbol{s} \ .$$

According to (13.8 and 10), we have $\theta\boldsymbol{j}\theta^{-1} = -\boldsymbol{j}$, so that

$$\theta R \theta^{-1} = R \ .$$

Commutability of inversion I with K (or $-\mathrm{i}\hat{\sigma}_y K$) may be confirmed directly by operating these upon a wavefunction. (Note that $I\alpha = \alpha$, $I\beta = \beta$ for spin functions.) The translation T that carries $\boldsymbol{r}$ to $\boldsymbol{r} + \boldsymbol{b}$ can be expressed as

$$T = \exp(-\mathrm{i}\boldsymbol{p}\cdot\boldsymbol{b}/\hbar)$$

It is easy to show by using (13.9) that

$$\theta T \theta^{-1} = T$$

holds also for this operator.

13.2 Nonunitary Groups and Corepresentations

When the Hamiltonian H is not only invariant with respect to the unitary operators u_j of the group G but also with respect to the time reversal θ, it is convenient to consider the set of antiunitary operators $\{\theta u_j\}$ together with the set of operators $\{u_j\}$ belonging to G and regard them as forming a group $\bar{G}$, the *nonunitary group*. The law of composition for this group reads as follows: When we have $u_i u_j = u_k$ for the elements of G, we put[2]

$$(\theta u_i)u_j = u_i(\theta u_j) = \theta u_k \ ,$$
$$(\theta u_i)(\theta u_j) = (-1)^N u_k \ . \tag{13.19}$$

Note that the group G is an invariant subgroup with index 2 of the nonunitary group $\bar{G}$:

$$\bar{G} = G + \theta G \ . \tag{13.20}$$

The Hamiltonian H is invariant under the operations of $\bar{G}$.

Let us generalize the problem a little more and consider a nonunitary group $\bar{G}$ consisting of unitary as well as antiunitary operators. The product of two unitary operators is unitary. So is the product of two antiunitary operators, whereas the product of unitary and antiunitary operators is antiunitary. This implies that $\bar{G}$ contains an equal number of unitary and antiunitary operators and the set of unitary operators form an invariant subgroup G of index 2. The set of antiunitary operators forms a coset of G in $\bar{G}$. If we pick out one representative from the set and denote it as a_0, we have the coset decomposition

$$\bar{G} = G + a_0 G \ . \tag{13.21}$$

We may put here $a_0 = \theta v_0$. The operator v_0 is unitary, but does not necessarily belong to G. (We shall later see such an example in the magnetic space group.) When v_0 belongs to G, (13.21) reduces to (13.20).

In this section, we consider representations of the nonunitary group $\bar{G}$ given by (13.21). If we denote the unitary operators by u and antiunitary operators by a, the following equations must hold for the representation matrices $\hat{D}$:

$$u_1 u_2 \phi_k = \sum_j \phi_j [\hat{D}(u_1)\hat{D}(u_2)]_{jk} \ ,$$

$$u a \phi_k = \sum_j \phi_j [\hat{D}(u)\hat{D}(a)]_{jk} \ , \tag{13.22}$$

$$a u \phi_k = \sum_j \phi_j [\hat{D}(a)\hat{D}(u)^*]_{jk} \ ,$$

$$a_1 a_2 \phi_k = \sum_j \phi_j [\hat{D}(a_1)\hat{D}(a_2)^*]_{jk} \ .$$

---

[2] It may be difficult to accept the set $\{u_j, \theta u_j\}$ as forming a group in the usual sense because of the presence of the factor $(-1)^N$ on the right-hand side of the second equation of (13.19). We may consider that this is the factor originating from the property of the wavefunctions and we are dealing here with a ray representation (Sect. 5.4) with such a factor system.

Note that the representation matrices of the products au and $a_1 a_2$ are not quite equal to the corresponding products of the matrices. Since complex conjugates of the matrices are involved, representations of this kind are called *corepresentations* of a nonunitary group. Two corepresentations D and D' are equivalent when there exists a matrix $\hat{T}$ that satisfies

$$\hat{D}'(u) = \hat{T}^{-1}\hat{D}(u)\hat{T} \ , \quad \hat{D}'(a) = \hat{T}^{-1}\hat{D}(a)\hat{T}^* \ . \tag{13.23}$$

Otherwise they are inequivalent.

If we are given an irreducible representation Δ with dimension d of the unitary subgroup G, we can construct an irreducible corepresentation D of $\bar{G}$ in the following way. Let us denote by ψ_j $(j = 1, 2, \ldots, d)$ the basis for the representation Δ. Then, we have

$$u\psi_k = \sum_{j=1}^{d} \psi_j \Delta_{jk}(u) \ . \tag{13.24}$$

If we put

$$\psi'_k = a_0 \psi_k \ , \tag{13.25}$$

we find that they satisfy

$$u\psi'_k = \sum_{j=1}^{d} \psi'_j \bar{\bar{\Delta}}_{jk}(u) \ , \quad \hat{\bar{\Delta}}(u) = \hat{\Delta}(a_0^{-1} u a_0)^* \ . \tag{13.26}$$

The representation $\bar{\Delta}$ is also an irreducible representation of G. It can be equivalent or inequivalent to Δ.

Let us first consider the case where Δ and $\bar{\Delta}$ are equivalent. In this case, we can find a unitary matrix $\hat{U}$ such that

$$\bar{\Delta}(u) = \hat{U}^{-1}\hat{\Delta}(u)\hat{U} \ , \quad u \in G \ . \tag{13.27}$$

If we change the basis of $\bar{\Delta}$ from ψ' to φ' by means of the transformation

$$a_0 \psi_k = \psi'_k = \sum_{j=1}^{d} \varphi'_j U_{jk} \ , \tag{13.28}$$

we obtain

$$u\varphi'_k = \sum_{j=1}^{d} \varphi'_j \Delta_{jk}(u) \ . \tag{13.29}$$

Operating a_0 on both sides of (13.28), we have

$$a_0^2 \psi_k = \sum_{j=1}^{d} a_0 \varphi'_j U_{jk}^* \ .$$

Since a_0^2 is unitary, we can solve these equations for $a_0\varphi'_k$ in the form

$$a_0\varphi'_k = \sum_{j=1}^{d} \psi_j[\hat{\Delta}(a_0^2)(\hat{U}^{-1})^*]_{jk} . \tag{13.30a}$$

Let us consider further the consequences of (13.27), which reads

$$\hat{\bar{\Delta}}(u) = \hat{\Delta}(a_0^{-1}ua_0)^* = \hat{U}^{-1}\hat{\Delta}(u)\hat{U} . \tag{13.31}$$

Replacing u by $a_0ua_0^{-1}$ here, we obtain

$$\hat{\Delta}(a_0^{-2}ua_0^2) = (\hat{U}^{-1})^*\hat{\Delta}(a_0^{-1}ua_0)^*\hat{U}^* .$$

If we note that the left-hand side is the product of three unitary matrices $\hat{\Delta}(a_0^{-2})\hat{\Delta}(u)\hat{\Delta}(a_0^2)$ and that the right-hand side can be expressed in terms of $\hat{U}$ and $\hat{\Delta}(u)$ by means of (13.31), it is not difficult to derive the relation

$$\hat{U}\hat{U}^*\hat{\Delta}(a_0^2)^{-1}\hat{\Delta}(u) = \hat{\Delta}(u)\hat{U}\hat{U}^*\hat{\Delta}(a_0^2)^{-1} .$$

Remembering that the representation Δ is irreducible, we may conclude that

$$\hat{U}\hat{U}^* = \lambda\hat{\Delta}(a_0^2) \tag{13.32}$$

because of Schur's lemma. Both sides of this equation are unitary matrices, so that $|\lambda| = 1$. Equation (13.32) enables us to put (13.30a) into the form

$$a_0\varphi'_k = \lambda^* \sum_{j=1}^{d} \psi_j U_{jk} . \tag{13.30b}$$

If we put $u = a_0^2$ in (13.31), we obtain

$$\hat{\Delta}(a_0^2)^* = \hat{U}^{-1}\hat{\Delta}(a_0^2)\hat{U} . \tag{13.33}$$

This together with (13.32) leads to $\lambda = \lambda^*$, so that

$$\lambda = \pm 1 .$$

We must further distinguish two cases.

(a) $\lambda = 1$. Taking the sum and difference of (13.28) and (13.30b) we have

$$a_0(\psi_k + \varphi'_k) = \sum_{j=1}^{d} (\psi_j + \varphi'_j)U_{jk} ,$$

$$a_0(\psi_k - \varphi'_k) = \sum_{j=1}^{d} (\psi_j - \varphi'_j)(-U_{jk}) .$$

If we put $\Phi_k = (\psi_k + \varphi'_k)/\sqrt{2}$, these together with (13.24 and 29) give us

$$u\Phi_k = \sum_{j=1}^{d} \Phi_j\Delta_{jk}(u) , \quad a_0\Phi_k = \sum_{j=1}^{d} \Phi_j U_{jk} . \tag{13.34}$$

The same set of equations hold[3] for Φ'_k defined by $\Phi'_k = \mathrm{i}(\psi_k - \varphi'_k)/\sqrt{2}$. This means that no new linearly independent bases of corepresentation are introduced by the inclusion of a_0. According to (13.34), the irreducible corepresentation D of $\bar{G}$ is given by

$$\hat{D}(u) = \hat{\Delta}(u) , \quad \hat{D}(a_0) = \hat{U} . \tag{13.35}$$

(b) $\lambda = -1$. From (13.28 and 30b), we obtain

$$(\psi_k, \varphi'_k) = (a_0 \varphi'_k, a_0 \psi_k) = \lambda(\psi_k, \varphi'_k) ,$$

where we have made use of the fact that ψ_k and φ'_k transform as the same partner in the basis for the irreducible representation Δ. Since $\lambda = -1$,

$$(\psi_k, \varphi'_k) = 0 ,$$

so that they are orthogonal to each other and linearly independent. This means that the irreducible corepresentation D of $\bar{G}$ is given by

$$\hat{D}(u) = \begin{bmatrix} \hat{\Delta}(u) & 0 \\ 0 & \hat{\Delta}(u) \end{bmatrix} , \quad \hat{D}(a_0) = \begin{bmatrix} 0 & -\hat{U} \\ \hat{U} & 0 \end{bmatrix} . \tag{13.36}$$

(c) When the representations Δ and $\bar{\Delta}$ are inequivalent, ψ_k and φ'_k are linearly independent and the irreducible corepresentation D of $\bar{G}$ is given by

$$\hat{D}(u) = \begin{bmatrix} \hat{\Delta}(u) & 0 \\ 0 & \hat{\bar{\Delta}}(u) \end{bmatrix} , \quad \hat{D}(a_0) = \begin{bmatrix} 0 & \hat{\Delta}(a_0^2) \\ \hat{1} & 0 \end{bmatrix} . \tag{13.37}$$

The following equation can be proved for any of the three cases (a), (b) and (c) considered above:

$$\hat{\Delta}(a_0 u a_0 u) = \hat{\Delta}(a_0^2)\hat{\bar{\Delta}}(u)^* \hat{\Delta}(u) . \tag{13.38}$$

In particular, in the cases (a) and (b), the right-hand side can be written as

$$\lambda^{-1} \hat{U}\hat{\Delta}(u)^* \hat{U}^* \hat{\Delta}(u) ,$$

using (13.27 and 32). By taking the trace of both sides and remembering the great orthogonality relation (4.46) for irreducible representations, we obtain

$$\lambda \sum_u \chi((a_0 u)^2) = \lambda \sum_j \sum_u \Delta_{jj}(a_0 u a_0 u) = g \tag{13.39}$$

for the character χ of the irreducible representation Δ, where g is the order of the

---

[3] If Φ_k vanishes identically, we have $a_0 \psi_k = -\sum \psi_j U_{jk}$. If Φ'_k vanishes, we have $a_0 \psi_k = \sum \psi_j U_{jk}$. In any case the conclusion is the same.

group G. In case (c) the left-hand side of (13.39) vanishes. Altogether, we have the *Wigner criterion* for the three cases:

$$\sum_a \chi(a^2) = \sum_u \chi((a_0 u)^2) = \begin{cases} g, & \text{case (a)} , \\ -g, & \text{case (b)} , \\ 0, & \text{case (c)} . \end{cases} \tag{13.40}$$

Suppose G is the unitary group which leaves the Hamiltonian H invariant and the basis functions ψ_k ($k = 1, 2, \ldots, d$) for its irreducible representation Δ are the eigenfunctions of H with the energy eigenvalue E. In case (a), there is no additional degeneracy due to the presence of the antiunitary operator a_0, because we may assume without loss of generality that the basis ψ_k is chosen from the start so as to satisfy (13.34). In case (b), the presence of the operator a_0 gives rise to additional degeneracy, because we obtain φ_k''s which are linearly independent of ψ_k though belonging to the same irreducible representation of G as the latter. In case (c), the basis ψ_k' which belongs to the representation $\bar{\Delta}$ inequivalent to Δ is obtained by applying a_0 and the degeneracy is doubled here as well.

Let us put $a_0 = \theta$ in the discussion given above. This is the case of time reversal leaving H invariant. Then the representation

$$\hat{\bar{\Delta}}(u) = \hat{\Delta}(\theta^{-1}u\theta)^* = \hat{\Delta}(u)^* \tag{13.41}$$

is the complex conjugate of the representation Δ. Since

$$\hat{\Delta}(a_0^2) = \hat{\Delta}(\theta^2) = (-1)^N \hat{1} , \tag{13.42}$$

we have

$$\hat{U}\hat{U}^* = \lambda \hat{1} = \begin{cases} \hat{1}, & \text{(a), case (a) of Sect. 4.12,} \\ -\hat{1}, & \text{(b), case (b) of Sect. 4.12,} \end{cases} \tag{13.43}$$

when spin is neglected or the number of electrons N is even. We have already discussed implications of (13.43) in Sect. 4.12. As a matter of fact, (a), (b) and (c) of (4.88) correspond to (a), (b) and (c) of the present section, respectively. In case (a), the representation matrices can be chosen real. In case (b), we have pseudoreal representations of even dimensions. Examples of these cases have been given in Sect. 4.12.

When N is odd, we have

$$\hat{U}\hat{U}^* = -\lambda \hat{1} = \begin{cases} -\hat{1}, & \text{(a), case (b) of Sect. 4.12,} \\ \hat{1}, & \text{(b), case (a) of Sect. 4.12,} \end{cases} \tag{13.44}$$

so that it is impossible to choose all the representation matrices real in case (a).

However, this is possible in case (b), thereby bringing $\hat{D}(\theta)$ of (13.36) into the form

$$\hat{D}(\theta) = \begin{bmatrix} 0 & -\hat{1} \\ \hat{1} & 0 \end{bmatrix} .$$

Examples of this case (N odd) are listed here for the rotation group and some point groups:

(a) $D^{(J)}$ with J half of an odd integer; double-valued representations of D_{2n} and O; the double-valued representation $E_{1/2}$ of D_{2n+1} and T.
(b) The double-valued representation $B_{3/2}$ of C_3.
(c) The double-valued representations of C_n ($n \geqq 2$), D_{2n+1}, and T, which appear as pairs.

13.3 Criteria for Space Groups and Examples

Let us write down the Wigner criterion (13.40) for extra degeneracy in the case of a space group G. The character of the irreducible representation Δ of G will be denoted as χ. Noting that $\chi(a^2)$ on the right-hand side of (13.40) is nonvanishing only when $a^2 \in \mathscr{G}(\boldsymbol{k}_j)$, $\boldsymbol{k}_j \in \{\boldsymbol{k}\} = (\boldsymbol{k}, \boldsymbol{k}_2, \boldsymbol{k}_3, \ldots, \boldsymbol{k}_s)$, we obtain

$$\sum_a \chi(a^2) = s \sum_{a^2 \in \mathscr{G}(\boldsymbol{k})} \chi^k(a^2) , \tag{13.45}$$

where s is the number of wavevectors constituting the star $\{\boldsymbol{k}\}$ and χ^k is the character of the small representation of the group $\mathscr{G}(\boldsymbol{k})$ of wavevector $\boldsymbol{k}$. When we write down the expression of the antiunitary operator a explicitly as

$$a = a_0 u = \theta v_0 u ,$$
$$v_0 = \{\varrho | \boldsymbol{\sigma}\} ,$$
$$u = \{\beta | \boldsymbol{b}\}\{\varepsilon | \boldsymbol{t}_n\} ,$$

we find that a^2 is given by

$$a^2 = (-1)^N (v_0\{\beta|\boldsymbol{b}\})^2 \{\varepsilon | \beta^{-1}\varrho^{-1}\boldsymbol{t}_n + \boldsymbol{t}_n\} ,$$

remembering θ commutes with $v_0 u$ and θ^2 may be replaced by $(-1)^N$. This then leads to

$$\chi^k(a^2) = (-1)^N \exp[\mathrm{i}(\varrho\beta\boldsymbol{k} + \boldsymbol{k})\cdot\boldsymbol{t}_n]\chi^k((v_0\{\beta|\boldsymbol{b}\})^2) .$$

Substituting this expression into the right-hand side of (13.45) and carrying out the summation over translations $\{\varepsilon|\boldsymbol{t}_n\}$, we can put the Wigner criterion (13.40) in the form

$$\sum_{\{\beta|\boldsymbol{b}\}} \chi^k((v_0\{\beta|\boldsymbol{b}\})^2) = \begin{cases} (-1)^N[\boldsymbol{k}], & \text{(a)} , \\ -(-1)^N[\boldsymbol{k}], & \text{(b)} , \\ 0, & \text{(c)} . \end{cases} \tag{13.46}$$

On the left-hand side, the summation is to be taken over the representative elements $\{\beta|\boldsymbol{b}\}$ of $G/\mathscr{T}$ satisfying

$$\varrho\beta\boldsymbol{k} \doteq -\boldsymbol{k} \ . \tag{13.47}$$

Because of this restriction, $(v_0\{\beta|\boldsymbol{b}\})^2$ belongs to $\mathscr{G}(\boldsymbol{k})$. The notation $[\boldsymbol{k}]$ stands for the order of the factor group $\mathscr{G}(\boldsymbol{k})/\mathscr{T}$.

The criterion (13.46) is valid in general for any non-unitary space group (such as magnetic space groups). If we put $v_0 = \{\varepsilon|0\}$, namely $a_0 = \theta$, we obtain the *Herring criterion* for space groups

$$\sum_{\{\beta|\boldsymbol{b}\}} \chi^{\boldsymbol{k}}(\{\beta|\boldsymbol{b}\}^2) = \begin{cases} (-1)^N[\boldsymbol{k}], & \text{(a)} \ , \\ -(-1)^N[\boldsymbol{k}], & \text{(b)} \ , \\ 0, & \text{(c)} \ . \end{cases} \tag{13.48}$$

The condition (13.47) turns into

$$\beta\boldsymbol{k} \doteq -\boldsymbol{k} \ . \tag{13.49}$$

In case (a), Δ and Δ^* are equivalent and there is no additional degeneracy. In case (b), Δ and Δ^* are equivalent, whereas they are inequivalent in case (c). In either case degeneracy is doubled. Corresponding to (13.43,44), we also have

$$\hat{U}\hat{U}^* = \begin{cases} (-1)^N\hat{1}, & \text{(a)} \ , \\ -(-1)^N\hat{1}, & \text{(b)} \ . \end{cases} \tag{13.50}$$

When the right-hand side of this equation is equal to $\hat{1}$, the representation matrices of the space group can be chosen real. However, when it is equal to $-\hat{1}$, this is impossible.

Let us be more specific about the case where spin is neglected. The basis functions of the irreducible representation Δ of the space group G (which are eigenfunctions of H) will be denoted as $\psi_{\boldsymbol{k}_j\mu}$, see (11.48).

Case (a): Since Δ and Δ^* are equivalent, the star of Δ^*, namely, $\{\boldsymbol{k}\}^* \equiv \{-\boldsymbol{k}\}$ and the star $\{\boldsymbol{k}\}$ are one and the same, so that $-\boldsymbol{k}$ belongs to $\{\boldsymbol{k}\}$. Let us first consider the case where $-\boldsymbol{k} \not\doteq \boldsymbol{k}$ (i.e., $-\boldsymbol{k}$ and $\boldsymbol{k}$ cannot be connected by a reciprocal lattice vector). Then there will be an element P belonging to G that will send the vector $\boldsymbol{k}$ into $-\boldsymbol{k}$, and we may put

$$\psi_{-\boldsymbol{k}\mu} = P\psi_{-\boldsymbol{k}\mu} \ , \qquad -\boldsymbol{k} \doteq P\boldsymbol{k} \ . \tag{13.51}$$

The functions $\psi_{-\boldsymbol{k}\mu}$ form the basis for an irreducible representation of $\mathscr{G}(-\boldsymbol{k})$ with its representation matrices $\hat{D}^{-\boldsymbol{k}}(R)$ related to those of $D^{\boldsymbol{k}}$ by

$$\hat{D}^{-\boldsymbol{k}}(R) = \hat{D}^{\boldsymbol{k}}(P^{-1}RP) \ , \qquad R \in \mathscr{G}(\boldsymbol{k}) \equiv \mathscr{G}(-\boldsymbol{k}) \ , \tag{13.52}$$

where $\hat{D}^k(R)$ are the representation matrices of $\mathscr{G}(\boldsymbol{k})$ based on $\psi_{k\mu}$. The representation D^{-k} is equivalent to D^{k*}, so that

$$\hat{D}^{-k}(R) = \hat{U}\hat{D}^k(R)^*\hat{U}^{-1} . \tag{13.53}$$

Since there is no additional degeneracy due to θ, we can express $\psi_{-k\mu}$ in terms of $\psi^*_{k\mu}$ as

$$\psi_{-k\mu} = \sum_{\mu'=1}^{d} \psi^*_{k\mu'}(\hat{U}^{-1})_{\mu'\mu} . \tag{13.54}$$

Especially, when $-\boldsymbol{k} \doteq \boldsymbol{k}$, D^k and D^{k*} are equivalent. Note that in this case we can further choose all the small representation matrices $\hat{D}^k(R)$ (as well as the basis functions) to be real.

Case (b): Since Δ and Δ^* are equivalent also here, (13.53) still holds when $-\boldsymbol{k} \not\doteq \boldsymbol{k}$. However, the functions $\varphi_{-k\mu}$ defined by

$$\varphi_{-k\mu} = \sum_{\mu'} \psi^*_{k\mu'}(\hat{U}^{-1})_{\mu'\mu} \tag{13.55}$$

are orthogonal to the functions $\psi_{-k\mu}$:

$$(\varphi_{-k\mu}, \psi_{-k\mu}) = 0$$

and linearly independent. When $-\boldsymbol{k} \doteq \boldsymbol{k}$, D^k and D^{k*} are equivalent. However, it is impossible here to make all the matrices real. Note also that the dimension of the small representation D^k is bound to be an even number.

Case (c): We have two possibilities here: either the stars $\{\boldsymbol{k}\}^* \equiv \{-\boldsymbol{k}\}$ and $\{\boldsymbol{k}\}$ are different, or $\{\boldsymbol{k}\}^* \equiv \{\boldsymbol{k}\}$. In the former event, there will be no element $\{\beta|\boldsymbol{b}\}$ that satisfies (13.49) and $\psi_{k\mu}$ and $\psi^*_{k\mu}$ form the bases of two inequivalent representations of G. In the latter event, the vector $-\boldsymbol{k}$ belongs to the star $\{\boldsymbol{k}\}$. When $-\boldsymbol{k} \not\doteq \boldsymbol{k}$, $\psi_{k\mu}$ and $\psi^*_{-k\mu}$ are the bases of two paired inequivalent representations of $\mathscr{G}(\boldsymbol{k})$. (If the character of the representation of $\mathscr{G}(\boldsymbol{k})$ based on $\psi_{k\mu}$ is denoted by $\chi^k(R)$, the character of the representation based on $\psi^*_{-k\mu}$ is given by $\chi^k(P^{-1}RP)^*$ according to (13.51).) When $-\boldsymbol{k} \doteq \boldsymbol{k}$, the representations D^k and D^{k*} are inequivalent.

Let us give below some simple examples.

Example 1. Suppose inversion $\{I|\tau\}$ (or $\{I|0\}$) is an element of G and consider a general point $\boldsymbol{k}$ in the Brillouin zone. For the star $\{\boldsymbol{k}\} = \{\boldsymbol{k}, -\boldsymbol{k}\}$, we set $\{\beta|\boldsymbol{b}\} = \{I|\tau\}$ in (13.48). Then we have

$$\sum \chi^k(\{\beta|\boldsymbol{b}\}^2) = \chi^k(\{I|\tau\}^2) = 1 ,$$

because the group $\mathscr{G}(\boldsymbol{k})/\mathscr{T}$ can have only the identity representation. This is case (a) and we may put $\psi_{-k} = \psi^*_k$. Because of the inversion symmetry, we have $E(\boldsymbol{k}) = E(-\boldsymbol{k})$. There is no extra degeneracy due to time reversal (Fig. 13.1). It is

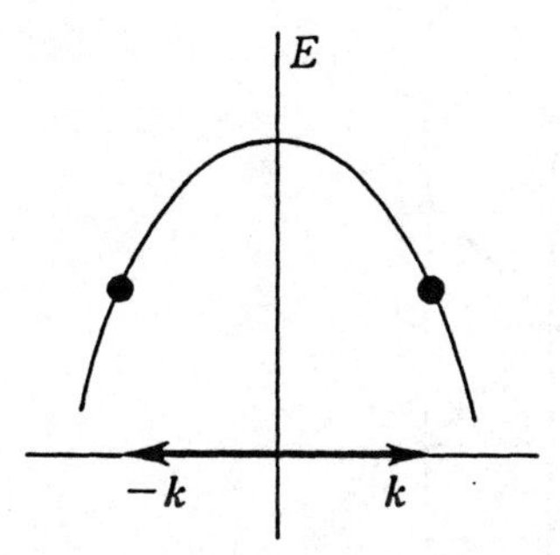

Fig. 13.1. Degeneracy due to inversion symmetry (without spin)

easy to confirm that both the representation matrices of G and its bases can be chosen real.

Example 2. When inversion is not included in G, no operation in G will carry a general point $\boldsymbol{k}$ to $-\boldsymbol{k}$, so that

$$\sum \chi^{k}(\{\beta|\boldsymbol{b}\}^2) = 0 \ .$$

This is case (c). The states $\psi_{\boldsymbol{k}}$ and $\psi_{\boldsymbol{k}}^*$ are degenerate because of time-reversal symmetry. In other words, we always have $E(-\boldsymbol{k}) = E(\boldsymbol{k})$ for energy bands, even when the crystal lacks inversion symmetry (Fig. 13.2).

Example 3. As an example of case (b), we consider the two-dimensional space group Pa and its representation at a point $\boldsymbol{k}_1 = (\pi/a, k_y)$ on the Brillouin zone boundary. This space group has the glide $g = \{m_y|\boldsymbol{\tau}\}$ with $\boldsymbol{\tau} = (a/2, 0)$ among its symmetry operations. The star of $\boldsymbol{k}_1$ consists of $\boldsymbol{k}_1$ and $\boldsymbol{k}_2 = m_y\boldsymbol{k}_1$ (Fig. 13.3). Since $\boldsymbol{k}_2 \doteq -\boldsymbol{k}_1$, we obtain

$$\sum \chi^{k_1}(\{\beta|\boldsymbol{b}\}^2) = \chi^{k_1}(g^2) = -1$$

by noting $g^2 = \{\varepsilon|2\boldsymbol{\tau}\}$, and $k_x = \pi/a$. If we denote the Bloch function at $\boldsymbol{k}_1$ as $\psi_{\boldsymbol{k}_1}$, its complex conjugate $\psi_{\boldsymbol{k}_1}^*$ will be orthogonal to $\psi_{\boldsymbol{k}_2} = g\psi_{\boldsymbol{k}_1}$:

$$(\psi_{\boldsymbol{k}_1}^*, \psi_{\boldsymbol{k}_2}) = (g\psi_{\boldsymbol{k}_1}^*, g\psi_{\boldsymbol{k}_2}) = -(\psi_{\boldsymbol{k}_2}^*, \psi_{\boldsymbol{k}_1}) = -(\psi_{\boldsymbol{k}_1}^*, \psi_{\boldsymbol{k}_2}) \ .$$

This means that in addition to the basis $\{\psi_{\boldsymbol{k}_1}, \psi_{\boldsymbol{k}_2}\}$ for an irreducible representation of Pa, we have an independent set of functions $\{\psi_{\boldsymbol{k}_2}^*, -\psi_{\boldsymbol{k}_1}^*\}$ which is the

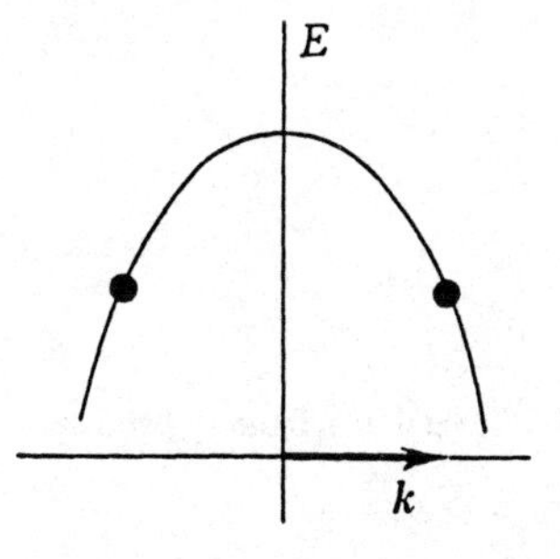

Fig. 13.2. Degeneracy due to time reversal symmetry (without spin)

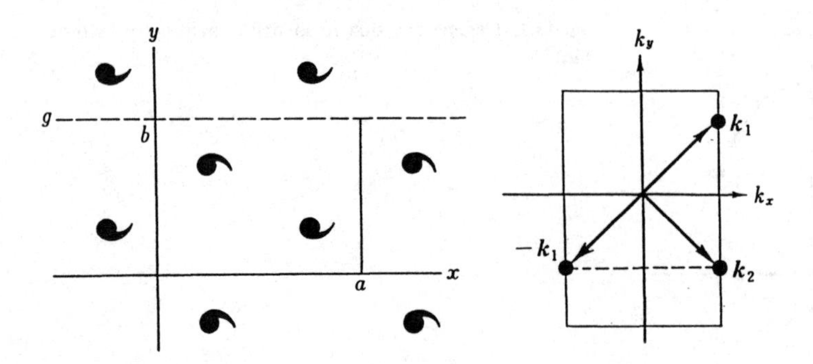

Fig. 13.3. Pattern with the two-dimensional symmetry Pa and the Brillouin zone for the space group Pa

basis for the same irreducible representation and has the same energy. At the points $\boldsymbol{k}_1$ and $\boldsymbol{k}_2$, we have twofold degeneracy due to the time-reversal symmetry (sticking together of two bands).

We will briefly sketch the cases with spin. In these cases it is necessary to consider the double space group G' as the symmetry group.

Example 4. Take a general point $\boldsymbol{k}$ in the Brillouin zone. When the inversion $\{I|\boldsymbol{\tau}\}$ belongs to G, we choose $\{I|\boldsymbol{\tau}\}$ and $\{\bar{I}|\boldsymbol{\tau}\}$ as $\{\beta|\boldsymbol{b}\}$ in (13.48). Since the square of either element is the identity, we have

$$\sum \chi^{k}(\{\beta|\boldsymbol{b}\}^2) = 2 \ ,$$

which corresponds to case (b). Needless to say, $\boldsymbol{k}$ and $-\boldsymbol{k}$ belong to the same star because of the presence of $\{I|\boldsymbol{\tau}\}$. If we write $\phi_{\boldsymbol{k}}$ for the Bloch function at $\boldsymbol{k}$, the one at $-\boldsymbol{k}$ is given by $\phi_{-\boldsymbol{k}} = \{I|\boldsymbol{\tau}\}\phi_{\boldsymbol{k}}$. The time reversal of the latter, $\theta\phi_{-\boldsymbol{k}}$, is orthogonal to $\phi_{\boldsymbol{k}}$. The two sets of functions $\{\phi_{\boldsymbol{k}}, \phi_{-\boldsymbol{k}}\}$ and $\{\theta\phi_{-\boldsymbol{k}}, \theta\phi_{\boldsymbol{k}}\}$ form a basis for the same irreducible representation of G', and are mutually orthogonal and independent. We thus have twofold degeneracy at each of $\boldsymbol{k}$ and $-\boldsymbol{k}$ due to time-reversal symmetry (Fig. 13.4). When the spin-orbit interaction is neglected, we may put $\phi_{\boldsymbol{k}} = \psi_{\boldsymbol{k}}\alpha$, so that $\theta\phi_{-\boldsymbol{k}} = \psi^*_{-\boldsymbol{k}}\beta$. The twofold degeneracy

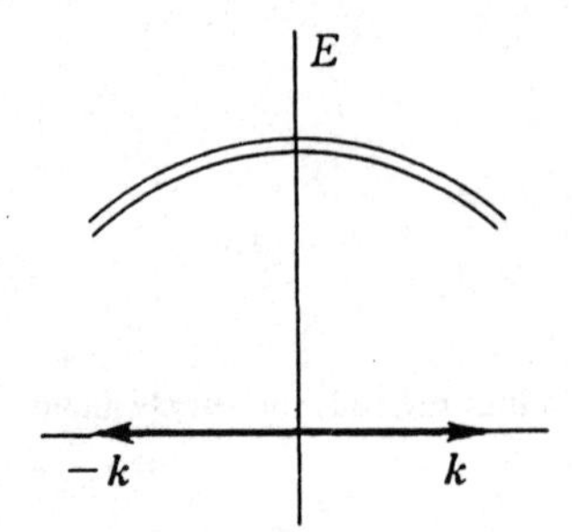

Fig. 13.4. Degeneracy due to inversion symmetry (with spin)

thus corresponds to that of the spin states α and β. When the spin-orbit interaction is taken into consideration, the eigenfunction ϕ_k turns into a linear combination of $\psi_k\alpha$ and $\psi^*_{-k}\beta$. However, the degeneracy of ϕ_k and $\theta\phi_{-k}$ will not be lifted. As remarked before, if we put $\phi'_1 = (\phi_k + i\phi_{-k})/\sqrt{2}$ and $\phi'_2 = (i\phi_k + \phi_{-k})/\sqrt{2}$, and use the bases $\{\phi'_1, \phi'_2\}$ and $\{\theta\phi'_1, \theta\phi'_2\}$, the representation matrices will be made real and take the same form for these two sets.

Example 5. When G does not contain inversion, we have

$$\sum \chi^k(\{\beta|\boldsymbol{b}\}^2) = 0$$

for a general point $\boldsymbol{k}$. This is case (c) and there is no degeneracy at $\boldsymbol{k}$. The state $\theta\phi_k$ has, however, the same energy as ϕ_k (Fig. 13.5).

Example 6. As an example of case (a), we consider again the irreducible representation of Pa, but this time at the point $\boldsymbol{k} = \{0, k_y\}$ on the k_y axis. The vector $m_y\boldsymbol{k} = -\boldsymbol{k}$ belongs to the same star as $\boldsymbol{k}$. Since $g^2 = \bar{g}^2 = \{\bar{\varepsilon}|0\}$, we find

$$\sum \chi^k(\{\beta|\boldsymbol{b}\}^2) = -2 .$$

The time reversal $\theta\phi_k$ of ϕ_k does not differ from $\phi_{-k} = g\phi_k$ apart from a phase factor. This implies that we simply obtain $\{\phi_{-k}, -\phi_k\}$ even if we operate θ upon the set $\{\phi_k, \phi_{-k}\}$ and there is no extra degeneracy due to time reversal. It is also impossible in this case to have real representation of G with the basis $\{\phi_k, \phi_{-k}\}$.

Example 7. As a more complicated example, we choose a double-valued representation at point X on the Brillouin zone boundary of the rutile structure. The character table for $\mathscr{G}'_X$ has been given in Table 11.6. Since $-\boldsymbol{k}$ is equivalent to $\boldsymbol{k}$ at X, we take all the representative elements of $\mathscr{G}_X/\mathscr{T}$ as $\{\beta|\boldsymbol{b}\}$. Application of Herring's criterion (13.48) to the representation X_3 then leads to

$$\sum \chi^{X_3}(\{\beta|\boldsymbol{b}\}^2) = 0$$

from Table 13.1. This is case (c), so X_3 and $X_3^* = X_4$ are degenerate. Altogether, we have fourfold degeneracy at X and this cannot be lifted even with the spin-orbit interaction.

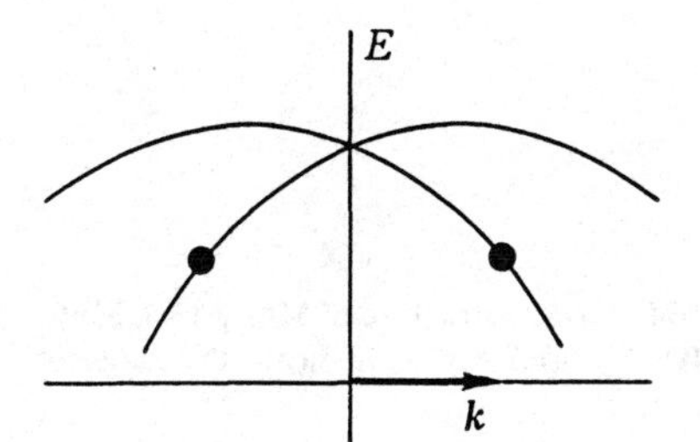

Fig. 13.5. Degeneracy due to time reversal symmetry (with spin)

Table 13.1. Herring test for X_3

$u \in \mathscr{G}_X/T$	u^2	$\chi^{X_3}(u^2)$
$\{\varepsilon\|0\}$	$\{\varepsilon\|0\}$	2
$\{C_2\|0\}$	$\{\bar{\varepsilon}\|0\}$	−2
$\{C_{2x}\|\boldsymbol{\tau}\}$	$\{\bar{\varepsilon}\|\boldsymbol{t}_1\}$	−2
$\{C_{2y}\|\boldsymbol{\tau}\}$	$\{\bar{\varepsilon}\|\boldsymbol{t}_2\}$	2
$\{I\|0\}$	$\{\varepsilon\|0\}$	2
$\{\sigma_h\|0\}$	$\{\bar{\varepsilon}\|0\}$	−2
$\{\sigma_{vx}\|\boldsymbol{\tau}\}$	$\{\bar{\varepsilon}\|\boldsymbol{t}_2+\boldsymbol{t}_3\}$	2
$\{\sigma_{vy}\|\boldsymbol{\tau}\}$	$\{\bar{\varepsilon}\|\boldsymbol{t}_1+\boldsymbol{t}_3\}$	−2

13.4 Magnetic Space Groups

We considered in Sect. 13.2 a nonunitary group G consisting of unitary and antiunitary operators. There exist indeed systems that permit symmetry operations $a_0 = \theta v_0$ involving time reversal θ together with the simultaneous application of translation-rotation v_0. Consider, for example, the antiferromagnetic compounds of rutile structure: MnF_2, FeF_2 and CoF_2. Below the Néel temperature T_N, ordering of the (average) magnetic moments (or spins, for simplicity) of ions of the iron group elements is observed. In such magnetically ordered phase, the spins at the corners of the unit cell are aligned upward along the c-axis of the crystal, whereas those at the body center point downward, antiparallel to the former (Fig. 13.6).

Let us consider the symmetry operations that keep this magnetic ordering invariant. The following eight operations are included in the space group D_{4h}^{14} ($P4_2/mnm$) appropriate to the system without spin or to the paramagnetic phase:

$$\begin{aligned}&\{\varepsilon|0\},\quad \{C_2|0\},\quad \{C_{2x}|\boldsymbol{\tau}\},\quad \{C_{2y}|\boldsymbol{\tau}\}\ ,\\ &\{I|0\},\quad \{\sigma_h|0\},\quad \{\sigma_{vx}|\boldsymbol{\tau}\},\quad \{\sigma_{vy}|\boldsymbol{\tau}\}\ ,\end{aligned} \tag{13.56}$$

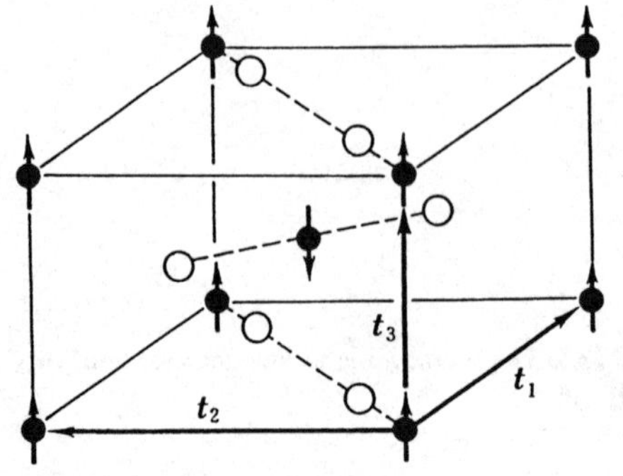

Fig. 13.6. Crystal structure of MnF_2. (●): Mn^{2+} ion, (z): F^- ion. Arrows indicate the magnetic moments

where the vector τ is defined in (11.10). When we multiply the N^3 translations by these operations, we obtain the space group D_{2h}^{12} (Pnnm) consisting of $8N^3$ elements.

Since spins are axial vectors, we may simply regard them as arrows under translation-rotations. Under reflections in a plane, the component of spin perpendicular to the plane remains invariant, while the component parallel to the plane gets reversed. (This behavior can be easily understood, if we replace the spin or the magnetic moment by an equivalent circular electric current.)

If we keep in mind that the direction of spin is reversed by the time reversal θ, we find that the following combinations with θ leave the spin alignments invariant:

$$
\begin{aligned}
&\theta\{C_4|\tau\}, \quad \theta\{C_4^{-1}|\tau\}, \quad \theta\{C_{2\bar{x}}|0\}, \quad \theta\{C_{2\bar{y}}|0\} , \\
&\theta\{S_4|\tau\}, \quad \theta\{S_4^{-1}|\tau\}, \quad \theta\{\sigma_{d\bar{x}}|0\}, \quad \theta\{\sigma_{d\bar{y}}|0\} .
\end{aligned}
\tag{13.57}
$$

In the system without spin, the operations (13.56) and those of (13.57) with θ deleted constituted the space group $P4_2/mnm$. For the spin alignments in Fig. 13.6, (13.56, 57) and the translations form a group $\bar{G}$ ($= P4_2'/mnm'$). The group $G = $ Pnnm is a unitary subgroup of $\bar{G}$, and (13.21) holds for $a_0 = \theta\{C_4|\tau\}$.

At temperatures above T_N, the average magnetic moments vanish and the group of operations leaving the system invariant is $G_0 = P4_2/mnm$. A non-unitary group $\bar{G}_0$ which also keeps the system invariant is obtained by augmenting G_0 by the time reversal θ:

$$
\bar{G}_0 = G_0 + \theta G_0 .
$$

This group $\bar{G}_0$ ($= P4_2/mnm1'$) is called the paramagnetic group.

Suppose we have arrived at the ordered phase described by $\bar{G}$ (or G) starting from the paramagnetic phase described by $\bar{G}_0$ (or G_0) by lowering the temperature through T_N (or T_C, the Curie temperature in the case of ferromagnets). We then have the following subgroup relations among them:

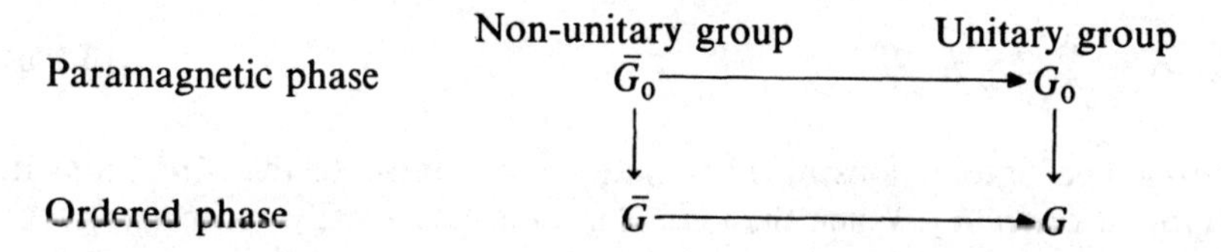

where the notation $A \rightarrow B$ means that B is a subgroup of A. Note that it may happen that no operation involving time reversal is permitted in the ordered phase (as, for example, in the ferromagnetic state) so that the only symmetry group is G. Note also that the unit cells of G and G_0 need not coincide and the magnetic unit cell can be larger than the original cell.

Space groups involving time reversal may be constructed from the 230 space groups in the following way. Let us pick out an invariant subgroup G with index

2 from a space group and form a product a_0 of θ with an element v_0 which does not necessarily belong to G. Then G together with $a_0 G$ forms a nonunitary space group $\bar{G}$. There are altogether 1421 such nonunitary space groups. Among them we have 230 paramagnetic space groups (gray groups[4]), which are simply space groups augmented with θ. The remaining 1191 groups are the essentially new *magnetic space groups* (black and white groups) which are of the type (13.21) with $a_0 \neq \theta$. These 1421 groups and the conventional 230 space groups together form the 1651 Shubnikov groups. These exhaust all the groups of symmetry operations of a crystal when time reversal is included besides translations and rotations. The symmetry group of a magnetically ordered state belongs either to the original 230 space groups or to the 1191 magnetic space groups.

In a similar way, the 32 point groups are extended to 122 Shubnikov point groups. Of these, 32 groups are the original point groups, another 32 groups are the paramagnetic groups of the type (13.20) and the remaining 58 groups are magnetic point groups of the type (13.21).

In the nomenclature of magnetic point and space groups, the international symbols follow those for the ordinary point and space groups and distinguish symmetry operations involving time reversal by adding a prime ($'$). In the paramagnetic groups, $1'$ denotes time reversal as a symmetry operation. (See the notation employed above in the example of the rutile structure.)

13.5 Excitons in Magnetic Compounds; Spin Waves

Since excitons in the ordered phases of magnetic compounds can be treated in almost the same way as in molecular crystals (Sect. 12.11), we will directly proceed to the application of the general theory to a real system, MnF_2 (FeF_2, CoF_2).

The interaction between the spins of Mn^{2+} ions in the crystal is described by the Hamiltonian

$$H_{ex} = -\sum_{i>j} 2J_{ij}\boldsymbol{S}_i \cdot \boldsymbol{S}_j \ . \tag{13.58a}$$

The "exchange" interactions in (13.58a) apply, of course, to the Mn^{2+} ions in their ground state $^6A_1$. When the ion i is in an electronically excited state, the interaction with the neighboring spins $\boldsymbol{S}_j$ may not necessarily be the same as in (13.58a). It may happen that the magnitude of J_{ij} changes greatly. It is even

---

[4] If we indicate the up spin by a black circle and the down spin by a white one, time reversal changes black into the white, and vice versa. The black and white average out in the paramagnetic state, leading to a gray color. The names of the gray group and the black and white groups originate from this description.

possible that the interaction energy can no longer be expressed as a simple scalar product of the spins. Let us assume here, however, that the symmetry of the environment around an ion in the ordered phase does not change even when the ion is excited.

To make the situation easier to understand, we replace the spins $\boldsymbol{S}_j$ around the spin $\boldsymbol{S}_i$ by their time averages $\langle \boldsymbol{S}_j \rangle$ and regard the spin $\boldsymbol{S}_i$ as placed in an effective magnetic field $\boldsymbol{H}(i)$ (the molecular field):

$$H_{\text{mol}} = 2\mu_{\text{B}} \sum_i \boldsymbol{S}_i \cdot \boldsymbol{H}(i) \; ,$$

$$\mu_{\text{B}} \boldsymbol{H}(i) = - \sum_j J_{ij} \langle \boldsymbol{S}_j \rangle \; . \tag{13.58b}$$

It need hardly be emphasized that the alignments of the average spin $\langle \boldsymbol{S}_j \rangle$ at each lattice point conform to the symmetry of the magnetic space group P$4'_2$/mnm$'$.

In the paramagnetic phase, the average spin $\langle \boldsymbol{S}_j \rangle$ vanishes and the site group $G_{0\text{s}}$ for each Mn ion is a subgroup of the space group $\text{D}_{4\text{h}}^{14}$. The site group is isomorphic to $\text{D}_{2\text{h}}$ and consists of

$$\{\varepsilon|0\}, \quad \{C_2|0\}, \quad \{C_{2\bar{x}}|0\}, \quad \{C_{2\bar{y}}|0\} \; ,$$
$$\{I|0\}, \quad \{\sigma_{\text{h}}|0\}, \quad \{\sigma_{\text{d}\bar{x}}|0\}, \quad \{\sigma_{\text{d}\bar{y}}|0\} \; .$$

In the ordered phase, the spins of Mn^{2+} order along the c-axis upward or downward as shown in Fig. 13.6. The site group G_{s}, which will be called the magnetic site group, now consists of only those symmetry operations that leave the molecular field acting on the ion unchanged:

$$\{\varepsilon|0\}, \quad \{C_2|0\}, \quad \{I|0\}, \quad \{\sigma_{\text{h}}|0\} \; .$$

This is the point group $\text{C}_{2\text{h}}$. The irreducible representations and the corresponding characters of $\text{C}_{2\text{h}}$ are given in Table 13.2. Since we are dealing with spins, we must include double-valued representations. The values M in the rightmost column indicate that the eigenfunctions ϕ_M of J_z form the basis for the irreducible representation given on the leftmost column. Compatibility relations between the irreducible representations of $\text{D}_{2\text{h}}$ and $\text{C}_{2\text{h}}$ are given in Table 13.3.

Table 13.2. Irreducible representations and characters of the double group $\text{C}'_2(\text{C}_{2\text{h}} = \text{C}_2 \times \text{C}_\text{i})$

		E	$\bar{E}$	C_2	$\bar{C}_2$	$M =$
A	Γ_1	1	1	1	1	$0, \pm 2, \pm 4, \ldots$
B	Γ_2	1	1	-1	-1	$\pm 1, \pm 3, \pm 5, \ldots$
$\text{E}_{1/2}^{+}$	Γ_3	1	-1	$-\text{i}$	i	$1/2, -3/2, 5/2, \ldots$
$\text{E}_{1/2}^{-}$	Γ_4	1	-1	i	$-\text{i}$	$-1/2, 3/2, -5/2, \ldots$

Table 13.3. Compatibility between the irreducible representations of D_{2h} and C_{2h}

D_{2h}	C_{2h}
$A_{g,u}$, $B_{1g,u}$	$A_{g,u}$
$B_{2g,u}$, $B_{3g,u}$	$B_{g,u}$
$E_{1/2g,u}$	$E^{+}_{1/2g,u}$, $E^{-}_{1/2g,u}$

The wavefunction for the ground state in the molecular field approximation is given by

$$\Psi_0 = \prod_n \psi_{n0} \prod_n \psi_{n1} = \prod_{n\beta} \psi_{n\beta} \,, \tag{13.59}$$

where the functions ψ_{n0} and ψ_{n1} stand, respectively, for the ground state wavefunction of the up-spin site (0) and down-spin site (1) of the nth cell.[5] In the case of Mn^{2+}, they represent the $M_S = 5/2$ and $M_S = -5/2$ states of ${}^6A_1$. The wavefunction for the excited state is given by

$$\Psi_{m\alpha\lambda} = \psi_{m\alpha\lambda} \prod_{n\beta}{}' \psi_{n\beta} \,, \tag{13.60}$$

where the function $\psi_{m\alpha\lambda}$ ($\alpha = 0, 1$) denotes the excited state λ of the ion $m\alpha$.

The ion 0 at the corner of the cell feels the same crystal field as the ion 1 at the body center, although the field is rotated by the angle $\pi/2$. It is also subject to the same molecular field as the ion 1, though in the reversed direction. These facts imply that both ions have the identical level structure and thus are completely equivalent. In fact, they are interchanged by symmetry operations such as $\{C_{2x}|\tau\}$ belonging to D^{12}_{2h}.

The symmetry of the wavefunction of the excited state can be found by examining the representation based on the functions (13.60) of D^{12}_{2h}, the unitary subgroup of the magnetic space group. In contrast to the case of molecular crystals, the ground state Ψ_0 will not always be totally symmetric, although it will certainly be translationally invariant (with $\mathbf{k} = 0$). This makes the matter complicated to a certain extent. Let us therefore always treat $\Psi_0^* \Psi_{m\alpha\lambda}$ instead of

[5] This is the so-called Néel state, where spins point definitely up and down along the c-axis. The correct ground state wavefunction will be more complicated and involve components where many spins are canted. As far as the symmetry is concerned, however, the Néel state inherits that of the exact wavefunction so that discussions based on (13.59 and 60) will be valid also for the correct eigenfunctions.

$\Psi_{m\alpha\lambda}$ itself when considering the symmetry of excited states hereafter. (In this way, we are actually considering the symmetry of the excitation.) This treatment is convenient, because now $\Psi_0^* \Psi_0$ describing the charge and spin density of the system is certainly invariant to the operations of the space group G and thus totally symmetric. The transformation properties of $\Psi_0^* \Psi_{m\alpha\lambda}$ under the operations of G_s are the same as that of $\psi_{m\alpha}^* \psi_{m\alpha\lambda}$. Another merit of it is that it makes the use of double-valued representations unnecessary.

As seen from Table 13.2, the symmetry of the product $\psi_{m\alpha}^* \psi_{m\alpha\lambda}$ becomes A_g when $\psi_{m\alpha}$ and $\psi_{m\alpha\lambda}$ have the same symmetry. Otherwise, it is B_g, A_u or B_u. The latter two result when the two factors have different parity.

Let us calculate the character of the representation of D_{2h}^{12} based on $\Psi_0^* \Psi_{k=0,\alpha\lambda}$ to examine the symmetry of the exciton at $\boldsymbol{k} = 0$. Remembering that there are two equivalent ions in a cell, we find, as in the case of molecular crystals,

$$\chi^{k=0}(\{C_2|0\}) = 2 , \quad \chi^{k=0}(\{I|0\}) = \chi^{k=0}(\{\sigma_h|0\}) = \pm 2 ,$$

when the symmetry of $\psi_{0\alpha}^* \psi_{0\alpha\lambda}$ is $A_{g,u}$. For the case of $B_{g,u}$, we have

$$\chi^{k=0}(\{C_2|0\}) = -2 , \quad \chi^{k=0}(\{I|0\}) = -\chi^{k=0}(\{\sigma_h|0\}) = \pm 2 .$$

The irreducible characters for D_{2h}^{12} at $\boldsymbol{k} = 0$ are given by those for the point group D_{2h} (Appendix B). Reducing the representation into the irreducible representations of D_{2h}, we obtain the result in Table 13.4. The wavefunctions of the excited states are given by (12.40) using (12.30, 31). The symmetry of the exciton given in Table 13.4 stands for that of $\Psi_0^* \Psi_{k=0,\nu\Gamma\gamma}$.

Application of the criterion (13.46) tells us about the extra degeneracy due to the inclusion of the time-reversal operation. In the present case, the magnetic space group is obtained by augmenting D_{2h}^{12} by $a_0 = \theta v_0$ with

$$v_0 = \{\varrho|\sigma\} = \{C_4|\tau\} .$$

Since we are considering the problem at $\boldsymbol{k} = 0$, the operations $\{\beta|\boldsymbol{b}\}$ satisfying (13.47) are all the representative elements of $D_{2h}^{12}/\mathscr{T}$, namely, those given in (13.56). Note also that we put $(-1)^N = 1$ on the right-hand side of (13.46). This is because we deal only with single-valued representations based on $\Psi_0^* \Psi_{k=0,\alpha\lambda}$. Calculation of the left-hand side of (13.46) is easy, since we need not worry about the translations at $\boldsymbol{k} = 0$. We have only to evaluate the sum of $\chi((C_4R)^2)$ over

Table 13.4. Relations between the symmetry of a local excitation and that of the exciton at $\boldsymbol{k} = 0$

Symmetry of $\psi_{0\alpha}^* \psi_{0\alpha\lambda}$	Symmetry of the exciton at $k = 0$
$A_{g(u)}$	$A_{g(u)}$, $B_{1g(u)}$
$B_{g(u)}$	$B_{3g(u)} + B_{2g(u)}$

the eight elements R of D_{2h}. The conclusion is: excitons $A_{g(u)}$ and $B_{1g(u)}$ belong to case (a) and they have no additional degeneracy. In general, A_g and B_{1g} have different energies at $\boldsymbol{k} = 0$. In other words, Davydov splitting should be observable.[6] In contrast, $B_{2g(u)}$ and $B_{3g(u)}$ belong to case (c). Their wavefunctions are related to each other by the antiunitary operator a_0 so that the two are degenerate (no Davydov splitting).

As regards the optical excitation, we note that the three components P_x, P_y, P_z of the electric dipole moment $\boldsymbol{P}$ transform respectively like B_{3u}, B_{2u}, and B_{1u} of D_{2h}, whereas the components M_x, M_y, and M_z of the magnetic dipole moment $\boldsymbol{M}$ transform like B_{3g}, B_{2g}, and B_{1g}, respectively. This leads to the selection rules given in Table 13.5. The $A_{g,u}$ excitons correspond to the forbidden transitions and will not be observed unless external stress or a magnetic field causes lowering of the symmetry.

Spin waves or magnons in the antiferromagnet MnF_2 are to be regarded as a kind of exciton as discussed so far. Local excitation corresponding to the canting of an up spin from the state $M_S = 5/2$ to $M_S = 3/2$ gives rise to a spin wave. The symmetry of this type of excitation $\psi^*\psi_\lambda$ is found from Table 13.2 to be B_g. The symmetry of the spin wave at $\boldsymbol{k} = 0$ is therefore $B_{3g} + B_{2g}$ according to Table 13.4. Creation of this kind of exciton is made possible by means of the antiferromagnetic resonance, namely by applying an oscillating magnetic field perpendicular to the c-axis.

Excitons having B_{1g} symmetry are observed in MnF_2 at energies of 18 418 cm^{-1} and 18 435 cm^{-1}. They correspond to magnetic dipole transitions from the ground state $^6A_{1g}$ (of O_h), $M_S = 5/2$, to the excited state $^4T_{1g}$, $M_L = \pm 1$, $M_S = 3/2$ [13.1].

Simultaneous creation of two excitons (including spin waves) by means of an electric dipole transition has been observed in MnF_2. Let us briefly discuss the

Table 13.5. Selection rules for optical transitions

Type of transition	Polarization of light	Symmetry of exciton
Electric dipole	Oscillating electric field	
	$// \, c \, (\pi)$	B_{1u}
	$\perp c \, (\sigma, \alpha)$	$B_{3u} + B_{2u}$
Magnetic dipole	Oscillating magnetic field	
	$// \, c \, (\sigma)$	B_{1g}
	$\perp c \, (\pi, \alpha)$	$B_{3g} + B_{2g}$

[6] No Davydov splitting has been observed for the B_{1g} exciton at 18 400 cm^{-1} of MnF_2 (to be discussed below). This is due to the fact that the actual symmetry of the Hamiltonian (13.58a) is higher than D_{2h}^{12}.

selection rules for this type of excitation. When we denote the state of each exciton by $kv\Gamma\gamma$ and $k'v'\Gamma'\gamma'$, respectively, the wavefunction representing the state with two excitons is

$$\Psi_2 = \Psi_{kv\Gamma\gamma,k'v'\Gamma'\gamma'}$$

$$= N^{-3}\sum_{m\alpha}\sum_{m'\alpha'} e^{i\boldsymbol{k}\cdot\boldsymbol{t}_m + i\boldsymbol{k}'\cdot\boldsymbol{t}_{m'}} C_\alpha(\Gamma\gamma)C_{\alpha'}(\Gamma'\gamma')\psi_{m\alpha v}\psi_{m'\alpha'v'}\prod_{n\beta}{}' \psi_{n\beta} ,$$

where we have used the coefficients $C_{\alpha\mu}(\Gamma\gamma)$ in (12.40), neglecting the interaction between the excitons.[7] If Ψ_2 is to be excited by light, it must have the same wavevector as the ground state Ψ_0, so that $\boldsymbol{k} + \boldsymbol{k}' = 0$. More precisely, optical excitation is possible when

$$\int d\tau \Psi_0^* P \Psi_2$$

does not vanish. This can be checked by examining whether or not the product representation $\{\boldsymbol{k}\}\Gamma \times \{-\boldsymbol{k}\}\Gamma'$ contains any one of the irreducible representations B_{1u}, B_{3u}, B_{2u} at $\boldsymbol{k} = 0$ when completely reduced. The procedures discussed in Sect. 12.12 are applicable for this purpose.

Table 13.6 gives the symmetry of excitons at several points of the Brillouin zone (BZ) [13.2]. For the irreducible characters of small representations of D_{2h}^{12} at points X, Z and A of the BZ, see Table 13.7. Selection rules for the two-exciton transitions at the points X, Z and A on the BZ boundary are given in Table 13.8. Note that electric dipole transitions are not allowed at Γ, M and R, because each exciton has definite (even) parity at these points.

Derivation of these results will be left to the reader as an exercise. Representations in braces { } in the left-hand column of Table 13.8 belong to the

Table 13.6. Symmetry of excitons at points of high symmetry in the Brillouin zone (BZ). (See Fig. 11.7d for the BZ of rutile structure.) For the irreducible representations at Γ, Bethe's notation for D_{2h} is adopted

	A_g excitation	B_g excitation
Γ	Γ_1^+, Γ_2^+	$\Gamma_3^+ + \Gamma_4^+$
X	X_1	X_2
M	M_1^+, M_2^+	M_3^+, M_4^+
Z	Z_1	Z_2
R	R_1^+	R_2^+
A	A_1	A_2

[7] The term with $m\alpha = m'\alpha'$ in the sum on the right-hand side will not appear in the correct wavefunction. This term is, however, included to let $\Psi_0^*\Psi_2$ behave like the basis of $\{\boldsymbol{k}\}\Gamma \times \{\boldsymbol{k}'\}\Gamma'$.

Table 13.7. Irreducible characters of small representations of D_{2h}^{12} at the Z and A points of the BZ. Those for the X point are given in Table 11.4

	Z_1 A_1	Z_2 A_2
$\{\varepsilon\|0\}$	2	2
$\{C_2\|0\}$	2	-2
$\{C_{2x}\|\tau\}$	0	0
$\{C_{2y}\|\tau\}$	0	0
$\{I\|0\}$	0	0
$\{\sigma_h\|0\}$	0	0
$\{\sigma_{vx}\|\tau\}$	0	0
$\{\sigma_{vy}\|\tau\}$	0	0

Table 13.8. Symmetry of the two-exciton state at points X, Z and A

Two A_g or B_g excitations	A_g excitation + B_g excitation
$\left.\begin{matrix} X_1 \times X_1 \\ X_2 \times X_2 \end{matrix}\right\} = \begin{cases} \Gamma_1^+ + \Gamma_2^+ + \Gamma_3^- + \{\Gamma_4^-\} \\ \Gamma_1^+ + \Gamma_2^+ + \{\Gamma_3^-\} + \Gamma_4^- \end{cases}$	$X_1 \times X_2 = \Gamma_1^- + \Gamma_2^- + \Gamma_3^+ + \Gamma_4^+$
$\left.\begin{matrix} Z_1 \times Z_1 \\ Z_2 \times Z_2 \end{matrix}\right\} = \Gamma_1^+ + \Gamma_2^+ + \Gamma_1^- + \{\Gamma_2^-\}$	$Z_1 \times Z_2 = \Gamma_3^+ + \Gamma_4^+ + \Gamma_3^- + \Gamma_4^-$
$\left.\begin{matrix} A_1 \times A_1 \\ A_2 \times A_2 \end{matrix}\right\} = \Gamma_1^+ + \Gamma_2^+ + \{\Gamma_1^-\} + \Gamma_2^-$	$A_1 \times A_2 = \Gamma_3^+ + \Gamma_4^+ + \Gamma_3^- + \Gamma_4^-$

antisymmetric representation. When the two excitons are of the same species and $v = v'$, $\boldsymbol{k} = \boldsymbol{k}'$, $\Gamma = \Gamma'$, we have only symmetric representations in the reduction of $\{\boldsymbol{k}\}\Gamma \times \{\boldsymbol{k}\}\Gamma$, because Ψ_2 is a symmetric function. This means that it is actually impossible to create two excitons at the Z point by optical means. The two results of decomposition for the product $X_1 \times X_1 = X_2 \times X_2$ corres-pond to the two choices $X = (0, \pi/a, 0)$ and $X = (\pi/a, 0, 0)$. Note that these two points do not belong to the same star.

*13.5.1. Symmetry of the Hamiltonian

Finally, a few words will be in order on the relation between the symmetry of the Hamiltonian H_{ex} given by (13.58a) for the spin system and its symmetry in the magnetically ordered phase. The former is described by the paramagnetic space group $\bar{G}_0$, and the latter by the magnetic space group $\bar{G}$ (or its unitary subgroup G). Since $\bar{G}$ is in general a subgroup of $\bar{G}_0$, this implies, for example, that if the

Néel state Ψ_0 were the correct ground state of H_{ex}, the state $\theta\Psi_0$ with all the spins reversed would also be an eigenstate degenerate with Ψ_0. This degeneracy will not be lifted even when the anisotropy energy is taken into consideration. In presuming Ψ_0 as the nondegenerate ground state, we are selecting only the spin configuration that is actually realized. The Hamiltonian which give Ψ_0 as its correct ground state is really the molecular-field Hamiltonian H_{mol}, that is, (12.58b) with its symmetry $\bar{G}$ (or G). In fact, with H_{mol}, the state $\theta\Psi_0$ has macroscopically higher energy, so that the state Ψ_0 has no degeneracy.

Suppose we have constructed a state Ψ_1 by exciting several ions from the ground state described by Ψ_0. The states $R\Psi_1$ derived from Ψ_1 by operating with R of G are all equivalent and eigenstates acquired by diagonalizing H_{ex} using them as bases provide us with the bases of the irreducible representation of G. This is why we treat the system as if the symmetry of the Hamiltonian has been lowered to $\bar{G}$ (or G) in the ordered phase.

In the next chapter we will discuss how the symmetry of $\bar{G}$ or G follows from that of G_0 on the grounds of symmetry.

14. Landau's Theory of Phase Transitions

Landau's theory concerns the kind of invariants consistent with the symmetry of the crystal that may appear in the expression for the free energy. In this sense, it should be viewed as one of the successful applications of the group theory in classical physics. Unlike ordinary classical applications, however, the representation theory for space groups plays an essential role here, because a change of crystal symmetry is the main issue. It is most interesting to note that possible types of phase transitions are determined definitely solely from symmetry considerations.

Since determination of possible spin configurations in a magnetic compound may be treated in a similar way, a brief section is devoted to the discussion of this problem.

14.1 Landau's Theory of Second-Order Phase Transitions

In the first-order phase transitions of crystals, an abrupt change from one phase to another occurs accompanied by a discontinuous change in lattice constants, and generation or absorption of latent heat. At the transition point, the two phases coexist. The condition for the possibility of such a first-order transition between the two phases is the equality of their free energies at the transition point and there is no restriction on their symmetry. In contrast to this, changes in second-order phase transitions are more continuous with no latent heat and no discontinuity in lattice parameters at the critical point. However, derivatives of thermodynamical quantities (with respect to temperature) such as specific heat and thermal expansion coefficient are discontinuous. In this case, there exists only one phase at each temperature (even at the critical temperature) and we may regard it as a transition within a single phase. Landau showed that this type of transition is possible only when there is a severe restriction on the symmetry of the crystal above and below the critical temperature [14.1]. Let us review the theory here, since it is most interesting as an application of the representation theory of space groups.

The symmetry of a crystal is described by the density $\varrho(\boldsymbol{r})$ of particles constituting it. (It is of course necessary to introduce $\varrho_j(\boldsymbol{r})$ with $j = 1, 2, \ldots$ when there are two or more kinds of particles. For simplicity, we assume only one ϱ here, since generalization will not be difficult.) The set of operations which leave ϱ invariant form the space group of the crystal.

In general, when pressure P and temperature T change, the density ϱ changes accordingly. The Gibbs free energy Φ is a function of P and T and is a functional of ϱ at the same time, so that we may put

$$\Phi = \Phi(P, T, \varrho) . \tag{14.1}$$

In the equilibrium state, the free energy has to take a minimum value against a slight change in ϱ with fixed values of P and T. This condition determines the function ϱ. Suppose (P_0, T_0) is a point on the line of second-order transitions in the phase diagram and let ϱ at this point be ϱ_0. The space group corresponding to ρ_0 will be denoted as G_0. If we take a point (P, T) close to (P_0, T_0) and write the density at that point as

$$\varrho = \varrho_0 + \Delta\varrho , \tag{14.2}$$

we have $\Delta\varrho \to 0$ as $(P, T) \to (P_0, T_0)$. The difference $\Delta\varrho$ represents the small change corresponding to the lowering of the symmetry (assuming ϱ_0 to have higher symmetry). The symmetry group G of the density ϱ cannot include any operation that is not included in G_0, which implies that G is a subgroup of G_0.

Now, it may be assumed that $\Delta\varrho$ can be expressed as a linear combination of the bases of irreducible representations of G_0 as

$$\Delta\varrho(\boldsymbol{r}) = \sum_{\alpha}\sum_{i} c_i^{(\alpha)}\phi_i^{(\alpha)}(\boldsymbol{r}) , \tag{14.3}$$

where $\phi_i^{(\alpha)}$ is the ith partner $(i = 1, 2, \ldots, n_\alpha)$ of the basis for the irreducible representation $D^{(\alpha)}$ of G_0. Since $\Delta\varrho$ is a real function, the coefficients $c_i^{(\alpha)}$ will be real if we choose $\phi_i^{(\alpha)}$ as real. This is possible when $D^{(\alpha)}$ is a real representation [case (a) of (4.88)]. When $D^{(\alpha)}$ is not real, we can use real combinations $\underline{\phi}_i^{(\alpha)} = \phi_i^{(\alpha)} + \phi_i^{(\alpha)*}$ and $\underline{\phi}_i^{(\alpha)\prime} = \mathrm{i}(\phi_i^{(\alpha)} - \phi_i^{(\alpha)*})$ instead of $\phi_i^{(\alpha)}$ and $\phi_i^{(\alpha)*}$, because the bases $\phi_i^{(\alpha)*}$ of $D^{(\alpha)*}$ are linearly independent of $\phi_i^{(\alpha)}$. The real representation $\underline{D}^{(\alpha)} = D^{(\alpha)} + D^{(\alpha)*}$ obtained with such $2n_\alpha$ real bases is called *physically irreducible.* If we are always to use such bases, we may assume the coefficients $c_i^{(\alpha)}$ to be real all along. The superscript α there, of course, refers to a physically irreducible representation.

Strictly speaking, we should write (14.3) as

$$\Delta\varrho(\boldsymbol{r}) = \sum_{n\alpha i} c_i^{n(\alpha)}\phi_i^{n(\alpha)}(\boldsymbol{r}) ,$$

using in place of a single $\phi_i^{(\alpha)}$ a complete set of functions $\{\phi_i^{n(\alpha)}\}$ with $n = 0, 1, 2, \ldots$ that transform like the ith partner of the irreducible representation $D^{(\alpha)}$. But then we must know how to set up such a complete set. This will be discussed in Sect. 14.3. We will see there that the functions $\phi_i^{n(\alpha)}$ are determined by an eigenvalue problem and that $\phi_i^{(\alpha)}$ in (14.3) actually corresponds to the lowest eigenvalue of that problem.

We may assume that the basis of the identity representation is not included in the summation of (14.3), because such a term is invariant to the operations of G_0 like ϱ_0 and may be viewed as absorbed into the latter from the beginning.

Substituting (14.2, 3) into (14.1), and expanding Φ into a power series of $c_i^{(\alpha)}$, assuming $\Delta\varrho$ sufficiently small, yields

$$\Phi(P, T, \varrho) = \Phi^{(0)} + \Phi^{(1)} + \Phi^{(2)} + \cdots, \tag{14.4}$$

$$\Phi^{(0)} = \Phi(P, T, \varrho_0),$$

$$\Phi^{(1)} = \sum_{\alpha i} c_i^{(\alpha)} \left(\frac{\partial \Phi}{\partial c_i^{(\alpha)}} \right)_0, \tag{14.5}$$

$$\Phi^{(2)} = \frac{1}{2} \sum_{\alpha i} \sum_{\alpha' i'} c_i^{(\alpha)} c_{i'}^{(\alpha')} \left(\frac{\partial^2 \Phi}{\partial c_i^{(\alpha)} \partial c_{i'}^{(\alpha')}} \right)_0.$$

Since $\phi_i^{(\alpha)}$ are the bases of the irreducible (physically irreducible) representation $D^{(\alpha)}$, they transform as

$$R\phi_i^{(\alpha)} = \sum_j \phi_j^{(\alpha)} D_{ji}^{(\alpha)}(R) \tag{14.6}$$

under the symmetry operations R of G_0. We then have

$$R\Delta\varrho = \sum_{\alpha i} c_i^{(\alpha)} R\phi_i^{(\alpha)} = \sum_\alpha \sum_j \left(\sum_i D_{ji}^{(\alpha)}(R) c_i^{(\alpha)} \right) \phi_j^{(\alpha)}, \tag{14.7}$$

when we operate R on $\Delta\varrho$. This shows that we may consider operation of R as inducing the transformation

$$c_j^{(\alpha)} \to c_j^{(\alpha)\prime}, \quad c_j^{(\alpha)\prime} \equiv \sum_i D_{ji}^{(\alpha)}(R) c_i^{(\alpha)} \tag{14.8}$$

on the coefficients $c_j^{(\alpha)}$ instead of acting upon the functions $\phi_i^{(\alpha)}$. From here on, we will take this viewpoint.

We note that both the density function $\Delta\varrho$ and $R\Delta\varrho$ give rise to geometrically the same density distribution, so that the free energy Φ takes the same value for both distributions. In other words, Φ is invariant under the symmetry operations belonging to G_0. This means that the coefficients $c_i^{(\alpha)}$ have to be included in Φ only in the form of invariants of G_0, or the bases of the identity representation, when Φ is expanded as in (14.4,5). Since we have assumed the identity representation is absent from the sum on the right-hand side of (14.3), $\Phi^{(1)}$ vanishes identically. We also know that the only second-order invariant made from $c_i^{(\alpha)}$ is given by $\sum_i c_i^{(\alpha)2}$, so that $\Phi^{(2)}$ must take the form

$$\Phi^{(2)} = \sum_\alpha A^{(\alpha)}(P, T, \varrho_0, \{\phi_i^{(\alpha)}\}) \sum_j c_j^{(\alpha)2}. \tag{14.9}$$

In the equilibrium state, we have $\Delta\varrho = 0$ at the critical point (P_0, T_0), so that (14.9) must take a minimum at $c_j^{(\alpha)} = 0$, which requires that none of the coefficients $A^{(\alpha)}$ $(P_0, T_0, \ldots)$ can be negative. If all $A^{(\alpha)}$'s were positive, they would be positive also at (P, T) in the neighborhood of (P_0, T_0), which in turn would lead to $c_j^{(\alpha)} = 0$ there for any α, in contradiction with the assumption that the point (P_0, T_0) lies on the line of transitions. Therefore, it must be that at least one of them, say $A^{(\beta)}$, vanishes at the transition points:

$$A^{(\beta)}(P_0, T_0, \ldots) = 0 \ , \tag{14.10}$$

while all the other coefficients are positive:

$$A^{(\alpha)}(P_0, T_0, \ldots) > 0 \ , \quad \alpha \neq \beta \ . \tag{14.11}$$

In fact, (14.10) may be regarded as an equation giving the line of transition points.

According to (14.10, 11), all the coefficients $A^{(\alpha)}$, including $A^{(\beta)}$, will be positive on one side of the line of transitions, leading to the vanishing of all $c_j^{(\alpha)}$ there. On the other side of the curve, however, we have $A^{(\beta)} < 0$ and the symmetry of the crystal there is described by

$$\varrho = \varrho_0 + \Delta\varrho = \varrho_0 + \sum_j c_j^{(\beta)} \phi_j^{(\beta)} \tag{14.12}$$

with nonvanishing coefficients $c_j^{(\beta)}$. (At this stage, their values are still undetermined.) Thus, second-order phase transitions correspond, in general, to a change in symmetry corresponding to one of the irreducible representations (i.e., $D^{(\beta)}$) of G_0. We will drop the superscript β hereafter, when we are concerned with only this particular symmetry or representation. Then we have

$$\Phi^{(2)} = A(P, T) \sum_j c_j^2 \ . \tag{14.13}$$

Since $A(P, T)$ vanishes at (P_0, T_0) as seen above, we must have

$$\Phi^{(3)} = 0$$

for the third-order term, because the crystal would not be stable there otherwise. This condition will be satisfied automatically when it is impossible to construct a third-order (homogeneous) invariant, or when the totally symmetric product representation $[D^3]$ does not contain the identity representation, where $D = D^{(\beta)}$ or $D^{(\beta)}$.

For the same reason, the fourth-order term with respect to c_j

$$\Phi^{(4)} = \sum_\lambda B_\lambda(P, T) f_\lambda^{(4)}(c_j) \tag{14.14}$$

must be positive definite, where $f_\lambda^{(4)}(c_j)$ are the (homogeneous) fourth-order invariants of c_j. The suffix λ has been introduced to distinguish between invariants of different type.

At point (P, T) in the neighborhood of (P_0, T_0), the c_j's or $\Delta\varrho$ can be determined by minimizing the expression

$$\begin{aligned}\Phi &= \Phi^{(0)} + \Phi^{(2)} + \Phi^{(4)} \\ &= \Phi(P, T, \varrho_0) + A(P, T)\sum_j c_j^2 + \sum_\lambda B_\lambda(P, T) f_\lambda^{(4)}(c_j) \,. \end{aligned} \tag{14.15}$$

As regards the representation D $(= D^{(\beta)}\ \underline{D}^{(\beta)})$ we have the condition (14.10, 11) besides the one that $[D^3]$ must not contain the identity representation. The irreducible representation $D^{(\beta)}$ of the space group G_0 is specified by star$\{\boldsymbol{k}\}$ and the (irreducible) small representation Γ of the group $\mathscr{G}(\boldsymbol{k})$ of the wavevector $\boldsymbol{k}$. Its partners are specified by the wavevector $\boldsymbol{k}$ belonging to the star and the partner ν of the representation Γ.[1] The condition (14.10,11) can be expressed as

$$A^{\{\boldsymbol{k}'\}\Gamma'}(P_0, T_0) > A^{\{\boldsymbol{k}\}\Gamma}(P_0, T_0) = 0 \tag{14.16}$$

for $\boldsymbol{k}'$ in the neighborhood of $\boldsymbol{k}$ and small representations Γ' compatible with Γ. In other words, $A^{\{\boldsymbol{k}'\}\Gamma'}$ takes its minimum at $\boldsymbol{k}' = \boldsymbol{k}$ $(\Gamma' = \Gamma)$. If we put $\boldsymbol{k}' = \boldsymbol{k} + \boldsymbol{\kappa}$ and expand A into a Maclaurin series with respect to the components of $\boldsymbol{\kappa}$, the coefficients of the first-order terms have to vanish. It may happen that they vanish at $\boldsymbol{k}$ for reasons other than symmetry. In such cases, $\boldsymbol{k}$ need not be limited to special points in the Brillouin zone (BZ). However, it will be shown below that $\boldsymbol{k}$ must be a point of high symmetry in the BZ if the coefficients of the first-order terms are to vanish on the grounds of symmetry. Putting off the detailed consideration to Sect. 14.3, we give here the result. In order that the coefficients vanish by symmetry,

the antisymmetric representation $\{D^2\}$ of G_0 with $D = \{\boldsymbol{k}\}\Gamma$ (or $D = \{\boldsymbol{k}\}\Gamma + (\{\boldsymbol{k})\Gamma)^*$ when $\{\boldsymbol{k}\}\Gamma$ is not real) should not contain the vector representation V of G_0, (14.17)

where V denotes the 3-dimensional representation (with $\boldsymbol{k} = 0$) of G_0 by the components of a vector. This is the condition corresponding to (14.10,11) or (14.16).

Let us summarize the above considerations. In general, second-order phase transitions induce a symmetry change corresponding to an irreducible representation $D = \{\boldsymbol{k}\}\Gamma$ (or $D = \{\boldsymbol{k}\}\Gamma + (\{\boldsymbol{k}\}\Gamma)^*$ when $\{\boldsymbol{k}\}\Gamma$ is not real) of G_0, so

[1] In general, the function $\phi_{\boldsymbol{k}\nu}$ itself may not be real with this choice of the basis function, even when the representation $\{\boldsymbol{k}\}\Gamma$ is real.

that

$$\varrho_0 \to \varrho = \varrho_0 + \sum_j c_j \phi_j \ , \tag{14.18}$$

where the last term is to be replaced by $\sum_j (c_j\phi_j + c_j^* \phi_j^*)$, when $\{\boldsymbol{k}\}\Gamma$ is not real. The representation D must satisfy the following conditions:

(i) $[D^3]$ does not contain the identity representation (the Landau criterion).

(ii) $\{D^2\}$ does not contain the vector representation (the Lifshitz criterion). We must keep in mind, however, that (14.16) may be satisfied for reasons other than symmetry, so that this condition is not absolutely necessary.

(iii) Since the density (14.18) is invariant under the symmetry group G of the low-temperature phase, the representation D necessarily contains the identity representation when it is decomposed into its irreducible components as a representation of G. This helps us to limit the possible types of group G for the low-temperature phase when G_0 and its irreducible representation $\{\boldsymbol{k}\}\Gamma$ are specified. That is to say, G is a subgroup of G_0 such that $\{\boldsymbol{k}\}\Gamma$ contains the identity representation (of G) when reduced as a representation of G. Of several possibilities, the one actually realized is naturally the structure that makes the fourth-order term smallest.

When it is impossible to select three vectors $\boldsymbol{k}_1, \boldsymbol{k}_2, \boldsymbol{k}_3$ belonging to the star $\{\boldsymbol{k}\}$ and satisfying $\boldsymbol{k}_1 + \boldsymbol{k}_2 + \boldsymbol{k}_3 \doteq 0$, the condition (i) is satisfied. However, when several sets of such vectors exist, we must construct a representation based on the functions

$$\phi_{\boldsymbol{k}_1\nu_1}\phi_{\boldsymbol{k}_2\nu_2}\phi_{\boldsymbol{k}_3\nu_3} \ , \quad \boldsymbol{k}_1 + \boldsymbol{k}_2 + \boldsymbol{k}_3 \doteq 0 \ ,$$

and confirm that its character χ satisfies $\sum_R \chi(R) = 0$. When the representation $\{\boldsymbol{k}\}\Gamma$ is not real, it is further necessary to check that $[(\{\boldsymbol{k}\}\Gamma)^2] \times (\{\boldsymbol{k}\}\Gamma)^*$ does not contain the identity representation, or equivalently, that $[(\{\boldsymbol{k}\}\Gamma)^2]$ does not contain $\{\boldsymbol{k}\}\Gamma$.

Explicitly, the condition (ii) is elaborated as follows. The first step is to calculate $\chi_0(Q) = \chi^{\boldsymbol{K}=0\{(\{\boldsymbol{k}\}\Gamma)^2\}}$, the character at $\boldsymbol{K} = 0$ of the antisymmetric representation $\{(\{\boldsymbol{k}\}\Gamma)^2\}$. It may be obtained by the procedure given in Sect. 12.12.1. If we write $V(Q)$ for the character of the representation V based on the three components of a vector (with $\boldsymbol{K} = 0$), we can express the requirement that $\boldsymbol{K} = 0\{(\{\boldsymbol{k}\}\Gamma)^2\}$ should not contain V as

$$\begin{aligned} S &\equiv \frac{1}{gN^3} \sum_{R \in G_0} \chi_0(R) V(R) \\ &= \frac{1}{2[\boldsymbol{k}]} \sum_Q [\chi^{\boldsymbol{k}(\Gamma)}(Q)\chi^{\boldsymbol{k}(\Gamma)}(P^{-1}QP)V(Q) - \chi^{\boldsymbol{k}(\Gamma)}((PQ)^2)V(PQ)] = 0 \ . \end{aligned} \tag{14.19a}$$

The summation over Q is to be taken over the representative elements of the factor group $\mathscr{G}(\boldsymbol{k})/\mathscr{T}$ (namely, those with the smallest translational parts). The notation $[\boldsymbol{k}]$ stands for the number of such elements. The quantity $\chi^{\boldsymbol{k}(\Gamma)}$ is the character of the small representation Γ of $\mathscr{G}(\boldsymbol{k})$ and P is an element of G_0 that satisfies $P\boldsymbol{k} \doteq -\boldsymbol{k}$. When the representation $\{\boldsymbol{k}\}\Gamma$ is real, S may be put in the form

$$S = \frac{1}{2[\boldsymbol{k}]} \sum_Q [|\chi^{\boldsymbol{k}(\Gamma)}(Q)|^2 V(Q) - \chi^{\boldsymbol{k}(\Gamma)}((PQ)^2)V(PQ)] \ . \tag{14.19b}$$

When $-\boldsymbol{k} \doteq \boldsymbol{k}$, we may use the following expression for S in place of (14.19a):

$$S = \frac{1}{2[\boldsymbol{k}]} \sum_{Q} \{(\chi^{\boldsymbol{k}(\Gamma)}(Q))^2 - \chi^{\boldsymbol{k}(\Gamma)}(Q^2)\} V(Q) \ . \tag{14.19c}$$

In this case, $S = 0$ means that $\{\Gamma^2\}$ does not contain V.

The condition that $(\{\boldsymbol{k}\}\Gamma) \times (\{\boldsymbol{k}\}\Gamma)^*$ should not contain V may be expressed as

$$S = \frac{1}{[\boldsymbol{k}]} \sum_{Q} |\chi^{\boldsymbol{k}(\Gamma)}(Q)|^2 V(Q) = 0 \ . \tag{14.20}$$

A representation D satisfying the conditions (i) and (ii) is said to be active. A point $\boldsymbol{k}$ in the BZ which yields an active representation will be called an active point. As seen below, active points are limited to several special points in the BZ. All these results follow from (14.19a, 20).

First of all, points $\boldsymbol{k}$ at which the representation of $\mathscr{G}(\boldsymbol{k})$ based on the components of a vector contains the identity representation cannot be active. Since $\mathscr{G}(\boldsymbol{k})/\mathscr{T}$ at general points $\boldsymbol{k}$ in the BZ consists only of the identity operation, they need not be considered further as active points. Similar situations hold for those points at which $\mathscr{G}(\boldsymbol{k})/\mathscr{T}$ consists only of rotations about the axis $\boldsymbol{k}$ and reflections in the symmetry planes containing $\boldsymbol{k}$. Active points will not, therefore, be present inside the BZ except for the zone center. Points $\boldsymbol{k}$ can be active when $\mathscr{G}(\boldsymbol{k})/\mathscr{T}$ includes, besides rotations about the axis $\boldsymbol{k}$, rotations about axes intersecting $\boldsymbol{k}$ or reflections in the planes crossed by $\boldsymbol{k}$ (i.e., in the cases where $\mathscr{G}(\boldsymbol{k})/\mathscr{T}$ contains any one of the groups D_2, D_3, C_{3h}, or S_4). Such points are limited to a few points of high symmetry on the BZ boundary.

Let us give a simple example. Consider an order–disorder transition of a binary alloy of elements A and B, and assume that the atoms occupy the sites of a body-centered cubic lattice (Im3m, O_h^9) in the disordered (high temperature) phase. We will examine what kinds of ordering (formation of superlattices) are permitted. Possible active points and their stars are found to be (see Fig. 11.7c for the BZ)

$$\text{N:} \quad \boldsymbol{k}_1 = \left(0, \frac{\pi}{a}, \frac{\pi}{a}\right) , \quad \boldsymbol{k}_2 = \left(\frac{\pi}{a}, 0, \frac{\pi}{a}\right) , \quad \boldsymbol{k}_3 = \left(\frac{\pi}{a}, \frac{\pi}{a}, 0\right) ,$$

$$\boldsymbol{k}_4 = \left(0, \frac{\pi}{a}, -\frac{\pi}{a}\right) , \quad \boldsymbol{k}_5 = \left(-\frac{\pi}{a}, 0, \frac{\pi}{a}\right) , \quad \boldsymbol{k}_6 = \left(\frac{\pi}{a}, -\frac{\pi}{a}, 0\right) ,$$

$$\Gamma: \quad \boldsymbol{k} = (0, 0, 0) \ ,$$

$$\text{H:} \quad \boldsymbol{k} = \left(0, 0, \frac{2\pi}{a}\right) .$$

They all satisfy the relation $-\boldsymbol{k} \doteq \boldsymbol{k}$. Besides them,

$$\text{P:} \quad \boldsymbol{k}_1 = \left(\frac{\pi}{a}, \frac{\pi}{a}, \frac{\pi}{a}\right) , . \quad \boldsymbol{k}_2 = \left(-\frac{\pi}{a}, -\frac{\pi}{a}, -\frac{\pi}{a}\right)$$

are also active. In the case of an order–disorder transition, we may consider $\Delta\varrho$ as invariant under all rotations (without accompanying translations) of $\mathscr{G}(\boldsymbol{k})$, because formation of the superlattice simply makes lattice sites belonging to different cells inequivalent. This means that it suffices to consider only the identity representations N_1, Γ_1, H_1, P_1 as the irreducible representation Γ of $\mathscr{G}(\boldsymbol{k})$. These are all real representations of G_0.

If we write ϕ_j for the basis function of N_1 at the six points $\boldsymbol{k}_j$ ($j = 1, 2, \ldots, 6$) given above for N, we can construct an invariant with the symmetry $[D^3]$,

$$\phi_1\phi_3\phi_5 + \phi_2\phi_3\phi_4 + \phi_1\phi_2\phi_6 + \phi_4\phi_5\phi_6 \ ,$$

using sets of three ϕ_j satisfying $\sum \boldsymbol{k}_j \doteq 0$. This means N_1 is not active.

For the same reason Γ_1 is not active.

In the case of H_1, the sum of three $\boldsymbol{k}$'s is not equivalent to 0. In addition, it is impossible to construct $\{H_1^2\}$. Therefore, H_1 will be active. The basis function ϕ of H_1 is invariant to all the rotations belonging to O_h and changes its sign under the translation ($a/2$, $a/2$, $a/2$). The density function $\varrho = \varrho_0 + \Delta\varrho$ describes a superlattice with A and B atoms on (0, 0, 0) and ($a/2$, $a/2$, $a/2$) sites of the crystal, respectively (CsCl structure, Pm3m, O_h^1).

No combination of three vectors chosen from $\boldsymbol{k}_1$ and $\boldsymbol{k}_2$ belonging to the star P leads to 0. Further, it can be verified that (14.19) holds for $\boldsymbol{k}(\Gamma) = P_1$ and $V = \Gamma_{15}$, so that P_1 is active. If we denote the basis functions of P_1 at $\boldsymbol{k}_1$ and $\boldsymbol{k}_2$ ($= -\boldsymbol{k}_1$) by ϕ_1 and ϕ_2, respectively, we may put $\phi_2 = \phi_1^*$, because the representation is real.[2] Accordingly, in the expression

$$\Delta\varrho = c_1\phi_1 + c_2\phi_2 \ ,$$

we have $c_2 = c_1^*$. Noting that $\phi_1^2 + \phi_2^2$ and $\phi_1\phi_2$ are the basis functions of H_1 and Γ_1, respectively, we find that the fourth-order invariants constructed from c_1 and c_2 are given by $(c_1^2 + c_2^2)^2$ and $c_1^2c_2^2$. This gives us the following form for (14.15):

$$\Phi = \Phi^{(0)} + 2Ac_1c_2 + B_1(c_1^2 + c_2^2)^2 + B_2c_1^2c_2^2 \ .$$

Substituting $c_1 = c\exp(\mathrm{i}\theta)$ ($= c_2^*$) here, we obtain

$$\Phi = \Phi^{(0)} + 2Ac^2 + 4B_1c^4\cos^2 2\theta + B_2c^4 \ .$$

Let us minimize this expression with respect to θ. For positive B_1, its minimum will be taken at $\theta = \pm\pi/4$, so that we have

$$\Delta\varrho = \frac{c}{\sqrt{2}}[(\phi_1 + \phi_2) \pm \mathrm{i}(\phi_1 - \phi_2)] \ .$$

---

[2] A typical form of ϕ_1 will be $\cos\frac{\pi x}{a}\cos\frac{\pi y}{a}\cos\frac{\pi z}{a} + \mathrm{i}\sin\frac{\pi x}{a}\sin\frac{\pi y}{a}\sin\frac{\pi z}{a}$.

For this $\Delta\varrho$, we have the equality

$$\Delta\varrho(a, 0, 0) = -\Delta\varrho(0, 0, 0)$$

together with

$$\Delta\varrho\left(\frac{a}{2}, \frac{a}{2}, \frac{a}{2}\right) = \mp\Delta\varrho(0, 0, 0)$$

because of the equality $\phi_1(0,0,0) = \phi_2(0,0,0)$ which follows from $\{I|0\}\phi_1 = \phi_2$. The superlattice here consists of two sublattices of NaCl structure shifted by the vector $(a/2, a/2, a/2)$, each having A (or B) atoms at the corner and face center and B (or A) atoms at the body and edge centers of a cubic unit cell with the lattice constant $2a$ (Fd3m, O_h^7).

When B_1 is negative, $\theta = 0$, π gives the minimum ($\theta = \pi/2$ leads to an equivalent result) and we have

$$\Delta\varrho = \pm c(\phi_1 + \phi_2) \ .$$

In this case, we find

$$\Delta\varrho(a, 0, 0) = -\Delta\varrho(0, 0, 0) \ ,$$

$$\Delta\varrho\left(\frac{a}{2}, \frac{a}{2}, \frac{a}{2}\right) = 0 \ .$$

The size of the cubic cell becomes eight times as large, with A atoms occupying its corner and face center and B atoms body and edge centers. The body center of the original cell is occupied by B (an alloy of composition AB_3) or by C (in the case of a ternary alloy ABC_2, Fm3m, O_h^5).

14.2 Crystal Structures and Spin Alignments

The problem of spin alignments in the ordered phase of magnetic compounds (at $T = 0$ K) may also be treated as an application of the representation theory of space groups [14.2].

Suppose that the interaction between spins is described by the Hamiltonian

$$H = -\sum_{mi\alpha,\, nj\beta} \mathcal{J}_{i\alpha, j\beta}(m, n) S_{mi\alpha} S_{nj\beta} \ , \qquad (14.21)$$

where m, n denote the number given to the cells and i, j distinguish between different ions in the unit cell. Suffixes α, β stand for the x, y, and z components. Summations over $mi\alpha$ and $nj\beta$ are to be taken independently

$[\mathcal{J}_{i\alpha,j\beta}(m,n) = \mathcal{J}_{j\beta,i\alpha}(n,m)]$. When the coefficients take the form

$$\mathcal{J}_{i\alpha,j\beta}(m,n) = \delta_{\alpha\beta} J_{ij}(m,n) \ , \tag{14.22}$$

the terms in (14.21) represent the ordinary isotropic exchange interaction. However, we assume here the more general form (14.21) to include the dipolar interaction as well as the anisotropy energy. The coefficients $\mathcal{J}_{i\alpha,j\beta}(m,n)$ depend only on the difference $\boldsymbol{R}_n - \boldsymbol{R}_m$ of the position vectors of the cells n and m, so that we have

$$\mathcal{J}_{i\alpha,j\beta}(m,n) = \mathcal{J}_{i\alpha,j\beta}(0, n-m) \ . \tag{14.23}$$

The components of spin $S_{mi\alpha}$ in (14.21) were originally quantum-mechanical operators. We treat them here, however, simply as components of classical vectors. (This is permissible when the magnitude $\boldsymbol{S}_i$ of the spin is sufficiently large.) In accordance with this approximation, we write E for H hereafter. At $T = 0$ K, the spin vectors will tend to align so that the total interaction energy is minimum. The problem of determining the spin alignment means looking for a set of values $\{S_{mi\alpha}\}$ that minimize E under the conditions

$$\sum_{\alpha} S_{mi\alpha}^2 = S_i^2 \ . \tag{14.24}$$

For this purpose, we introduce undetermined multipliers λ_{mi} and solve

$$-\sum_{nj\beta} \mathcal{J}_{i\alpha,j\beta}(m,n) S_{nj\beta} + \lambda_{mi} S_{mi\alpha} = 0 \tag{14.25}$$

for $S_{mi\alpha}$ as functions of λ_{mi} and determine the latter variables from the conditions (14.24). When we have several sets of values for λ_{mi} so determined, the one that makes the value of

$$E = -\sum_{mi\alpha,nj\beta} \mathcal{J}_{i\alpha,j\beta}(m,n) S_{mi\alpha} S_{nj\beta} = -\sum_{mi} \lambda_{mi} S_i^2 \tag{14.26}$$

smallest will correspond to the alignment actually realized. Since the first term on the left-hand side of (14.25) is the molecular field $\boldsymbol{H}_{\text{eff}}(mi)$ (times $g\mu_{\text{B}}$) acting upon the spin $\boldsymbol{S}_{mi}$, equation (14.25) represents the requirement that $\boldsymbol{S}_{mi}$ must be parallel (more exactly, antiparallel, as $\lambda_{mi} > 0$) to the molecular field.

If we write λ_{mi}^0 for the λ_{mi} that makes (14.26) take its minimum value E_0 and denote the corresponding values of $S_{mi\alpha}$ as $S_{mi\alpha}^0$, we must naturally have the inequality

$$\delta E = E - E_0 = -\sum_{\substack{mi\alpha\\nj\beta}} \mathcal{J}_{i\alpha,j\beta}(m,n) \delta S_{mi\alpha} \delta S_{nj\beta} + \sum_{mi\alpha} \lambda_{mi}^0 (\delta S_{mi\alpha})^2 > 0 \tag{14.27a}$$

for arbitrary (nonzero) vectors $\delta \boldsymbol{S}_{mi}$ satisfying

$$\boldsymbol{S}^0_{mi} \cdot \delta \boldsymbol{S}_{mi} = 0 \ . \tag{14.27b}$$

As mentioned above, λ_{mi} are the parameters to be determined so as to satisfy (14.24) after the solutions $S_{mi\alpha}(\lambda)$ of (14.25) have been found. We do not know anything about their symmetry by then, except that they satisfy

$$\det| - \mathscr{J}_{i\alpha, j\beta}(m, n) + \lambda_{mi} \delta_{m,n} \delta_{i,j} \delta_{\alpha,\beta}| = 0 \ . \tag{14.28}$$

For simplicity, however, we make here the following assumption:

The values of λ^0_{mi} are all equal for all equivalent magnetic ions in the crystal. (14.29)

In other words, we put

$$\lambda^0_{mi} = \lambda^0_i$$

and assume λ^0_i for all the equivalent ions in the unit cell to be equal.[3] It must be borne in mind that we are looking in the following for λ^0_{mi} and $S^0_{mi\alpha}$ under such restrictions.

With the assumption (14.29), solving (14.25) is equivalent to solving the stationary value problem for the quantity

$$W = - \sum_{\substack{mi\alpha \\ nj\beta}} \mathscr{J}_{i\alpha, j\beta}(m, n) S_{mi\alpha} S_{nj\beta} + \sum_{mi\alpha} \lambda_i (S_{mi\alpha})^2 \ , \tag{14.30}$$

which is invariant under the symmetry operations of the crystal. Note that the stationary value of W is equal to zero because of (14.26). To take advantage of the crystal symmetry, we introduce the Fourier transform $\sigma_i(\boldsymbol{k})$ of $\boldsymbol{S}_{mi}$:

$$S_{mi\alpha} = N^{-3/2} \sum_{\boldsymbol{k}} \exp(\mathrm{i}\boldsymbol{k} \cdot \boldsymbol{R}_m) \sigma_{i\alpha}(\boldsymbol{k}) \ , \tag{14.31}$$

where

$$\sigma_{i\alpha}(\boldsymbol{k})^* = \sigma_{i\alpha}(-\boldsymbol{k}) \ ,$$

because $S_{mi\alpha}$ is real. The interaction parameter $\mathscr{J}$ may also be transformed as

$$\begin{aligned} \mathscr{J}_{i\alpha, j\beta}(\boldsymbol{k}) &= \sum_n \mathscr{J}_{i\alpha, j\beta}(m, n) \exp[\mathrm{i}\boldsymbol{k} \cdot (\boldsymbol{R}_n - \boldsymbol{R}_m)] \\ &= \mathscr{J}_{j\beta, i\alpha}(\boldsymbol{k})^* = \mathscr{J}_{j\beta, i\alpha}(-\boldsymbol{k}) \ . \end{aligned} \tag{14.32}$$

[3] We have an example in which (14.29) is violated: the metastable vortex configuration in the two-dimensional XY model [14.3].

In terms of these quantities, W can be expressed in the form

$$W = -\sum_k \sum_{i\alpha j\beta} \mathscr{J}_{i\alpha, j\beta}(\boldsymbol{k})\sigma_{i\alpha}(\boldsymbol{k})^* \sigma_{j\beta}(\boldsymbol{k}) + \sum_{ki\alpha} \lambda_i \sigma_{i\alpha}(\boldsymbol{k})^* \sigma_{i\alpha}(\boldsymbol{k}) \ , \tag{14.33}$$

so that its stationary value will be attained by $\sigma_i(\boldsymbol{k})$ obtained from

$$-\sum_{j\beta} \mathscr{J}_{i\alpha, j\beta}(\boldsymbol{k})\sigma_{j\beta}(\boldsymbol{k}) + \lambda_i \sigma_{i\alpha}(\boldsymbol{k}) = 0$$

with λ_i determined so as to satisfy

$$\sum_k \boldsymbol{\sigma}_i(\boldsymbol{k})^* \cdot \boldsymbol{\sigma}_i(\boldsymbol{k} - \boldsymbol{q}) = N^3 S_i^2 \delta_{q,0} \ , \tag{14.34}$$

which corresponds to (14.24).

Suppose we apply a unitary transformation

$$\sigma_{i\alpha}(\boldsymbol{k}) = \sum_\mu U_{i\alpha,\mu} \sigma_\mu(\boldsymbol{k}) \tag{14.35}$$

to diagonalize the Hermitian form on the right-hand side of (14.33) and obtain

$$W = \sum_{\mu k} \Lambda(\mu, \boldsymbol{k}) |\sigma_\mu(\boldsymbol{k})|^2 \ . \tag{14.36}$$

Then the eigenvector $\sigma_\mu(\boldsymbol{k})$ will be characterized by an irreducible representation of the space group G_0 whose symmetry operations leave the Hamiltonian (14.21) invariant. Let us therefore write $\mu = \iota\Gamma\nu$. The symbol Γ denotes the small representation of $\mathscr{G}(\boldsymbol{k})$, ν its basis function and ι distinguishes between different eigenvectors with the same Γ and ν.

In general, it is necessary to know the explicit expression for $\mathscr{J}(\boldsymbol{k})$ in order to determine the set $\{\boldsymbol{k}\}\iota\Gamma$ which makes W (and $\Lambda(\{\boldsymbol{k}\}\iota\Gamma)$ as a function of $\boldsymbol{k}$) stationary with its value zero. However, if we are only to look for $\{\boldsymbol{k}\}\Gamma$ which makes Λ stationary for reasons of symmetry, we can apply the result obtained in the preceding section. Our task then is to search for $\{\boldsymbol{k}\}\Gamma$ which satisfies the condition (14.17), where Γ is the irreducible representation obtained by reducing the representation of $\mathscr{G}(\boldsymbol{k})$ based on the components of spin vectors $\sigma_{i\alpha}(\boldsymbol{k})$. Naturally, this method will not tell us anything about the possibility of Λ taking its stationary (or minimum) value at a general point $\boldsymbol{k}$ inside the BZ (for reasons other than symmetry) as in the case of the helical structure observed, for example, in MnO_2.

It should be pointed out as a particular feature of the spin-alignment problem that the eigenvalue $\Lambda = 0$ often has degeneracy higher than required by the crystal symmetry. a) This is especially true with the isotropic exchange interaction given by (14.22). Note in this case that the spin system has symmetry much higher than that described by the space group G_0 [14.4]. b) When there are more than two inequivalent ions in a unit cell, several λ_i will be involved. It can happen for a certain set of

λ_i that Λ vanishes simultaneously for different representations Γ_1 and Γ_2 at $\boldsymbol{k}$. Then $\sigma_{\Gamma_1}(\boldsymbol{k})$ as well as $\sigma_{\Gamma_2}(\boldsymbol{k})$ will be nonvanishing. It is, of course, necessary in such a case that $D_1 \times D_2$ does not contain V. (More generally, Λ with different stars $\{\boldsymbol{k}_1\}$ and $\{\boldsymbol{k}_2\}$ can vanish simultaneously.)

When a single $D = \imath\{\boldsymbol{k}\}\Gamma$ is active, we examine if (14.34) can be satisfied by (14.35) assuming only the components $\sigma_\mu(\boldsymbol{k})$ of the representation D to be nonvanishing. If the answer is yes, there is a possibility that the corresponding spin alignment given by (14.31) will be realized.

Let us discuss a simple example. Consider the body-centered cubic lattice of the preceding section. We put spin vectors at each lattice site and study possible spin structures. We have only one spin per unit cell here, and this makes the problem easy to handle.[4] Besides, active representations are all real, as seen below.

At the point N, we note that $\boldsymbol{k}_j \doteq -\boldsymbol{k}_j$ and that the point group of the wavevector is D_{2h}. Decomposition of the representation at N based on the components of an axial vector gives N_2, N_3 and N_4. Clearly, it is impossible to construct from any one of them an antisymmetric representation that contains the vector representation Γ_{15}. This means that an alignment as given, for example, in Fig. 14.1 corresponding to $\boldsymbol{k}_1$ is possible (ordering of the third kind in the bcc lattice). This is an antiferromagnetic ordering where the black and white circles represent ions with antiparallel spins in a direction along one of the three cubic axes.

At the Γ point, Γ'_{15} is active because $\{\Gamma'^2_{15}\}$ does not contain Γ_{15}. This yields ferromagnetic ordering, where all the spins point in the same direction.

The situation is similar at H, where $\boldsymbol{k} \doteq -\boldsymbol{k}$ and $\{H'^2_{15}\}$ does not contain Γ_{15}. The corresponding ordering is antiferromagnetic with the spin at the body center pointing upward and the spin at the corner downward (ordering of the first kind in the bcc lattice).

In the case of the P point, we have to examine whether (14.19a) holds for $\boldsymbol{k}(\Gamma) = P_5$ and $V = \Gamma_{15}$, because $\boldsymbol{k}_2 = -\boldsymbol{k}_1$. Choosing $P = \{I|0\}$, we find that it does not. Therefore, the point P is inactive for a general form of $\mathscr{J}_{\alpha\beta}(m, n)$. In the case of an isotropic exchange interaction (14.22), however, the symmetry of the Hamiltonian becomes much higher than that of the actual crystal. Only magnitudes of the vectors $\boldsymbol{\sigma}(\boldsymbol{k})$ are involved in (14.33) and as a result we may put $\boldsymbol{k}(\Gamma) = P_1$ to treat this case. Then, we find that (14.19a) holds and P_1 is active. Assuming only $\sigma_z(\boldsymbol{k}_1)$ and $\sigma_z(\boldsymbol{k}_2) = \sigma_z(\boldsymbol{k}_1)^*$ to be nonvanishing, we find

$$\sigma_z(\boldsymbol{k}_1) \propto (1 + \mathrm{i})$$

from (14.34). Putting this into the right-hand side of (14.31), we obtain

$$S_{mz} \propto \cos(\boldsymbol{k}_1 \cdot \boldsymbol{R}_m) - \sin(\boldsymbol{k}_1 \cdot \boldsymbol{R}_m) \ .$$

[4] If we choose (14.22) for $\mathscr{J}$ in this case, we find that λ is given by the maximum value of $J(\boldsymbol{k}) = \sum_m J(m, n) \exp[\mathrm{i}\boldsymbol{k} \cdot (\boldsymbol{R}_n - \boldsymbol{R}_m)]$. The results given below can also be obtained by determining $\boldsymbol{k}$ directly from $\mathrm{grad}_{\boldsymbol{k}} J(\boldsymbol{k}) = 0$.

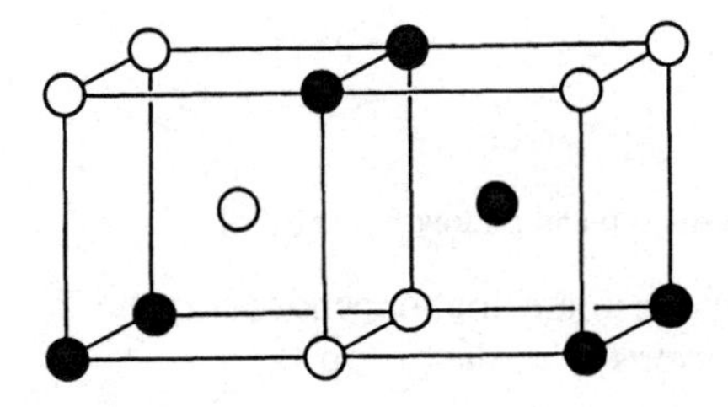

Fig. 14.1. Ordering of the third kind in the bcc lattice

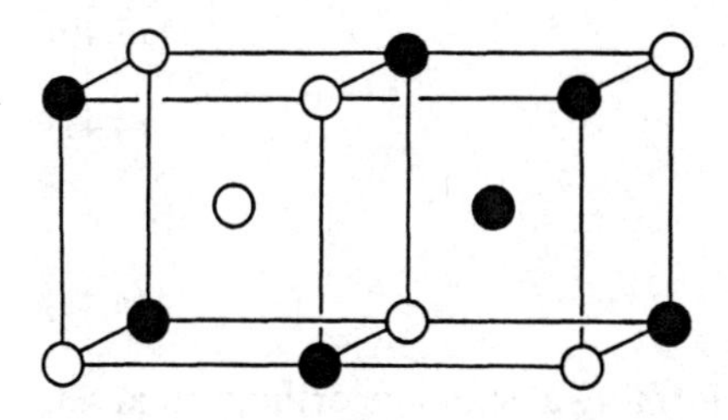

Fig. 14.2. Ordering of the second kind in the bcc lattice

Noting that $\boldsymbol{k}_1 = (\pi/a, \pi/a, \pi/a)$, we have the ordering given in Fig. 14.2 (ordering of the second kind in the bcc lattice).

We cannot decide which of the above four kinds of ordering is realized until after we have compared the magnitudes of the corresponding values of λ. The relative values of λ depend naturally on the form of $\mathscr{J}$. However, we shall not go into further details of this problem.

*14.3 Derivation of the Lifshitz Criterion

In this section, we derive the criterion for the coefficient $A^{\{k'\}\Gamma'}(P_0, T_0)$ to take its stationary value at a point $\boldsymbol{k}$ for reasons of symmetry.

In (14.4, 5), we expanded the free energy Φ as a function of $c_i^{(\alpha)}$. If we expand it directly as a *functional* of $\Delta\varrho(r)$, the second-order term $\Phi^{(2)}$ will take the form

$$\Phi^{(2)} = \int d\boldsymbol{r} \int d\boldsymbol{r}' h(\boldsymbol{r}, \boldsymbol{r}') \Delta\varrho(\boldsymbol{r}) \Delta\varrho(\boldsymbol{r}') \ . \tag{14.37}$$

This result may be understood as follows. We divide the whole crystal into a very large number of small volume elements and number them from 1 to N. Let us denote the volume of the ith element by Δv_i and represent its position by $\boldsymbol{r}_i$. By a sufficiently fine division, we may let $\varrho_i = \varrho(\boldsymbol{r}_i)$ stand for the values of the density ϱ in the ith element. This will allow us to regard the free energy Φ as a function of N variables ϱ_i. If we vary ϱ_i from the equilibrium value $\varrho_{0i} = \varrho_0(r_i)$ to $\varrho_i = \varrho_{0i} + \Delta\varrho_i$, the second-order term in the increment of Φ will be

$$\Phi^{(2)} = \sum_{i,j} h_{ij} \Delta\varrho_i \Delta\varrho_j (\Delta v_i \Delta v_j) \tag{14.38}$$

with

$$h_{ij} = \frac{1}{2} \frac{\partial^2 \Phi}{\partial(\varrho_i \Delta v_i)\partial(\varrho_j \Delta v_j)}\bigg|_{\varrho=\varrho_0} .$$

Equation (14.37) will be obtained by taking the limit $\Delta v_i \to 0$ and putting $h_{ij} \to h(r_i, r_j)$.

In (14.37), h is a real symmetric function of $\boldsymbol{r}$ and $\boldsymbol{r}'$ that depends on P and T and is invariant under the symmetry operations R of the group G_0:

$$h(\boldsymbol{r}', \boldsymbol{r}) = h(\boldsymbol{r}, \boldsymbol{r}') , \tag{14.39}$$

$$h(\boldsymbol{r}, \boldsymbol{r}')^* = h(\boldsymbol{r}, \boldsymbol{r}') , \tag{14.40}$$

$$h(R^{-1}\boldsymbol{r}, R^{-1}\boldsymbol{r}') = h(\boldsymbol{r}, \boldsymbol{r}') . \tag{14.41}$$

Then, the eigenvalue problem

$$[h\phi](\boldsymbol{r}) \equiv \int h(\boldsymbol{r}, \boldsymbol{r}')\phi(\boldsymbol{r}')d\boldsymbol{r}' = A\phi(\boldsymbol{r}) \tag{14.42}$$

yields real eigenvalues with eigenfunctions forming a complete orthonormal set. The eigenfunctions ϕ corresponding to a (degenerate) eigenvalue A are characterized as the basis functions $\phi_{\boldsymbol{k}\nu}^{n(D)}$ $(n = 0, 1, 2, \ldots)$ of an irreducible representation $D = \{\boldsymbol{k}\}\Gamma$ of the group G_0. Note also that there may be additional degeneracy for the eigenvalue $A^{n(D)}$ due to the time-reversal symmetry.

Let us employ this set of eigenfunctions in the expansion of $\Delta\varrho(\boldsymbol{r})$ and retain there for each $\boldsymbol{k}$ only those functions $\phi_{\boldsymbol{k}\nu}$ $(\boldsymbol{k} \in \{\boldsymbol{k}\}, \nu \in \Gamma)$ that correspond to the lowest eigenvalue $A^{n=0(D)}$, where we have dropped the superscripts 0 and (D) to ϕ for simplicity. This then gives us (14.3), or we may as well say that we consider only such $\Delta\varrho(\boldsymbol{r})$ there. If we rewrite (14.3) in terms of the original (not necessarily real) basis functions, we have

$$\Delta\varrho = \sum_D^{\mathrm{r}} \sum_{\boldsymbol{k}\nu} c_{\boldsymbol{k}\nu}\phi_{\boldsymbol{k}\nu} + \sum_D^{\mathrm{n.r.}} \sum_{\boldsymbol{k}\nu} (c_{\boldsymbol{k}\nu}\phi_{\boldsymbol{k}\nu} + c_{\boldsymbol{k}\nu}^* \phi_{\boldsymbol{k}\nu}^*)/\sqrt{2} . \tag{14.43}$$

The first sum is to be taken over real representations D and the second sum over nonreal representations. When D is nonreal, we need not include its conjugate representation D^* in the sum, because its basis ϕ^* is already taken care of in the second sum, paired with ϕ.

If we put (14.43) into (14.37), we obtain

$$\Phi^{(2)} = \sum_D A^{(D)}(P, T) \sum_{\boldsymbol{k}\nu} |c_{\boldsymbol{k}\nu}|^2 \tag{14.44}$$

corresponding to (14.9), where the summation is taken over both real and nonreal representations. As seen below in (14.47), we have

$$c_{\boldsymbol{k}\nu}^* = \sum_{\nu'} c_{-\boldsymbol{k}\nu'}(\hat{U}^{-1})_{\nu\nu'}$$

for the coefficients $c_{\boldsymbol{k}\nu}$ corresponding to real representations D.

In the following, we shall always mean by $D = \{k\}\Gamma$ the representation that corresponds to the lowest eigenvalue $A^{(D)}(P, T)$ among all the eigenvalues A. This is the representation $D^{(\beta)}$ in (14.12,13). Let us write D' for the representation with a wavevector $\boldsymbol{k}' = \boldsymbol{k} + \boldsymbol{\kappa}$ slightly different from $\boldsymbol{k}$. Our problem is to find the condition for $A_{\boldsymbol{k}+\boldsymbol{\kappa}} \equiv A^{(D')}(P, T)$ to take the stationary value $A_{\boldsymbol{k}} = A^{(D)}(P, T)$ at $\boldsymbol{\kappa} = 0$. The quantity $A_{\boldsymbol{k}+\boldsymbol{\kappa}}$ can be obtained by diagonalizing the matrix of h in terms of the functions $\exp(-\mathrm{i}\boldsymbol{\kappa}\cdot\boldsymbol{r})\phi^{n(D)}_{\boldsymbol{k}\nu}$ with the wavevector $\boldsymbol{k} + \boldsymbol{\kappa}$, where the $\phi^{n(D)}_{\boldsymbol{k}\nu}$'s are the unperturbed functions at $\boldsymbol{k}$ (the $\boldsymbol{k}\cdot\boldsymbol{p}$ perturbation method in band theory). As long as we are interested in the correction to $A_{\boldsymbol{k}}$ to first order in $\boldsymbol{\kappa}$, we may confine ourselves to the manifold of the lowest degenerate eigenvalue $A^{n=0(D)}$ as in the perturbation theory for degenerate states (so that $n = 0$ and (D) will not be explicitly written out hereafter). The point $\boldsymbol{k} + \boldsymbol{\kappa}$ will be assumed to be a general point near $\boldsymbol{k}$ in the following.

(i) Real D: When the dimension of the representation Γ is d, $A_{\boldsymbol{k}+\boldsymbol{\kappa}}$ can be obtained as the lowest eigenvalue of the $d \times d$ matrix

$$\begin{aligned}\langle \mathrm{e}^{-\mathrm{i}\boldsymbol{\kappa}\cdot\boldsymbol{r}}\phi_{\boldsymbol{k}\nu}|h|\mathrm{e}^{-\mathrm{i}\boldsymbol{\kappa}\cdot\boldsymbol{r}}\phi_{\boldsymbol{k}\nu'}\rangle &= \iint d\boldsymbol{r}\,d\boldsymbol{r}'\,\phi_{\boldsymbol{k}\nu}(\boldsymbol{r})^*\mathrm{e}^{\mathrm{i}\boldsymbol{\kappa}\cdot(\boldsymbol{r}-\boldsymbol{r}')}h(\boldsymbol{r},\boldsymbol{r}')\phi_{\boldsymbol{k}\nu'}(\boldsymbol{r}')\\ &= A_{\boldsymbol{k}}\delta_{\nu\nu'} + \boldsymbol{\kappa}\cdot\boldsymbol{a}_{\nu\nu'}\ .\end{aligned} \tag{14.45}$$

where the second line corresponds to the expansion of the exponential function to first order in $\boldsymbol{\kappa}$, and $\boldsymbol{a}_{\nu\nu'}$ is given by

$$\begin{aligned}\boldsymbol{a}_{\nu\nu'} &= \mathrm{i}\iint d\boldsymbol{r}\,d\boldsymbol{r}'\,\phi_{\boldsymbol{k}\nu}(\boldsymbol{r})^*(\boldsymbol{r}-\boldsymbol{r}')h(\boldsymbol{r},\boldsymbol{r}')\phi_{\boldsymbol{k}\nu'}(\boldsymbol{r}')\\ &= \mathrm{i}\iint d\boldsymbol{r}\,d\boldsymbol{r}'\boldsymbol{r}h(\boldsymbol{r},\boldsymbol{r}')\{\phi_{\boldsymbol{k}\nu}(\boldsymbol{r})^*\phi_{\boldsymbol{k}\nu'}(\boldsymbol{r}') - \phi_{\boldsymbol{k}\nu'}(\boldsymbol{r})\phi_{\boldsymbol{k}\nu}(\boldsymbol{r}')^*\}\ .\end{aligned} \tag{14.46}$$

For $A_{\boldsymbol{k}+\boldsymbol{\kappa}}$ to be stationary at $\boldsymbol{\kappa} = 0$, it is necessary that all the vectors $\boldsymbol{a}_{\nu\nu'}$ vanish. As is seen from (14.46), this will be the case when $D^* \times D$ does not contain the representation V. However, since we have

$$\phi^*_{\boldsymbol{k}\nu} = \sum_{\nu'}\phi_{-\boldsymbol{k}\nu'}U_{\nu'\nu} \tag{14.47}$$

in this case according to (13.54), $\boldsymbol{a}_{\nu\nu'}$ are linear combinations of the following quantity:

$$\boldsymbol{\beta}_{\nu\nu'} - \mathrm{i}\iint d\boldsymbol{r}\,d\boldsymbol{r}'\boldsymbol{r}h(\boldsymbol{r},\boldsymbol{r}')\{\phi_{-\boldsymbol{k}\nu}(\boldsymbol{r})\phi_{\boldsymbol{k}\nu'}(\boldsymbol{r}') - \phi_{\boldsymbol{k}\nu'}(\boldsymbol{r})\phi_{-\boldsymbol{k}\nu}(\boldsymbol{r}')\}\ . \tag{14.48}$$

These vectors will vanish identically when $\{D^2\}$ does not contain V, because the expression in the braces is the basis function of the antisymmetric representation.

(ii) Nonreal D: (a) When $-\boldsymbol{k}$ does not belong to $\{\boldsymbol{k}\}$, it is necessary that $D^* \times D$ should not contain V. (b) When we have $-\boldsymbol{k} \in \{\boldsymbol{k}\}$, $\phi^*_{-\boldsymbol{k}\nu}$ and $\phi_{\boldsymbol{k}\nu}$ are orthogonal and belong to the same eigenvalue $A_{\boldsymbol{k}}$. We obtain $A_{\boldsymbol{k}+\boldsymbol{\kappa}}$ by diagonalizing the $2d \times 2d$ matrix of h in terms of $\exp(-\mathrm{i}\boldsymbol{\kappa}\cdot\boldsymbol{r})\phi^*_{-\boldsymbol{k}\nu}$ and

$\exp(-\mathrm{i}\boldsymbol{\kappa}\cdot\boldsymbol{r})\phi_{\boldsymbol{k}\nu}$, where we have the following type of matrix elements:

$$\langle \mathrm{e}^{-\mathrm{i}\boldsymbol{\kappa}\cdot\boldsymbol{r}}\phi^*_{-\boldsymbol{k}\nu}|h|\mathrm{e}^{-\mathrm{i}\boldsymbol{\kappa}\cdot\boldsymbol{r}}\phi_{\boldsymbol{k}\nu'}\rangle = \iint d\boldsymbol{r}\, d\boldsymbol{r}'\phi_{-\boldsymbol{k}\nu}(\boldsymbol{r})\mathrm{e}^{\mathrm{i}\boldsymbol{\kappa}\cdot(\boldsymbol{r}-\boldsymbol{r}')}h(\boldsymbol{r},\boldsymbol{r}')\phi_{\boldsymbol{k}\nu'}(\boldsymbol{r}')$$
$$= \boldsymbol{\kappa}\cdot\boldsymbol{\beta}_{\nu\nu'} \tag{14.49}$$

in addition to (14.45). The terms of first order in $\boldsymbol{\kappa}$ appearing in (14.45) as well as (14.49) have to vanish, and this leads to the condition that neither $D^* \times D$ nor $\{D^2\}$ should contain V. Since we know that

$$\{\underline{D}^2\} = \{(D + D^*)^2\} = \{D^2\} + \{D^{*2}\} + D^* \times D \ , \tag{14.50}$$

we may simply put the requirement in case (ii) as "$\{\underline{D}^2\}$ should not contain V".

*14.3.1 Lifshitz's Derivation of the Lifshitz Criterion

Lifshitz derived (14.17) from a more direct and physical argument. We require that the structure described by the wavevector $\boldsymbol{k}$ be stable compared to the one with the vector $\boldsymbol{k} + \boldsymbol{\kappa}$. The latter structure is obtained by modulating the former by a wavevector $\boldsymbol{\kappa}$ and has spatial inhomogeneity of the order of κ^{-1} compared to it. This inhomogeneity can be taken into consideration macroscopically by regarding the coefficients c_i in (14.18) as slowly varying functions of the spatial coordinates. This implies that Φ will depend not only on c_i but also on its derivatives such as $\partial c_i/\partial x$. Therefore, the first-order terms with respect to the latter derivatives must also vanish when Φ is expanded as a power series in c_i and $\partial c_i/\partial x$. Now, the total free energy of the crystal is given by a volume integral like $\int \Phi dV$. When integrated, the terms such as $\partial c_i/\partial x$ and the symmetric part of the product $c_i\partial c_j/\partial x$, i.e.,

$$c_i\frac{\partial c_j}{\partial x} + \frac{\partial c_i}{\partial x}c_j = \frac{\partial}{\partial x}(c_i c_j) \ ,$$

will contribute to the total free energy as the surface integral, so that a significant contribution comes only from the antisymmetric part

$$c_i\frac{\partial c_j}{\partial x} - \frac{\partial c_i}{\partial x}c_j \ .$$

If such terms are invariant under the operations of G_0, they will be present in the expansion of Φ and the structure described by the wavevector $\boldsymbol{k}$ will become unstable. Stability is assured only when no linear combination of the antisymmetric products of c_i transforms like a component of a vector, and this is the criterion given in (14.17).

15. The Symmetric Group

In this chapter, fundamentals of the symmetric group, namely, classes of permutations, Young diagrams, irreducible characters, and the construction of irreducible representations and their bases, are reviewed and summarized.

The relation between the irreducible representations of the symmetric group and tensor representations of the unitary group is essential in understanding the wavefunctions of a many-electron atom with a definite magnitude of spin S. We have also seen that the concepts of symmetric and antisymmetric product representations play important roles in other fields covered in this book.

15.1 The Symmetric Group (Permutation Group)

Suppose that n objects are placed on n seats numbered from 1 to n. A permutation P is a move to shift the object on the seat p_i to a new seat i, which is expressed as

$$P = \begin{pmatrix} 1 & 2 & \cdots & n \\ p_1 & p_2 & \cdots & p_n \end{pmatrix} \quad \text{or} \quad P = (k \to p_k) \ . \tag{15.1}$$

According to this definition, the permutation P rearranges the array of figures $1, 2, \ldots, n$ into the array $p_1, p_2, \ldots, p_n$. If we perform a permutation $Q = (k \to q_k)$ successively after P, the result is another permutation R, which is written as QP. The rule of composition is given by

$$\begin{aligned} R = QP &= (k \to q_k)(k \to p_k) = (k \to q_k)(q_k \to p_{q_k}) \\ &= (k \to p_{q_k}) \ . \end{aligned} \tag{15.2}$$

Details of this rule have already been given in Sect. 2.3.

The symmetric group $\mathfrak{S}_n$ of degree n is the set of permutations of n objects. Its order is $n!$.

Permutations such as $C = \begin{pmatrix} 1 & 2 \ldots & p-1 & p \\ 2 & 3 \ldots & p & 1 \end{pmatrix}$, which produces a cyclic replacement $1 \to 2, 2 \to 3, \ldots, p-1 \to p, p \to 1$, are called *cycles* of length (order) p and are denoted as

$$C = (1\ 2 \ldots p) \ . \tag{15.3}$$

Any permutation can be put into the form of a product of cycles that share no common numbers. For example, we have

$$\begin{pmatrix} 1 & 2 & 3 & 4 & 5 & 6 \\ 2 & 5 & 6 & 4 & 1 & 3 \end{pmatrix} = (4)(3\ \ 6)(1\ \ 2\ \ 5) \equiv (3\ \ 6)(1\ \ 2\ \ 5)\ , \tag{15.4}$$

where we have followed the convention of writing out only cycles of length greater than unity. If repetitions are allowed for, cycles may be expressed as products of *transpositions* (ij) as, for example,

$$(1\ \ 2 \cdots p) = (1\ \ 2)(1\ \ 3) \cdots (1\ \ p)\ . \tag{15.5}$$

When a permutation is expressed as a product of transpositions, one requiring an even number of transpositions is called an *even permutation*, while one requiring an odd number is called an *odd permutation*. Every permutation is either even or odd.

The set of all even permutations of n objects forms the *alternating group* $\mathfrak{A}_n$ of order $n!/2$, which is an invariant subgroup of $\mathfrak{S}_n$.

Exercise 15.1. Confirm $\mathfrak{S}_4 = \mathfrak{A}_4 + (12)\mathfrak{A}_4$ as shown in Table 15.1.

If we transform a permutation $P = (i \to i')$ by $Q = (j \to i) = (j' \to i')$ to obtain a permutation R conjugate to P, the result is

$$R = QPQ^{-1} = (j \to i)(i \to i')(i' \to j') = (j \to j')\ , \tag{15.6}$$

so that R is obtained by replacing i, i' of P by j, j', respectively. For example, we find

$$(1\ \ 2\ \ 3)(2\ \ 4\ \ 5\ \ 1\ \ 6)(1\ \ 2\ \ 3)^{-1} = (1\ \ 4\ \ 5\ \ 3\ \ 6)\ .$$

Table 15.1. The alternating group $\mathfrak{A}_4$ and $(1\ \ 2)\mathfrak{A}_4$

$\mathfrak{A}_4$	$(1\ \ 2)\mathfrak{A}_4$
E	$(1\ \ 2)$
$(1\ \ 3\ \ 2) = (1\ \ 2)(2\ \ 3)$	$(2\ \ 3)$
$(1\ \ 4\ \ 2) = (1\ \ 2)(2\ \ 4)$	$(2\ \ 4)$
$(1\ \ 2\ \ 3) = (1\ \ 2)(1\ \ 3)$	$(1\ \ 3)$
$(1\ \ 2\ \ 4) = (1\ \ 2)(1\ \ 4)$	$(1\ \ 4)$
$(1\ \ 2)(3\ \ 4)$	$(3\ \ 4)$
$(2\ \ 3\ \ 4) = (2\ \ 3)(2\ \ 4)$	$(1\ \ 3\ \ 4\ \ 2)$
$(2\ \ 4\ \ 3) = (2\ \ 4)(2\ \ 3)$	$(1\ \ 4\ \ 3\ \ 2)$
$(1\ \ 3)(2\ \ 4)$	$(1\ \ 4\ \ 2\ \ 3)$
$(1\ \ 3\ \ 4) = (1\ \ 3)(1\ \ 4)$	$(1\ \ 2\ \ 3\ \ 4)$
$(1\ \ 4)(2\ \ 3)$	$(1\ \ 3\ \ 2\ \ 4)$
$(1\ \ 4\ \ 3) = (1\ \ 4)(1\ \ 3)$	$(1\ \ 2\ \ 4\ \ 3)$

Accordingly, transforming a cycle does not change its length. In other words, permutations that can be obtained by transforming a permutation have the same cycle structure. Conversely, permutations with the same cycle structure belong to the same class. This is because cycles with equal lengths $C = (i_1\, i_2 \ldots i_s)$ and $C' = (j_1\, j_2 \ldots j_s)$ can be correlated by the transformation $C' = QCQ^{-1}$ with $Q = (j_m \to i_m)$, as seen above.

These considerations show that a conjugate class of the symmetric group is specified by a set of nonnegative integers $k_1, k_2, \ldots, k_n$, when the elements of the class have k_s cycles of length s. Let us write (k) for this set of numbers and use (k) to denote a class of $\mathfrak{S}_n$:

$$(k) = (k_1, k_2, \ldots, k_n) \tag{15.7}$$

with

$$k_1 + 2k_2 + \cdots + nk_n = n\ . \tag{15.8}$$

The number of elements $n(k)$ in the class (k) is given by

$$n(k) = \frac{n!}{1^{k_1}\, k_1!\, 2^{k_2}\, k_2! \cdots n^{k_n}\, k_n!}\ . \tag{15.9}$$

This may be understood as follows. The number of ways of assigning n different figures $1, 2, \ldots, n$ to n seats is $n!$. Among these $n!$ ways, we do not distinguish between the $k_j!$ ways of arranging k_j cycles of length j, so that $n!$ is to be divided by $\Pi_j k_j!$. We note that the j cyclic changes of figures in a cycle of length j simply lead to the same permutation. Since we have k_j such cycles, further division by $\Pi_j j^{k_j}$ is necessary and the result is (15.9).

We may also use a sequence of integers $(\mu_1, \mu_2, \ldots, \mu_n)$ with $\mu_j \geqq \mu_{j+1} \geqq 0$ to specify a class, each μ_j representing the length of a cycle. For example, we find

$$(\mu_1, \mu_2, \ldots, \mu_6) = (3, 2, 1, 0, 0, 0)\ , \qquad (k) = (1, 1, 1, 0, 0, 0)$$

for the class which the permutation (15.4) belongs to. The sequence $\mu_1, \mu_2, \ldots, \mu_n$ is obtained by reading (k) from the right and putting k_μ times the figure μ (from left to right) every time we meet a nonzero value of k_μ, so that it is characterized as a partition of n:

$$\left.\begin{array}{l} \mu_1 + \mu_2 + \cdots + \mu_n = n \\ \mu_1 \geqq \mu_2 \geqq \cdots \geqq \mu_n \geqq 0 \end{array}\right\}\ . \tag{15.10}$$

This means that we have as many classes as different partitions of n.

15.2 Irreducible Characters

We will now quote without proof various theorems on the irreducible representations of the symmetric group [15.1].

The number of inequivalent irreducible representations of the symmetric group $\mathfrak{S}_n$ is given by the number of classes in $\mathfrak{S}_n$, which is equal to the number of partitions of n. In fact, each irreducible representation of the symmetric group is characterized by a *partition* of n:

$$[\lambda] = [\lambda_1, \lambda_2, \ldots, \lambda_n] \ ,$$
$$\lambda_1 + \lambda_2 + \cdots + \lambda_n = n \ ,$$
$$\lambda_1 \geqq \lambda_2 \geqq \cdots \geqq \lambda_n \geqq 0 \ .$$

In the symbol $[\lambda_1, \lambda_2, \ldots, \lambda_n]$, it is customary to write $[\lambda_1, \lambda_2, \ldots, \lambda_\nu]$ when $\lambda_{\nu+1}$ and the rest are equal to zero. When λ_j appears p times in succession, the abbreviated notation $[\cdots \lambda_j^p \cdots]$ will also be used. The partition $[\lambda]$ may be expressed as a diagram (the *Young diagram*) which consists of ν rows of squares, with λ_1 squares in the first row, λ_2 squares in the second and so on, aligned at the left end. For instance, we have the following three diagrams for the irreducible representations of $\mathfrak{S}_3$:

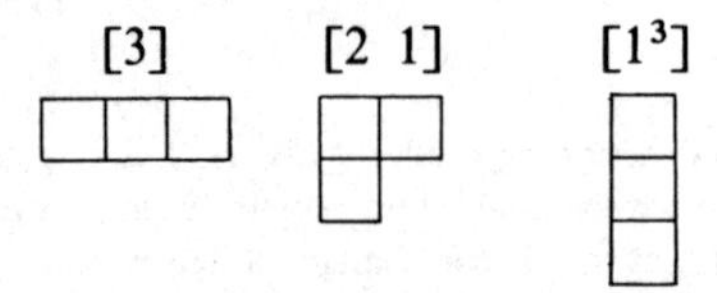

The character $\chi^{[\lambda]}(k)$ for the class (k) of the irreducible representation $D^{[\lambda]}$ can be obtained from the identity

$$s_1^{k_1} s_2^{k_2} \cdots s_n^{k_n} |x_1^{m-1}, x_2^{m-2}, \ldots, x_{m-1}^1, x_m^0|$$
$$= \sum_{[\lambda]} \chi^{[\lambda]}(k) |x_1^{l_1}, x_2^{l_2}, \ldots, x_m^{l_m}| \qquad (15.11)$$

as the coefficient of $x_1^{l_1} x_2^{l_2} \ldots x_m^{l_m}$, choosing $m \geqq \nu$, where

$$s_q = x_1^q + x_2^q + \cdots + x_m^q \ , \qquad q = 1, 2, \ldots, n \ ,$$
$$l_j = \lambda_j + m - j \ , \qquad j = 1, 2, \ldots, m \ ,$$

$$|x_1^{l_1}, x_2^{l_2}, \ldots, x_m^{l_m}| = \begin{bmatrix} x_1^{l_1} & x_1^{l_2} \ldots x_1^{l_m} \\ x_2^{l_1} & x_2^{l_2} \ldots x_2^{l_m} \\ \cdots & \cdots\cdots\cdots \\ x_m^{l_1} & x_m^{l_2} \ldots x_m^{l_m} \end{bmatrix} \ . \qquad (15.12)$$

The summation on the right-hand side of (15.11) is taken over the partitions of n:

$$\lambda_1 + \lambda_2 + \cdots + \lambda_m = n \ , \quad \lambda_1 \geqq \lambda_2 \geqq \cdots \geqq \lambda_m \geqq 0 \ .$$

The dimension $g[\lambda]$ of the representation $D^{[\lambda]}$ is equal to the character of the identity element, $\chi^{[\lambda]}(n, 0, \ldots)$, which reads

$$g[\lambda] = \frac{n!\, D(l_1, l_2, \ldots, l_m)}{l_1!\, l_2! \cdots l_m!}\,, \tag{15.13}$$

$$D(l_1, l_2, \ldots, l_m) = |l_1^{m-1}, l_2^{m-2}, \ldots, l_m^0| = \prod_{i<j}(l_i - l_j)\,. \tag{15.14}$$

For the class of transpositions $(k) = (n-2, 1, 0, \ldots)$, we find

$$\frac{1}{2}n(n-1)\chi^{[\lambda]}(n-2, 1, 0, \ldots) = g[\lambda]\sum_{j=1}^{m}\left(\frac{1}{2}\lambda_j(\lambda_j+1) - j\lambda_j\right).$$

It is also to be remarked that we have a recursion formula for $\chi^{[\lambda]}$ which can be derived from (15.11). When (k) includes a cycle of order p, we denote by (k') the class obtained by deleting this cycle from (k). Then, the following equation holds:

$$\begin{aligned}\chi^{[\lambda_1, \ldots, \lambda_m]}(k) &= \sum_{[\lambda']}\chi^{[\lambda']}(k') \\ &= \chi^{[\lambda_1 - p, \lambda_2, \ldots, \lambda_m]}(k') + \chi^{[\lambda_1, \lambda_2 - p, \ldots, \lambda_m]}(k') + \cdots\end{aligned} \tag{15.15}$$

For $\chi^{[\lambda']}$ with unnatural order $\lambda'_{j+1} - \lambda'_j \geqq 2$, which occurs when λ_j is substituted by $\lambda_j - p$ on the right-hand side of this equation, the following replacement is repeated until a natural order of λ_j's is reached:

$$\chi^{[\lambda'_1, \ldots, \lambda'_j, \lambda'_{j+1}, \ldots, \lambda'_m]} \to -\chi^{[\lambda'_1, \ldots, \lambda'_{j+1} - 1, \lambda'_j + 1, \ldots, \lambda'_m]}$$

It may happen during this process (or on the right-hand side of (15.15)) that we meet $\chi^{[\lambda']}$'s with $\lambda'_{j+1} - \lambda'_j = 1$. They are put equal to zero, as are $\chi^{[\lambda']}$'s with negative λ'_m. In Table 15.2, the irreducible characters $\chi^{[\lambda]}$ for $n = 2$–5 are presented.

15.3 Construction of Irreducible Representation Matrices

Irreducible unitary representations $D^{[\lambda]}$ of $\mathfrak{S}_n$ may be constructed by the following procedure. We first note that any permutation of $\mathfrak{S}_n$ can be expressed as a product of $n-1$ transpositions (12), (23), ..., $(n-1, n)$. For example, we find

$$\begin{aligned}&(1\ 3) = (1\ 2)(2\ 3)(1\ 2)\,, \quad (1\ 3\ 2) = (1\ 2)(2\ 3)\,, \\ &(1\ 2\ 3) = (2\ 3)(1\ 2)\,.\end{aligned} \tag{15.16}$$

Table 15.2. Tables of irreducible characters for the symmetric groups $\mathfrak{S}_n$ ($n = 2$–5). Classes (k) are given in the first column of each table with partitions $[\lambda]$ in the first row

$\mathfrak{S}_2$	[2]	[1²]
(2 0)	1	1
(0 1)	1	−1

$\mathfrak{S}_3$	[3]	[2 1]	[1³]
(3 0 0)	1	2	1
3(1 1 0)	1	0	−1
2(0 0 1)	1	−1	1

$\mathfrak{S}_4$	[4]	[3 1]	[2²]	[2 1²]	[1⁴]
(4 0 0 0)	1	3	2	3	1
6(2 1 0 0)	1	1	0	−1	−1
3(0 2 0 0)	1	−1	2	−1	1
8(1 0 1 0)	1	0	−1	0	1
6(0 0 0 1)	1	−1	0	1	−1

$\mathfrak{S}_5$	[5]	[4 1]	[3 2]	[3 1²]	[2² 1]	[2 1³]	[1⁵]
(5 0 0 0 0)	1	4	5	6	5	4	1
10(3 1 0 0 0)	1	2	1	0	−1	−2	−1
15(1 2 0 0 0)	1	0	1	−2	1	0	1
20(2 0 1 0 0)	1	1	−1	0	−1	1	1
20(0 1 1 0 0)	1	−1	1	0	−1	1	−1
30(1 0 0 1 0)	1	0	−1	0	1	0	−1
24(0 0 0 0 1)	1	−1	0	1	0	−1	1

Accordingly, the representation matrix $\hat{D}^{[\lambda]}(P)$ for any permutation P is obtainable once we know the matrices $\hat{D}^{[\lambda]}(i, i+1)$ for transpositions $(i, i+1)$ $(i = 1, 2, \ldots, n-1)$.

To specify the rows and columns of the representation matrices, we draw *standard tableaux* (or simply *tableaux*) by filling in the squares of the Young diagram $[\lambda]$ with the figures $1, 2, \ldots, n$. The figures are to be arranged so that they increase to the right in each row and always increase downward in each column. The number $g[\lambda]$ of different tableaux constructed from a single Young diagram $[\lambda]$ is equal to the dimension of the representation $D^{[\lambda]}$. For example, we have three tableaux for the irreducible representation [31]:

1	2	3
4		

(2 1 1 1)

1	2	4
3		

(1 2 1 1)

1	3	4
2		

(1 1 2 1) (15.17)

Instead of the tableaux, we can use the Yamanouchi symbol (or simply *Y-symbol* [15.2]), which is a sequence of figures showing the number of the row in which the figures $n, n-1, \ldots, 1$ appear in the tableau in this order. In (15.17) we have also given the corresponding Y-symbol under each tableau. Note that, if the figure j appears in the square of rth row and μth column, then the jth to last figure of the Y-symbol will be r. This particular figure r is the μth to last one among the figures r in the Y-symbol. When we simply write α or β below, we will mean by it a tableau or Y-symbol.

When we specify the rows and columns of the representation matrices of $D^{[\lambda]}$ by tableaux, the rule for constructing $\hat{D}^{[\lambda]}(i, i+1)$ is given as follows.

(1) When i and $i+1$ appear in the same row of the tableau α, we set $\hat{D}^{[\lambda]}_{\alpha\alpha}(i, i+1) = 1$ with all the other elements of the row α and column α put to zero.
(2) When i and $i+1$ appear in the same column of α, we set $D^{[\lambda]}_{\alpha\alpha}(i, i+1) = -1$ and put all the other elements of the row α and column α equal to zero.
(3) When i and $i+1$ are located respectively in the rth row, μth column and the r'th row, μ'th column of the tableau α and $r \neq r'$, $\mu \neq \mu'$ (which means that i and $i+1$ are neither in the same row nor in the same column), $\hat{D}^{[\lambda]}(i, i+1)$ has an off-diagonal element between α and β, the latter tableau being obtained by exchanging i and $i+1$ in the tableau α. Nonvanishing elements in the rows α, β and columns α, β are given by[1]

$$
\begin{aligned}
D^{[\lambda]}_{\alpha\alpha}(i, i+1) &= -D^{[\lambda]}_{\beta\beta}(i, i+1) = \varrho_\alpha = \frac{1}{\mu' - r' - \mu + r}\,, \\
D^{[\lambda]}_{\alpha\beta}(i, i+1) &= D^{[\lambda]}_{\beta\alpha}(i, i+1) = \sqrt{1 - \varrho_\alpha^2}\;.
\end{aligned}
\tag{15.18}
$$

Let us put down, as an example, some matrices of the irreducible representation [31] of $\mathfrak{S}_4$ derived according to the rule given above (the rows and columns are specified by the tableaux of (15.17) in the order given there):

$$
\hat{D}(1\ 2) = \begin{bmatrix} 1 & 0 & 0 \\ 0 & 1 & 0 \\ 0 & 0 & -1 \end{bmatrix}, \quad
\hat{D}(2\ 3) = \begin{bmatrix} 1 & 0 & 0 \\ 0 & -1/2 & \sqrt{3}/2 \\ 0 & \sqrt{3}/2 & 1/2 \end{bmatrix},
$$

$$
\hat{D}(3\ 4) = \begin{bmatrix} -1/3 & \sqrt{8}/3 & 0 \\ \sqrt{8}/3 & 1/3 & 0 \\ 0 & 0 & 1 \end{bmatrix}.
$$

[1] In the representation $[\lambda]^*$ conjugate to $[\lambda]$, we choose the right-hand side of the second line as $-\sqrt{1-\varrho_\alpha^2}$ following (15.19).

Representation matrices for other permutations of $\mathfrak{S}_4$ may be obtained by making use of relations like (15.16):

$$\hat{D}(1\ 3) = \begin{bmatrix} 1 & 0 & 0 \\ 0 & -1/2 & -\sqrt{3}/2 \\ 0 & -\sqrt{3}/2 & 1/2 \end{bmatrix} ,$$

$$\hat{D}(1\ 4) = \begin{bmatrix} -1/3 & -\sqrt{2}/3 & -\sqrt{2/3} \\ -\sqrt{2}/3 & 5/6 & -1/2\sqrt{3} \\ -\sqrt{2/3} & -1/2\sqrt{3} & 1/2 \end{bmatrix} ,$$

$$\hat{D}(2\ 4) = \begin{bmatrix} -1/3 & -\sqrt{2}/3 & \sqrt{2/3} \\ -\sqrt{2}/3 & 5/6 & 1/2\sqrt{3} \\ \sqrt{2/3} & 1/2\sqrt{3} & 1/2 \end{bmatrix} .$$

The diagram $[\lambda]^*$ obtained by transposing the rows and columns of a diagram $[\lambda]$ is said to be *conjugate* to the latter. The representation corresponding to $[\lambda]^*$ is also characterized as conjugate to $D^{[\lambda]}$. For example, $D^{[31]}$ and $D^{[211]}$ are conjugate to each other. The representation matrices of $D^{[\lambda]^*}$ are related to those of $D^{[\lambda]}$ by

$$D^{[\lambda]^*}_{\alpha^*\beta^*}(P) = (-1)^P D^{[\lambda]}_{\alpha\beta}(P) , \tag{15.19}$$

where $(-1)^P$ takes the value $+1$ or -1 according as the permutation P is even or odd, and α^* denotes the transpose of the tableau α. Equation (15.19) leads to the irreducible character for $D^{[\lambda]^*}$ given by

$$\chi^{[\lambda]^*}(P) = (-1)^P \chi^{[\lambda]}(P) , \tag{15.20}$$

where * is not to be confused with the symbol for a complex conjugate.

15.4 The Basis for Irreducible Representations

In the unitary (or orthogonal) representation given in the preceding section, $n-1$ mutually commuting operators

$$A_j = \sum_{i=1}^{j-1} (ij) , \quad j = 2, 3, \ldots, n , \tag{15.21}$$

are diagonalized simultaneously. This can be verified directly for the example $D^{[31]}$ of $\mathfrak{S}_4$. If we put $(A_j)_{\alpha\alpha} = a_j(\alpha)$, the eigenvalue $a_j(\alpha)$ is given by

$$a_j(\alpha) = \mu(j\alpha) - r(j\alpha) , \tag{15.22}$$

where we assume the figure j occupies the square at the $r(j\alpha)$th row and $\mu(j\alpha)$th column of the tableau α.

The eigenvector $Q_\alpha^{[\lambda]}$ corresponding to the eigenvalue $a_j(\alpha)$ may be constructed in the following way: Let us imagine that we take successive steps to build up the tableau α of the Young diagram $[\lambda]$. We first write down the tableau α as $\alpha(2)$ for two figures 1 and 2. We then add $\boxed{3}$ to make it a tableau $\alpha(3)$ for three figures. Further addition of $\boxed{4}$ will then lead to $\alpha(4)$. We may regard the tableau α as $\alpha(n)$ derived in this way. We next define *Young's symmetrizer* $Q_{\alpha(s)}$ by

$$Q_{\alpha(s)} = H_{\alpha(s)} K_{\alpha(s)} \tag{15.23}$$

for each tableau $\alpha(s)$ $(s = 2, 3, \ldots, n)$ that appears during this process, where $H_{\alpha(s)}$ is the symmetrizer of the rows of the tableau $\alpha(s)$ and $K_{\alpha(s)}$ is the antisymmetrizer of its columns. For example, Young's symmetrizer for the tableau $\alpha(3) = \begin{array}{|c|c|}\hline 1 & 2 \\ \hline 3 \\ \cline{1-1}\end{array}$ is given by $Q_{\alpha(3)} = (E + (12))(E - (13))$. In general, H_α is expressed as $\sum\limits_M M$ in terms of the permutations M of the figures in the same row, whereas K_α is given by $\sum\limits_N (-1)^N N$ with permutations N of the figures in the same column of the tableau α.

Finally, we construct a product of $Q_{\alpha(s)}$,

$$Q_\alpha^{[\lambda]} = Q_{\alpha(2)} Q_{\alpha(3)} \cdots Q_{\alpha(n)} , \tag{15.24}$$

which provides us with the simultaneous eigenvector of A_j $(j = 2, 3, \ldots, n)$, which has the eigenvalues $a_j(\alpha)$:[2]

$$A_j Q_\alpha^{[\lambda]} = a_j(\alpha) Q_\alpha^{[\lambda]} . \tag{15.25}$$

This means that we can also use the set of eigenvalues $(a_2(\alpha), a_3(\alpha), \ldots, a_n(\alpha))$ instead of the tableau α or Y-symbol to label the rows and columns of the representation matrices.

As an example, the basis set for the representation [21] of $\mathfrak{S}_3$ is given below. Besides the tableaux, the Y-symbols and the corresponding set of eigenvalues of A_j, (a_2, a_3) are also shown.

$\begin{array}{|c|c|}\hline 1 & 2 \\ \hline 3 \\ \cline{1-1}\end{array}$ (2 1 1)

$$Q_{(211)}^{[21]} = (E + (1\ 2))(E + (1\ 2))(E - (1\ 3)) = 2E + 2(1\ 2) - 2(1\ 3) - 2(1\ 2\ 3), \quad a_2 = 1, a_3 = -1 ,$$

$\begin{array}{|c|c|}\hline 1 & 3 \\ \hline 2 \\ \cline{1-1}\end{array}$ (1 2 1)

$$Q_{(121)}^{[21]} = (E - (1\ 2))(E + (1\ 3))(E - (1\ 2)) = 2E + (1\ 3) + (2\ 3) - 2(1\ 2) - (1\ 2\ 3) - (1\ 3\ 2), \quad a_2 = -1, a_3 = 1 .$$

[2] We may equally well put $Q_{\alpha(s)}$ equal to $K_{\alpha(s)} H_{\alpha(s)}$ and define $Q_\alpha^{[\lambda]}$ by (15.24) in terms of these $Q_{\alpha(s)}$. The result will also satisfy (15.25).

Corresponding to the rule given in the preceding section, we have the following equations:

(1) When i and $i+1$ appear in the same row of the tableau α,

$$(i, i+1)Q_\alpha^{[\lambda]} = Q_\alpha^{[\lambda]} .$$

(2) When i and $i+1$ appear in the same column of α,

$$(i, i+1)Q_\alpha^{[\lambda]} = -Q_\alpha^{[\lambda]} .$$

(3) When i and $i+1$ appear neither in the same row nor in the same column of α,

$$(i, i+1)Q_\alpha^{[\lambda]}/\|Q_\alpha^{[\lambda]}\| = \varrho_\alpha Q_\alpha^{[\lambda]}/\|Q_\alpha^{[\lambda]}\| + \sqrt{1-\varrho_\alpha^2}\, Q_\beta^{[\lambda]}(i, i+1)/\|Q_\beta^{[\lambda]}\| ,$$

$$(i, i+1)Q_\beta^{[\lambda]}/\|Q_\beta^{[\lambda]}\| = \sqrt{1-\varrho_\alpha^2}\, Q_\alpha^{[\lambda]}(i, i+1)/\|Q_\alpha^{[\lambda]}\| - \varrho_\alpha Q_\beta^{[\lambda]}/\|Q_\beta^{[\lambda]}\| ,$$

with

$$\frac{1}{\varrho_\alpha} = a_{i+1}(\alpha) - a_i(\alpha) = -\frac{1}{\varrho_\beta} ,$$

denoting by β the tableau derived from α by exchanging the figures i and $i+1$.

The norm $\|Q^{[\lambda]}\|(>0)$ of $Q^{[\lambda]} = \sum_P c(P)P$ is defined by

$$\|Q^{[\lambda]}\|^2 = \sum_P (c(P))^2 ,$$

so that $\|Q^{[21]}_{(211)}\| = 4$, $\|Q^{[21]}_{(121)}\| = 2\sqrt{3}$ in the example given above.

15.5 The Unitary Group and the Symmetric Group

The set of m-dimensional unitary matrices $\hat{u} = (u_{ij})$ forms the unitary group $\mathrm{U}(m)$ of dimension m. If we introduce the basis set X_i in the m-dimensional vector space and define a unitary transformation U of the set as

$$UX_j = \sum_{i=1}^{m} X_i u_{ij} , \tag{15.26}$$

the matrix $\hat{u}$ is the representation matrix $\hat{\mathscr{D}}(U)$ of U. As is well known, unitary matrices can be diagonalized by an appropriate unitary matrix $\hat{T}$:

$$\hat{T}\hat{u}\hat{T}^{-1} = \begin{bmatrix} x_1 & & & \\ & x_2 & & \\ & & \ddots & \\ & & & x_m \end{bmatrix} , \tag{15.27}$$

where x_j are complex numbers with modulus unity. The classes of the group $\mathrm{U}(m)$ are thus characterized by the set of numbers $(x_1, x_2, \ldots, x_n)$.

Let us introduce m^n basis functions $X_{j_1 j_2 \cdots j_n}$ which transform like a direct product of X_j as given by (15.26):

$$UX_{j_1 j_2 \cdots j_n} = \sum_{\{i\}} X_{i_1 i_2 \cdots i_n} u_{i_1 j_1} u_{i_2 j_2} \ldots u_{i_n j_n} \; .$$

They provide us with the basis for the product representation $(\mathscr{D})^n \equiv \mathscr{D} \times \mathscr{D} \times \cdots \times \mathscr{D}$ of $\mathrm{U}(m)$ (representation by a covariant tensor of rank n). The representation $(\mathscr{D})^n$ is not irreducible in general as a representation of $\mathrm{U}(m)$. It is possible to show, however, that representations of $\mathrm{U}(m)$ by tensors with definite symmetry are irreducible [15.2]. We examine this situation in the case of $n = 2$.

Let us define the result of operating the transposition (12) upon a tensor $X_{j_1 j_2}$ by

$$(1\;\;2) X_{j_1 j_2} = X_{j_2 j_1} \; .$$

Then, it follows from the equation

$$UX_{j_1 j_2} = \sum_{i_1 i_2} X_{i_1 i_2} u_{i_1 j_1} u_{i_2 j_2} \tag{15.28}$$

that we have

$$\begin{aligned} (1\;\;2) UX_{j_1 j_2} &= \sum (1\;\;2) X_{i_1 i_2} u_{i_1 j_1} u_{i_2 j_2} \\ &= \sum X_{i_2 i_1} u_{i_1 j_1} u_{i_2 j_2} \\ &= UX_{j_2 j_1} = U(1\;\;2) X_{j_1 j_2} \; . \end{aligned} \tag{15.29}$$

This leads to the consequence that the components of a symmetric tensor $\mathbf{X}(\mathrm{S})$ and those of an antisymmetric tensor $\mathbf{X}(\mathrm{A})$

$$X_{j_1 j_2}(\mathrm{S}) = (E + (1\;\;2)) X_{j_1 j_2} \; , \qquad X_{j_1 j_2}(\mathrm{A}) = (E - (1\;\;2)) X_{j_1 j_2}$$

transform, respectively, among themselves as

$$\begin{aligned} UX_{j_1 j_2}(\mathrm{S}) &= \sum X_{i_1 i_2}(\mathrm{S}) \{ u_{i_1 j_1} u_{i_2 j_2} + u_{i_2 j_1} u_{i_1 j_2} \} / 2 \; , \\ UX_{j_1 j_2}(\mathrm{A}) &= \sum X_{i_1 i_2}(\mathrm{A}) \{ u_{i_1 j_1} u_{i_2 j_2} - u_{i_2 j_1} u_{i_1 j_2} \} / 2 \; , \end{aligned}$$

under U and mixing of different symmetry never occurs. It can further be shown that the representation $\mathscr{D}^{\mathrm{S}} = [\mathscr{D}^2] = \mathscr{D}^{[2]}$ (the symmetric product representation) of $\mathrm{U}(m)$ by $\mathbf{X}(\mathrm{S})$ and $\mathscr{D}^{\mathrm{A}} = \{\mathscr{D}^2\} = \mathscr{D}^{[11]}$ (the antisymmetric product representation) by $\mathbf{X}(\mathrm{A})$ are irreducible. The dimensions of these representations are given by $f[2] = m(m+1)/2$ and $f[11] = m(m-1)/2$, respectively, and we have

$$(\mathscr{D})^2 = \mathscr{D}^{[2]} + \mathscr{D}^{[1^2]} \; . \tag{15.30}$$

See page 348 for their characters. They are derived easily from the transformation properties given above.

In the general case of tensors with rank n, operation of the permutation $P = \begin{pmatrix} 1 & 2 & \dots & n \\ 1' & 2' & \dots & n \end{pmatrix}$ on $X_{j_1 j_2 \dots j_n}$ is defined by

$$PX_{j_1 j_2 \dots j_n} = X_{j'_1 j'_2 \dots j'_n} . \tag{15.31}$$

When the tensor is given by the product of $X_j(i)$ as a function of the coordinates $1, 2, \dots, n$,

$$X_{j_1 j_2 \dots j_n}(1\ 2 \cdots n) = X_{j_1}(1) \cdots X_{j_n}(n) , \tag{15.32}$$

(15.31) is equivalent to defining P as an operator acting on the wavefunction $\Psi(1, 2, \dots, n)$ of the electron coordinates $1, 2, \dots, n$ to give Ψ' in the following way:

$$\Psi' = P\Psi ,$$

$$\Psi'(1'\ 2' \dots n') = \Psi(1\ 2 \dots n) . \tag{15.33}$$

For example, we find according to (15.31)

$$\begin{pmatrix} 1 & 2 & 3 \\ 3 & 2 & 1 \end{pmatrix} X_{j_1 j_2 j_3} = X_{j_3 j_2 j_1} , \quad \begin{pmatrix} 1 & 2 & 3 \\ 1 & 3 & 2 \end{pmatrix} X_{j_3 j_2 j_1} = X_{j_3 j_1 j_2} \tag{15.34}$$

corresponding to (2.12). Now (15.33) leads to the set of equations

$$\begin{pmatrix} 1 & 2 & 3 \\ 3 & 2 & 1 \end{pmatrix} \Psi(1\ 2\ 3) = \Psi(3\ 2\ 1) ,$$

$$\Psi'(1\ 2\ 3) = \begin{pmatrix} 1 & 2 & 3 \\ 3 & 1 & 2 \end{pmatrix} \Psi(1\ 2\ 3) = \left[\begin{pmatrix} 1 & 2 & 3 \\ 1 & 3 & 2 \end{pmatrix} \begin{pmatrix} 1 & 2 & 3 \\ 3 & 2 & 1 \end{pmatrix} \Psi \right] (1\ 2\ 3)$$

$$= \begin{pmatrix} 1 & 2 & 3 \\ 3 & 2 & 1 \end{pmatrix} \Psi(1\ 3\ 2) = \Psi(2\ 3\ 1).$$

Note that this result agrees with (15.34) if we put $\Psi(1\ 2\ 3)$ as in (15.32).

We observe that, with this definition, the components of the tensor $X_{j_1 j_2 \dots j_n}$ form a basis for a (m^n-dimensional) representation $\Delta^{(n)}$ of the symmetric group $\mathfrak{S}_n$, which is reducible in general. Corresponding to (15.29), we find

$$PU = UP , \quad P \in \mathfrak{S}_n , \quad U \in U(m) , \tag{15.35}$$

so that we should be able to choose the basis sets for an irreducible representation of the unitary group $U(m)$ in such a way that they serve simultaneously as the basis sets for a particular irreducible representation of $\mathfrak{S}_n$.

In fact, such basis sets are constructed in the following way. We operate $Q_\alpha^{[\lambda]}$ given by (15.24) on the components of tensor $X_{j_1 j_2 \dots j_n} = X_{\{j\}}$ to obtain $Q_\alpha^{[\lambda]} X_{\{j\}}$. Various choices of the set $\{j\}$ will then give rise to different linear combinations of tensor components. If we have $f[\lambda]$ linearly independent ones among them, appropriate linear combinations of them will provide us with an orthonormal basis set $X_\sigma([\lambda]\alpha)$ with $\sigma = 1, 2, \ldots, f[\lambda]$.[3] For the tableau β derived from α by exchanging i and $i+1$, we construct $X_\sigma([\lambda]\beta)$ from $Q_\beta^{[\lambda]}(i, i+1)X_{\{j\}}$ by the same procedure as for α (see the example given below for the three-electron problem).

The set $X_\sigma([\lambda]\alpha)$ forms the basis for the $g[\lambda]$-dimensional irreducible representation $D^{[\lambda]}$ of the symmetric group $\mathfrak{S}_n$:

$$PX_\sigma([\lambda]\alpha) = \sum_{\alpha'} X_\sigma([\lambda]\alpha')D_{\alpha'\alpha}^{[\lambda]}(P) \ . \tag{15.36}$$

On the other hand, we have from (15.35)

$$UQ_\alpha^{[\lambda]}X_{j_1 j_2 \cdots j_n} = \sum_{\{i\}} Q_\alpha^{[\lambda]} X_{i_1 i_2 \cdots i_n} u_{i_1 j_1} u_{i_2 j_2} \cdots u_{i_n j_n}$$

which will turn into the equation

$$UX_\sigma([\lambda]\alpha) = \sum_{\sigma'} X_{\sigma'}([\lambda]\alpha)\mathscr{D}_{\sigma'\sigma}^{[\lambda]}(U) \ , \tag{15.37}$$

when the $Q_\alpha^{[\lambda]}X_{\{j\}}$'s are expressed in terms of the $X_\sigma([\lambda]\alpha)$'s. Equation (15.37) of course represents that the $X_\sigma([\lambda]\alpha)$'s ($\sigma = 1, 2, \ldots, f[\lambda]$) are the basis set for the representation $\mathscr{D}^{[\lambda]}$ of the unitary group U(m). Actually, it can be shown that $\mathscr{D}^{[\lambda]}$ is irreducible. We are, however, not going to prove it here [15.3].

The functions of the simultaneous basis set for the representations corresponding to a given partition $[\lambda]$ of U(m) and $\mathfrak{S}_n$ may be arranged as in Table 15.3. Each row of the table gives the basis for the irreducible representation $D^{[\lambda]}$ of $\mathfrak{S}_n$, and each column the basis for the irreducible representation $\mathscr{D}^{[\lambda]}$ of U(m). This result shows that the representation $(\mathscr{D})^n$ is decomposed into its irreducible components as

$$(\mathscr{D})^n = \sum_{[\lambda]} g[\lambda]\mathscr{D}^{[\lambda]} \ , \qquad \lambda_1 + \lambda_2 + \cdots + \lambda_m = n \ , \tag{15.38}$$

$$m^n = \sum_{[\lambda]} g[\lambda]f[\lambda] \ , \tag{15.39}$$

where $g[\lambda]$ denotes the dimension of the representation $D^{[\lambda]}$. It must be kept in mind that, for $n > m$, we have to drop Young diagrams $[\lambda]$ with columns longer than m on the right-hand sides of (15.38, 39). This is because we have at least two equal figures among $j_1, j_2, \ldots, j_n$ ($1 \leqq j_p \leqq m$) in such a case and this makes the antisymmetrized tensor vanish identically. It is hardly necessary to point out that we also have

$$\Delta^{(n)} = \sum_{[\lambda]} f[\lambda]D^{[\lambda]} \tag{15.40}$$

---

[3] Define the inner product by $(X_{i_1 i_2 \ldots i_n}, X_{j_1 j_2 \ldots j_n}) = \delta_{i_1 j_1}\delta_{i_2 j_2}\ldots\delta_{i_n j_n}$.

Table 15.3. Simultaneous basis set $X_\sigma([\lambda]\alpha)$ for the irreducible representation characterized by $[\lambda]$ of U(m) and $\mathfrak{S}_n$

$\mathfrak{S}_n$ / U(m)		Basis sets for $D^{[\lambda]}$ ($g[\lambda]$ functions)			
		α	β	$\cdots$	ν
Basis sets for $\mathscr{D}^{[\lambda]}$	$\sigma = 1$	$X_1([\lambda]\alpha)$	$X_1([\lambda]\beta)$	$\cdots$	$X_1([\lambda]\nu)$
	$\sigma = 2$	$X_2([\lambda]\alpha)$	$X_2([\lambda]\beta)$	$\cdots$	$X_2([\lambda]\nu)$
	$\vdots$	$\vdots$	$\vdots$		$\vdots$
	$\sigma = f[\lambda]$	$X_f([\lambda]\alpha)$	$X_f([\lambda]\beta)$	$\cdots$	$X_f([\lambda]\nu)$

for the representation $\Delta^{(n)}$ of $\mathfrak{S}_n$. The number $f[\lambda]$ represents the dimension of the representation $\mathscr{D}^{[\lambda]}$.

The irreducible character of $\mathscr{D}^{[\lambda]}$ for the class of elements U having eigenvalues $(x_1, x_2, \ldots, x_m)$ as (15.27) may be obtained from

$$\chi^{[\lambda]}(U) = \frac{|x_1^{l_1}, x_2^{l_2}, \ldots, x_m^{l_m}|}{D(x_1, x_2, \ldots, x_m)} \tag{15.41}$$

in terms of the determinants given in (15.12 and 14), with $l_j = \lambda_j + m - j$ $(j = 1, 2, \ldots, m)$.

Using the character $\chi(U)$ of the representation $\mathscr{D}$

$$\chi(U) = \sum_{j=1}^{m} x_j \, , \quad \chi(U^p) = \sum_{j=1}^{m} x_j^p \equiv s_p \, , \quad p = 1, 2, \ldots, n \, , \tag{15.42}$$

and (15.11), the character (15.41) may also be put in the following form:

$$\chi^{[\lambda]}(U) = \frac{1}{n!} \sum_{(k)} n(k) \chi^{[\lambda]}(k) s_1^{k_1} s_2^{k_2} \ldots s_n^{k_n} \tag{15.43}$$

The summation on the right-hand side is to be taken over the classes (k) of $\mathfrak{S}_n$ with $n(k)$ given by (15.9). The right-hand side of (15.43) may be calculated easily using the character table for $\mathfrak{S}_n$ (Table 15.4).

The dimension $f[\lambda]$ of $\mathscr{D}^{[\lambda]}$ is calculated by

$$f[\lambda] = \frac{D(l_1, l_2, \ldots, l_m)}{D(m-1, m-2, \ldots, 0)}$$

which is derived from (15.41) by putting $x_j = \exp[\mathrm{i}(m-j)\varphi]$ and taking the limit of φ tending to zero.

Let us derive, as an example, the basis set for the irreducible representation $\mathscr{D}^{[2\ 1]}$ $(n = 3)$ of U(2), the two-dimensional unitary group. We use α, β in place of the basis functions X_1, X_2 for the representation $\mathscr{D}$, and choose the components of the tensor in the form of a product as in (15.32). Applying $Q^{[2\ 1]}_{(2\ 1\ 1)}$ and $Q^{[2\ 1]}_{(1\ 2\ 1)}$ to the functions $\alpha(1)\,\alpha(2)\,\beta(3)$, $\alpha(1)\,\beta(2)\,\alpha(3)$ and so on, we find that only the following functions are nonvanishing:

$$\begin{aligned}
Q^{[2\ 1]}_{(2\ 1\ 1)}\,\alpha(1)\,\alpha(2)\,\beta(3) &= 2\{2\alpha(1)\,\alpha(2)\,\beta(3) - \alpha(1)\,\beta(2)\,\alpha(3) - \beta(1)\,\alpha(2)\,\alpha(3)\}\\
&= 2\sqrt{6}\,X_1([2\ 1](2\ 1\ 1))\ ,\\
Q^{[2\ 1]}_{(2\ 1\ 1)}\,\beta(1)\,\alpha(2)\,\alpha(3) &= -2\sqrt{6}\,X_1([2\ 1](2\ 1\ 1))\ ,\\
Q^{[2\ 1]}_{(2\ 1\ 1)}\,\beta(1)\,\beta(2)\,\alpha(3) &= 2\{2\beta(1)\,\beta(2)\,\alpha(3) - \beta(1)\,\alpha(2)\,\beta(3) - \alpha(1)\,\beta(2)\,\beta(3)\}\\
&= -2\sqrt{6}\,X_2([2\ 1](2\ 1\ 1))\ ,\\
Q^{[2\ 1]}_{(2\ 1\ 1)}\,\alpha(1)\,\beta(2)\,\beta(3) &= 2\sqrt{6}\,X_2([2\ 1](2\ 1\ 1))\ ,\\
Q^{[2\ 1]}_{(1\ 2\ 1)}\,\alpha(1)\,\beta(2)\,\alpha(3) &= 3\{\alpha(1)\,\beta(2)\,\alpha(3) - \beta(1)\,\alpha(2)\,\alpha(3)\}\\
&= 3\sqrt{2}\,X_1([2\ 1](1\ 2\ 1))\ ,\\
Q^{[2\ 1]}_{(1\ 2\ 1)}\,\beta(1)\,\alpha(2)\,\alpha(3) &= -3\sqrt{2}\,X_1([2\ 1](1\ 2\ 1))\ ,\\
Q^{[2\ 1]}_{(1\ 2\ 1)}\,\beta(1)\,\alpha(2)\,\beta(3) &= 3\{\beta(1)\,\alpha(2)\,\beta(3) - \alpha(1)\,\beta(2)\,\beta(3)\}\\
&= -3\sqrt{2}\,X_2([2\ 1](1\ 2\ 1))\ ,\\
Q^{[2\ 1]}_{(1\ 2\ 1)}\,\alpha(1)\,\beta(2)\,\beta(3) &= 3\sqrt{2}\,X_2([2\ 1](1\ 2\ 1))\ .
\end{aligned}$$

With $f[2\ 1] = 2$ and $g[2\ 1] = 2$, the basis functions are arranged as in Table 15.5.

Suppose we consider in this example only those U that belong to the subgroup SU(2) of U(2). Since irreducible representations of U(2) are also irreducible with respect to SU(2) and SU(2) $\sim$ SO(3), the representation $\mathscr{D}$ is precisely the irreducible representation $D^{(1/2)}$ of the rotation group SO(3) and the functions α and β are nothing but the spin functions that form the basis for it. Note that the basis for $\mathscr{D}^{[21]}$ provides us with that for $D^{(1/2)}$ as confirmed easily from the expressions given above. The functions $X_\sigma([2\ 1]\alpha)$ represent the states of a system of three electrons with its total spin $S = 1/2$, and the suffixes 1 and 2 correspond respectively to $M_s = +1/2$ and $-1/2$. As shown above, we can construct two independent sets of such functions corresponding to the tableaux $\alpha = (2\ 1\ 1)$ and $\alpha = (1\ 2\ 1)$. If we put $\boldsymbol{s}_1 + \boldsymbol{s}_2 = \boldsymbol{S}_{12}$, we see that the former set is obtained by coupling the spin $S_{12} = 1$ with $s_3 = 1/2$, while the latter is derived from the coupling of $S_{12} = 0$ and $s_3 = 1/2$.

When the elements of U(m) are restricted to those of the subgroup SU(m), the subduced representation $\mathscr{D}^{[\lambda]} \downarrow \mathrm{SU}(m)$ is irreducible in general. Remember, however, that irreducible representations corresponding to the partitions $[\lambda]$ and $[\lambda']$ become equivalent in SU(m) in the following cases:

$$(1)\ [\lambda] = [\lambda_1, \lambda_2, \ldots, \lambda_m]\ ,\quad \lambda_m \geqq 1\ ,\quad [\lambda'] = [\lambda_1 - \lambda_m, \ldots, \lambda_{m-1} - \lambda_m, 0]\ ,$$

$$(2)\ [\lambda] = [\lambda_1, \lambda_2, \ldots, \lambda_m]\ ,\quad \lambda_m \geqq 0\ ,\quad [\lambda'] = [\lambda_1 - \lambda_m, \lambda_1 - \lambda_{m-1}, \ldots, 0]\ .$$

Table 15.4. Irreducible characters $\chi^{[\lambda]}(U)$ for the representation $\mathscr{D}^{[\lambda]}$ of U(m). The definition of s_p is given by (15.42)

$n = 2$: $\chi^{[2]} = \frac{1}{2}(s_1^2 + s_2)$

$\chi^{[1^2]} = \frac{1}{2}(s_1^2 - s_2)$

$n = 3$: $\left.\begin{matrix}\chi^{[3]}\\ \chi^{[1^3]}\end{matrix}\right\} = \frac{1}{6}(s_1^3 \pm 3s_1 s_2 + 2s_3)$

$\chi^{[21]} = \frac{1}{3}(s_1^3 - s_3)$

$n = 4$: $\left.\begin{matrix}\chi^{[4]}\\ \chi^{[1^4]}\end{matrix}\right\} = \frac{1}{24}(s_1^4 \pm 6s_1^2 s_2 + 3s_2^2 + 8s_1 s_3 \pm 6s_4)$

$\left.\begin{matrix}\chi^{[31]}\\ \chi^{[21^2]}\end{matrix}\right\} = \frac{1}{24}(3s_1^4 \pm 6s_1^2 s_2 - 3s_2^2 \mp 6s_4)$

$\chi^{[2^2]} = \frac{1}{12}(s_1^4 + 3s_2^2 - 4s_1 s_3)$

$n = 5$: $\left.\begin{matrix}\chi^{[5]}\\ \chi^{[1^5]}\end{matrix}\right\} = \frac{1}{120}(s_1^5 \pm 10s_1^3 s_2 + 15s_1 s_2^2 + 20s_1^2 s_3 \pm 20s_2 s_3 \pm 30s_1 s_4 + 24s_5)$

$\left.\begin{matrix}\chi^{[41]}\\ \chi^{[21^3]}\end{matrix}\right\} = \frac{1}{60}(2s_1^5 \pm 10s_1^3 s_2 + 10s_1^2 s_3 \mp 10s_2 s_3 - 12s_5)$

$\left.\begin{matrix}\chi^{[32]}\\ \chi^{[2^2 1]}\end{matrix}\right\} = \frac{1}{120}(5s_1^5 \pm 10s_1^3 s_2 + 15s_1 s_2^2 - 20s_1^2 s_3 \pm 20s_2 s_3 \mp 30s_1 s_4)$

$\chi^{[31^2]} = \frac{1}{20}(s_1^5 - 5s_1 s_2^2 + 4s_5)$

Table 15.5. Simultaneous basis set for the representation $D^{[21]}$ of $\mathfrak{S}_3$ and $\mathscr{D}^{[21]}$ of U(2). (See Table 15.3)

	(2 1 1)	(1 2 1)
$\sigma = 1$	$X_1([2\ 1](2\ 1\ 1))$	$X_1([2\ 1](1\ 2\ 1))$
$\sigma = 2$	$X_2([2\ 1](2\ 1\ 1))$	$X_2([2\ 1](1\ 2\ 1))$

These results may be derived from the expression (15.41) for the character of $\mathscr{D}^{[\lambda]}$ by noting that $\det(\hat{u}) = x_1 x_2 \ldots x_m = 1$ in SU(m). Examples of the rules (1) and (2) for the case of $m = 3$ are shown in Fig. 15.1.

The basis wavefunctions for an n-electron system with its resultant spin S can be expressed generally as $X_\sigma([(n/2) + S, (n/2) - S]\alpha)$. Corresponding to differ-

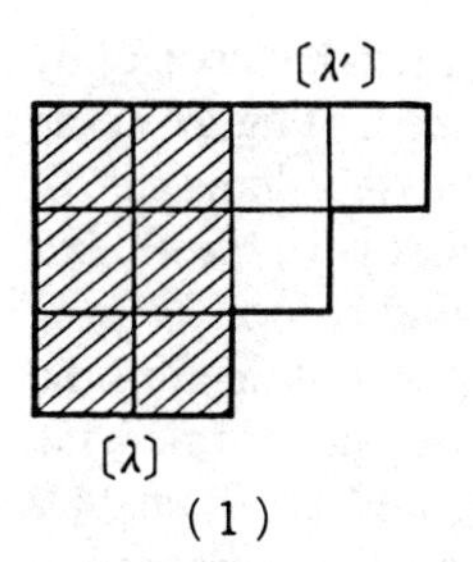

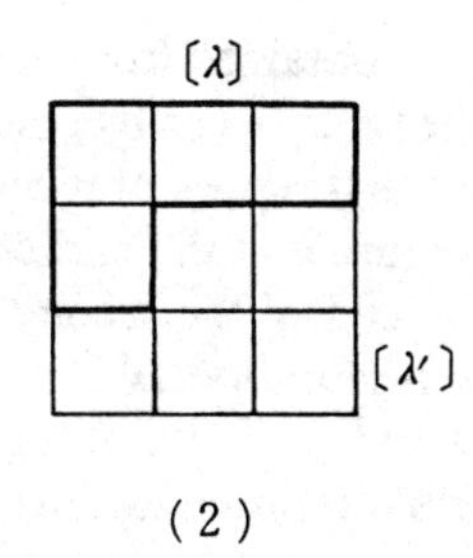

Fig. 15.1. Equivalence of $\mathscr{D}^{[\lambda]}$ in SU(3)

ent choices of α, we have as many such sets of functions as

$$g\left[\frac{n}{2}+S, \frac{n}{2}-S\right] = \frac{(2S+1)n!}{\left(\frac{n}{2}+S+1\right)!\left(\frac{n}{2}-S\right)!}, \tag{15.44}$$

each set constituting the basis for the representation $D^{(S)}$ of SO(3) with dimension $f[(n/2)+S, (n/2)-S] = 2S+1$.

15.6 The Branching Rule

The irreducible representation $D^{[\lambda]}$ of the symmetric group $\mathfrak{S}_n$ with $[\lambda] = [\lambda_1, \lambda_2, \ldots, \lambda_n]$ is in general reducible as a representation of $\mathfrak{S}_{n-1}$. As shown below, (15.15) enables us to find what kind of irreducible representations $D^{[\lambda']}$ of $\mathfrak{S}_{n-1}$ with $[\lambda'_1, \lambda'_2, \ldots, \lambda'_{n-1}]$ appear in the decomposition of $D^{[\lambda]} \downarrow \mathfrak{S}_{n-1}$. We first note that we will be dealing with the elements of $\mathfrak{S}_{n-1}$ if we consider among the elements of $\mathfrak{S}_n$ only such permutations as $\begin{pmatrix} 1 & 2 & \ldots & n-1 & n \\ 1' & 2' & \ldots & (n-1)' & n \end{pmatrix}$. This amounts to obtaining a class (k') of $\mathfrak{S}_{n-1}$ from a class (k) of $\mathfrak{S}_n$ by deleting from the latter a particular cycle of order one, i.e., (n). If $(k) = (k_1, k_2, \ldots)$, we have $(k') = (k_1 - 1, k_2, \ldots)$. According to (15.15), it then follows that

$$\begin{aligned} \chi^{[\lambda]}(k) &= \sum_{[\lambda']} \chi^{[\lambda']}(k') \\ &= \chi^{[\lambda_1 - 1, \lambda_2, \ldots, \lambda_n]}(k') + \chi^{[\lambda_1, \lambda_2 - 1, \ldots, \lambda_n]}(k') + \cdots . \end{aligned} \tag{15.45}$$

On the right-hand side, the characters $\chi^{[\lambda']}$ for which $\lambda'_1, \lambda'_2, \ldots$ are not arranged in natural order or any one of them is negative are to be discarded. This leads immediately to the following result:

$$D^{[\lambda]} \downarrow \mathfrak{S}_{n-1} = \sum_{[\lambda']} D^{[\lambda']} = D^{[\lambda_1 - 1, \lambda_2, \ldots]} + D^{[\lambda_1, \lambda_2 - 1, \ldots]} + \cdots . \tag{15.46}$$

The basis set for $D^{[\lambda]} \downarrow \mathfrak{S}_{n-1}$ may be obtained from each of the tableaux $[\lambda]$ by removing the square containing the figure n (or by deleting the first figure in the Y-symbol). Note that we obtain the first representation on the right-hand side of (15.46) from the tableau with the figure n at the end of the first row, the second one from the tableau with n at the end of the second row and so on.

Conversely, the irreducible representation $D^{[\lambda']}$ of $\mathfrak{S}_{n-1}$ is contained in the representations $D^{[\lambda'_1+1,\,\lambda'_2,\ldots]}, D^{[\lambda'_1,\,\lambda'_2+1,\ldots]}, \ldots$ of $\mathfrak{S}_n$ as seen in (15.46). The relation between the irreducible representations of $\mathfrak{S}_n$ and $\mathrm{U}(m)$ enables us to write down the rule of decomposition for the product representation $\mathscr{D}^{[\lambda']} \times \mathscr{D}^{[1]}$ of $\mathrm{U}(m)$ as

$$\mathscr{D}^{[\lambda']} \times \mathscr{D}^{[1]} = \sum_{[\lambda]} \mathscr{D}^{[\lambda]} = \mathscr{D}^{[\lambda'_1+1,\,\lambda'_2,\ldots]} + \mathscr{D}^{[\lambda'_1,\,\lambda'_2+1,\ldots]} + \cdots . \tag{15.47}$$

The summation on the right-hand side is to be taken only over $[\lambda]$'s with λ_j's in natural order. Representations corresponding to Young diagrams with more rows than m are also dropped.

Example 1. A simple example is given in Fig. 15.2 using Young diagrams. When $m = 4$, the last diagram on the right-hand side is excluded.

Example 2. Applying (15.47) for the case of $m = 2$, we obtain

$$\begin{aligned}\mathscr{D}^{\left[\frac{n}{2}+s,\,\frac{n}{2}-s\right]} \times \mathscr{D}^{[1]} &= \mathscr{D}^{\left[\frac{n+1}{2}+s+\frac{1}{2},\,\frac{n+1}{2}-s-\frac{n}{2}\right]} \\ &\quad + \mathscr{D}^{\left[\frac{n+1}{2}+s-\frac{1}{2},\,\frac{n+1}{2}-s+\frac{1}{2}\right]} \qquad \text{for } S \neq 0 , \\ &= \mathscr{D}^{\left[\frac{n+1}{2}+\frac{1}{2},\,\frac{n+1}{2}-\frac{1}{2}\right]} \qquad \text{for } S = 0 .\end{aligned}$$

If we go over to SU(2), this gives the *branching rule* for the spin functions. See Fig. 15.3.

In the preceding section the basis sets for $\mathscr{D}$ and $\mathscr{D}^{[\lambda]}$ were written as X_j and $X_\sigma([\lambda]\alpha)$. Let us denote them as $\phi(\varrho)$ and $\psi_\alpha(\Sigma\sigma)$ in this section, where Σ stands for the representation $\mathscr{D}^{[\lambda]}$ of $\mathrm{U}(m)$. Corresponding to (15.47), we have

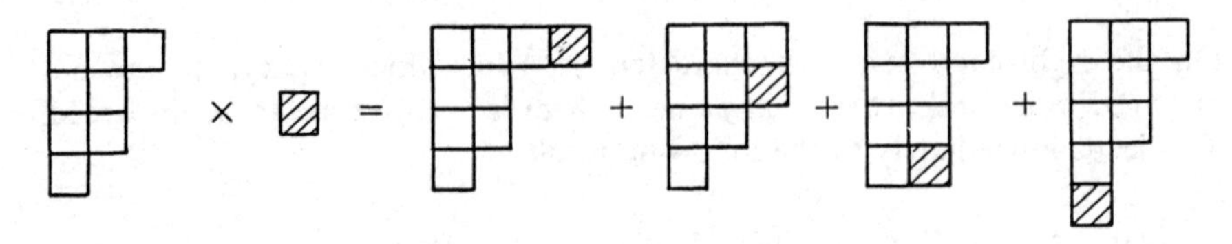

Fig. 15.2. Reduction of $\mathscr{D}^{[\lambda]} \times \mathscr{D}^{[1]}$

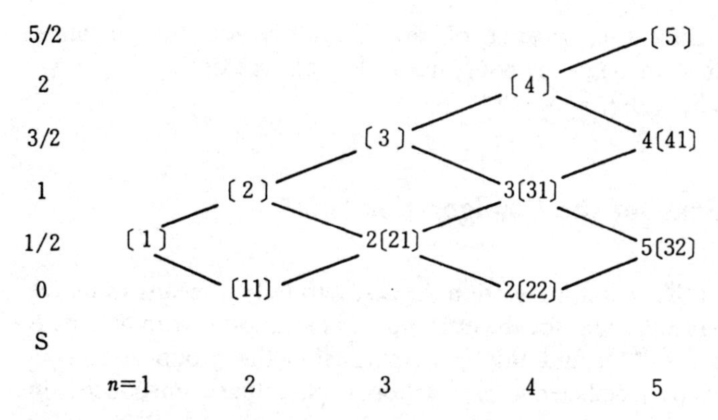

Fig. 15.3. Branching rule for spin functions. The figure before the partition $[\lambda]$ represents $g[\lambda]$, the number of independent spin functions with spin S

the following relation between the basis sets:

$$\psi_\alpha(\Sigma\sigma) = \sum_{\sigma_1 \varrho} \Psi_{\alpha_1}(\Sigma_1\sigma_1)\phi(\varrho)\langle \Sigma_1\sigma_1, [1]\,\varrho|\Sigma\sigma\rangle \tag{15.48}$$

with the *Clebsch–Gordan coefficients* (abbreviated as the *CG coefficients*) $\langle \Sigma_1\sigma_1, [1]\,\varrho|\Sigma\sigma\rangle$. The CG coefficients are the elements of a unitary matrix whose rows are specified by $\sigma_1\varrho$ and columns by $\Sigma\sigma$. The Greek characters α and α_1 are the Y-symbols with $\alpha_1 = (r_{n-1} \ldots r_2 r_1)$ for $\alpha = (r_n r_{n-1} \ldots r_2 r_1)$. When α_1 (or α) is specified, the representation $\Sigma_1 = \mathscr{D}^{[\lambda_1]}$ is determined accordingly. If we know the values of the CG coefficients, we can derive the basis set for Σ from that for Σ_1 through (15.48). In other words, we can obtain the basis set $\psi_\alpha(\Sigma\sigma) = X_\sigma([\lambda]\alpha)$ for the representation $D^{[\lambda]}$ of the symmetric group $\mathfrak{S}_n$ from that for $D^{[\lambda_1]}$ of $\mathfrak{S}_{n-1}$. Conversely, we can use (15.48) to determine the CG coefficients in such a way that the functions $\psi_\alpha(\Sigma\sigma)$ form the basis set for $D^{[\lambda]}$ of the symmetric group.

Specifically, for the spin functions $\phi(\varrho) \equiv \chi(m)$, $\psi_\alpha(\Sigma\sigma) \equiv \Theta_\alpha(SM)$ that form the basis set for the representations of SU(2), (15.48) turns into

$$\Theta_{\alpha*}(SM) = \sum_{M_1 m} \Theta_{\alpha_1^*}(S_1M_1)\chi(m)\langle S_1M_1\tfrac{1}{2}m|SM\rangle \tag{15.49}$$

with the Wigner coefficients of Sect. 7.10 appearing here as the CG coefficients on the right-hand side. As noted at the end of the preceding section, the relation between S and $\Sigma^* = D^{[\lambda]^*}$ is given by

$$[\lambda]^* = \left[\frac{n}{2} + S, \frac{n}{2} - S\right]. \tag{15.50}$$

In this case, we may use, instead of the Y-symbol α^*, the sequence $S_2, S_3, \ldots, S_n$ defined through the coupling of the spin vectors $s_1 + s_2 = S_2$, $S_2 + s_3 = S_3, \ldots, S_{n-1} + s_n = S_n$.

15.7 Wavefunctions for the Configuration $(nl)^x$

The results given in the preceding section suggest naturally a means to derive antisymmetrized wavefunctions for the electron configuration l^n with definite S and L from those for l^{n-1}, and this in turn clarifies the group-theoretical meaning of the c.f.p. (coefficients of fractional parentage) introduced in Sect. 7.14.

Let us regard the $(2l+1)$ orbitals $\phi(m_l)$ $(m_l = -l, -l+1, \ldots, l)$ for an l electron as a basis for the group $U(2l+1)$ or $SU(2l+1)$. The function $\psi_\alpha(\Sigma\sigma)$ given by (15.48) is then the basis function for the representation Σ of $U(2l+1)$. If we limit now the elements of $U(2l+1)$ to those belonging to SO(3), Σ will be reducible in general as a representation of SO(3) so that

$$\Sigma \downarrow \mathrm{SO}(3) = \sum_L \nu_L D^{(L)} , \tag{15.51}$$

where ν_L denotes the number of times the representation $D^{(L)}$ appears in the decomposition. In accordance with this reduction, we can choose the basis functions so that they form simultaneously the basis for $D^{(L)}$. This means that we put $\sigma = \xi L M_L$, where ξ has been introduced to distinguish between the basis functions with the same L and M_L when $\nu_L \geqq 2$. According to *Racah* [15.4], the CG coefficients can then be factorized as

$$\langle \Sigma_1\sigma_1, [1]\varrho|\Sigma\sigma\rangle = \langle L_1 M_{L_1} l m_l|L M_L\rangle \langle \Sigma_1\xi_1 L_1, [1]l|\Sigma\xi L\rangle \tag{15.52}$$

using the Wigner coefficients. As a result, (15.48) now takes the form

$$\psi_\alpha(\Sigma\xi L M_L) = \sum_{\xi_1 L_1} \sum_{M_{L_1} m_l} \psi_{\alpha_1}(\Sigma_1\xi_1 L_1 M_{L_1})\phi(m_l) \times \langle L_1 M_{L_1} l m_l|L M_L\rangle \langle \Sigma_1\xi_1 L_1, [1]l|\Sigma\xi L\rangle . \tag{15.53}$$

Although we have not given the arguments of functions explicitly, $\psi_{\alpha_1}(\Sigma_1\xi_1 L_1 M_{L_1})$ here is a function of the coordinates of electrons $1, 2, \ldots, n-1$, while $\psi(m_l)$ is a function of the electron n.

If we denote by $[\lambda]^*$ and α^* respectively the representation conjugate to $[\lambda]$ and the Y-symbol conjugate to α, we find that the function $\sum_\alpha \Theta_{\alpha^*}\psi_\alpha$ is totally antisymmetric with respect to the permutations of the electron coordinates (including the spin coordinates). To verify this, we first observe that the functions ψ_α, Θ_{α^*} transform respectively as the bases for the representations $D^{[\lambda]}$ and $D^{[\lambda]^*}$ under the permutation P of the coordinates. By an appropriate choice of phase factors for the functions, it is always possible to make the correlation (15.19) hold between the representation matrices $\hat{D}^{[\lambda]}(P)$ and $\hat{D}^{[\lambda]*}(P)$. We then have

$$P\sum_{\alpha}\Theta_{\alpha*}\psi_{\alpha}=\sum_{\alpha}P\Theta_{\alpha*}P\psi_{\alpha}$$

$$=\sum_{\alpha\beta\beta'}\Theta_{\beta^*}\psi_{\beta'}D^{[\lambda]*}_{\beta*\alpha*}(P)D^{[\lambda]}_{\beta'\alpha}(P)$$

$$=(-1)^P\sum_{\beta\beta'}\Theta_{\beta^*}\psi_{\beta'}\sum_{\alpha}D^{[\lambda]}_{\beta\alpha}(P)D^{[\lambda]}_{\beta'\alpha}(P)=(-1)^P\sum_{\beta}\Theta_{\beta^*}\psi_{\beta}\ ,$$

where we have made use of the fact that $\hat{D}^{[\lambda]}(P)$ is actually an orthogonal matrix. The totally antisymmetric (normalized) wavefunction for l^n with definite S and L is thus given by

$$\Psi(l^n\xi SLM_SM_L)=\frac{1}{\sqrt{g[\lambda]}}\sum_{\alpha}\Theta_{\alpha*}(SM_S)\psi_{\alpha}(\Sigma\xi LM_L)\ . \tag{15.54}$$

Let us substitute the expressions (15.49 and 53) into the right-hand side of (15.54). Since summing over α is equivalent to carrying out summation simultaneously over α_1 and Σ_1 (which corresponds to α_1) or S_1, we may put the result as

$$\Psi(l^n\xi SLM_SM_L)=\sum_{S_1\xi_1L_1}\Phi(l^{n-1}(\xi_1S_1L_1)lSLM_SM_L)$$
$$\times\langle l^{n-1}(\xi_1S_1L_1)lSL|\}l^n\xi SL\rangle\ , \tag{15.55}$$

where the function $\Phi(l^{n-1}\ldots)$ is defined by (7.150), and

$$\langle l^{n-1}(\xi_1S_1L_1)lSL|\}l^n\xi SL\rangle=\langle\Sigma_1\xi_1L_1,[1]l|\Sigma\xi L\rangle\langle l^{n-1}\Sigma_1,l[1]||l^n\Sigma\rangle\ , \tag{15.56}$$

$$\langle l^{n-1}\Sigma_1,l[1]||l^n\Sigma\rangle=\sqrt{\frac{g[\lambda_1]}{g[\lambda]}}\ . \tag{15.57}$$

Equation (15.56) gives the relation between the c.f.p. for l^n and the second factor on the right-hand side of (15.52), or the CG coefficient defined by (15.48).

These considerations show that the values of L allowed for a given value of S in a configuration l^n are limited to those for which the representation $D^{(L)}$ appears at least once in the reduction (15.51) of $\mathscr{D}^{[\lambda]}\downarrow SO(3)$ with $[\lambda]=[(n/2)+S,(n/2)-S]^*$. To carry out the reduction (15.51), we need the characters $\chi^{[\lambda]}(U)$ of the representation $\mathscr{D}^{[\lambda]}$. They are given by (15.43) and are tabulated in Table 15.4 for $n\leqq 5$. If we take the rotation $R(\varphi)$ about the

z-axis through an angle φ, we obtain

$$\chi(R(\varphi)) = \frac{\sin\left(l + \frac{1}{2}\right)\varphi}{\sin\frac{1}{2}\varphi}$$

with

$$s_p = \chi((R(\varphi))^p) = \chi(R(p\varphi)) = \frac{\sin\left(l + \frac{1}{2}\right)p\varphi}{\sin\frac{1}{2}p\varphi} .$$

For the irreducible characters $\chi^{(L)}$ for $D^{(L)}$ of SO(3), we have

$$\chi^{(L)}(R(\phi)) = \frac{\sin\left(L + \frac{1}{2}\right)\varphi}{\sin\frac{1}{2}\varphi} .$$

From

$$\chi^{[\lambda]}(R(\varphi)) = \sum_L \nu_L \chi^{(L)}(R(\varphi)) ,$$

we can determine the number of times ν_L the representation $D^{(L)}$ appears in the reduction.

So far, we have dealt with the case of *LS* (Russell–Saunders) coupling. The case of *jj* coupling is simpler. Corresponding to (15.55, 56), we now have

$$\Psi(j^n\xi J) = \sum_{\xi_1 J_1} \Phi(j^{n-1}(\xi_1 J_1)jJ)\langle j^{n-1}(\xi_1 J_1)jJ|\}j^n\xi J\rangle ,$$

$$\langle j^{n-1}(\xi_1 J_1)jJ|\}j^n\xi J\rangle = \langle \Sigma_1\xi_1 J_1, [1]j|\Sigma\xi J\rangle ,$$

where $\Sigma_1 = \mathscr{D}^{[\lambda_1]}$ and $\Sigma = \mathscr{D}^{[\lambda]}$ as before, but with $[\lambda_1] = [1^{n-1}]$ and $[\lambda] = [1^n]$, so that both are totally antisymmetric.

In the *jj* coupling scheme of the nuclear shell model, isospin τ is introduced to treat the proton and neutron on an equal footing. In this case, we have only to put $l \to j$, $L \to J$, $s \to \tau$, $S \to T$ in (15.55) and in the treatments leading up to this equation.

Exercise 15.2. Show by the method described above that the allowed combinations of S and L of multiplet terms in the p^3 configuration ($l = 1$) are given by

$$(S, L) = (\tfrac{1}{2}, 1) , \quad (\tfrac{1}{2}, 2) , \quad (\tfrac{3}{2}, 0) .$$

In this case, it is wiser to use the original expression for the character

$$\frac{\sin\left(l+\frac{1}{2}\right)\varphi}{\sin\frac{1}{2}\varphi} = e^{il\varphi} + e^{i(l-1)\varphi} + \cdots + e^{-il\varphi} .$$

*15.8 $D^{(J)}$ as Irreducible Representations of SU(2)

In Sect. 7.5, we derived the expression for representation matrices of $D^{(J)}$ by an analytic method. As shown below, we can obtain it by an algebraic means regarding $D^{(J)}$ as irreducible representations of SU(2).

In the general theory (Sect. 15.5), we used the notation X_i for the basis functions in the m-dimensional complex vector space. Here we put $m = 2$ and write ξ, η for X_1, X_2. We first note that any two-dimensional special unitary matrix $\hat{u}$ may be expressed as

$$\hat{u} = \begin{bmatrix} a & -b \\ b^* & a^* \end{bmatrix} , \quad |a|^2 + |b|^2 = 1 . \tag{15.58}$$

Let us put a 2×2 matrix with determinant 1 in the form

$$\hat{u} = \begin{bmatrix} a & -b \\ c & d \end{bmatrix} , \quad ad + bc = 1 .$$

For this matrix to be unitary, the following two matrices have to be equal:

$$\hat{u}^{-1} = \begin{bmatrix} d & b \\ -c & a \end{bmatrix} , \quad \hat{u}^\dagger = \begin{bmatrix} a^* & c^* \\ -b^* & d^* \end{bmatrix} ,$$

which leads to

$$c = b^* , \quad d = a^* .$$

We then define the transformation U through

$$\begin{aligned} U\xi &= a\xi + b^*\eta , \\ U\eta &= -b\xi + a^*\eta , \end{aligned} \tag{15.59}$$

or

$$[U\xi, U\eta] = [\xi, \eta]\hat{u} , \tag{15.60}$$

using the matrix $\hat{u}$. The set of such transformations forms the group SU(2) and $\hat{u}$ may be viewed as the representation matrix $\hat{\mathscr{D}}(U)$ corresponding to the element U.

In (15.58), a and b are complex numbers whose absolute values are not greater than 1, and they may be written as

$$a = \mathrm{e}^{-\mathrm{i}(\alpha+\gamma)/2} \cos\frac{\beta}{2} \ , \qquad b = \mathrm{e}^{-\mathrm{i}(\alpha-\gamma)/2} \sin\frac{\beta}{2} \tag{15.61}$$

without loss of generality. Then the matrix $\hat{u}$ coincides with the representation matrix $\hat{D}^{(1/2)}(R(\alpha, \beta, \gamma))$ of the rotation group given by (7.52).

Let us consider a set of monomials of ξ, η of degree n. As seen below, representations of SU(2) based on these monomials provide us with the known expressions for the matrices of $D^{(J)}$. We begin with simple cases.

When $n = 1$, ξ and η form the basis set for the (irreducible) representation $\mathscr{D}$.

When $n = 2$, we have three monomials of degree 2:

$$f_1 = \frac{\xi^2}{\sqrt{2}} \ , \quad f_0 = \xi\eta \ , \quad f_{-1} = \frac{\eta^2}{\sqrt{2}} \ , \tag{15.62}$$

where appropriate numerical factors have been introduced to make the resulting representation unitary.

If we apply U to, say, f_1, the result will be a homogeneous polynomial of ξ and η of degree 2, because of (15.59) together with $U(\xi^2) = (U\xi)(U\xi)$. This means that Uf_1 is a linear combination of f_1, f_0 and f_{-1}. In fact, we find that the transformation is given by

$$[Uf_1, Uf_0, Uf_{-1}] = [f_1, f_0, f_{-1}] \begin{bmatrix} a^2 & -\sqrt{2}ab & b^2 \\ \sqrt{2}ab^* & |a|^2 - |b|^2 & -\sqrt{2}a^*b \\ b^{*2} & \sqrt{2}a^*b^* & a^{*2} \end{bmatrix} . \tag{15.63}$$

The representation obtained in this way is irreducible. It is indeed the symmetric product representation $[\mathscr{D} \times \mathscr{D}]$ or $\mathscr{D}^{[2]}$ in the notation we have been using since Sect. 15.5, as is expected from the choice of the basis (15.62).

For a general integer n, we put

$$F_\nu \equiv \frac{\xi^{n-\nu}\eta^\nu}{\sqrt{(n-\nu)!\,\nu!}} \ , \qquad \nu = 0, 1, \ldots, n \ ,$$

and find that UF_ν can be expressed as a linear combination of $(n + 1)$ such F_ν's. They form a basis set for the $(n + 1)$-dimensional irreducible representation $\mathscr{D}^{[n]}$ of SU(2).

To make the correspondence with the result for the rotation group clear, it is more convenient to put

$$n = 2J \ , \quad \nu = J - K$$

and employ

$$f_K^{(J)} = \frac{\xi^{J+K}\eta^{J-K}}{\sqrt{(J+K)!(J-K)!}}, \qquad \begin{array}{l} 2J = 0, 1, 2, \ldots, \\ K = J, J-1, \ldots, -J . \end{array}$$

Then we have

$$Uf_K^{(J)} = \frac{(a\xi + b^*\eta)^{J+K}(-b\xi + a^*\eta)^{J-K}}{\sqrt{(J+K)!(J-K)!}} .$$

Expanding the right-hand side, we obtain

$$Uf_K^{(J)} = \sqrt{(J+K)!(J-K)!}\sum_{s=0}^{J+K}\sum_{r=0}^{J-K}\frac{(a\xi)^{J+K-s}(b^*\eta)^s(-b\xi)^{J-K-r}(a^*\eta)^r}{(J+K-s)!s!(J-K-r)!r!} .$$

If we assume $0! = 1$ and $1/n! = 0$ for negative integers, we may drop the restriction on the range for the nonnegative s and r. (In other words, the summations are to be carried out over all the possible values of s and r as long as the factorials make sense.) Putting

$$r + s = J - M ,$$

we replace the sum over s by a sum over M. The result is given by

$$Uf_K^{(J)} = \sum_{M=-J}^{J} f_M^{(J)} D_{MK}^{(J)}(U) \tag{15.64}$$

with

$$D_{MK}^{(J)}(U) = \sum_r \frac{a^{M+K+r}(b^*)^{J-M-r}(-b)^{J-K-r}(a^*)^r}{(M+K+r)!(J-M-r)!(J-K-r)!r!} \times \sqrt{(J+M)!(J-M)!(J+K)!(J-K)!} . \tag{15.65}$$

This is the representation matrix for U of the $(2J+1)$-dimensional irreducible representation $\mathscr{D}^{[2J]}$ of SU(2), for which the $f_K^{(J)}$'s form the basis set.

If we use (15.61) in (15.65), we obtain

$$D_{MK}^{(J)}(U) = \sum_r \frac{(-1)^{J-K-r}\left(\cos\frac{\beta}{2}\right)^{M+K+2r}\left(\sin\frac{\beta}{2}\right)^{2J-M-K-2r}}{(M+K+r)!(J-M-r)!(J-K-r)!r!} \times \mathrm{e}^{-iM\alpha}\mathrm{e}^{-iK\gamma}\sqrt{(J+M)!(J-M)!(J+K)!(J-K)!} ,$$

which agrees with the results (7.40, 41) and (7.49) for the representations of the rotation group. In this way, we have derived the irreducible representations $D^{(J)}$ ($J = 0, 1/2, 1, 3/2, 2, \ldots$) of the rotation group SO(3) as those of the two-dimensional special unitary group SU(2).

Exercise 15.3. By using (15.61), show that (15.63) turns into the representation matrix of $D^{(1)}$ of the rotation group as follows:

$$D^{(1)}(R(\alpha,\beta,\gamma)) = \begin{bmatrix} e^{-i\alpha}\cos^2\frac{\beta}{2}\,e^{-i\gamma} & -e^{-i\alpha}\frac{1}{\sqrt{2}}\sin\beta & e^{-i\alpha}\sin^2\frac{\beta}{2}\,e^{i\gamma} \\ \frac{1}{\sqrt{2}}\sin\beta\, e^{-i\gamma} & \cos\beta & -\frac{1}{\sqrt{2}}\sin\beta\, e^{i\gamma} \\ e^{i\alpha}\sin^2\frac{\beta}{2}\,e^{-i\gamma} & e^{i\alpha}\frac{1}{\sqrt{2}}\sin\beta & e^{i\alpha}\cos^2\frac{\beta}{2}\,e^{i\gamma} \end{bmatrix} .$$

Exercise 15.4. Show that if we use the basis set f_x, f_y and f_z defined by

$$f_1 = -\frac{1}{\sqrt{2}}(f_x + \mathrm{i}f_y)\ , \quad f_0 = f_z\ , \quad f_{-1} = \frac{1}{\sqrt{2}}(f_x - \mathrm{i}f_y)\ ,$$

instead of f_1, f_0 and f_{-1} given by (15.62), we have

$$[Uf_x, Uf_y, Uf_z] = [f_x, f_y, f_z]\hat{R}\ ,$$

where $\hat{R}$ is the 3×3 orthogonal matrix defined by (7.7).

*15.9 Irreducible Representations of U(m)

In Sect. 15.5, we saw that representations $\mathscr{D}^{[\lambda]}$ based on tensors with the symmetry corresponding to the partition $[\lambda] = [\lambda_1, \lambda_2, \ldots, \lambda_m]$ $(\lambda_1 \geqq \lambda_2 \geqq \cdots \geqq \lambda_m \geqq 0)$ are irreducible. A complete set of irreducible representations of U(m) is obtained by supplementing them with the following ones derived from $\mathscr{D}^{[\lambda]}$.

Let us multiply the representation matrices of $\mathscr{D}^{[\lambda]}$ by the eth power of det $\hat{u}$, where e is an integer which may be either positive or negative. The matrix obtained in this way

$$\hat{\mathscr{D}}'(U) = (\det \hat{u})^e \hat{\mathscr{D}}^{[\lambda]}(U) \tag{15.66}$$

yields an irreducible representation of U(m). The character $\chi'(U)$ for it is given by

$$\chi'(U) = x_1^e x_2^e \cdots x_m^e \chi^{[\lambda]}(U) \tag{15.67}$$

in terms of $\chi^{[\lambda]}(U)$ of (15.41). We then note that the right-hand side may also be expressed by (15.41) as $\chi^{[\lambda']}(U)$ with $[\lambda'] = [\lambda_1 + e, \lambda_2 + e, \ldots, \lambda_m + e]$. In the case $\lambda_m + e \geqq 0$, $\mathscr{D}'$ does not lead to any new representations. However, $\mathscr{D}'$ will provide us with a new representation that was not considered in Sect. 15.5 when e is negative and at least one $\lambda_j + e$ becomes negative.

When we include the representations considered above, we may say that the irreducible representations of U(m) are characterized in general by a *signature* $[\lambda] = [\lambda_1, \lambda_2, \cdots, \lambda_m]$ $(\lambda_1 \geqq \lambda_2 \geqq \cdots \geqq \lambda_m)$. The term signature is used instead of *partition* to indicate that any λ_j may be negative. The corresponding representation will also be denoted as $\mathscr{D}^{[\lambda]}$. It can be shown that there are no irreducible representations of U(m) other than $\mathscr{D}^{[\lambda]}$ extended in this way [15.3]. As seen above, (15.41) is also valid for the irreducible characters of these representations.

Let us give the simplest example. In the case of the one-dimensional unitary group U(1), we obtain representations $\mathscr{D}^{[\lambda]}(U) = \exp(\mathrm{i}\lambda\varphi)$ $(\lambda \geqq 1)$ for $u = \exp(\mathrm{i}\varphi)$ through the process given in Sect. 15.5. Only after supplementing them with $\mathscr{D}^{[-\lambda]}(U) = \exp(-\mathrm{i}\lambda\varphi)$ $(\lambda \geqq 0)$ obtained as above will the set of irreducible representations of U(1) be complete.

Exercise 15.5. To confirm the irreducibility and completeness of the representations, show in the case of $m = 1$ and 2 that the characters given by (15.41) satisfy the orthogonality relations of the first and second kinds.

Appendices

A. The Thirty-Two Crystallographic Point Groups

Table A.1. Stereograms of the 32 crystallographic point groups

Triclinic	(Monoclinic) (1st setting)	Tetragonal
1	2	4
———	$m(=\bar{2})$	$\bar{4}$
$\bar{1}$	$2/m$	$4/m$
Monoclinic (2nd setting)	Orthorhombic	
2	222	422
m	$2mm$	$4mm$
———	———	$\bar{4}2m$
$2/m$	mmm	$4/mmm$

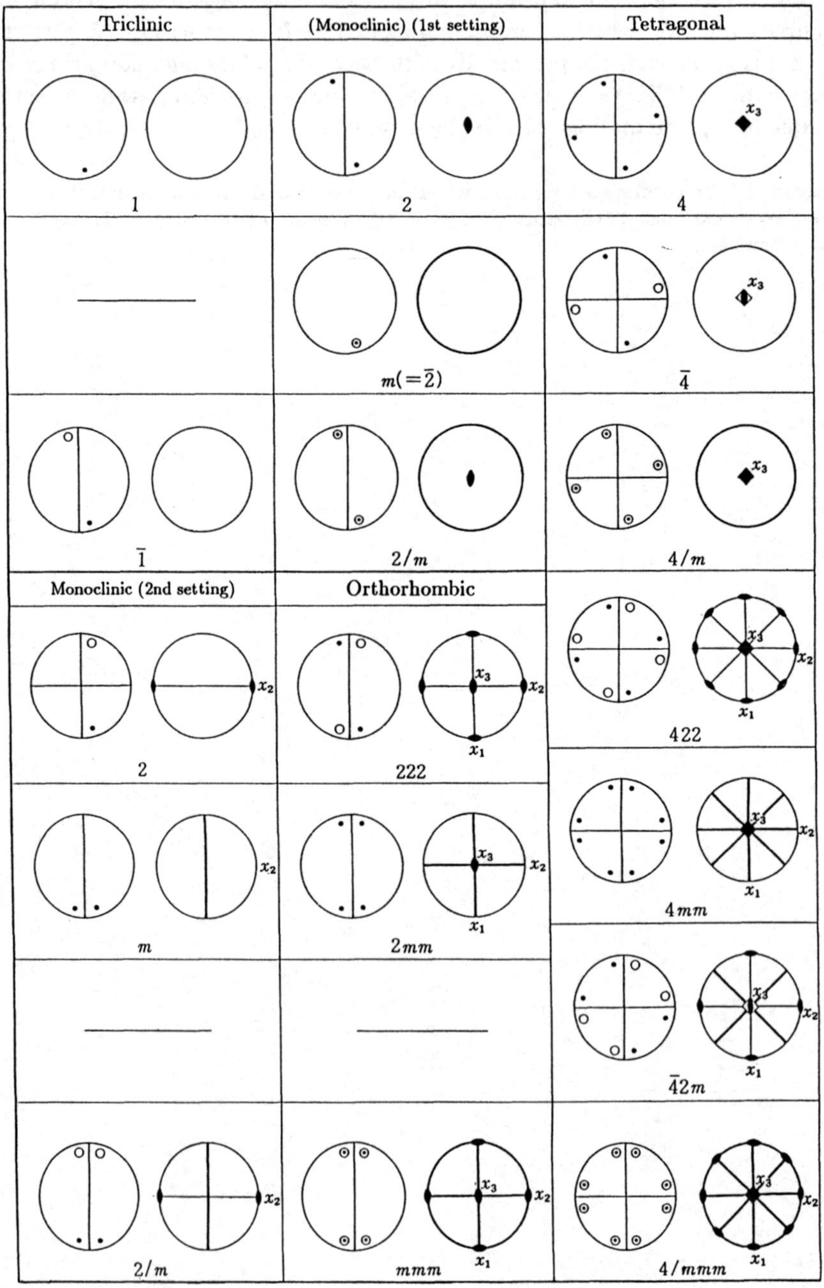

Table A.1. (continued)

Trigonal	Hexagonal	Cubic
x_3 3	x_3 6	x_3 x_2 x_1 23
—	x_3 $\bar{6}$	—
x_3 $\bar{3}$	x_3 $6/m$	x_3 x_2 x_1 $m3$
x_3 x_2 x_1 32	x_3 x_2 x_1 622	x_3 x_2 x_1 432
x_3 x_2 x_1 $3m$	x_3 x_2 x_1 $6mm$	—
—	x_3 x_2 x_1 $\bar{6}m2$	x_3 x_2 x_1 $\bar{4}3m$
x_3 x_2 x_1 $\bar{3}m$	x_3 x_2 x_1 $6/mmm$	x_3 x_2 x_1 $m3m$

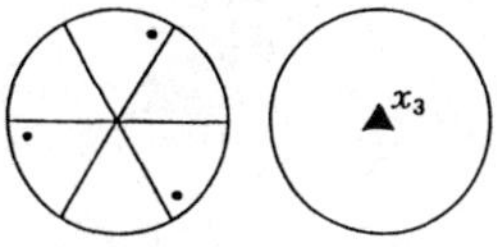

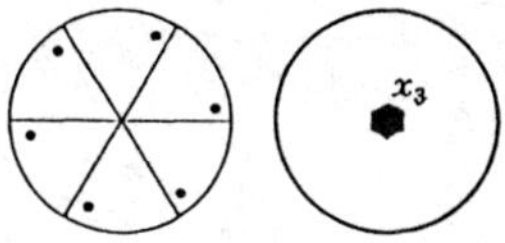

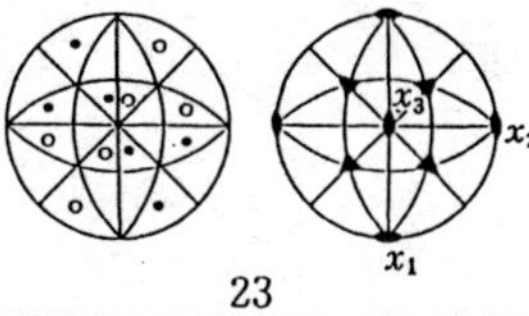

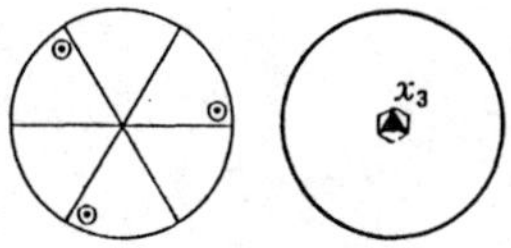

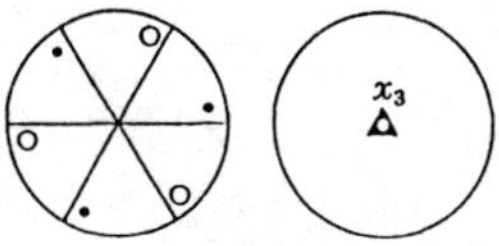

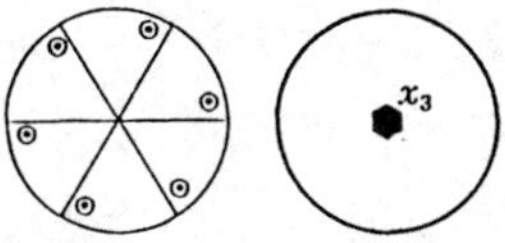

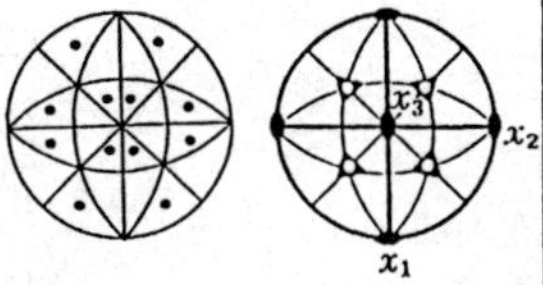

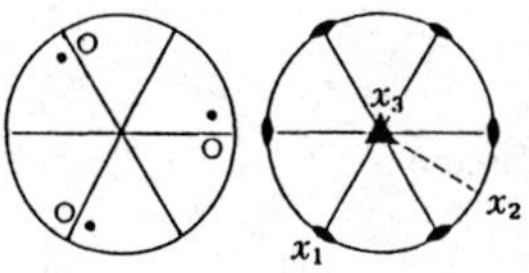

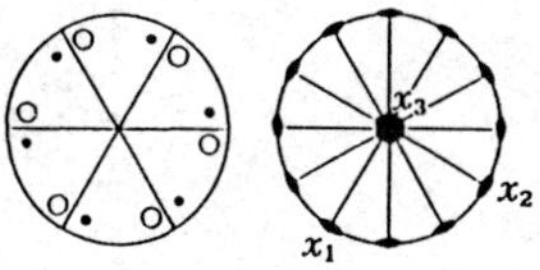

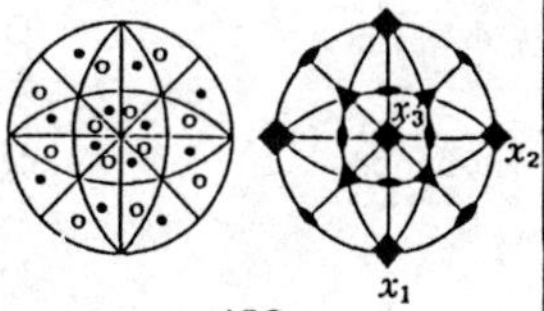

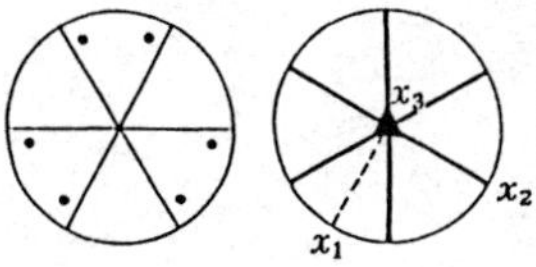

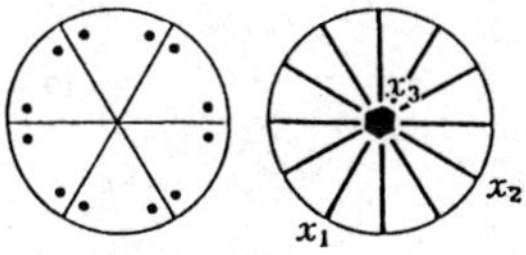

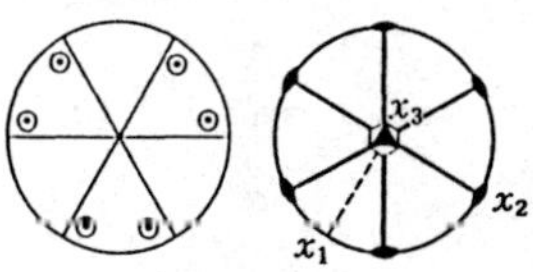

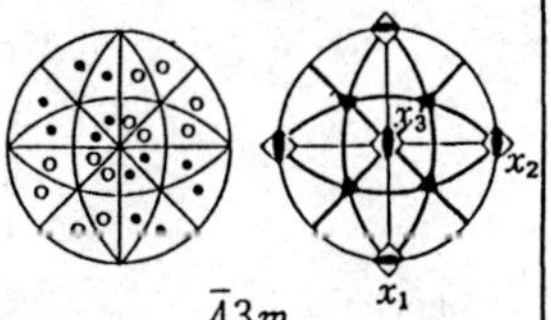

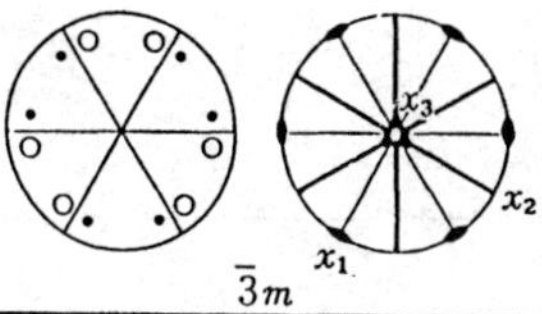

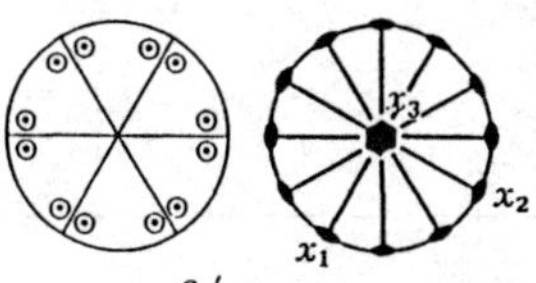

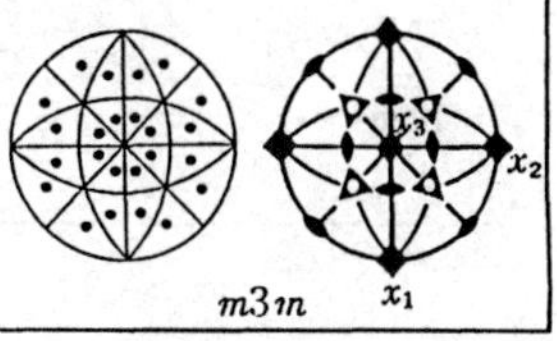

Table A.2. Schönflies and international symbols for the 32 crystallographic point groups

Crystal system	Schönflies symbol	International symbol (abbreviated)
Cubic	O_h	$\frac{4}{m}\bar{3}\frac{2}{m}$ (m3m)
	O	432
	T_d	$\bar{4}3m$
	T_h	$\frac{2}{m}\bar{3}$ (m3)
	T	23
Tetragonal	D_{4h}	$\frac{4}{m}\frac{2}{m}\frac{2}{m}$ (4/mmm)
	D_4	422
	D_{2d}	$\bar{4}2m$
	C_{4v}	4mm
	C_{4h}	$\frac{4}{m}$ (4/m)
	S_4	$\bar{4}$
	C_4	4
Orthorhombic	D_{2h}	$\frac{2}{m}\frac{2}{m}\frac{2}{m}$ (mmm)
	D_2	222
	C_{2v}	2mm
Hexagonal	D_{6h}	$\frac{6}{m}\frac{2}{m}\frac{2}{m}$ (6/mmm)
	D_6	622
	D_{3h}	$\bar{6}m2$
	C_{6v}	6mm
	C_{6h}	$\frac{6}{m}$ (6/m)
	C_{3h}	$\bar{6}$
	C_6	6
Trigonal	D_{3d}	$\bar{3}\frac{2}{m}$ ($\bar{3}m$)
	D_3	32
	C_{3v}	3m
	$C_{3i}(S_6)$	$\bar{3}$
	C_3	3
Monoclinic	C_{2h}	$\frac{2}{m}$ (2/m)
	$C_{1h}(C_s)$	m
	C_2	2
Triclinic	C_i	$\bar{1}$
	C_1	1

B. Character Tables for Point Groups

Characters and basis functions are tabulated below for the irreducible representations of point groups. For a proper rotation group and its direct-product group with C_i, only one of the two groups is considered. Both Mulliken and Bethe notations are written in parallel to denote the irreducible representations. For an explanation of the nomenclature, see Sects. 8.4 and 8.5. The representations below the broken line are double-valued. When barred and unbarred operations have different characters, they appear simultaneously in the same place. Spin functions α and β used in the basis functions are quantized along the z-axis. See Sect. 8.6 for transformation properties of the spin functions.

Complex conjugate representations are degenerate by virtue of time-reversal symmetry. The Mulliken notation regards such a pair of representations as a single (physically irreducible) representation. The other single-valued representations are all real and do not have additional degeneracy. Most double-valued representations are pseudoreal, having no additional degeneracy. Real representations, which appear twice because of time-reversal symmetry, are noted in the tables. See Sects. 4.12, 13.1 and 13.2 for degeneracy due to time-reversal symmetry.

For the symmorphic space groups O_h^1(sc), O_h^5(fcc) and O_h^9(bcc), the labels of small representations [B.1] are given in the rightmost columns of the corresponding point groups. For convenience of reference, we summarize in Table B.1 the point groups of $\boldsymbol{k}$ for high-symmetry points in the Brillouin zone.

Table B.1. Point group of $\boldsymbol{k}$ for high-symmetry points in the BZ

	O_h	T_d	D_{4h}	D_{2d}	C_{4v}	D_{2h}	C_{2v}	D_{3d}	C_{3v}	Fig. 11.7
O_h^1	ΓR		MX		ΔT		ΣZS		Λ	(a)
O_h^5	Γ		X	W	Δ		ΣZS	L	Λ	(b)
O_h^9	ΓH	P			Δ	N	ΣDG		ΛF	(c)

B.1 Cubic Point Groups

Table B.2. O_h

O_h		E	$6C_4$	$3C_4^2$	$6C_2'$	$8C_3$	I	$6IC_4$	$3\sigma_h$	$6\sigma_d$	$8IC_3$	Γ, R, H
A_{1g}	Γ_1^+	1	1	1	1	1	1	1	1	1	1	Γ_1
A_{2g}	Γ_2^+	1	−1	1	−1	1	1	−1	1	−1	1	Γ_2
E_g	Γ_3^+	2	0	2	0	−1	2	0	2	0	−1	Γ_{12}
T_{1g}	Γ_4^+	3	1	−1	−1	0	3	1	−1	−1	0	Γ_{15}'
T_{2g}	Γ_5^+	3	−1	−1	1	0	3	−1	−1	1	0	Γ_{25}'
A_{1u}	Γ_1^-	1	1	1	1	1	−1	−1	−1	−1	−1	Γ_1'
A_{1u}	Γ_2^-	1	−1	1	−1	1	−1	1	−1	1	−1	Γ_2'
E_u	Γ_3^-	2	0	2	0	−1	−2	0	−2	0	1	Γ_{12}'
T_{1u}	Γ_4^-	3	1	−1	−1	0	−3	−1	1	1	0	Γ_{15}
T_{2u}	Γ_5^-	3	−1	−1	1	0	−3	1	1	−1	0	Γ_{25}
$E_{1/2g}$	Γ_6^+	2 −2	$\sqrt{2}$ $-\sqrt{2}$	0	0	1 −1	2 −2	$\sqrt{2}$ $-\sqrt{2}$	0	0	1 −1	Γ_6^+
$E_{5/2g}$	Γ_7^+	2 −2	$-\sqrt{2}$ $\sqrt{2}$	0	0	1 −1	2 −2	$-\sqrt{2}$ $\sqrt{2}$	0	0	1 −1	Γ_7^+
$G_{3/2g}$	Γ_8^+	4 −4	0 0	0	0	−1 1	4 −4	0 0	0	0	−1 1	Γ_8^+
$E_{1/2u}$	Γ_6^-	2 −2	$\sqrt{2}$ $-\sqrt{2}$	0	0	1 −1	−2 2	$-\sqrt{2}$ $\sqrt{2}$	0	0	−1 1	Γ_6^-
$E_{5/2u}$	Γ_7^-	2 −2	$-\sqrt{2}$ $\sqrt{2}$	0	0	1 −1	−2 2	$\sqrt{2}$ $-\sqrt{2}$	0	0	−1 1	Γ_7^-
$G_{3/2u}$	Γ_8^-	4 −4	0 0	0	0	−1 1	−4 4	0 0	0	0	1 −1	Γ_8^-

A_{1g}: $r^2,\ x^4 + y^4 + z^4 - \frac{3}{5}r^4$

A_{2g}: $x^4(y^2 - z^2) + y^4(z^2 - x^2) + z^4(x^2 - y^2)$

E_g: $\{u, v\}$, $u \equiv 2z^2 - x^2 - y^2$, $v \equiv \sqrt{3}(x^2 - y^2)$

T_{1g}: $\{yz(y^2 - z^2), zx(z^2 - x^2), xy(x^2 - y^2)\}$

T_{2g}: $\{yz, zx, xy\}$ Γ_6^+: $\{\alpha, \beta\}$ Γ_7^+: $\{xy\alpha + (yz + \mathrm{i}zx)\beta,\ -xy\beta + (yz - \mathrm{i}zx)\alpha\}$

Γ_8^+: $\{v\beta,\ -u\alpha,\ u\beta,\ -v\alpha\}$, $\{(zx + \mathrm{i}yz)\alpha + 2\mathrm{i}xy\beta,\ -\sqrt{3}(zx + \mathrm{i}yz)\beta, -\sqrt{3}(zx - \mathrm{i}yz)\alpha, (zx - \mathrm{i}yz)\beta + 2\mathrm{i}xy\alpha\}$

Γ_6^-: $\{z\alpha + (x + \mathrm{i}y)\beta,\ -z\beta + (x - \mathrm{i}y)\alpha\}$

Γ_7^-: $\{xyz\alpha, xyz\beta\}$ Γ_8^-: $\{-\sqrt{3}(x + \mathrm{i}y)\alpha, 2z\alpha - (x + \mathrm{i}y)\beta, 2z\beta + (x - \mathrm{i}y)\alpha, \sqrt{3}(x - \mathrm{i}y)\beta\}$ $O_h = O \times C_i$

A_{1u}: $(A_{2g}) \times xyz$

A_{2u}: xyz

Eu: $\{xyzv,\ -xyzu\}$

T_{1u}: $\{x, y, z\}$

T_{2u}: $\{x(y^2 - z^2), y(z^2 - x^2), z(x^2 - y^2)\}$

Note that the above basis functions for the double-valued representations are not eigenfunctions of the total angular momentum $\boldsymbol{j} = \boldsymbol{l} + \boldsymbol{s}$, and do not diagonalize the spin-orbit interaction. Basis functions expressed in terms of the eigenfunctions of $\boldsymbol{j}$ are given in [B.2].

Table B.3. T_d

T_d		E		$6IC_4$		$3C_2$	$6\sigma_d$	$8C_3$		Compatibility with O_h	P
A_1	Γ_1	1		1		1	1	1		A_{1g}, A_{2u}	P_1
A_2	Γ_2	1		-1		1	-1	1		A_{2g}, A_{1u}	P_2
E	Γ_3	2		0		2	0	-1		E_g, E_u	P_3
T_1	Γ_4	3		1		-1	-1	0		T_{1g}, T_{2u}	P_5
T_2	Γ_5	3		-1		-1	1	0		T_{2g}, T_{1u}	P_4
$E_{1/2}$	Γ_6	2	-2	$\sqrt{2}$	$-\sqrt{2}$	0	0	1	-1	Γ_6^+, Γ_7^-	P_6
$E_{5/2}$	Γ_7	2	-2	$-\sqrt{2}$	$\sqrt{2}$	0	0	1	-1	Γ_7^+, Γ_6^-	P_7
$G_{3/2}$	Γ_8	4	-4	0	0	0	0	-1	1	Γ_8^+, Γ_8^-	P_8

Basis functions for T_d may be readily obtained from those for O_h by noting that xyz is invariant in T_d.

Table B.4. T

T		E		$3C_2$	$4C_3$		$4C_3^2$	
A	Γ_1	1		1	1		1	
E	Γ_2	1		1	ω		ω^2	
	Γ_3	1		1	ω^2		ω	
T	Γ_4	3		-1	0		0	
$E_{1/2}$	Γ_5	2	-2	0	1	-1	-1	1
$G_{3/2}$	Γ_6	2	-2	0	ω	$-\omega$	$-\omega^2$	ω^2
	Γ_7	2	-2	0	ω^2	$-\omega^2$	$-\omega$	ω

$\omega = \exp(-2\pi i/3)$, $T_h = T \times C_i$

B.2 Tetragonal Point Groups

Table B.5. D_{4h}

D_{4h}		E	$2C_4$	C_4^2	$2C_2'$	$2C_2''$	I	$2IC_4$	σ_h	$2\sigma_v$	$2\sigma_d$	M, X
A_{1g}	Γ_1^+	1	1	1	1	1	1	1	1	1	1	X_1
A_{2g}	Γ_2^+	1	1	1	−1	−1	1	1	1	−1	−1	X_4
B_{1g}	Γ_3^+	1	−1	1	1	−1	1	−1	1	1	−1	X_2
B_{2g}	Γ_4^+	1	−1	1	−1	1	1	−1	1	−1	1	X_3
E_g	Γ_5^+	2	0	−2	0	0	2	0	−2	0	0	X_5
A_{1u}	Γ_1^-	1	1	1	1	1	−1	−1	−1	−1	−1	X_1'
A_{2u}	Γ_2^-	1	1	1	−1	−1	−1	−1	−1	1	1	X_4'
B_{1u}	Γ_3^-	1	−1	1	1	−1	−1	1	−1	−1	1	X_2'
B_{2u}	Γ_4^-	1	−1	1	−1	1	−1	1	−1	1	−1	X_3'
E_u	Γ_5^-	2	0	−2	0	0	−2	0	2	0	0	X_5'
$E_{1/2g}$	Γ_6^+	2 −2	$\sqrt{2}$ $-\sqrt{2}$	0	0	0	2 −2	$\sqrt{2}$ $-\sqrt{2}$	0	0	0	X_6^+
$E_{3/2g}$	Γ_7^+	2 −2	$-\sqrt{2}$ $\sqrt{2}$	0	0	0	2 −2	$-\sqrt{2}$ $\sqrt{2}$	0	0	0	X_7^+
$E_{1/2u}$	Γ_6^-	2 −2	$\sqrt{2}$ $-\sqrt{2}$	0	0	0	−2 2	$-\sqrt{2}$ $\sqrt{2}$	0	0	0	X_6^-
$E_{3/2u}$	Γ_7^-	2 −2	$-\sqrt{2}$ $\sqrt{2}$	0	0	0	−2 2	$\sqrt{2}$ $-\sqrt{2}$	0	0	0	X_7^-

A_{1g}: z^2
A_{2g}: $xy(x^2 - y^2)$
B_{1g}: $x^2 - y^2$
B_{2g}: xy
E_g: $\{-zx, zy\}$
Γ_6^+: $\{\alpha, \beta\}$
Γ_7^+: $\{xy\alpha, xy\beta\}$
$D_{4h} = D_4 \times C_i$

A_{1u}: $xyz(x^2 - y^2)$
A_{2u}: z
B_{1u}: xyz
B_{2u}: $z(x^2 - y^2)$
E_u: $\{x, y\}$
Γ_6^-: $\{z\alpha, -z\beta\}, \{(x + iy)\beta, (x - iy)\alpha\}$
Γ_7^-: $\{(x - iy)\beta, -(x + iy)\alpha\}$

Table B.6. D_{2d}

D_{2d}		Basis	E		$2IC_4$		C_2	$2C'_2$	$2\sigma_d$	W
A_1	Γ_1	z^2, xyz	1		1		1	1	1	W_1
A_2	Γ_2	$z(x^2-y^2)$	1		1		1	-1	-1	W_2
B_1	Γ_3	x^2-y^2	1		-1		1	1	-1	W'_1
B_2	Γ_4	z, xy	1		-1		1	-1	1	W'_2
E	Γ_5	$\{x, y\}$	2		0		-2	0	0	W_3
$E_{1/2}$	Γ_6	$\{\alpha, \beta\}$	2	-2	$\sqrt{2}$	$-\sqrt{2}$	0	0	0	W_6
$E_{3/2}$	Γ_7	$\{z\alpha, z\beta\}$	2	-2	$-\sqrt{2}$	$\sqrt{2}$	0	0	0	W_7

When considering the group of W, note that the C'_2 axis of the D_{2d} group is in the [110] direction in Fig. 11.7b. Therefore, one has to read $xy \to x^2 - y^2$ and $x^2 - y^2 \to xy$ in the above table.

Table B.7. C_{4v}

C_{4v}		E		$2C_4$		C_4^2	$2\sigma_v$	$2\sigma_d$	Compatibility with D_{4h}	T, Δ
A_1	Γ_1	1		1		1	1	1	A_{1g}, A_{2u}	Δ_1
A_2	Γ_2	1		1		1	-1	-1	A_{2g}, A_{1u}	Δ'_1
B_1	Γ_3	1		-1		1	1	-1	B_{1g}, B_{2u}	Δ_2
B_2	Γ_4	1		-1		1	-1	1	B_{2g}, B_{1u}	Δ'_2
E	Γ_5	2		0		-2	0	0	E_g, E_u	Δ_5
$E_{1/2}$	Γ_6	2	-2	$\sqrt{2}$	$-\sqrt{2}$	0	0	0	Γ_6^+, Γ_6^-	Δ_6
$E_{3/2}$	Γ_7	2	-2	$-\sqrt{2}$	$\sqrt{2}$	0	0	0	Γ_7^+, Γ_7^-	Δ_7

Table B.8. C_4, S_4

C_4		E		C_4		C_4^2		C_4^3		Basis	
S_4		E		IC_4		C_2		IC_4^3			Basis
A	Γ_1	1		1		1		1		z	xyz
B	Γ_2	1		-1		1		-1		xy	z
E	Γ_3	1		$-i$		-1		i		$x+iy$	$x-iy$
	Γ_4	1		i		-1		$-i$		$x-iy$	$x+iy$
$E_{1/2}$	Γ_5	1	-1	ρ	$-\rho$	$-i$	i	$-\rho^*$	ρ^*	α	α
	Γ_6	1	-1	ρ^*	$-\rho^*$	i	$-i$	$-\rho$	ρ	β	β
$E_{3/2}$	Γ_7	1	-1	$-\rho$	ρ	$-i$	i	ρ^*	$-\rho^*$	$(x-iy)\beta$	$(x+iy)\beta$
	Γ_8	1	-1	$-\rho^*$	ρ^*	i	$-i$	ρ	$-\rho$	$(x+iy)\alpha$	$(x-iy)\alpha$

$\rho = \exp(-\pi i/4)$, $C_{4h} = C_4 \times C_i$

B.3 Orthorhombic Point Groups

Table B.9. D_{2h}

D_{2h}		Basis	E	C_{2z}	C_{2y}	C_{2x}	I	σ_z	σ_y	σ_x	N
A_g	Γ_1^+	x^2, y^2, z^2	1	1	1	1	1	1	1	1	N_1
B_{1g}	Γ_2^+	xy	1	1	−1	−1	1	1	−1	−1	N_2
B_{2g}	Γ_3^+	xz	1	−1	1	−1	1	−1	1	−1	N_4
B_{3g}	Γ_4^+	yz	1	−1	−1	1	1	−1	−1	1	N_3
A_u	Γ_1^-	xyz	1	1	1	1	−1	−1	−1	−1	N_2'
B_{1u}	Γ_2^-	z	1	1	−1	−1	−1	−1	1	1	N_1'
B_{2u}	Γ_3^-	y	1	−1	1	−1	−1	1	−1	1	N_3'
B_{3u}	Γ_4^-	x	1	−1	−1	1	−1	1	1	−1	N_4'
$E_{1/2g}$	Γ_5^+	$\{\alpha, \beta\}$	2 −2	0	0	0	2 −2	0	0	0	N_5^+
$E_{1/2u}$	Γ_5^-	$\{z\alpha, -z\beta\}$	2 −2	0	0	0	−2 2	0	0	0	N_5^-

$D_{2h} = D_2 \times C_i$
In the group of N, the x-, y-, and z-axes of the symmetry operations are directed $z \parallel [110]$, $y \parallel [001]$, $x \parallel [\bar{1}10]$. The names of the small representations obey the connectivity relations with Σ.

Table B.10. C_{2v}

C_{2v}		Basis	E	C_2	σ_y	σ_x	Z
A_1	Γ_1	z, x^2, y^2, z^2	1	1	1	1	Z_1
A_2	Γ_2	xy	1	1	−1	−1	Z_2
B_1	Γ_3	x, xz	1	−1	1	−1	Z_3
B_2	Γ_4	y, yz	1	−1	−1	1	Z_4
$E_{1/2}$	Γ_5	$\{\alpha, \beta\}$	2 −2	0	0	0	Z_5

Table B.11. ΣKUS

ΣKUS	Basis	E	C_2'	σ_d	IC_{4z}^2
Σ_1	$x + y, xy$	1	1	1	1
Σ_2	$z(x - y)$	1	1	−1	−1
Σ_3	z	1	−1	1	−1
Σ_4	$x - y$	1	−1	−1	1
Σ_5	$\{\alpha, \beta\}$	2 −2	0	0	0

B.6 Hexagonal Point Groups

Table B.12. D_{3h}, C_{6v}, D_6

D_{3h}			E	$2IC_6$	$2C_3$	σ_h	$3C_{2y}$	$3\sigma_x$			
	C_{6v}		E	$2C_6$	$2C_3$	C_2	$3\sigma_y$	$3\sigma_x$		C_{6v}	D_{3h}
	D_6		E	$2C_6$	$2C_3$	C_2	$3C_{2y}$	$3C_{2x}$	Basis	Basis	Basis
A_1'	A_1	Γ_1	1	1	1	1	1	1	z^2	z	$y^3 - 3x^2y$
A_2'	A_2	Γ_2	1	1	1	1	-1	-1	z		$x^3 - 3xy^2$
A_1''	B_1	Γ_3	1	-1	1	-1	$+1$	-1	$y^3 - 3x^2y$	$x^3 - 3xy^2$	
A_2''	B_2	Γ_4	1	-1	1	-1	-1	1	$x^3 - 3xy^2$	$y^3 - 3x^2y$	z
E''	E_1	Γ_6	2	1	-1	-2	0	0	$\{x, y\}$	$\{x, y\}$	$\{zx, zy\}$
E'	E_2	Γ_5	2	-1	-1	2	0	0	$\{xy, x^2 - y^2\}$	$\{xy, x^2 - y^2\}$	$\{x, y\}$
	$E_{1/2}$	Γ_7	2 -2	$\sqrt{3}$ $-\sqrt{3}$	1 -1	0	0	0		$\{\alpha, \beta\}$	
	$E_{5/2}$	Γ_8	2 -2	$-\sqrt{3}$ $\sqrt{3}$	1 -1	0	0	0			
	$E_{3/2}$	Γ_9	2 -2	0 0	-2 2	0	0	0		$\{(x + iy)\alpha, (x - iy)\beta\}$	

$D_{6h} = D_6 \times C_i$

Some authors [11.5–6] interchange Γ_5 and Γ_6.

Table B.13. C_{3h}, C_6

C_{3h}			E		IC_6		C_3		σ_h		C_3^2		IC_6^5		Basis
	C_6		E		C_6		C_3		C_2		C_3^2		C_6^5		
A′	A	Γ_1	1		1		1		1		1		1		z^2
A″	B	Γ_2	1		-1		1		-1		1		-1		$z(x \pm iy)^3$
E′	E_2	Γ_3	1		ω		ω^2		1		ω		ω^2		$(x + iy)^2$
		Γ_4	1		ω^2		ω		1		ω^2		ω		$(x - iy)^2$
E″	E_1	Γ_5	1		$-\omega$		ω^2		-1		ω		$-\omega^2$		$z(x - iy)$
		Γ_6	1		$-\omega^2$		ω		-1		ω^2		$-\omega$		$z(x + iy)$
	$E_{1/2}$	Γ_7	1	-1	ϱ	$-\varrho$	ϱ^2	$-\varrho^2$	$-i$	i	ϱ^4	$-\varrho^4$	ϱ^5	$-\varrho^5$	α
		Γ_8	1	-1	$-\varrho^5$	ϱ^5	$-\varrho^4$	ϱ^4	i	$-i$	$-\varrho^2$	ϱ^2	$-\varrho$	ϱ	β
	$E_{5/2}$	Γ_9	1	-1	$-\varrho$	ϱ	ϱ^2	$-\varrho^2$	i	$-i$	ϱ^4	$-\varrho^4$	$-\varrho^5$	ϱ^5	$(x - iy)^2\beta$
		Γ_{10}	1	-1	ϱ^5	$-\varrho^5$	$-\varrho^4$	ϱ^4	$-i$	i	$-\varrho^2$	ϱ^2	ϱ	$-\varrho$	$(x + iy)^2\alpha$
	$E_{3/2}$	Γ_{11}	1	-1	i	$-i$	-1	1	$-i$	i	1	-1	i	$-i$	$z(x - iy)\beta$
		Γ_{12}	1	-1	$-i$	i	-1	1	i	$-i$	1	-1	$-i$	i	$z(x + iy)\alpha$

$\varrho = \exp(-\pi i/6)$, $\omega = \varrho^4 = \exp(-2\pi i/3)$
$C_{6h} = C_6 \times C_i$

B.5 Trigonal Point Groups

Table B.14. D_{3d}

D_{3d}		Basis	E	$2C_3$	$3C_2'$	I	$2IC_3$	$3\sigma_v$	L
A_{1g}	Γ_1^+	z^2	1	1	1	1	1	1	L_1
A_{2g}	Γ_2^+	$x_1y_2 - y_1x_2$	1	1	−1	1	1	−1	L_2
E_g	Γ_3^+	$\{zx, zy\}$	2	−1	0	2	−1	0	L_3
A_{1u}	Γ_1^-	$3x^2y - y^3$	1	1	1	−1	−1	−1	L_1'
A_{2u}	Γ_2^-	z	1	1	−1	−1	−1	1	L_2'
E_u	Γ_3^-	$\{x, y\}$	2	−1	0	−2	1	0	L_3'
$E_{1/2g}$	Γ_6^+	$\{\alpha, \beta\}$	2 −2	1 −1	0	2 −2	1 −1	0	L_6^+
$E_{3/2g}$	Γ_4^+	$(zx + izy)\alpha \pm$	1 −1	−1 1	i −i	1 −1	−1 1	i −i	L_4^+
	Γ_5^+	$i(zx - izy)\beta$	1 −1	−1 1	−i i	1 −1	−1 1	−i i	L_5^+
$E_{1/2u}$	Γ_6^-	$\{z\alpha, z\beta\}$	2 −2	1 −1	0	−2 2	−1 1	0	L_6^-
$E_{3/2u}$	Γ_4^-	$(x + iy)\alpha \pm$	1 −1	−1 1	i −i	−1 1	1 −1	−i i	L_5^-
	Γ_5^-	$i(x - iy)\beta$	1 −1	−1 1	−i i	−1 1	1 −1	i −i	L_4^-

$D_{3d} = D_3 \times C_i$

Table B.15. C_{3v}

C_{3v}		E	$2C_3$	$3\sigma_v$	F, Λ
A_1	Γ_1	1	1	1	Λ_1
A_2	Γ_2	1	1	−1	Λ_2
E	Γ_3	2	−1	0	Λ_3
$E_{1/2}$	Γ_6	2 −2	1 −1	0	Λ_6
$E_{3/2}$	Γ_4	1 −1	−1 1	i −i	Λ_4
	Γ_5	1 −1	−1 1	−i i	Λ_5

Table B.16. C_3

C_3		E		C_3		C_3^2		Basis
A	Γ_1	1		1		1		z
E	Γ_2	1		ω		ω^2		$x + iy$
	Γ_3	1		ω^2		ω		$x - iy$
$E_{1/2}$	Γ_4	1	-1	$-\omega^2$	ω^2	ω	$-\omega$	α
	Γ_5	1	-1	$-\omega$	ω	ω^2	$-\omega^2$	β
$B_{3/2}$	Γ_6	1	-1	-1	1	1	-1	$(x + iy)\alpha$

$\omega = \exp(-2\pi i/3)$, $C_{3i}(= S_6) = C_3 \times C_i$. $B_{3/2}$ is a real representation. It appears in a pair because of time-reversal symmetry.

B.6 Monoclinic Point Groups

Table B.17. $C_{1h}(= C_s)$, C_2

$C_{1h}(= C_s)$			E		σ	
C_2			E		C_2	
A′	A	Γ_1	1		1	
A″	B	Γ_2	1		-1	
$E_{1/2}$		Γ_3	1	-1	$-i$	i
		Γ_4	1	-1	i	$-i$

$C_{2h} = C_2 \times C_i$

B.7 Triclinic Point Groups

Table B.18. C_1

C_1		E	
A	Γ_1	1	
$B_{1/2}$	Γ_2	1	-1

$C_i = C_1 \times C_i$
$B_{1/2}$ is a real representation. It appears in a pair because of time-reversal symmetry.

B.8 Axial Rotation Groups

Table B.19. $D_{\infty h}$

$D_{\infty h}$		Basis	E	$2C(\phi)$	$\infty\sigma_v$	I	$2IC(\phi)$	$\infty C'_2$
A_{1g}	Σ_g^+	z^2	1	1	1	1	1	1
A_{1u}	Σ_u^-		1	1	-1	-1	-1	1
A_{2g}	Σ_g^-	$x_1y_2 - y_1x_2$	1	1	-1	1	1	-1
A_{2u}	Σ_u^+	z	1	1	1	-1	-1	-1
E_{1g}	Π_g	$\{zx, zy\}$	2	$2\cos\phi$	0	2	$2\cos\phi$	0
E_{1u}	Π_u	$\{x, y\}$	2	$2\cos\phi$	0	-2	$-2\cos\phi$	0
E_{2g}	Δ_g	$\{xy, x^2 - y^2\}$	2	$2\cos 2\phi$	0	2	$2\cos 2\phi$	0
E_{2u}	Δ_u		2	$2\cos 2\phi$	0	-2	$-2\cos 2\phi$	0
$\vdots$								
E_{ng}			2	$2\cos n\phi$	0	2	$2\cos n\phi$	0
E_{nu}			2	$2\cos n\phi$	0	-2	$-2\cos n\phi$	0
$E_{1/2g}$		$\{\alpha, \beta\}$	2	$2\cos\phi/2$	0	2	$2\cos\phi/2$	0
$E_{1/2u}$		$\{z\alpha, z\beta\}$	2	$2\cos\phi/2$	0	-2	$-2\cos\phi/2$	0
$E_{3/2g}$		$\{\alpha^3, \beta^3\}$	2	$2\cos 3\phi/2$	0	2	$2\cos 3\phi/2$	0
$E_{3/2u}$		$\{(x + iy)\alpha, (x - iy)\beta\}$	2	$2\cos 3\phi/2$	0	-2	$-2\cos 3\phi/2$	0
$\vdots$								
$E_{n+1/2g}$			2	$2\cos(n + \frac{1}{2})\phi$	0	2	$2\cos(n + \frac{1}{2})\phi$	0
$E_{n+1/2u}$			2	$2\cos(n + \frac{1}{2})\phi$	0	-2	$-2\cos(n + \frac{1}{2})\phi$	0

$D_{\infty h} = C_{\infty v} \times C_i$

The symbols Σ_g^+, Σ_u^-, etc. are preferred in linear molecules.

Answers and Hints to the Exercises

Chapter 2

2.10 The four aces (SHDC) are rearranged as follows:

$$P(\text{SHDC}) = (\text{DSCH}), \quad Q(\text{DSCH}) = (\text{DCHS}) .$$

The multiplication rule (2.11) gives

$$QP = \begin{pmatrix} 1 & 2 & 3 & 4 \\ 1 & 3 & 4 & 2 \end{pmatrix} \begin{pmatrix} 1 & 2 & 3 & 4 \\ 3 & 1 & 4 & 2 \end{pmatrix} = \begin{pmatrix} 1 & 2 & 3 & 4 \\ 1 & 3 & 4 & 2 \end{pmatrix} \begin{pmatrix} 1 & 3 & 4 & 2 \\ 3 & 4 & 2 & 1 \end{pmatrix} = \begin{pmatrix} 1 & 2 & 3 & 4 \\ 3 & 4 & 2 & 1 \end{pmatrix} .$$

2.11 Putting $G_i = E$ in (2.16), we obtain

$$f(E)f(G_j) = f(G_j) .$$

Since this relation holds for any $f(G_j)$, the element $f(E)$ must be the unit element of the group $\mathscr{G}'$. Next put $G_j = G_i^{-1}$, then

$$f(G_i)f(G_i^{-1}) = f(E) .$$

Since $f(E)$ is the unit element, $f(G_i^{-1})$ is inverse to $f(G_i)$.

2.14 Consider some element C (other than the unit element E) of the group $\mathscr{G}$ of order g. Then we have $C^m = E$ for some integer m, and the cyclic group $\{C, C^2, \ldots, C^{m-1}, E\}$ is a subgroup of $\mathscr{G}$. Now, $\mathscr{G}$ cannot have a proper subgroup because g is a prime number. Consequently, m must be equal to g.

2.15 Use the geometrical considerations explained in the text, or else use (2.23) and Table 2.3.

2.18

$\mathscr{C}_1$	$\mathscr{C}_2$	$\mathscr{C}_3$	$\mathscr{C}_4$	$\mathscr{C}_5$
$\mathscr{C}_2$	$\mathscr{C}_1$	$\mathscr{C}_3$	$\mathscr{C}_4$	$\mathscr{C}_5$
$\mathscr{C}_3$	$\mathscr{C}_3$	$2\mathscr{C}_1 + 2\mathscr{C}_2$	$2\mathscr{C}_5$	$2\mathscr{C}_4$
$\mathscr{C}_4$	$\mathscr{C}_4$	$2\mathscr{C}_5$	$2\mathscr{C}_1 + 2\mathscr{C}_2$	$2\mathscr{C}_3$
$\mathscr{C}_5$	$\mathscr{C}_5$	$2\mathscr{C}_4$	$2\mathscr{C}_3$	$2\mathscr{C}_1 + 2\mathscr{C}_2$

2.19 If A is conjugate to B, then A^{-1} is conjugate to B^{-1}. So, the elements inverse to the ones belonging to a class $\mathscr{C}_j$ will form a class $\mathscr{C}_{j'}$. If $\mathscr{C}_i \neq \mathscr{C}_{j'}$, then c_{ij}^1 vanishes since $\mathscr{C}_i\mathscr{C}_j$ does not contain the unit element. If $\mathscr{C}_i = \mathscr{C}_{j'}$, products of the form GG^{-1} appear h_i times in the class product $\mathscr{C}_i\mathscr{C}_j$.

2.20 From (2.26), $G\mathscr{C}_k = \mathscr{C}_k G$. Sum both sides over the elements G of the class $\mathscr{C}_j$.

2.23 $C_{4v} = C_2 + C_2 C_4 + C_2 \sigma_x + C_2 \sigma_d$.

2.24 For two elements K_i and K_j of the set $\mathscr{K}$, we have $f(K_i) = f(K_j) = E'$. Then $K_i K_j$ is an element of $\mathscr{K}$, because

$$f(K_i K_j) = f(K_i) f(K_j) = E'E' = E' ,$$

using the definition (2.16) of homomorphism. Furthermore, from Exercise 2.11,

$$f(K_i^{-1}) = f(K_i)^{-1} = (E')^{-1} = E' ,$$

indicating that K_i^{-1} is an element of $\mathscr{K}$. Therefore, $\mathscr{K}$ is a subgroup of $\mathscr{G}$.
Next, for an arbitrary element G of $\mathscr{G}$,

$$\begin{aligned} f(GK_iG^{-1}) &= f(G)f(K_i)f(G^{-1}) \\ &= f(G)E'f(G)^{-1} = E' , \end{aligned}$$

showing that GK_iG^{-1} belongs to $\mathscr{K}$. Since the elements of the set $G\mathscr{K}G^{-1}$ are distinct, $G\mathscr{K}G^{-1}$ coincides with $\mathscr{K}$ as a set.

2.25 The group D_{3h} has six classes: $\{E\}$, $\{C_3, C_3^{-1}\}$, $\{\sigma_1, \sigma_2, \sigma_3\}$, $\{\sigma_h\}$, $\{C_3\sigma_h, C_3^{-1}\sigma_h\}$, $\{U_1, U_2, U_3\}$.

Chapter 3

3.1 The eigenvalue λ of an eigenvector $\boldsymbol{v}$ of the Hermitian operator A is given by $\lambda = (\boldsymbol{v}, A\boldsymbol{v})$. Take the complex conjugate of this equation:

$$\lambda^* = (\boldsymbol{v}, A\boldsymbol{v})^* = (A\boldsymbol{v}, \boldsymbol{v}) ,$$

and use the hermiticity (3.45).

3.2 Because of the unitarity of A, we have in general $(A\boldsymbol{v}, A\boldsymbol{v}) = (\boldsymbol{v}, \boldsymbol{v})$. For an eigenvector $\boldsymbol{v}$ of A with eigenvalue λ, we have $(A\boldsymbol{v}, A\boldsymbol{v}) = \lambda^*\lambda(\boldsymbol{v}, \boldsymbol{v})$ so that $|\lambda| = 1$.
If $\hat{A}$ is a symmetric unitary matrix, then $\hat{A}^{-1} = \hat{A}^*$. Take the complex conjugate of $A\boldsymbol{v} = \lambda\boldsymbol{v}$, and show that

$$A^{-1}\boldsymbol{v}^* = \lambda^*\boldsymbol{v}^* = \lambda^{-1}\boldsymbol{v}^* ,$$

which means $A\boldsymbol{v}^* = \lambda\boldsymbol{v}^*$. Since $\boldsymbol{v}^*$ is an eigenvector with the same eigenvalue as $\boldsymbol{v}$, real vectors $\boldsymbol{v} + \boldsymbol{v}^*$ or $\mathrm{i}(\boldsymbol{v} - \boldsymbol{v}^*)$ may be chosen as the eigenvector.

Chapter 4

4.1 For $\sigma_y v$ to be a multiple of v, c_2 has to be equal to c_1 or $-c_1$. Neither choice can bring $R(\alpha)v$ into a multiple of v, unless $\lambda = 0$.

4.3 Put $x = r\cos\phi$, $y = r\sin\phi$ in polar coordinates. Then $l_z = -\mathrm{i}\partial/\partial\phi$, and the right-hand side of (4.30) becomes $f(r\cos(\phi-\alpha), r\sin(\phi-\alpha))$.

4.4 You have only to verify the transformation for the generating elements C_3 and σ_1:

$$C_3 f_1^{(\mathrm{E})} = (2f_2 - f_3 - f_1)/\sqrt{6}$$
$$= -\frac{1}{2}f_1^{(\mathrm{E})} + \frac{\sqrt{3}}{2}f_2^{(\mathrm{E})} \ .$$

4.6 The matrix $\hat{M}$ commutes with any matrix of the representation if it commutes with the matrices for the generating elements. Show that the matrix $\hat{M}$, which commutes with $\hat{D}^{(\mathrm{E})}(C_3)$ and $\hat{D}^{(\mathrm{E})}(\sigma_1)$ of (4.3), is necessarily a multiple of the unit matrix. For the three-dimensional representation (4.37),

$$\hat{M} = \begin{bmatrix} a & b & b \\ b & a & b \\ b & b & a \end{bmatrix}$$

satisfies (4.44).

4.9 Use (4.61) and (4.56) to evaluate the left-hand side.

4.10 Irreducibility of the direct-product representation is evident from

$$\sum_A \sum_B |\chi^{(a\times b)}(AB)|^2 = \sum_A |\chi^{(a)}(A)|^2 \sum_B |\chi^{(b)}(B)|^2$$
$$= g(\mathscr{A})g(\mathscr{B}) = g(\mathscr{A}\times\mathscr{B}) \ .$$

Furthermore, the number $n_\mathrm{r}(\mathscr{A}\times\mathscr{B})$ of direct-product representations constructed in this way is equal to the number $n_\mathrm{c}(\mathscr{A}\times\mathscr{B})$ of classes of the direct-product group $\mathscr{A}\times\mathscr{B}$, because

$$n_\mathrm{r}(\mathscr{A}) = n_\mathrm{c}(\mathscr{A}) \ , \qquad n_\mathrm{r}(\mathscr{B}) = n_\mathrm{c}(\mathscr{B}) \ ,$$
$$n_\mathrm{r}(\mathscr{A}\times\mathscr{B}) = n_\mathrm{r}(\mathscr{A})\,n_\mathrm{r}(\mathscr{B}) \ , \qquad n_\mathrm{c}(\mathscr{A}\times\mathscr{B}) = n_\mathrm{c}(\mathscr{A})\,n_\mathrm{c}(\mathscr{B}) \ .$$

4.11

$$\sum_j [\hat{D}^{(\mathrm{reg})}(G)]_{ij}[\hat{D}^{(\mathrm{reg})}(G')]_{jk} = \sum_j \delta(G_i^{-1}GG_j)\,\delta(G_j^{-1}G'G_k)$$
$$= \delta(G_i^{-1}GG'G_k) \ .$$

4.12 Use the result from Exercise 2.16.

4.13 The group $\mathrm{C}_{4\mathrm{v}}$ has five classes and hence five irreducible representations, whose dimensions can be obtained from $4\times 1^2 + 2^2 = 8$. The characters may be determined in the same way as mentioned in the text for the C_{3v} group. Here, we show another method to determine the characters using the relation (5). If we write x_i for $h_i\, \chi^{(\alpha)}(\mathscr{C}_i)$, we have

$$x_i x_j = d_\alpha \sum_k c_{ij}^k x_k \ .$$

The class constants may be found in the table of Exercise 2.18. For example, $\mathscr{C}_2\mathscr{C}_3 = \mathscr{C}_3$ and $\mathscr{C}_3\mathscr{C}_3 = 2\mathscr{C}_1 + 2\mathscr{C}_2$ give respectively $x_2 x_3 = d_\alpha x_3$ and $x_3^2 = d_\alpha(2d_\alpha + 2x_2)$, from which we find $\{x_2 = d_\alpha,\ x_3 = \pm 2d_\alpha\}$ or $\{x_2 = -d_\alpha,\ x_3 = 0\}$. Proceeding in this way and using the orthogonality (2), complete the character table.

4.15 The character χ of the direct-product representation $\tilde{D}_1 \times D_2$ is

$$\chi(G) = \chi_1(G^{-1})\, \chi_2(G) = \chi_1(G)^*\, \chi_2(G) \ .$$

Its sum over G is equal to g if $\chi_1 = \chi_2$ and vanishes otherwise.
Invariance of Ψ_0 is verified as follows:

$$G\Psi_0 = \sum_j (G\tilde{\psi}_j)(G\phi_j)$$

$$= \sum_j \sum_k \sum_i \tilde{\psi}_k \phi_i \tilde{D}_{kj}(G)\, D_{ij}(G)$$

$$= \sum_k \sum_i \tilde{\psi}_k\, \phi_i \sum_j D_{ij}(G)\, D_{jk}(G^{-1}) \ .$$

$$= \sum_k \sum_i \tilde{\psi}_k \phi_i [\hat{D}(G)\, \hat{D}(G^{-1})]_{ik} \ .$$

Chapter 5

5.1 Evaluate $w_{\lambda i}$ using (5.7, 8), to obtain

$$w_{\lambda i} = \frac{1}{g} \sum_P \sum_{j=1}^{k} \delta_{ii}(P)\, \chi^{(\lambda)}(R_j^{-1} P R_j)\, \chi^{(i)}(P)^* \ .$$

On the right-hand side, you have only to sum over $P = R_j S R_j^{-1} (S \in \mathscr{H})$. Note that $\chi^{(i)}(P) = \chi^{(i)}(S)$, which yields

$$w_{\lambda i} = \frac{1}{h} \sum_S \chi^{(\lambda)}(S)\, \chi^{(i)}(S)^* \ ,$$

which proves the Frobenius reciprocity theorem.

5.2 The number of times the representation $D^{(\lambda)}$ appears in $(D^{(\lambda)} \uparrow \mathscr{G}) \downarrow \mathscr{H}$ is calculated using (5.8, 9):

$$\frac{1}{h} \sum_{S \in \mathscr{H}} \chi^{(\lambda)}(S)^* \, \bar{\chi}^{(\lambda)}(S) = \sum_i w_{i\lambda} w_{\lambda i} = \sum_i w_{\lambda i}^2 \ .$$

5.3 Sum (5.7) over P belonging to the class $\mathscr{C}_i$, and use (2.26).

5.4 Reduction of the subduced representations $D^{(i)} \downarrow \mathscr{H}$ is obvious from Table 5.1. As for the induction $D_6 \to D_{6h}$, the operation R_2 in (5.7) is here the inversion I. Since $IPI = P$, we obtain

$$\bar{\chi}^{(\lambda)}(P) = \begin{cases} 2\chi^{(\lambda)}(P) & \text{if } P \in D_6 \ , \\ 0 & \text{otherwise} \ . \end{cases}$$

This character gives, for example, $A \uparrow D_{6h} = A_{1g} + A_{1u}$. As for the induction $C_6 \to D_6$, take $R_2 = C_{2y}$. Then $C_{2y} C_6^m C_{2y} = C_6^{-m}$ will give

$$\bar{\chi}^{(\lambda)}(C_6^m) = \chi^{(\lambda)}(C_6^m) + \chi^{(\lambda)}(C_6^m)^* \ .$$

$$A \uparrow D_6 = A_1 + A_2 \ , \quad \Gamma_5(E_1) \uparrow D_6 = E_1 \ , \quad \Gamma_3(E_2) \uparrow D_6 = E_2$$

$$B \uparrow D_6 = B_1 + B_2 \ , \quad \Gamma_6(E_1) \uparrow D_6 = E_1 \ , \quad \Gamma_4(E_2) \uparrow D_6 = E_2$$

5.5 From the text above this problem, subduction of D and $D \uparrow \mathscr{G}$ onto $\mathscr{H}$ gives

$$D \downarrow \mathscr{H} = n\Delta^{(11)} \ , \qquad (D \uparrow \mathscr{G}) \downarrow \mathscr{H} = \sum_{i=1}^{s} n\Delta^{(i1)} \ .$$

Since the representations $\Delta^{(i1)}$ with different i are inequivalent, $(D \uparrow \mathscr{G}) \downarrow \mathscr{L}$ contains D only once.

5.6 The group C_{3v} is decomposed as (2.37) with respect to the invariant subgroup C_3. Irreducible representations $\Gamma_1, \Gamma_2, \Gamma_3$ of the cyclic group C_3 are readily obtained (Appendix B). $g = 6$ and $h = 3$ give $k = g/h = 2$, a prime number.

Choose first $\Delta = \Gamma_1$, and show

$$\Delta(\sigma_1 C_3 \sigma_1) = \Delta(C_3^{-1}) = 1 \ , \quad \Delta(\sigma_1 C_3^{-1} \sigma_1) = \Delta(C_3) = 1 \ .$$

Then $\Delta(\sigma_1 \mathscr{H} \sigma_1)$ is equivalent to $\Delta(\mathscr{H})$, which corresponds to case (b) of the text ($\mathscr{L} = \mathscr{G}$). Determine $U = D(\sigma_1)$ such that $U^2 = \Delta(\sigma_1^2) = \Delta(E) = 1$. Then $U = \pm 1$ will give the A_1 and A_2 representations of C_{3v}.

Choose next $\Delta = \Gamma_2$, and show

$$\Delta(\sigma_1 C_3 \sigma_1) = \omega^2 \ , \quad \Delta(\sigma_1 C_3^{-1} \sigma_1) = \omega \ .$$

This is case (a) ($\mathscr{L} = \mathscr{H}$). $\Gamma_2 \uparrow C_{3v}$ gives the irreducible representation E of C_{3v}.

5.7 (i) Single-valued representations: Begin by evaluating p_{ii} of (5.44) for the given factor system to get two ray classes E and σ_h, and hence $r = 2$. Now, $2^2 + 2^2 = m = 8$. The two 2-dimensional irreducible representations are constructed from the irreducible representations of the invariant subgroup D_2. The group D_2 has only one ray class E, so that it has one 2-dimensional representation ($2^2 = 4$). Its representation matrices $\hat{\Delta}(S)$ are determined so as to satisfy the multiplication table

$$\begin{aligned}
\hat{\Delta}(C_{2z})\hat{\Delta}(C_{2x}) &= -\hat{\Delta}(C_{2x})\hat{\Delta}(C_{2z}) = -\hat{\Delta}(C_{2y}) \ , \\
\hat{\Delta}(C_{2x})\hat{\Delta}(C_{2y}) &= -\hat{\Delta}(C_{2y})\hat{\Delta}(C_{2x}) = \quad \hat{\Delta}(C_{2z}) \ , \\
\hat{\Delta}(C_{2y})\hat{\Delta}(C_{2z}) &= -\hat{\Delta}(C_{2z})\hat{\Delta}(C_{2y}) = \quad \hat{\Delta}(C_{2x}) \ , \\
\hat{\Delta}(C_{2x})^2 &= -\hat{\Delta}(C_{2y})^2 = \hat{\Delta}(C_{2z})^2 = \hat{1} \ .
\end{aligned}$$

Such matrices are given by

$$\hat{\Delta}(E) = \hat{1} \ , \quad \Delta(C_{2x}) = \hat{\sigma}_x \ , \quad \Delta(C_{2y}) = -\mathrm{i}\hat{\sigma}_y \ , \quad \Delta(C_{2z}) = \hat{\sigma}_z$$

in terms of the Pauli spin matrices (7.34). Next, use (5.56) to calculate

$$\hat{\Delta}^{(I)}(S) = \frac{\alpha^{SI}_{S,I}}{\alpha^{SI}_{I,ISI}} \hat{\Delta}(ISI) \ , \quad S \in D_2$$

and find $\hat{\Delta}^{(I)}(S)$ is equivalent to $\hat{\Delta}(S)$ [case (b)]. Determine the representation matrix for inversion $\hat{U} = \hat{D}(I)$ such that

$$\begin{aligned}
&\hat{\Delta}(C_{2x})\hat{U} = -\hat{U}\hat{\Delta}(C_{2x}) \ , \quad \hat{\Delta}(C_{2z})\hat{U} = \hat{U}\hat{\Delta}(C_{2z}) \ , \\
&\hat{\Delta}(C_{2y})\hat{U} = -\hat{U}\hat{\Delta}(C_{2y}) \ , \quad \hat{U}^2 = \hat{1} \ .
\end{aligned}$$

You will have two such $\hat{U}$'s, namely, $\hat{U} = \pm\hat{\sigma}_z$, each corresponding to an irreducible representation.

	E	C_{2z}	C_{2x}	C_{2y}	I	σ_h	σ_{vx}	σ_{vy}
X_1	$\hat{1}$	$\hat{\sigma}_z$	$\hat{\sigma}_x$	$-\mathrm{i}\hat{\sigma}_y$	$\hat{\sigma}_z$	$\hat{1}$	$-\mathrm{i}\hat{\sigma}_y$	$\hat{\sigma}_x$
X_2	1	$\hat{\sigma}_z$	$\hat{\sigma}_x$	$-\mathrm{i}\hat{\sigma}_y$	$-\hat{\sigma}_z$	$-\hat{1}$	$\mathrm{i}\hat{\sigma}_y$	$-\hat{\sigma}_x$

(ii) Double-valued representations: p_{ii} is nonvanishing for E and C_{2z}. Notice the difference from (i). To find the two 2-dimensional irreducible representations, use again the invariant subgroup D_2, which now has four 1-dimensional representations given below. Induced representations $X_3 = \Gamma_1 \uparrow D_{2h}$ and $X_4 = \Gamma_3 \uparrow D_{2h}$ are irreducible [case (a)].

D_2	E	C_{2z}	C_{2x}	C_{2y}
Γ_1	1	i	i	1
Γ_2	1	i	−i	−1
Γ_3	1	−i	−i	1
Γ_4	1	−i	i	−1

	E	C_{2z}	C_{2x}	C_{2y}	I	σ_h	σ_{vx}	σ_{vy}
X_3	$\hat{1}$	$i\hat{1}$	$i\hat{\sigma}_z$	$\hat{\sigma}_z$	$\hat{\sigma}_x$	$i\hat{\sigma}_x$	$-\hat{\sigma}_y$	$i\hat{\sigma}_y$
X_4	$\hat{1}$	$-i\hat{1}$	$-i\hat{\sigma}_z$	$\hat{\sigma}_z$	$\hat{\sigma}_x$	$-i\hat{\sigma}_x$	$\hat{\sigma}_y$	$i\hat{\sigma}_y$

In Sects. 11.8 and 11.10, the above representations are obtained by means of Herring's method, which does not make use of ray representations. Although Herring's method is straightforward, it requires lengthy calculations as compared with the above method because of the doubled (and further doubled in the case of double-valued representations) number of group elements.

Chapter 6

6.1 $R\psi(\boldsymbol{r}) = \psi(R^{-1}\boldsymbol{r}) = \psi(\boldsymbol{r}\hat{R})$ requires (6.6). The proof can also be given explicitly using the operator expression $R = \exp(-i\alpha l_z)$ (see Exercise 4.3). Evaluate RxR^{-1} and Rp_xR^{-1} using commutation relations like $[l_z, x] = iy$.

6.3 Use (6.32, 33) and (6.20) to find

$$P^{(\beta)}_{l(m)} f = \sum_\alpha \sum_\mu c^{(\alpha)}_\mu \cdot \frac{d_\beta}{g} \sum_R D^{(\beta)}_{lm}(R)^* R\phi^{(\alpha)}_\mu$$

$$= \sum_\alpha \sum_\mu \sum_\nu c^{(\alpha)}_\mu \phi^{(\alpha)}_\nu \cdot \frac{d_\beta}{g} \sum_R D^{(\beta)}_{lm}(R)^* D^{(\alpha)}_{\nu\mu}(R).$$

Use here the great orthogonality theorem (4.46).

6.5 Use the representation matrices defined in (4.5, 6, 3).

6.10 $$(P^{(\alpha)}_{m'(m)} f, g) = \frac{d_\alpha}{g} \sum_R D^{(\alpha)}_{m'm}(R)\,(Rf, g)$$

$$= \frac{d_\alpha}{g} \sum_R D^{(\alpha)}_{m'm}(R)\,(f, R^{-1}g)$$

$$= \frac{d_\alpha}{g} \sum_R D^{(\alpha)}_{mm'}(R^{-1})^*\,(f, R^{-1}g)$$

$$= (f, P^{(\alpha)}_{m(m')} g)\ .$$

9.1 Let $\varphi_1, \varphi_2, \ldots, \varphi_6$ denote the six $2s$ orbitals of the carbon atoms (or six 1s orbitals of the hydrogen atoms). The problem is to reduce the representation based on these six orbitals which transform into themselves according to (9.15) except that all the negative signs are dropped. Since the s orbitals themselves remain the same under symmetry operations, you have only to count the number of s orbitals that do not move.

D_{6h}	E	$2C_6$	$2C_3$	C_2	$3C_{2y}$	$3C_{2x}$	I	$2S_3$	$2S_6$	σ_h	$3\sigma_d$	$3\sigma_v$
$\chi(R)$	6	0	0	0	2	0	0	0	0	6	0	2

Reduction of this character gives $A_{1g} + B_{1u} + E_{1u} + E_{2g}$. The corresponding wavefunctions are obtained by means of the methods mentioned in the text:

$$\psi(a_{1g}) = (\varphi_1 + \varphi_2 + \varphi_3 + \varphi_4 + \varphi_5 + \varphi_6)/\sqrt{6}\ ,$$
$$\psi(b_{1u}) = (\varphi_1 - \varphi_2 + \varphi_3 - \varphi_4 + \varphi_5 - \varphi_6)/\sqrt{6}\ ,$$
$$\psi(e_{1u}1) = (-\varphi_2 - \varphi_3 + \varphi_5 + \varphi_6)/2\ ,$$
$$\psi(e_{1u}2) = (2\varphi_1 + \varphi_2 - \varphi_3 - 2\varphi_4 - \varphi_5 + \varphi_6)/\sqrt{12}\ ,$$
$$\psi(e_{2g}1) = (-\varphi_2 + \varphi_3 - \varphi_5 + \varphi_6)/2\ ,$$
$$\psi(e_{2g}2) = (-2\varphi_1 + \varphi_2 + \varphi_3 - 2\varphi_4 + \varphi_5 + \varphi_6)/\sqrt{12}\ .$$

Here $\psi(e_{1u}1)$, $\psi(e_{1u}2)$ transform like x, y and $\psi(e_{2g}1)$, $\psi(e_{2g}2)$ like xy, and $x^2 - y^2$. a_{1g} and e_{1u} are bonding orbitals, while b_{1u} and e_{2g} are anti-bonding orbitals.

Consider the p_x and p_y orbitals in the following way: For each carbon atom, choose the y'-axis such that it goes through the center of the molecule and the x'-axis perpendicular to it. For the x'- and y'-axes defined in this way, the six $p_{y'}$ orbitals and the six $p_{x'}$ orbitals are closed within themselves under the symmetry operations. The six $p_{y'}$ orbitals, transforming like the s orbitals, give $A_{1g} + B_{1u} + E_{1u} + E_{2g}$, while the six $p_{x'}$ orbitals give the following character:

D_{6h}	E	$2C_6$	$2C_3$	C_2	$3C_{2y}$	$3C_{2x}$	I	$2S_3$	$2S_6$	σ_h	$3\sigma_d$	$3\sigma_v$
$\chi(R)$	6	0	0	0	-2	0	0	0	0	6	0	-2

9.2 Upon symmetry descent from O_h to C_4, we have the compatibility relations,

$$E_g \downarrow C_4 = A + B\ , \quad T_{2g} \downarrow C_4 = B + E\ .$$

As indicated in Table 9.6, we have the following symmetry-adapted functions:

$$\begin{aligned}
&\text{A:} \quad r^2 Y_{20} \sim 3z^2 - r^2 \,, \\
&\text{B:} \quad r^2 Y_{2,\pm 2} \sim (x + \mathrm{i}y)^2, (x - \mathrm{i}y)^2 \,, \\
&\text{E:} \quad r^2 Y_{2,\pm 1} \sim \mp (x \pm \mathrm{i}y)z
\end{aligned}$$

for the representations A, B and E of C_4 from the five d functions. Since the representations A and E derive respectively from E_g and T_{2g}, we find that $3z^2 - r^2$ and $(x \pm \mathrm{i}y)z$ respectively belong to E_g and T_{2g}.

On the other hand, B can derive from E_g as well as from T_{2g}, so that the functions $(x + \mathrm{i}y)^2$ and $(x - \mathrm{i}y)^2$ cannot be assigned uniquely to E_g or T_{2g} by means of the above compatibility relations alone. The assignment calls for a somewhat higher symmetry than C_4. Restrict the symmetry O_h down to C_{4v}, and so find

$$E_g \downarrow C_{4v} = A_1 + B_1 \,, \quad T_{2g} \downarrow C_{4v} = B_2 + E \,.$$

The distinction becomes possible at this level. The characters of B_1 and B_2 of C_{4v} differ for mirror reflections σ_v and σ_d. The symmetric linear combination of $(x + \mathrm{i}y)^2$ and $(x - \mathrm{i}y)^2$, being invariant under σ_v, belongs to B_1 and hence to E_g, while the antisymmetric one belongs to T_{2g}.

9.3 For σ orbitals, just count the number of atoms left unmoved by the symmetry operations.

O_h	E	$6C_4$	$3C_4^2$	$6C_2'$	$8C_3$	I	$6S_4$	$3\sigma_h$	$6\sigma_d$	$8S_6$
$\chi(R)$	6	2	2	0	0	0	0	4	2	0

Reduction of this character gives $A_{1g} + E_g + T_{1u}$. The basis function for the totally symmetric representation A_{1g} is $(s_1 + s_2 + s_3 + s_4 + s_5 + s_6)/\sqrt{6}$, while the three functions of T_{1u} that behave like x, y, z are $(s_4 - s_1)/\sqrt{2}$, $(s_5 - s_2)/\sqrt{2}$, $(s_6 - s_3)/\sqrt{2}$. For $p\pi$ orbitals, examine also how the orbitals are transformed.

O_h	E	$6C_4$	$3C_4^2$	$6C_2'$	$8C_3$	I	$6S_4$	$3\sigma_h$	$6\sigma_d$	$8S_6$
$\chi(R)$	12	0	-4	0	0	0	0	0	0	0

9.5 The wavefunctions for the $t_2^2 t_2'$ configuration, which has an inequivalent third electron, are constructed from (9.38). The allowed terms are

$$^2A_1 \quad ^2A_2 \quad 2\,^2E \quad 3\,^2T_1 \quad 4\,^2T_2 \quad ^4A_2 \quad ^4E \quad ^4T_1 \quad ^4T_2 \,.$$

For the t_2^3 configuration, use Table 15.4 to evaluate the character, and find

$$T_2^{[1^3]} = A_2 \ , \quad T_2^{[21]} = E + T_1 + T_2 \ .$$

O	E	$6C_4$	$3C_4^2$	$6C_2'$	$8C_3$
s_1	3	−1	−1	1	0
s_2	3	−1	3	3	0
s_3	3	−1	−1	1	3
$T_2^{[1^3]}$	1	−1	1	−1	1
$T_2^{[21]}$	8	0	0	0	−1

10.1 $Q(B_{2u}) = u_{1z} + u_{3z} - u_{2z} - u_{4z}$,

$$Q(B_{2g}) = u_{1y} - u_{3y} + u_{2x} - u_{4x} \ .$$

For the E_u representation, the projection operator constructed from the diagonal elements of the representation matrices becomes

$$P_{x(x)}^{(E)} = E - C_4^2 + C_{2x} - C_{2y} - I + \sigma_h - \sigma_{vx} + \sigma_{vy} \ ,$$

which may be factorized as

$$P_{x(x)}^{(E)} = (E - I)(E - C_4^2)(E + C_{2x}) \ .$$

Operate this on u_{1x}, and orthogonalize the result to the translational mode $u_{1x} + u_{2x} + u_{3x} + u_{4x}$ to find

$$Q(E_u x) = u_{1x} + u_{3x} - u_{2x} - u_{4x} \ .$$

10.2 All modes are allowed.

A_1: xx, yy, zz $\quad$ B_1: xz

A_2: xy $\quad$ B_2: yz

10.4 The proof rests on the time-reversal degeneracy of odd-electron wavefunctions Ψ. $\{\Psi, \theta\Psi\}$ will form a basis for the double-valued representation D. If a symmetry operation R transforms Ψ as

$$R\Psi = a\Psi + b\theta\Psi \ ,$$

then show

$$R\theta\Psi = \theta R\Psi = -b^*\Psi + a^*\theta\Psi \ .$$

The representation matrix $D(R)$ then becomes

$$\hat{D}(R) = \begin{bmatrix} a & -b^* \\ b & a^* \end{bmatrix} .$$

Unitarity of this matrix requires $|a|^2 + |b|^2 = 1$. Now evaluate the character of the antisymmetric product $\{D \times D\}$ using

$$\chi^{\{D \times D\}}(R) = \tfrac{1}{2}((\mathrm{Tr}\{D(R)\})^2 - \mathrm{Tr}\{D(R)^2\}) .$$

10.5 The center atom M gives $\mathrm{T_{1u}}$. Displacements of the surrounding six atoms give the following character $\chi(R)$.

$\mathrm{O_h}$	E	$6C_4$	$3C_4^2$	$6C_2'$	$8C_3$	I	$6S_4$	$3\sigma_h$	$6\sigma_d$	$8S_6$
N_R	6	2	2	0	0	0	0	4	2	0
$\chi^{(1)}(R)$	3	1	-1					1	1	
$\chi(R)$	18	2	-2	0	0	0	0	4	2	0

Find the normal coordinates

$$Q(\mathrm{A_{1g}}) = u_{1x} - u_{4x} + u_{2y} - y_{5y} + u_{3z} - u_{6z} ,$$

$$Q(\mathrm{E_g}\, u) = -u_{1x} + u_{4x} - u_{2y} + u_{5y} + 2u_{3z} - 2u_{6z} ,$$

$$Q(\mathrm{E_g}\, v) = \sqrt{3}\,(u_{1x} - u_{4x} - u_{2y} + u_{5y}) ,$$

$$Q(\mathrm{T_{2g}}\, xy) = u_{1y} + u_{2x} - u_{4y} - u_{5x} ,$$

$$Q(\mathrm{T_{2u}}\, z(x^2 - y^2)) = u_{1z} - u_{2z} + u_{4z} - u_{5z} ,$$

$$Q(\mathrm{T_{1u}}\, z) = \begin{cases} u_{1z} + u_{2z} + u_{4z} + u_{5z} - 2u_{3z} - 2u_{6z} , \\ u_{1z} + u_{2z} + u_{3z} + u_{4z} + u_{5z} + u_{6z} - 6U_z , \end{cases}$$

using the factorized projection operators given below. Here U_z stands for the displacement of the center atom M.

Projection operators for the irreducible representations of the octahedral group $\mathrm{O_h}$ are as follows [for "u" representations, replace $(E + I)$ by $(E - I)$]:

$$\mathrm{A_{1g}}\text{:}\quad (E + I)(E + C_2'[1\bar{1}0])(E + C_3[111] + C_3^{-1}[111]) \times (E + C_{4z}^2 + C_{4z} + C_{4z}^{-1}) ,$$

$$\mathrm{A_{2g}}\text{:}\quad (E + I)(E - C_2'[1\bar{1}0])(E + C_3[111] + C_3^{-1}[111]) \times (E + C_{4z}^2 - C_{4z} - C_{4z}^{-1}) ,$$

$$\mathrm{E_g}\, u\text{:}\quad (E + I)(E + C_2'[1\bar{1}0])(2E - C_3[111] - C_3^{-1}[111]) \times (E + C_{4z}^2 + C_{4z} + C_{4z}^{-1}) ,$$

$$\mathrm{E_g}\,v:\quad (E+I)(E-C_2'[1\bar{1}0])(2E-C_3[111]-C_3^{-1}[111])$$
$$\times\,(E+C_{4z}^2-C_{4z}-C_{4z}^{-1})\ ,$$

$$\mathrm{T_{1g}}\,\gamma:\quad (E+I)(E-C_2'[110])(E+C_{4z}^2+C_{4z}+C_{4z}^{-1})\ ,$$

$$\mathrm{T_{2g}}\,\zeta:\quad (E+I)(E+C_2'[110])(E+C_{4z}^2-C_{4z}-C_{4z}^{-1})\ .$$

The above projection operators were constructed from the diagonal elements of the representation matrices using the basis functions given in Appendix B.

11.1 The lower-right block of the representation matrix for $\{C_{2y}|\boldsymbol{\tau}\}$ should be, from (11.52), the small representation matrix for

$$\{\sigma_{\mathrm{d}\bar{y}}|0\}^{-1}\{C_{2y}|\boldsymbol{\tau}\}\{\sigma_{\mathrm{d}\bar{y}}|0\}=\{C_{2x}|\boldsymbol{\tau}\}\ .$$

For $\{C_4|\boldsymbol{\tau}\}$, use (11.55) to find

$$\{\sigma_{\mathrm{d}\bar{y}}|0\}^{-1}\{C_4|\boldsymbol{\tau}\}\{\varepsilon|0\}=\{\sigma_{\mathrm{v}y}|\boldsymbol{\tau}\}\ ,$$

$$\{\varepsilon|0\}^{-1}\{C_4|\boldsymbol{\tau}\}\{\sigma_{\mathrm{d}\bar{y}}|0\}=\{\sigma_{\mathrm{v}x}|\boldsymbol{\tau}\}\ .$$

An example of the basis of this four-dimensional representation is

$$\{\cos(\pi y/a),\ \sin(\pi y/a),\ \cos(\pi x/a),\ \sin(\pi x/a)\}\ .$$

You can also construct the representation matrices from the transformation of this basis.

11.2 For instance, $\{C_{2y}|\boldsymbol{\tau}\}$ has the following effect on orbital functions $f(x, y, z)$ and spin functions α, β.

$$\{C_{2y}|\boldsymbol{\tau}\}\,f(x,y,z)=f\left(\frac{a}{2}-x,\ y-\frac{a}{2},\frac{c}{2}-z\right)$$

$$\alpha\to\beta\ ,\quad \beta\to-\alpha$$

so that

$$\alpha\sin\frac{\pi y}{a}\ ,\quad \beta\cos\frac{\pi y}{a}\ ,\quad \beta\cos\frac{\pi y}{a}\ ,\quad \alpha\sin\frac{\pi y}{a}\ .$$

	$\{\varepsilon\|0\}$	$\{C_{2y}\|\boldsymbol{\tau}\}$	$\{C_2\|0\}$	$\{C_{2x}\|\boldsymbol{\tau}\}$	$\{\sigma_h\|0\}$	$\{\sigma_{\mathrm{v}x}\|\boldsymbol{\tau}\}$	$\{I\|0\}$	$\{\sigma_{\mathrm{v}y}\|\boldsymbol{\tau}\}$
X_3	$\begin{bmatrix}1 & \\ & 1\end{bmatrix}$	$\begin{bmatrix} & -1\\ -1 & \end{bmatrix}$	$\begin{bmatrix}\mathrm{i} & \\ & \mathrm{i}\end{bmatrix}$	$\begin{bmatrix} & -\mathrm{i}\\ -\mathrm{i} & \end{bmatrix}$	$\begin{bmatrix}-\mathrm{i} & \\ & \mathrm{i}\end{bmatrix}$	$\begin{bmatrix} & -\mathrm{i}\\ \mathrm{i} & \end{bmatrix}$	$\begin{bmatrix}-1 & \\ & 1\end{bmatrix}$	$\begin{bmatrix} & -1\\ 1 & \end{bmatrix}$
X_4	$\begin{bmatrix}1 & \\ & 1\end{bmatrix}$	$\begin{bmatrix} & 1\\ 1 & \end{bmatrix}$	$\begin{bmatrix}-\mathrm{i} & \\ & -\mathrm{i}\end{bmatrix}$	$\begin{bmatrix} & -\mathrm{i}\\ -\mathrm{i} & \end{bmatrix}$	$\begin{bmatrix}-\mathrm{i} & \\ & \mathrm{i}\end{bmatrix}$	$\begin{bmatrix} & \mathrm{i}\\ -\mathrm{i} & \end{bmatrix}$	$\begin{bmatrix}1 & \\ & -1\end{bmatrix}$	$\begin{bmatrix} & -1\\ 1 & \end{bmatrix}$

12.1 The representation M_4' at $\boldsymbol{k} = (\pi/a)(110)$ has a basis function like $\cos(\pi x/a)$ $\cos(\pi y/a)$ $\sin(2\pi z/a)$. Displace the origin through $(a/2, a/2, a/2)$. The phase factor (12.16) takes on -1 for $\alpha = C_4, C_2', IC_4$ and σ_v.

12.3 The rotation $\{C_{4z}^2|0\}$ sends $\psi_1, \psi_2, \psi_3, \psi_4$ to

$$\psi_1 \to \psi_3 , \quad \psi_3 \to \psi_1 , \quad \psi_2 \to \psi_4 , \quad \psi_4 \to \psi_2 .$$

For the $\{C_{2xy}'|\tau\}$ operation, show using (11.9) that

$$\{C_{2xy}'|\tau\} f(x, y, z) = f(y - a/4, x - a/4, a/4 - z),$$

and hence

$$\psi_1 \to -\psi_3 , \quad \psi_3 \to \psi_1 , \quad \psi_2 \to -\psi_2 , \quad \psi_4 \to -\psi_4 .$$

In addition,

$$\{C_{2\bar{x}y}'|\tau\} f(x, y, z) = f(a/4 - y, a/4 - x, a/4 - z) ,$$

$$\psi_1 \to \psi_1 , \quad \psi_3 \to -\psi_3 , \quad \psi_2 \to -\psi_4 , \quad \psi_4 \to -\psi_2 .$$

Comparing these transformation properties with the characters of Table 12.1, observe that

$$\{\psi_1(\boldsymbol{r})\psi_3(\boldsymbol{r}') - \psi_3(\boldsymbol{r})\psi_1(\boldsymbol{r}'), \psi_2(\boldsymbol{r})\psi_4(\boldsymbol{r}') - \psi_4(\boldsymbol{r})\psi_2(\boldsymbol{r}')\}$$

is a basis for the X_3 representation.

The same result can be obtained from the general theory of Sect. 12.12. Characters of the L_1 representation at $\boldsymbol{k}_1$, $\boldsymbol{k}_2$, $\boldsymbol{k}_3$, and $\boldsymbol{k}_4$ are given by [see (12.48)]

$$\chi^{\boldsymbol{k}_p}(Q) = \chi^{(L_1)}(S_{p-1}^{-1} Q S_{p-1}) , \quad S_\nu = \{IC_4|0\}^\nu .$$

For instance, for $Q = \{I|\tau\}$ and $p = 2$, we have

$$\{IC_4|0\}^{-1} \{I|\tau\} \{IC_4|0\} = \{I|IC_4^{-1}\tau\} = \{I|\tau\} \{\varepsilon| -\boldsymbol{t}_2\} ,$$

using the primitive vector $\boldsymbol{t}_2$ defined in (11.2), which yields

$$\chi^{\boldsymbol{k}_2}(\{I|\tau\}) = \chi^{(L_1)}(\{I|\tau\} \{\varepsilon| -\boldsymbol{t}_2\}) = -1 .$$

Use (12.53) and find the character of representation at X contained in $L_1 \times L_1$:

$$\chi^{X(L_1 \times L_1)}(Q) = 2\chi^{\boldsymbol{k}_1}(Q)\chi^{\boldsymbol{k}_3}(Q) + 2\chi^{\boldsymbol{k}_2}(Q)\chi^{\boldsymbol{k}_4}(Q) .$$

Use further (12.69) to evaluate

$$\chi^{X(L_1)}(Q^2) = \sum_{p=1}^{4} \delta(\boldsymbol{k}_p + Q\boldsymbol{k}_p, \boldsymbol{k}_X)\chi^{\boldsymbol{k}_p}(Q^2) .$$

The table below shows the process of character evaluation. The columns $k_1 \sim k_4$ give the values of the characters $\chi^{k_p}(Q)$ for k_p. Blank means that Q does not belong to the $\mathcal{G}(k_p)$. From this table and (12.68), we find $[L_1 \times L_1] = X_1$, and $\{L_1 \times L_1\} = X_3$.

$Q \in \mathcal{G}_X/T$	k_1	k_2	k_3	k_4	$\chi^{X(L_1 \times L_1)}(Q)$	$\chi^{X(L_1)}(Q^2)$
$\{\varepsilon\|0\}$	1	1	1	1	4×1	0
$\{C_{4x}^2, C_{4y}^2\|0\}$					0	0
$\{C_4^2\|0\}$					0	4
$2\{C_4\|\tau\}$					0	0
$\{C'_{2xy}\|\tau\}$		-1		-1	2×1	-2
$\{C'_{2\bar{x}y}\|\tau\}$	1		-1		$2 \times (-1)$	2
$\{I\|\tau\}$	1	-1	-1	-1	0	0
$\{\sigma_x, \sigma_y\|\tau\}$					0	0
$\{\sigma_h\|\tau\}$					0	0
$2\{S_4\|0\}$					0	0
$\{\sigma_{dxy}\|0\}$		1		1	2×1	2
$\{\sigma_{d\bar{x}y}\|0\}$	1		1		2×1	2

12.4

	Γ_{15}	X_4	$\Gamma_{15} \times X_4$
$\{\varepsilon\|0\}$	3	2	6
$\{C_{4x}^2, C_{4y}^2\|0\}$	-1	0	0
$\{C_4^2\|0\}$	-1	-2	2
$2\{C_4\|\tau\}$	1	0	0
$\{C'_{2xy}\|\tau\}$	-1	-2	2
$\{C'_{2\bar{x}y}\|\tau\}$	-1	2	-2
$\{I\|\tau\}$	-3	0	0
$\{\sigma_x, \sigma_y\|\tau\}$	1	0	0
$\{\sigma_h\|\tau\}$	1	0	0
$2\{S_4\|0\}$	-1	0	0
$2\{\sigma_d\|0\}$	1	0	0

12.5 Only those vibrations that appear in the reduction of $R_{15} \times \Gamma_{15} \times X_1$ can give rise to indirect transitions. Show that

$$\Gamma_{15} \times X_1 = X'_4 + X'_5 ,$$

$$R_{15} \times X'_4 = M_1 + M_5 ,$$

$$R_{15} \times X'_5 = M_1 + M_2 + M_3 + M_4 + M_5 .$$

Now, observe from Table 12.4 that the normal modes at the M point are M'_2, M'_4, and $2M'_5$.

12.6

$\{\Gamma\}$T: $\{\sin 2Y \sin 2Z, \sin 2Z \sin 2X, \sin 2X \sin 2Y\}$
$\{$M$\}$A: $\cos X \cos Y \sin^2 Z$,
$\{$M$\}$B$_1$: $\sin X \sin Y \cos^2 Z$,
where $X = \pi x/a$, $Y = \pi y/a$, $Z = \pi z/a$.

Exercise to Table 13.8. The group of $\boldsymbol{k}$ at Γ, X, Z and A is identical to the space group D_{2h}^{12}. In such cases, the general theory of Sect. 12.12 simplifies greatly: Characters of product representations may be evaluated as in the case of point groups without invoking complicated summations. Just one point to watch out for is that Q^2 can contain primitive lattice translations, as seen in the table below.

Q	Z_1 / A_1	$Z_1 \times Z_1$ / $A_1 \times A_1$	Q^2	Z_1	A_1	$\{Z_1 \times Z_1\}$	$\{A_1 \times A_1\}$
$\{\varepsilon \| 0\}$	2	4	$\{\varepsilon \| 0\}$	2	2	1	1
$\{C_2 \| 0\}$	−2	4	$\{\varepsilon \| 0\}$	2	2	1	1
$\{C_{2x} \| \tau\}$	0	0	$\{\varepsilon \| t_1\}$	2	−2	−1	1
$\{C_{2y} \| \tau\}$	0	0	$\{\varepsilon \| t_2\}$	2	−2	−1	1
$\{I \| 0\}$	0	0	$\{\varepsilon \| 0\}$	2	2	−1	−1
$\{\sigma_h \| 0\}$	0	0	$\{\varepsilon \| 0\}$	2	2	−1	−1
$\{\sigma_{vx} \| \tau\}$	0	0	$\{\varepsilon \| t_2 + t_3\}$	−2	2	1	−1
$\{\sigma_{vy} \| \tau\}$	0	0	$\{\varepsilon \| t_1 + t_3\}$	−2	2	1	−1

15.2 For an $l = 1$ electron, s_p of (15.42) is

$$s_p = e^{ip\varphi} + 1 + e^{-ip\varphi} .$$

In the p^3 configuration, the total spin angular momentum S takes on the values 1/2 and 3/2. The orbital part corresponding to the $S = 1/2$ spin state should belong to

$$[n/2 + S, n/2 - S]^* = [21]^* = [21]$$

of the product representation $D^{(1)} \times D^{(1)} \times D^{(1)}$. Evaluate its character using Table 15.4,

$$\begin{aligned}\chi^{[21]} &= (s_1^3 - s_3)/3 \\ &= (e^{2i\varphi} + e^{i\varphi} + 1 + e^{-i\varphi} + e^{-2i\varphi}) + (e^{i\varphi} + 1 + e^{-i\varphi}) ,\end{aligned}$$

which consists of $L = 2$ and $L = 1$ orbital states. For $S = 3/2$, show

$$[n/2 + S, n/2 - S]^* = [3]^* = [1^3] ,$$
$$\chi^{[1^3]} = (s_1^3 - 3s_1 s_2 + 2s_3)/6 = 1 .$$

Motifs of the family crests

(*found at the beginning of each chapter*)

1: Crane made from folded paper
2: Bound sheaves of the rice plant
3: Umbrellas
4: Tea berry
5: Hollyhock
6: Chinese flower
7: Arrow feathers
8: Feathers of the falcon
9: Cloves
10: Chinese balloon flower
11: Swallowtail butterfly
12: Small pouches
13: Scissors
14: Sword blades and leaves of the wood sorrel
15: Bell flower

References

Chapter 5

5.1 The following paper gives a slightly different treatment: W. G. Harter: J. Math. Phys. **10**, 739 (1969)
5.2 T. Janssen: *Crystallographic Groups* (North-Holland, Amsterdam 1973) pp. 243–254

Chapter 7

7.1 L. D. Landau, E. M. Lifshitz: *Quantum Mechanics*, 3rd ed. (Pergamon, Oxford 1976) pp. 410–419
7.2 Tables of the Racah coefficients are available, for example, in: T. Ishidzu, H. Horie, M. Sato, Y. Tanabe, S. Yanagawa: *Tables of Racah Coefficients* (Pan Pacific, Tokyo 1960); M. Rotenberg, N. Metropolis, R. Bivins, J. K. Wooten, Jr.: *The 3–j and 6–j Symbols* (MIT, Cambridge, MA, 1959)
7.3 R. Kubo, T. Nagamiya (eds.): *Solid State Physics* (McGraw-Hill, New York 1969) pp. 446–448
7.4 Tables of the c.f.p. for p, d, and f electrons have been given by G. Racah: Phys. Rev. **63**, 367 (1943); ibid. **76**, 1352 (1949)
7.5 For details of the treatment, see, for example, J. C. Slater: *Quantum Theory of Atomic Structure*, Vol. 2 (McGraw-Hill, New York 1960) pp. 95–157

Chapter 8

8.1 G. Herzberg: *Molecular Spectra and Molecular Structure, Vol. 3: Electronic Spectra and Electronic Structure of Polyatomic Molecules* (Van Nostrand, New York 1966)

Chapter 9

9.1 M. Kotani: J. Phys. Soc. Jpn. **19**, 2150–2156 (1964)
9.2 J. S. Griffith: *Irreducible Tensorial Method for Molecular Symmetry Groups* (Prentice-Hall, Englewood Cliffs, NJ 1962) pp. 4–31

Chapter 10

10.1 T. P. Wilson: J. Chem. Phys. **11**, 369 (1943)
10.2 H. A. Jahn, E. Teller: Proc. R. Soc. London A **161**, 220 (1937)
10.3 E. Ruch, A. Schönhofer: Theor. Chim. Acta **3**, 291 (1965)
10.4 L. D. Landau, E. M. Lifshitz: *Quantum Mechanics*, 3rd ed. (Pergamon, Oxford 1976) p. 408
10.5 E. I. Blount: J. Math. Phys. **12**, 1890 (1971)

Chapter 11

11.1 L. P. Bouckaert, R. Smoluchowski, E. Wigner: Phys. Rev. **50**, 58 (1936)
11.2 C. Herring: J. Franklin Inst. **233**, 525 (1942). An error at the W point is corrected in [11.3]
11.3 R. J. Elliott: Phys. Rev. **96**, 280 (1954)
11.4 R. H. Parmenter: Phys. Rev. **100**, 573 (1955)
11.5 R. C. Casella: Phys. Rev. **114**, 1514 (1959)
11.6 J. J. Hopfield: J. Phys. Chem. Solids **10**, 110 (1959)

11.7 E. I. Rashba: Soviet Phys.–Solid State **1**, 368 (1959)
11.8 R. Knox, A. V. Gold: *Symmetry in the Solid State* (Benjamin, New York 1964). This book contains reprints of the following original papers: [10.2], [11.1–4], [12.5].
11.9 G. F. Koster: "Space Groups and Their Representations", in *Solid State Physics*, Vol. 5 (Academic, New York 1957) pp. 173–256
11.10 J. L. Warren: Rev. Mod. Phys. **40**, 38 (1968)
11.11 C. J. Bradley, A. P. Cracknell: *The Mathematical Theory of Symmetry in Solids* (Oxford University Press, Oxford 1972)

Chapter 12

12.1 J. L. Birman: *Theory of Crystal Space Groups and Lattice Dynamics* (Springer, Berlin, Heidelberg 1984)
12.2 A. A. Maradudin, E. W. Montroll, G. H. Weiss, I. P. Ipatova: *Theory of Lattice Dynamics in the Harmonic Approximation*, 2nd ed. (Academic, New York 1971) Chap. 3
12.3 J. L. Warren: Rev. Mod. Phys. **40**, 38 (1968)
12.4 G. Nilsson, G. Nelin: Phys. Rev. B **3**, 364 (1971)
12.5 M. Lax, J. J. Hopfield: Phys. Rev. **124**, 115 (1961)

Chapter 13

13.1 R. E. Dietz, A. Misetich, H. J. Guggenheim: Phys. Rev. Lett. **16**, 841 (1966)
13.2 Tables of the irreducible characters are due to J. O. Dimmock, R. G. Wheeler: Phys. Rev. 127, 391 (1962)

Chapter 14

14.1 L. D. Landau, E. M. Lifshitz: *Statistical Physics*, Part 1, 3rd ed. (Pergamon, Oxford 1980) pp. 446–471
14.2 A detailed treatment without using group theory is given by T. Nagamiya: "Helical Spin Ordering" in *Solid State Physics*, Vol. 20, ed. by F. Seitz, D. Turnbull, H. Ehrenreich (Academic, New York 1967) pp. 305–411
14.3 J. M. Kosterlitz, D. J. Thouless: J. Phys. C **6**, 1181 (1973)
14.4 W. F. Brinkman, R. J. Elliott: Proc. R. Soc. London **294A**, 343 (1966)

Chapter 15

15.1 D. E. Rutherford: *Substitutional Analysis* (Edinburgh University Press, Edinburgh 1948) pp. 23–44
15.2 T. Yamanouchi: Proc. Phys.-Math. Soc. Jpn, **19**, 436 (1937)
15.3 H. Weyl: *The Classical Groups* (Princeton University Press, Princeton, 1939) pp. 115–136
15.4 G. Racah: Phys. Rev. **76**, 1352 (1949)

Appendix

B.1 L. P. Bouckaert, R. Smoluchowski, E. Wigner: Phys. Rev. **50**, 58 (1936)
B.2 Y. Onodera, M. Okazaki: J. Phys. Soc. Jpn. **21**, 2400 (1966)

Subject Index

Italic numbers indicate primary references.

Lightning Source UK Ltd.
Milton Keynes UK
UKOW041531050912

198506UK00002B/28/P